Number Theory

Volume 36

CRM
PROCEEDINGS &
LECTURE NOTES

Centre de Recherches Mathématiques
Université de Montréal

Number Theory

Canadian Number Theory Association VII, Montréal, May 19–25, 2002

Hershy Kisilevsky
Eyal Z. Goren
Editors

The Centre de Recherches Mathématiques (CRM) of the Université de Montréal was created in 1968 to promote research in pure and applied mathematics and related disciplines. Among its activities are special theme years, summer schools, workshops, postdoctoral programs, and publishing. The CRM is supported by the Université de Montréal, the Province of Québec (FCAR), and the Natural Sciences and Engineering Research Council of Canada. It is affiliated with the Institut des Sciences Mathématiques (ISM) of Montréal, whose constituent members are Concordia University, McGill University, the Université de Montréal, the Université du Québec à Montréal, and the Ecole Polytechnique. The CRM may be reached on the Web at www.crm.umontreal.ca.

American Mathematical Society
Providence, Rhode Island USA

The production of this volume was supported in part by the Fonds pour la Formation de Chercheurs et l'Aide à la Recherche (Fonds FCAR) and the Natural Sciences and Engineering Research Council of Canada (NSERC).

2000 *Mathematics Subject Classification.* Primary 11-XX.

Library of Congress Cataloging-in-Publication Data

Number theory/ Hershy Kisilevsky, Eyal Z. Goren, editors.
p. cm.–(CRM proceedings & lecture notes; v. 36)
English with one contribution in French.
Includes bibliographical references.
ISBN 0-8218-3331-6 (acid-free paper)
1. Number theory–Congresses. I. Kisilevsky, H. II. Goren, Eyal Z. (Eyal Zvi), 1963– III. Canadian Number Theory Association. Conference (7th: 2002: Montréal, Québec) IV. Series.
QA241.N8645 2004
512.7–dc22 2004046301

∞ The paper used in this book is acid-free and falls within the guidelines
established to ensure permanence and durability.
This volume was submitted to the American Mathematical Society
in camera ready form by the Centre de Recherches Mathématiques.
Visit the AMS home page at http://www.ams.org/

10 9 8 7 6 5 4 3 2 1 09 08 07 06 05 04

Contents

List of Participants

Agashe, Amod
University of Texas
amod@math.utexas.edu

Agboola, Adebisi
University of California
agboola@math.ucsb.edu

Ahlgren, Scott
University of Illinois,
Urbana-Champaign
ahlgren@math.uiuc.edu

Akbary, Amir
University of Lethbridge
akbary@cs.uleth.ca

Alaca, Saban
Carleton University
salaca@math.carleton.ca

Almeida, Paulo
University of Georgia
palmeida@math.uga.edu

Arbour, Benoît
McGill University
arbour@math.mcgill.ca

Archinard, Natalia
McGill University
archinard@math.mcgill.ca

Beck, Michael
University of Georgia
mbeck@math.uga.edu

Bennett, Michael
University of British Columbia
bennett@math.ubc.ca

Berger, Laurent
Brandeis University
laurent@brandeis.edu

Bertin, Marie-José
Université Pierre et Marie Curie
bertin@math.jussieu.fr

Besser, Amnon
Ben Gurion University
bessera@math.bgu.ac.il

Bhargava, Manjul
Princeton University
bhargava@math.princeton.edu

Borwein, Jonathan
Simon Fraser University
jborwein@cecm.sfu.ca

Boxall, John
Université de Caen
boxall@math.unicaen.fr

Boyd, David W.
University of British Columbia
boyd@math.ubc.ca

Boylan, Matthew
University of Wisconsin
boylan@math.wisc.edu

Brownawell, Dale
Penn State University
wdb@math.psu.edu

Bruin, Nils
Simon Fraser University
bruin@cecm.sfu.ca

Bugeaud, Yann
Université Louis Pasteur
bugeaud@math.u-strasbg.fr

Burns, David
King's College London
david.burns@kcl.ac.uk

Carter, James E.
College of Charleston
carteje@cocf.edu

Chang, Sungkon
University of Georgia
schang@math.uga.edu

Chen, Imin
Simon Fraser University
ichen@math.sfu.ca

Choi, Kwok-Kwong Stephen
Simon Fraser University
choi@cecm.sfu.ca

Cohen, Henri
Université de Bordeaux
cohen@math.u-bordeaux.fr

Cojocaru, Alina Carmen
Princeton University
cojocaru@math.princeton.edu

Consani, Caterina
University of Toronto
kc@math.toronto.edu

Cosma, Ioana
McGill University
ioanac@hotmail.com

Cremona, John
University of Nottingham
john.cremona.nottingham.ac.uk

David, Chantal
Concordia University
cdavid@mathstat.concordia.ca

Dèbes, Pierre
Université de Lille 1
pierre.debes@univ-lille1.fr

Del Corso, Ilaria
University of Pisa
delcorso@dm.unipi.it

Deninger, Christopher
Universität Münster
deninge@math.uni-muenster.de

Deshouillers, Jean-Marc
Université de Bordeaux
jean-marc.deshouillers@math.u-bordeaux.fr

Dilcher, Karl
Dalhousie University
dilcher@mathstat.dal.ca

Dorais, François
Concordia University
dendpa@look.ca

Doré, Vincent
dorenavant.r@videotron.ca

Dousmanis, Gerasimos
University of Wisconsin, Madison
dousmani@math.wisc.edu

Duke, William
University of California
duke@math.ucla.edu

Duursma, Iwan
University of Illinois,
Urbana-Champaign
duursma@math.uiuc.edu

Dvornicich, Roberto
Università di Pisa
dvornic@dm.unipi.it

Edgar, Hugh
San Jose State University

Edixhoven, Sebastiaan J.
Université de Rennes
edix@maths.univ-rennes1.fr

Elder, G. Griffith
University of Nebraska, Omaha
elder@unomaha.edu

Faber, Xander
University of Georgia
faber@math.uga.edu

Fastenberg, Lisa
Pace University
lfastenberg@pace.edu

Fearnley, Jack
Concordia University
jack@alcor.concordia.ca

Ferguson, Ronald
Simon Fraser University
rferguson@pims.math.ca

Ford, David
Concordia University
ford@cs.concordia.ca

Frechette, Sharon
College of the Holy Cross
sfrechet@mathcs.holycross.edu

Friedlander, John
University of Toronto
frdlndr@math.utoronto.ca

Friedman, Menahem
Nuclear Research Center-Negev
mlfrid@netvision.net.il

Fukshansky, Lenny
University of Texas at Austin
lenny@math.utexas.edu

Goren, Eyal Z.
McGill University
goren@math.mcgill.ca

Goss, David
The Ohio State University
goss@math.ohio-state.edu

Gournay, Antoine
McGill University
gournay@globetrotter.qc.ca

Gouvêa, Fernando
Colby College
fgouvea@colby.edu

Graham, Sid
Central Michigan University
sidney.w.graham@cmich.edu

Grantham, Jon
Center for Computing Sciences
grantham@super.org

Granville, Andrew
Université de Montréal
andrew@dms.umontreal.ca

Greenberg, Matthew
McGill University
greenberg@math.mcgill.ca

Gurak, Stan
University of San Diego
gurak@sandiego.edu

Guy, Michael G.
University of Georgia
guy@math.uga.edu

Hamblen, Spencer
Cornell University
srh23@cornell.edu

Hanke, Jonathan
Rutgers University
jonhanke@math.rutgers.edu

Haristoy, Julien
Université Louis Pasteur
haristoy@math.u-strasbg.fr

Herrmann, Emanuel
Universität des Saarlandes
herrmann@math.uni-sb.de

Hirabayashi, Mikihito
Kanazawa Institute
of Technology
hira@neptune.kanazawa-it.ac.jp

Hooper, Jeff
Acadia University
jeff.hopper@acadiau.ca

Huard, James G.
Canisius College
huard@canisius.edu

Iovita, Adrian
University of Washington
iovita@math.washington.edu

James, Kevin
Clemson University
kevja@clemson.edu

Jimbo, Henri Claver
Adv. Res. Center for Sci. & Engr.
yukimuki@ja2.so-net.ne.jp

Johnston, Henri
Cornell University
henri@math.cornell.edu

Jones, Lenny
Shippensburg University
lkjone@ship.edu

Juge, Nicolas
Université Laval
njuge@bioinfo.ulaval.ca

Jung Hwan, Ahn
Korea University
haitai@semi.korea.ac.kr

Kamgarpour, Masoud
University of Waterloo
mkamgarp@student.math.
uwaterloo.ca

Kedlaya, Kiran
MIT
kedlaya@math.mit.edu

Khan, Mizan
Eastern Connecticut
State University
khanm@easternct.edu

Khare, Chandrashekhar
University of Utah
shekhar@math.utah.edu

Kim, Henry
University of Toronto
henrykim@math.toronto.edu

Kisilevsky, Hershy
Concordia University
kisilev@mathstat.concordia.ca

Kitaek, Kim
Korea University
choypm1@hotmail.com

Knafo, Emmanuel
University of Toronto
knafo@math.toronto.edu

Knapp, Michael
University of Rochester
mknapp@math.rochester

Kolster, Manfred
McMaster University
kolster@mcmaster.ca

Kumchev, Angel
University of Toronto
kumchev@math.toronto.edu

Kuwata, Masato
Kanagawa Institute of Technology
kuwata@gen.kanagawa-it.ac.jp

Kwon, Soun-Hi
Korea University
shkwon@semi.korea.ac.kr

Laba, Izabella
University of British Columbia
ilaba@math.ubc.ca

Lagarias, Jeffrey
AT&T Labs – Research
jcl@research.att.com

Laliberté, Frédéric
McGill University
flalib@po-box.mcgill.ca

Laurent, Michel
Université de la Méditerranée
laurent@iml.univ-mrs.fr

Lecacheux, Odile
Université Paris 6
lecacheu@math.jussieu.fr

Lee, Geon No
Korea University
thisknow@hanmail.net

Lemire, Mathieu
Carleton University
mathieulem@yahoo.ca

Lévesque, Claude
Université Laval
Claude.Levesque@mat.ulaval.ca

Logan, Adam
McGill University
logan@math.mcgill.ca

Longhi, Ignazio
Universität Münster
longhi@uni-muenster.de

Lovejoy, Jeremy
University of Wisconsin
lovejoy@math.jussieu.fr

Luca, Florian
Universidad Nacional
Autónoma de México
fluca@matmor.unam.mx

MacHenry, Trueman
York University
machenry@mathstat.yorku.ca

Mahlburg, Karl
University of Wisconsin
mahlburg@math.wisc.edu

Marshall, David
McMaster University
marshall@math.mcmaster.ca

Marshall, S. Hammond
University of Texas at Austin
smarshall@math.utexas.edu

Martin, Greg
University of British Columbia
gerg@math.ubc.ca

McCallum, William
University of Arizona
wmc@math.arizona.edu

McGraw, William
University of Wisconsin
mcgraw@math.wise.edu

McKinnon, David
University of Waterloo
dmckinnon@math.uwaterloo.ca

Meyer, Jeffrey
Syracuse University
jlmeye01@syr.edu

Mihailescu, Preda
Universitaet Paderborn
preda@upd.de

Mochizuki, Asae
Chubu University
mczk@isc.chubu.ac.jp

Mortenson, Eric
University of Wisconsin, Madison
mort@math.wisc.edu

Murty, Ram
Queen's University
murty@mast.queensu.ca

Murty, V. Kumar
University of Toronto
murty@math.toronto.edu

Muzaffar, Habib
University of Waterloo
hbmuzaff@herod.uwaterloo.ca

Nafari, Manizheh
The Abdus Salam ICTP
manijnafari@yahoo.com

Nakahara, Toru
Saga University
nakahara@ms.saga-u.ac.jp

Ng, Nathan
Institut for Advanced Study
nathan@ias.edu

Nichifor, Alexandra
University of Washington
nichifor@math.washington.edu

Nicole, Marc-Hubert
McGill University
nicole@math.mcgill.ca

Nowak, Werner Georg
Universität für Bodenkultur
nowak@mail.boku.ac.at

Omar, Kihel
The University of Lethbridge
kihel@cs.uleth.ca

Onn, Uri
Technion
urion@tx.technion.ac.il

Ono, Ken
University of Wisconsin, Madison
ono@math.wisc.edu

Osburn, Robert
McMaster University
osburnr@icarus.math.mcmaster.ca

Pacheco, Amilcar
University Federal do Rio
de Janeiro
amilcar@impa.br

Papanikolas, Matthew
Texas A&M University
map@math.tamu.edu

Pappalardi, Francesco
Università degli Studi Roma Tre
pappa@mat.uniroma3.it

Pappas, George
Michigan State University
pappas@math.msu.edu

Park, Daeshik
University of Georgia
dpark@math.uga.edu

Park, Sun Mi
Korea University
smpark@semi.korea.ac.kr

Parsell, Scott
Penn State University
parsell@math.psu.edu

Patel, Deepam
Brandeis University
sreya@brandeis.edu

Penniston, David
Furman University
david.penniston@furmann.edu

Petsche, Clayton
University of Texas, Austin
clayton@math.utexas.edu

Pinch, Richard
Government Communications Headquarters
rgep@chalcedon.demon.co.uk

Pine, Eric
University of Georgia
epine@math.uga.edu

Pintér, Ákos
University of Debrecen
apinter@math.klte.hu

Pollack, Robert
University of Chicago
pollack@math.chicago.edu

Pomerance, Carl
Dartmouth College
carl.pomerance@dartmouth.edu

Pooh, Charles
University of Georgia
pooh@math.uga.edu

Poonen, Bjorn
University of California
poonen@math.berkeley.edu

Popescu, Christian
Johns Hopkins University
cpopescu@math.jhu.edu

Prasad, Amritanshu
The Institute of Mathematical Sciences
amri@imsc.res.in

Rainone, Timothy
McGill University
t_rainone@hotmail.com

Ramakrishna, Ravi
Cornell University
ravi@math.mcgill.ca

Rapoport, Michael
Köln University
rapoport@mi.uni-koeln.de

Rasmussen, Christopher
University of Arizona
chras@grad.math.arizona.edu

Ribenboim, Paulo
Queen's University
mathstat@mast.queensu.ca

Roberts, David
University of Minnesota
roberts@mrs.umn.edu

Roblot, Xavier
Université Lyon I
roblot@euler.univ-lyon1.fr

Roy, Damien
Université d'Ottawa
droy@acadvm1.uottawa.ca

Rubinstein, Michael
University of Waterloo
mrubinstein@math.uwaterloo.ca

Rudnick, Zeev
Tel Aviv University
rudnick@post.tau.ac.il

Saidak, Filip
University of Calgary
filips@math.ucalgary.ca

Saikia, Anupam
McGill University
saikia@math.mcgill.ca

Samet, Redha
Université de la Méditerranée
samet@iml.univ-mrs.fr

Savitt, David
McGill University
savitt@math mcgill ca

Simons, Lloyd
Saint Michael's College
lsimons@smcvt.edu

Sinha, Kaneenika
Queen's University
skaneen@mast.queensu.ca

Smyth, Chris
Edinburgh University
chris@maths.ed.ac.uk

Soundararajan, Kannan
University of Michigan
ksound@umich.edu

Sun, Xinyun
University of Waterloo
x5sun@math.uwaterloo.ca

Thaine, Francisco
Concordia University
ftha@vax2.concordia.ca

Tucker, Thomas
University of Georgia
ttucker@math.uga.edu

Vatsal, Vinayak
University of British Columbia
vatsal@math.ubc.ca

Wan, Daqing
University of California, Irvine
dwan@math.uci.edu

Williams, Kenneth
Carleton University
williams@math.carleton.ca

Sands, Jonathan
University of Vermont
sands@math.uvm.edu

Simon, Denis
Université de Caen
simon@math.unicaen.fr

Sinclair, Christopher
University of Texas, Austin
sinclair@math.utexas.edu

Smithline, Lawren
Cornell University
lawren@math.cornell.edu

Sondow, Jonathan
Princeton University
jsondow@alumni.princeton.edu

Stewart, Cameron L.
University of Waterloo
cstewart@watserv1.uwaterloo.ca

Talamanca, Valerio
Università di Roma Tre
valerio@matrm3.mat.uniroma3.it

Tsishchanka, Kiryl
Kansas State University
kit@math.ksu.edu

van der Waall, Alexa
Simon Fraser University
awaall@cecm.sfu.ca

Walsh, Gary
University of Ottawa
gwalsh@mathstat.uottawa.ca

Weiss, Al
University of Alberta
aweiss@math.ualberta.ca

Wooley, Trevor
University of Michigan
wooley@math.lsa.umich.edu

Yeung, David
University of British Columbia
dyeungca@yahoo.com

Zhu, Hui June
McMaster University
zhuh@mcmster.ca

List of Speakers

Ahlgren, Scott (University of Illinois)
Weierstrass points on modular curves and supersingular j-invariants

Akbary, Amir (University of Lethbridge)
Large divisors of Fourier coefficients of modular forms

Alaca, Saban (Carleton University)
p-integral bases of a quartic field defined by a trinomial

Bennett, Michael (University of British Columbia)
Classical Diophantine problems via Galois representations and modular forms

Berger, Laurent (Brandeis University)
p-adic representations

Bertin, Marie-José (Université Pierre et Marie Curie)
Mesure de Mahler et regulateur elliptique

Besser, Amnon (Ben-Gurion University)
p-adic Arakelov theory on curves

Bhargava, Manjul (Princeton University)
The representation of integers by quadratic forms

Borwein, Jonathan (Simon Fraser University)
Dirichlet series of squares of sums of squares

Boxall, John (Université de Caen)
Singular torsion on elliptic curves

Boyd, David W. (University of British Columbia)
A-polynomials and Mahler's measure

Boylan, Matthew (University of Wisconsin)
Exceptional congruences for the coefficients of certain eta-product newforms

Brownawell, Dale (Penn State University)
Algebraic independence over function fields

Bruin, Nils (Simon Fraser University)
Using towers of 2-covers of hyperelliptic curves to find rational points

Bugeaud, Yann (Université Louis Pasteur)
Diophantine approximation and Hausdorff dimension

Burns, David (King's College London)
On the equivariant Tamagawa number conjecture for Tate motives

Carter, James E. (College of Charleston)
Normal integral bases in quadratic and cyclic cubic extensions of quadratic fields

Chen, Imin (Simon Fraser University)
Elliptic curves with non-split $\bmod p^2$ *representations*

Choi, Kwok-Kwong Stephen (Simon Fraser University)
On Dirichlet series for sums of squares

Cohen, Henri (Université de Bordeaux)
Counting discriminants of number fields

Cojocaru, Alina Carmen (Princeton University)
Square-free orders for CM *elliptic curves modulo* p
and
The square sieve and the Lang–Trotter conjecture

Conrad, Brian (University of Michigan)
Multivariate Bateman–Horn conjecture

Consani, Caterina (University of Toronto)
Archimedean fibers and non-commutative geometry

Cremona, John (University of Nottingham)
Rational points on elliptic curves

Darmon, Henri (McGill University)
A fourteenth lecture on Fermat's last theorem

Debes, Pierre (Université de Lille 1)
A moduli form of the Shafarevich conjecture

Del Corso, Ilaria (Università di Pisa)
On the index of a number field

Deninger, Christopher (Universität Münster)
Elliptic curves over finite fields and transversal index theory

Duke, William (University of California, Los Angeles)
Extreme values of Artin L-functions and class numbers

Dvornicich, Roberto (Università di Pisa)
On polynomials taking small values at integral arguments

Edixhoven, Sebastiaan J. (Université de Rennes)
On $\bmod p$ *modular forms of weight one*

Elder, G. Griffith (University of Nebraska, Omaha)
Twists and Galois structure in bicyclic extensions

Fastenberg, Lisa (Pace University)
Computing Mordell–Weil ranks of cyclic covers of elliptic surfaces

Frechette, Sharon (College of the Holy Cross)
Nonvanishing results for L-functions associated to quadratic twists of Newforms

Friedlander, John (University of Toronto)
On some exponential sums over Z_m

Fukshansky, Lenny (University of Texas, Austin)
Points of small height outside of a collection of subspaces

Goss, David (The Ohio State University)
Modularity for Drinfeld modules

Granville, Andrew (Université de Montréal)
2002—Year of the successfully proven old conjectures ....

Gurak, Stan (University of San Diego)
Period polynomials for finite fields of fixed small degree

Hanke, Jonathan (Rutgers University)
Representing numbers by ternary quadratic forms

Herrmann, Emanuel (Universität des Saarlandes)
Computing all S-integral solution in a family of two simultaneous Pell equations

Hirabayashi, Mikihito (Kanazawa Institute of Technology)
A relative class number formula for an imaginary abelian number field by Dedekind sum

Hooper, Jeff (Acadia University)
Chinburg invariants and generalized quaternion group

Huard, James G. (Canisius College)
Sums of sixteen and twenty-four triangular numbers

Iovita, Adrian (Concordia University)
On the Iwasawa main conjecture for elliptic curves over $\mathbb{Q}$ at supersingular primes

James, Kevin (Clemson University)
The Lang–Trotter conjecture for elliptic curves with rational 3-torsion

Jimbo, Henri Claver (Advanced Research Center for Science & Engineering)
On one application of the Jacobi symbol

Jones, Lenny (Shippensburg University)
An extension of a theorem of D. H. Lehmer

Khare, Chandrashekhar (University of Utah)
Compatible systems of morphisms and rigidity

Kim, Henry (University of Toronto)
Applications of the functorial symmetric cube and symmetric fourth

Knafo, Emmanuel (University of Toronto)
Variance of distribution of almost primes in arithmetic progressions

Knapp, Michael (University of Rochester)
Systems of additive equations over p-adic field

Kumchev, Angel (University of Toronto)
Distribution of primes in arithmetic progressions and short intervals

Kuwata, Masato (Kanagawa Institute of Technology)
Vanishing of twisted L-functions of an elliptic curve and rational points on K3 *surfaces*

Lagarias, Jeffrey (AT&T Labs—Research)
Wavelets, tilings and number theory

Laurent, Michel (Université de la Méditerranée)
Inequalities of Liouville's type on a curve

Lecacheux, Odile (Université Paris 6)
Rang de familles de courbes elliptiques

Levesque, Claude (Université Laval)
Fundamental system of units of some family of quartic fields

Logan, Adam (McGill University)
Deformations of Galois representations associated to degenerate extensions

Lovejoy, Jeremy (University of Wisconsin)
Lacunary partition functions

Luca, Florian (UNAM)
A diophantine approximation property characterizing both algebraic and Mahler's numbers

Mahlburg, Karl (University of Wisconsin)
Partition identities and a theorem of Zagier

Marshall, David (McMaster University)
Cup product computations in the Galois cohomology of number fields

Marshall, S. Hammond (Affiliation University of Texas, Austin)
On Galois representations arising from Abelian varieties with bad reduction

Martin, Greg (University of British Columbia)
Inequities in the Shanks–Rényi prime number race

McCallum, William (University of Arizona)
The structure of fundamental groups of number fields

McGraw, William (University of Wisconsin)
The rationality of vector valued modular forms and implications for a theorem of Borcherds

McKinnon, David (University of Waterloo)
Vojta's main conjecture for blow-up surfaces

Meyer, Jeffrey (Syracuse University)
Some applications of a representation of the Dedekind sum

Mihailescu, Preda (Universitaet Paderborn)
Primary units and a proof of Catalan's conjecture

Mortenson, Eric (University of Wisconsin–Madison)
Rodriguez–Villegas supercongruences: The elliptic curves

Murty, V. Kumar (University of Toronto)
On the Sato–Tate conjecture

Murty, Ram (Queen's University)
Non-abelian generalizations of the Erdős–Kac theorem

Muzaffar, Habib (University of Waterloo)
A restricted Epstein zeta function and the evaluation of some definite integrals

Nakahara, Toru (Saga University)
On a problem of Hasse for certain imaginary Abelian fields

Ng, Nathan (Université de Montréal)
Discrete moments of the Riemann zeta function

Nicole, Marc-Hubert (McGill University)
F-crystals with real multiplication

Nowak, Werner Georg (Universität für Bodenkultur)
On the divisor problem: Moments of $D(x)$ over short intervals

Onn, Uri (Technion)
On interpolation between zonal spherical functions of $\mathrm{GL}_2(\mathbb{Q}_p)$ and $\mathrm{GL}_2(\mathbb{R})$

Ono, Ken (University of Wisconsin, Madison)
Arithmetic of the divisors of modular forms and values of modular functions

Osburn, Robert (McMaster University)
Tame kernels and 4-rank densities

Pacheco, Amilcar (Universidade Federal do Rio de Janeiro)
Distribution of the traces of Frobenius on elliptic curves over function fields

Papanikolas, Matthew (Brown University)
Extensions of elliptic curves

Pappas, George (Michigan State University)
Cubic structures and lattices of cusp forms

Parsell, Scott (Penn State University)
The values of irrational linear forms at prime arguments

Penniston, David (Furman University)
Ramanujan congruences for m-regular partition functions

Pinter, Akos (University of Debrecen)
On the equation $1^k + 2^k + \cdots + (x-1)^k = y^z$

Pollack, Robert (University of Chicago)
The main conjecture for CM *elliptic curves at supersingular primes*

Pomerance, Carl (Dartmouth College)
Mersenne numbers

Poonen, Bjorn (University of California)
Bertini theorems over finite fields

Popescu, Christian (Johns Hopkins University)
1-*motives and Iwasawa theory*

Ramakrishna, Ravi (Cornell University)
Deforming Galois representations

Rapoport, Michael (Köln University)
Derivatives of Eisenstein series and Faltings heights

Roberts, David (University of Minnesota, Morris)
An ABC *construction of number fields*

Rubinstein, Michael (University of Texas, Austin)
Moments of L-functions

Rudnick, Zeev (Tel-Aviv University)
Torsion points on curves and common divisors of $a^k - 1$ *and* $b^k - 1$

Saidak, Filip (University of Calgary)
On a theorem of Erdős

Savitt, David (Institut des Hautes Études Scientifiques)
The maximum number of points on a curve of genus 4 *over* $\mathbb{F}_8$ *is* 25

Shalalfeh, M. (Bilkent University, Ankara, Turkey)
Fibre product of Artin–Schreier and Kummer extensions and curves with many points

Sinha, Kaneenika (Queen's University)
Generalized Hurwitz zeta functions

Smithline, Lawren (Cornell University)
Computing compact operators with recursive generation

Smyth, Chris (Edinburgh University)
A new arithmetical function and a new height function

Sondow, Jonathan (Princeton University)
Criteria for irrationality of Euler's constant

Soundararajan, Kannan (University of Michigan)
An "uncertainty principle" for arithmetic sequences

Stewart, Cameron L. (University of Waterloo)
On the number of solutions of Diophantine equations

Talamanca, Valerio (Università di Roma Tre)
Heights on finite étale K-algebras

Tsishchanka, Kiryl (Kansas State University)
On approximation of real, complex, and p-adic numbers by algebraic numbers

Tucker, Thomas (University of Georgia)
Exceptional covers and bijections of rational points

Tupan, Alexandru (Johns Hopkins University)
Congruences for weight $\frac{3}{2}$ forms

Vatsal, Vinayak (University of British Columbia)
Multiplicative-type subgroups of $J_0(N)$

Wan, Daqing (University of California, Irvine)
Moment zeta functions in arithmetic geometry

Weiss, Al (University of Alberta)
Equivariant Iwasawa theory

Wooley, Trevor (University of Michigan)
Slim exceptional sets for stout representation problems

Preface

The seventh meeting of the Canadian Number Theory Association was held in Montréal, Québec, Canada during the week of May 19–25, 2002. It took place at the Centre de Recherches Mathématiques (CRM) on the campus of the Université de Montréal (UdeM).

The Canadian Number Theory Association (CNTA) was founded in 1987 at the International Number Theory Conference at Laval University. The purpose of the CNTA is to enhance and promote learning and research in Number Theory, particularly in Canada. To advance these goals the CNTA organizes major international conferences, with the aim of exposing Canadian students and researchers to the latest developments in number theory worldwide. The previous meetings have been held in Banff (1988), Vancouver (1989), Kingston (1991), Halifax (1994), Ottawa (1996) and Winnipeg (1999). These have been high quality conferences which have had wide participation of the international number theory community.

The conference was organized into several themes comprising the main directions of current research in number theory. There were sessions in Algebraic Number Theory, Analytic Number Theory, Arithmetic Algebraic Geometry, Computational Number Theory and Diophantine Analysis and Approximation. Each session had lectures exposing the recent main developments in the area, and others presenting new research results.

There were 186 participants representing 15 different countries, and 115 speakers. This volume is a collection of refereed articles selected from those presented at CNTA VII and is a cross section of current research in number theory. It contains expository articles as well as research articles and is aimed at an audience of active researchers as well as those entering the field.

A highlight of the meeting was the award of the second Ribenboim Prize to Henri Darmon. We were very pleased that this award was presented by Andrew Granville, the only previous recipient, and by Paulo Ribenboim.

We would like to thank the CRM and the Number Theory Foundation for their generous support. We also want to express our gratitude to the staff of the CRM, in particular Louis Pelletier, Louise Letendre and André Montpetit, for ensuring the smooth functioning of the meeting. Finally we thank the students and session organizers M. Kolster (Algebraic Number Theory), K. S. Williams (Analytic Number Theory), E. Z. Goren (Arithmetic Algebraic Geometry), G. Walsh (Computational Number Theory), and D. Roy (Diophantine Analysis and Approximation) for their selfless work in making the conference a success.

Hershy Kisilevsky and Eyal Z. Goren
CNTA VII organizers
Concordia University and McGill University, Montréal
March 2004

Centre de Recherches Mathématiques
CRM Proceedings and Lecture Notes
Volume **36**, 2004

Mesure de Mahler et Régulateur Elliptique : Preuve de Deux Relations *Exotiques*

Marie José Bertin

RÉSUMÉ. Les relations entre la mesure de Mahler de certains polynômes associés aux courbes elliptiques 11A et 14A d'une part et leur régulateur elliptique d'autre part, nous permettent de prouver des relations *exotiques* du dilogarithme elliptique sur des points de torsion, relations conjecturées par Bloch et Grayson.

1. Introduction

Introduite par Mahler pour mesurer la taille de polynômes, la mesure de Mahler a suscité par deux fois l'attention particulière des mathématiciens.

À l'origine se trouve sans doute la fameuse question de Lehmer [**10**] qui demeure toujours sans réponse : existe-t-il des entiers algébriques α, $\alpha \geq 1$, non racines de l'unité pour lesquels

$$M(\alpha) = \prod_{|\alpha^\sigma| \geq 1} |\alpha^\sigma| \geq 1,1762\ldots$$

où α^σ désigne les conjugués de α et $1,1762\ldots$ est l'unique racine > 1 du polynôme de Lehmer, de degré 10 et réciproque.

Grâce à la formule de Jensen, le problème peut se reformuler ainsi : existe-t-il des polynômes unitaires et irréductibles P, non cyclotomiques, ayant une mesure de Mahler logarithmique $m(P) \leq \log(1,1762)$ avec

$$m(P) := \log\bigl(M(P)\bigr) := \frac{1}{2\pi i}\int_{|x|=1} \log|P(x)|\frac{dx}{x}?$$

On définit de même la mesure de Mahler logarithmique $m(P)$ d'un polynôme de Laurent $P \in \mathbb{Z}[X^\pm, Y^\pm]$ par

$$m(P) := \frac{1}{(2\pi i)^2}\int_{\mathbb{T}^2} \log|P(x,y)|\frac{dx}{x}\frac{dy}{y}$$

où $\mathbb{T}^2$ désigne le tore complexe

$$\mathbb{T}^2 = \{(x,y) \in \mathbb{C}^2 / |x| = 1, |y| = 1\}.$$

2000 *Mathematics Subject Classification.* Primary : 11C08, 14H52 ; Secondary : 14G10, 11G05.

This is the final form of the paper.

L'intérêt pour le problème de Lehmer d'une meilleure connaissance de la mesure de Mahler de polynômes de plusieurs variables a été mis en évidence par un résultat de Boyd [**4**] prouvant que

$$\lim_{n\to\infty} M\big(P(X,X^n)\big) = M\big(P(X,Y)\big).$$

À cette époque remontent les valeurs approchées suivantes [**4, 5**]

$$M\left(X+\frac{1}{X}+Y+\frac{1}{Y}+1\right) = 1,28573\ldots,$$
$$M\big((X+1)Y^2+(X^2+X+1)Y+X(X+1)\big) = 1,25542\ldots$$

Une vingtaine d'années plus tard, la mesure de Mahler de polynômes de deux variables suscita un regain d'intérêt grâce à une conjecture de Deninger [**9**]

$$m\left(X+\frac{1}{X}+Y+\frac{1}{Y}+1\right) = L'(E,0) = \frac{15}{4\pi^2}L(E,2) := b_{15}$$

où $L(E,s)$ est la fonction L associée à la courbe elliptique de conducteur 15, clôture projective de la courbe d'équation affine

$$X+\frac{1}{X}+Y+\frac{1}{Y}+1=0,$$

isomorphe à la courbe 15A $=[1,1,1,0,0]$ des tables de Cremona et possédant de la 4-torsion.

Suivirent alors les travaux de Boyd [**6**] et Rodriguez–Villegas [**11, 12**], livrant, tel Boyd, de très nombreuses formules espérées comme

$$m(P_1) = m\big((x+y+1)(x+1)(y+1)+xy\big) \stackrel{?}{=} 7b_{11},$$
$$m(Q_1) = m(y^2+(x^2+2x-1)y+x^3) \stackrel{?}{=} 5b_{11}$$

où $b_{11} = L'(E_{11\mathrm{A}},0)$ et 11A $=[0,-1,1,0,0]$ est la courbe elliptique des tables de Cremona avec de la 5-torsion, ou bien

$$m(P_2) = m(y^2-xy+y-x^3) \stackrel{?}{=} 2b_{14},$$
$$m(Q_2) = m((x+1)y^2+(x^2+x+1)y+x(x+1)) \stackrel{?}{=} b_{14}$$

avec $b_{14} = L'(E_{14\mathrm{A}},0)$ et 14A $=[1,0,1,-1,0]$ avec de la 6-torsion.

Selon une méthode initiée par Rodriguez–Villegas [**13**], nous prouvons dans ce papier les égalités entre mesures de Mahler, la preuve des ? étant un exercice semble-t-il plus difficile car lié aux conjectures de Beilinson [**1**].

Un corollaire particulièrement intéressant de ces résultats est la preuve de relations *exotiques* de dilogarithmes elliptiques conjecturées par Bloch [**7**], à savoir

$$3D^E(P) - 2D^E(2P) = 0$$

où $P=(0,0)$ est le point de 5-torsion de la courbe 11A et D^E le dilogarithme elliptique ainsi que

$$2D^E(P) + 5D^E(2P) = 0$$

où $P=(1,0)$ désigne le point de 6-torsion de la courbe elliptique $E=14\mathrm{A}$.

2. Rappels

Soit E une courbe elliptique définie sur $\mathbb{Q}$. On a deux représentations de $E(\mathbb{C})$

$$E(\mathbb{C}) \xrightarrow{\sim} \mathbb{C}/(\mathbb{Z}\tau + \mathbb{Z}) \xrightarrow{\sim} \mathbb{C}^*/q^{\mathbb{Z}}$$
$$\big(\wp(u), \wp'(u)\big) \to u \pmod{\Lambda} \to z = e^{2\pi i u}$$

où $\wp$ désigne la fonction $\wp$ de Weierstrass, Λ le réseau $\Lambda = \{\tau, 1\}$, $\tau \in \mathbb{H}$ et $q = e^{2\pi i \tau}$.

Le dilogarithme elliptique D^E sur $E(\mathbb{C})$, introduit par Bloch (cf. Irvine Lectures [**3**]) est le symétrisé du dilogarithme D de Bloch et Wigner,

$$D\colon \mathbb{C}\backslash\{0,1\} \to \mathbb{R}$$
$$z \mapsto \Im\big(Li_2^{[c]}(z) + \log|z| \log^{[c]}(1-z)\big)$$

où c est un chemin allant de $\frac{1}{2}$ à z dans $\mathbb{C}\backslash\{0,1\}$ et

$$D^E(P) = \sum_{n=-\infty}^{+\infty} D(q^n z),$$

pour P point de $E(\mathbb{C})$ paramétré par $z \in \mathbb{C}^*$ et $q = e^{2\pi i \tau}$.

Le dilogarithme elliptique admet une autre représentation donnée par Bloch [**14**, **3**], en termes de série d'Eisenstein–Kronecker

$$D^E(P) = \frac{(\Im\tau)^2}{\pi} \Re\left(\sum_{m,n\in\mathbb{Z}}{}' \frac{\exp(2\pi i(n\xi - m\eta))}{(m\tau+n)^2(m\bar{\tau}+n)} \right)$$

avec $z = e^{2\pi i u}$ et $u = \xi\tau + \eta$.

1. Le groupe $K_2(E)$. Soit K un corps. Par le théorème de Matsumoto, le deuxième groupe de K-théorie $K_2(K)$ admet la représentation suivante par les symboles $\{f, g\}$

$$K_2(K) = \langle\{f,g\}\rangle/\{\text{relations}\}$$

avec $f, g \in K^*$ et les relations

- $\{f_1 f_2, g\} = \{f_1, g\} + \{f_2, g\}$,
- $\{f, g_1 g_2\} = \{f, g_1\} + \{f, g_2\}$,
- $\{1-f, f\} = 0$.

Si A est un groupe abélien, on obtient une bijection entre l'ensemble des symboles sur K à valeurs dans A et $\mathrm{Hom}\big(K_2(K), A\big)$.

Par exemple le symbole modéré de Tate

$$(x,y)_v \equiv (-1)^{v(x)v(y)} \frac{x^{v(y)}}{y^{v(x)}} \pmod{\mathcal{M}}$$

où v désigne une valuation discréte sur un corps F d'idéal maximal $\mathcal{M}$, de corps résiduel k, définit un homomorphisme λ_v

$$\lambda_v\colon K_2(F) \to k^*.$$

Par suite, si $\mathbb{Q}(E)$ est le corps des fonctions rationnelles de la courbe elliptique E tout $P \in E(\overline{\mathbb{Q}})$ définit une valuation sur $\mathbb{Q}(E)$ de corps résiduel noté $\mathbb{Q}(P)^*$. On obtient ainsi un homomorphisme

$$\lambda_P\colon K_2\big(\mathbb{Q}(E)\big) \to \mathbb{Q}(P)^*;$$

d'où la suite exacte définissant le groupe $K_2(E)$ modulo la torsion comme le noyau de l'homomorphisme $\lambda = \sqcup_P \lambda_P$

$$\lambda\colon K_2\big(\mathbb{Q}(E)\big) \otimes \mathbb{Q} \to \bigsqcup_{P \in E(\overline{\mathbb{Q}})} \big(\mathbb{Q}(P)^* \otimes \mathbb{Q}\big).$$

Un critère dû à Rodriguez–Villegas **[12]** permet alors de reconnaître si un symbole $\{x, y\}$ appartient à $K_2(E)$, lorsque $P(x, y) \in \mathbb{Q}[x, y]$ définit une courbe elliptique E.

THÉORÈME 2.1 ([**12**]). *Soit Q un polynôme de Laurent, $Q \in \mathbb{Q}[x^{\pm}, y^{\pm}]$, définissant une courbe lisse et projective C de $\mathbb{P}^2$.*

Le symbole $\{x, y\}^n$ appartient à $K_2(C)$ pour un $n \in \mathbb{N}$ si et seulement si Q est un polynôme tempéré (i.e. si les polynômes associés aux faces de son polygone de Newton n'ont pour racines que des racines de l'unité).

Par exemple, le polynôme

$$Q(x, y) = y^2 + (x^2 + 2x - 1)y + x^3$$

est un polynôme tempéré.

Car on associe à la face τ de Q un polynôme d'une seule variable t **[12]**,

$$Q_\tau(t) = \sum_{k=0}^{\infty} c_{\tau(k)} t^k \in \mathbb{C}[t]$$

où

$$Q(x, y) = \sum_{(m,n) \in \mathbb{Z}^2} c_{(m,n)} x^m y^n$$

et $\tau(0), \tau(1), \ldots$ sont les points consécutifs du réseau des entiers situés sur la face τ. On utilisera alors dans nos exemples le corollaire suivant.

COROLLAIRE 2.2. *Si les racines des polynômes en t des faces sont ± 1, alors $\{x, y\}^2 \in K_2(E)$.*

2. L'application régulateur. Cette application admet deux expressions l'une intégrale et l'autre analytique.

(1) Expression intégrale: L'application régulateur r s'exprime sous forme intégrale

$$r\colon K_2(E) \to \mathbb{R}$$
$$\{f, g\} \mapsto \frac{1}{2\pi} \int_\gamma \eta(f, g)$$

avec

$$\eta(f, g) = \log|f| d(\arg g) - \log|g| d(\arg f),$$

$f, g \in \mathbb{Q}(E)^*$ et γ chemin fermé ne passant ni par les zéros ni par les pôles de f et g et engendrant le sous-groupe $H_1(E, \mathbb{Z})^-$ des cycles sur lesquels la conjugaison complexe agit par -1, i.e. associés aux périodes imaginaires pures.

(2) Expression analytique: Soit $(f) = \sum_i a_i[P_i]$ et $(g) = \sum_j b_j[P_j]$, a_i et $b_j \in \mathbb{Q}$, les points P_i et P_j étant des points de la courbe elliptique E de paramètres réels t_i et t_j tels que $\{f, g\} \in K_2(E)$. Alors $r(\{f, g\})$ s'exprime au moyen de la série d'Eisenstein–Kronecker du dilogarithme elliptique

$$r(\{f,g\}) = \frac{(\Im\tau)^2}{\pi^2} \sum_{i,j} a_i b_j K_{2,1}(t_i - t_j)$$

avec

$$K_{2,1}(t) := \sum_{\gamma\in\Lambda\gamma\neq 0} \frac{\langle t,\gamma\rangle}{\gamma^2\overline{\gamma}}$$

et

$$\langle t,\gamma\rangle := \exp\left(\pi\frac{t\overline{\gamma} - \bar{t}\gamma}{\Im\tau}\right).$$

3. L'application losange.

Définition 1. Soit $\mathbb{Z}[\langle P\rangle]$ le sous-groupe de $\mathbb{Z}[E(\mathbb{Q})]$ engendré par un point $P \in E(\mathbb{Q})$. Soit $\mathbb{Z}[E(\mathbb{Q})]^-$ le groupe quotient de $E(\mathbb{Q})$ par la relation d'équivalence $\mathrm{cl}(-P) = -\,\mathrm{cl}(P)$.

L'application losange est définie par

$$\begin{aligned} \diamond\colon \mathbb{Z}[\langle P\rangle] \times \mathbb{Z}[\langle P\rangle] &\to \mathbb{Z}[E(\mathbb{Q})]^- \\ \big((f),(g)\big) &\mapsto (f)\diamond(g) = \sum_{m,n} a_n b_m \,\mathrm{cl}\big((n-m)P\big) \end{aligned}$$

si $(f) = \sum_{n\in\mathbb{Z}} a_n[nP]$ et $(g) = \sum_{n\in\mathbb{Z}} b_n[nP]$.

On rappelle le théorème suivant.

Théorème 2.3 (**[3]**). *Le dilogarithme elliptique D^E s'étend par linéarité en un morphisme de $\mathbb{Z}[E(\mathbb{Q})]^-$ dans $\mathbb{R}$. Si f et g sont des fonctions sur E dont les diviseurs associés sont éléments de $\mathbb{Z}[\langle P\rangle]$ et si $\{f, g\} \in K_2(E)$, alors*

$$\pi r(\{f,g\}) = D^E\big((f)\diamond(g)\big).$$

En particulier

$$D^E\big((f)\diamond(1-f)\big) = 0.$$

3. De la mesure de Mahler au régulateur

Le lien est donné par la formule de Jensen. On peut se référer à **[13]**.

Prenons l'exemple suivant :

$$P(x,y) = \big(y - y_1(x)\big)\big(y - y_2(x)\big)$$

avec $|y_1(x)y_2(x)| = 1$ et $|y_1(x)| \leq 1$, $|y_2(x)| \geq 1$.

Alors

$$\begin{aligned} m(P) &= \frac{1}{(2\pi i)^2}\int_{\mathbb{T}^2} \log\big(\big|\big(y-y_1(x)\big)\big(y-y_2(x)\big)\big|\big)\frac{dx}{x}\frac{dy}{y} \\ &= -\frac{1}{2\pi i}\int_{|x|=1} \log\big(|y_1(x)|\big)\frac{dx}{x} \\ &= \frac{1}{2\pi i}\int_{|x|=1} \log\big(|y_2(x)|\big)\frac{dx}{x} \end{aligned}$$

grâce à la formule de Jensen.

Notons

$$\sigma_1 = \{(x, y_1(x)) \in E(\mathbb{C})/|x| = 1\}, \quad \sigma_2 = \{(x, y_2(x)) \in E(\mathbb{C})/|x| = 1\}$$

les chemins étant orientés par x parcourant le cercle unité dans le sens direct.

Si σ_1 convenablement orienté, par exemple, engendre $H_1(E, \mathbb{Z})^-$ et si $\{x, y\} \in K_2(E)$, on obtient

$$\begin{aligned} r(\{x, y\}) &= \frac{1}{2\pi} \int_{\sigma_1} \eta(x, y) \\ &= -\frac{1}{2\pi} \int_{\sigma_1} \log |y| d(\arg x) \\ &= -\frac{1}{2\pi i} \int_{|x|=1} \log(|y_1(x)|) \frac{dx}{x} \\ &= \pm m(P). \end{aligned}$$

Dans ce qui suit la situation ne sera pas aussi simple et nécessitera une étude plus spécifique.

4. Les résultats

Lorsque les courbes elliptiques E, de conducteur N ont même classe d'isogénie, on notera b_N la quantité

$$b_N = \frac{N}{4\pi^2} L(E, 2) = \varepsilon L'(E, 0)$$

où L désigne la série L et ε le signe de l'équation fonctionnelle de la courbe elliptique E.

Théorème 4.1.

(1) $$7m(Q_1) = 5m(P_1).$$

En outre, si $X_1(11)$ est la courbe modulaire

$$Y^2 + Y = X^3 - X^2$$

munie de son point de 5-torsion $P = (0, 0)$, alors

$$\pi r(\{X, Y\}) = 5D^E(P) - 5D^E(2P).$$

D'où

$$r(\{X, Y\}) \stackrel{?}{=} b_{11}.$$

(2) $$m(P_2) = 2m(Q_2).$$

En outre, si E est la courbe elliptique 14A

$$Y^2 + XY + Y = X^3 - X$$

munie de son point de 6-torsion $P = (1, 0)$, alors

$$\pi r(\{X, Y\}) = 3D^E(P) + 3D^E(2P)$$

et

$$r(\{X, Y\}) \stackrel{?}{=} b_{14}.$$

Preuve de (1). La preuve de (1) se trouve dans **[2]**. Elle sera donc omise. □

PREUVE DE (2). Soit

$$P_2(x_2, y_2) = y_2^2 - x_2y_2 + y_2 - x_2^3.$$

Notons E_2 la courbe elliptique associée à P_2. Puique P_2 vérifie le Corollaire 2.2, $\{x_2, y_2\}$ appartient à $K_2(E_2)$. On a également

$$y_1(x_2) = \frac{x_2 - 1 + \varepsilon\sqrt{4x_2^3 + (x_2-1)^2}}{2}$$

où $\varepsilon = \pm 1$ est choisi de façon à avoir $|y_1(x_2)| \leq 1$ pour $|x_2| = 1$.

La courbe elliptique E_2 peut être considérée comme un revêtement double de $\mathbb{P}_1$

$$\pi_{x_2} \colon E_2 \longrightarrow \mathbb{P}_1$$

ramifié aux racines de $4x_2^3 + (x_2-1)^2$ soit en -1, β, $\bar{\beta}$ avec $\beta = 3/8 + i\sqrt{7}/8$.

Alors $\sigma = \pi_{x_2}^{-1}([-1, \beta]) - \pi_{x_2}^{-1}([-1, \bar{\beta}])$ engendre $H_1(E_2, \mathbb{Z})^-$.

On va montrer que $\sigma_1 = \{(x_2, y_1(x_2)) \in E_2(\mathbb{C})/|x_2| = 1\}$ est dans la classe d'homologie de σ.

Remarquons que P_2 ne s'annule sur le tore qu'au point $(-1, -1)$ et projetons σ_1 et σ à l'aide de la seconde projection

$$\pi_y \colon E_2 \longrightarrow \mathbb{P}^1.$$

Grâce à `MAPLE`, on peut montrer que $\pi_y(\sigma_1)$ est formé de deux boucles contenues dans le disque unité et ne le coupant qu'en -1, parcourues dans le sens direct, la première boucle contenant $\pi_y(\pi_x^{-1}([-1, \beta]))$ parcouru dans le sens direct, la seconde contenant $\pi_y(\pi_x^{-1}([-1, \bar{\beta}]))$ parcouru dans le sens indirect. Par suite σ_1 engendre $H_1(E_2, \mathbb{Z})^-$ et

$$m(P_2) = \pm r(\{x_2, y_2\}).$$

Soit

$$Q_2(x_1, y_1) = (x_1+1)y_1^2 + (x_1^2 + x_1 + 1)y_1 + x_1(x_1+1)$$

et notons E_1 la courbe elliptique associée à Q_2.

D'après le Corollaire 2.2, $\{x_1, y_1\} \in K_2(E_1)$.

En outre

$$y_1(x_1) = \frac{-(x_1^2 + x_1 + 1) + \varepsilon\sqrt{(x_1^2 + x_1 + 1)^2 - 4x_1(x_1+1)^2}}{2(x_1+1)}$$

avec ε choisi tel que $|y_1(x_1)| \leq 1$.

La courbe elliptique E_1 peut être considérée comme un revêtement double de $\mathbb{P}^1$ par

$$\pi_{x_1} \colon E_1 \longrightarrow \mathbb{P}^1$$

ramifié aux racines de $(x_1^2 + x_1 + 1)^2 - 4x_1(x_1+1)^2$, c'est-à-dire en $\alpha = \frac{1}{2} + \sqrt{2} + \frac{1}{2}\sqrt{5 + 4\sqrt{2}}, 1/\alpha, \beta = \frac{1}{2} - \sqrt{2} + \frac{1}{2}\sqrt{5 - 4\sqrt{2}}, \bar{\beta}$ où $\arg\beta = 2,7243\ldots$ et $|\beta| = 1$.

Par suite, $\sigma = \pi_{x_1}^{-1}([\bar{\beta}, \beta])$ est une courbe fermé de E_1 engendrant $H_1(E_1, \mathbb{Z})^-$.

D'autre part, on a

$$m(Q_2) = -\frac{1}{2\pi}\int_{\arg\beta}^{\arg\bar{\beta}} \log|y_1(e^{i\phi})|\, d\phi = \frac{1}{2\pi}\int_{\arg\beta}^{\arg\bar{\beta}} \log|y_2(e^{i\phi})|\, d\phi$$

soit

$$2m(Q_2) = -\frac{1}{2\pi}\int_{\sigma_1\cup\tilde{\sigma}_2} \eta(x_1, y_1)$$

où

$$\sigma_1 = \{(x_1, y_1(x_1)) \in E_1(\mathbb{C})/|x_1| = 1 \arg\beta \leq \arg x_1 \leq \arg\bar{\beta}\}$$

et

$$\tilde{\sigma}_2 = \big\{\big(x_1, y_2(x_1)\big) \in E_1(\mathbb{C})/|x_1| = 1 \arg\bar{\beta} \leq \arg x_1 \leq \arg\beta\big\}.$$

On voit alors que $\sigma_1 \cup \tilde{\sigma}_2$ est une courbe fermée de E_1 ayant même classe d'homologie que σ. D'où

$$2m(Q_2) = \pm r(\{x_1, y_1\}).$$

Considérons maintenant la courbe elliptique E notée 14A dans les tables de Cremona et possédant l'équation affine

$$Y^2 + XY + Y = X^3 - X.$$

D'après le Corollaire 2.2, $\{X, Y\} \in K_2(E)$.

On a alors les isomorphismes

$$E_1 \simeq E$$

donné par les formules

$$x_1 = -\frac{Y}{X-1}, \quad y_1 = \frac{Y}{X+1}\frac{X^2-2Y-1}{(X-1)(X-Y-1)}$$

et

$$E_2 \simeq E$$

donné par les formules

$$x_2 = X, \quad y_2 = -Y-1.$$

Notons $P = (1, 0)$ le point de 6-torsion de E.

Comme $\{x_1, y_1\}$, $\{x_2, y_2\}$, $\{X, Y\}$ appartiennent respectivement à $K_2(E_1)$, $K_2(E_2)$ et $K_2(E)$, on allons exprimer leur régulateur en fonction du dilogarithme elliptique D^E de points rationnels de E.

On calcule d'abord les diviseurs suivants :

$$\begin{aligned}
(X) &= [2P] + [-2P] - 2[0],\\
(Y) &= [P] + [2P] + [3P] - 3[0],\\
(X+1) &= 2[3P] - 2[0],\\
(X-1) &= [P] + [-P] - 2[0],\\
(X-Y-1) &= 2[P] + [-2P] - 3[0],\\
(X^2-2Y-1) &= [3P] + 3[P] - 4[0],\\
(Y+1) &= 3[-2P] - 3[0],\\
(X-Y) &= 3[2P] - 3[0].
\end{aligned}$$

On en déduit

$$\begin{aligned}
(X) \diamond (Y) &= 3\,\mathrm{cl}([P]) + 3\,\mathrm{cl}([2P]),\\
(x_1) \diamond (y_1) &= 6\,\mathrm{cl}([P]) + 6\,\mathrm{cl}([2P]),\\
(x_2) \diamond (y_2) &= -9\,\mathrm{cl}([2P]),\\
(Y) \diamond (1+Y) &= -6\,\mathrm{cl}([P]) - 15\,\mathrm{cl}([2P]).
\end{aligned}$$

On montre alors aisément les relations

$$-4(X) \diamond (Y) + 2(x_2) \diamond (y_2) = 2(Y) \diamond (1+Y), \quad 2(X) \diamond (Y) = (x_1) \diamond (y_1),$$

d'où

$$2r(\{X,Y\}) = r(\{x_2,y_2\}), \quad 2r(\{X,Y\}) = r(\{x_1,y_1\})$$

et par suite

$$r(\{x_1,y_1\}) = r(\{x_2,y_2\})$$

c'est-à-dire

$$m(P_2) = 2m(Q_2),$$

puisque les mesures de Mahler sont des quantités positives.

On vérifie enfin à l'aide de `PARI` que

$$r(\{X,Y\}) \stackrel{?}{=} h_{14} \qquad \square$$

Le dilogarithme elliptique D^E vérifie pour certaines courbes elliptiques E des relations autres que les relations de distribution. Ces relations ont été appelées *exotiques* par Bloch et Grayson **[7]**.

La conjecture de Bloch et Grayson est la suivante.

Supposons $E(\mathbb{Q})_{\text{tors}}$ cyclique ayant un cardinal $d > 2$. Désignons par Σ le nombre de fibres de type I_ν pour $\nu \geq 3$, dans le modèle de Néron de E et supposons $[(d-1)/2] - \Sigma > 1$. (Ces conditions sont vérifiées en particulier pour les courbes 11A et 14A.)

Alors il existe au moins $[(d-1)/2] - \Sigma - 1$ relations *exotiques*

$$\sum_{r=1}^{[(d-1)/2]} a_r D^E(rP) = 0$$

où P est un point de d-torsion et $a_r \in \mathbb{Z}$.

Le corollaire suivant prouve la conjecture de Bloch et Grayson pour les courbes 11A et 14A.

COROLLAIRE 4.2.

(1) *Si E est la courbe* 11A *définie par*

$$Y^2 + Y = X^3 - X^2$$

et $P = (0,0)$ le point de 5-torsion de cette courbe, on a la relation exotique

$$3D^E(P) - 2D^E(2P) = 0.$$

(2) *Si E est la courbe elliptique* 14A *définie par*

$$Y^2 + XY + Y = X^3 - X$$

et $P = (1,0)$ le point de 6-torsion, on a la relation exotique

$$2D^E(P) + 5D^E(2P) = 0.$$

PREUVE DE (1). La preuve (1) est donnée dans **[2]**. Elle sera donc omise. $\square$

PREUVE DE (2). Dans la preuve du théorème nous avons obtenu les trois relations

$$2r(\{X,Y\}) = r(\{x_2,y_2\}),$$
$$(X) \diamond (Y) = 3\,\mathrm{cl}([P]) + 3\,\mathrm{cl}([2P]),$$
$$(x_2) \diamond (y_2) = -9\,\mathrm{cl}([2P]).$$

Grâce à l'expression analytique du régulateur, on en déduit

$$\pi r(\{X,Y\}) = 3D^E(P) + 3D^E(2P), \quad \pi r(\{x_2, y_2\}) = -9D^E(2P).$$

La première relation donne alors la relation *exotique* annoncée. □

Nous pouvons également prouver les théorèmes suivants.

THÉORÈME 4.3.

$$m\big(y^2(x+1)^2 + y(2x^2+3x+2) + (x+1)^2\big) = 2m\big(y^2 + y(x^2+x+1) + x^2\big).$$

Soit la courbe elliptique $15A = [1,1,1,0,0]$ *d'équation*

$$Y^2 + XY + Y = X^3 + X^2.$$

On a alors les relations

$$\pi r(\{X,Y\}) = 4D^E(P)$$

où P *désigne le point de* 4*-torsion* $P = (0,0)$ *et*

$$r(\{X,Y\}) \stackrel{?}{=} b_{15}.$$

DÉMONSTRATION. Les deux courbes elliptiques associées à ces polynômes sont isomorphes à la courbe 15A.

La preuve est semblable à celle des théorèmes précédents. □

THÉORÈME 4.4.

$$3m\big(y^2(x+1)^2 + y\big(2(x+1)^2 - 9x\big) + (x+1)^2\big)$$
$$=?8m\big((x+1)^2y^2 + xy + (x+1)^2\big) \stackrel{?}{=} 12b_{21}.$$

Si E_1 *est la courbe elliptique, isomorphe à la courbe* 21A*, d'équation*

$$Y_1(Y_1 + 3X_1) = X_1(X_1-1)^2$$

et $P = (1,0) \in E_1(\mathbb{C})$*, on a les relations*

$$\pi r(\{X_1, Y_1\}) = 8D^E(P)$$

et

$$\frac{r(\{X_1,Y_1\})}{4} \stackrel{?}{=} b_{21}.$$

DÉMONSTRATION. On montre que les courbes elliptiques associées aux deux polynômes sont isomorphes à la courbe 21A puis à la courbe E_1.

A l'aide de `PARI` on obtient la relation

$$K_{2,1}(\theta_1) + K_{2,1}(\theta_2) \stackrel{?}{=} \tfrac{3}{4}K_{2,1}(t)$$

où θ_1, θ_2, t sont respectivement les paramètres des points $P_1 = (i,-i)$, $P_2 = (-i,i)$ et $P = (1,0)$ de $E_1(\mathbb{C})$.

On termine alors comme précédemment. □

REMARQUE 1. Tous les résultats précédents ont été obtenus à partir de "bons" modèles tempérés des courbes des tables de Cremona **[8]**. On peut alors se demander si pour chaque courbe elliptique définie sur $\mathbb{Q}$, il existe un modèle de Weierstrass tempéré dont les racines des polynômes des faces valent ± 1. La famille étudiée par Rodriguez–Villegas dans **[11]** est alors particulièrement intéressante.

On montre aisément que la famille tempérée

$$y^2 + y(x^2+kx+1) + x^2 = 0$$

est isomorphe à la famille de Weierstrass E_k de "bons" modèles tempérés

$$Y^2 + kXY = X(X-1)^2$$

par l'isomorphisme

$$x = \frac{Y}{X-1}, \quad y = \frac{kY-(X-1)^2}{(X-1)^2}.$$

On en déduit que les courbes de conducteur inférieur à 100 notées 15A, 15B, 15D, 21A, 24A, 24D, 42D et 48F ont un "bon" modèle de Weierstrass tempéré.

Si l'on note

$$m_k = m\big(y^2 + y(x^2+kx+1) + x^2\big),$$

Boyd [**6**] a montré les relations étonnantes suivantes

$$m_1 \stackrel{?}{=} b_{15}, \quad m_5 \stackrel{?}{=} 6b_{15}, \quad m_{16} \stackrel{?}{=} 11b_{15}.$$

En fait on peut montrer que ces relations ne sont que la traduction de relations semblables sur les régulateurs

$$r(\Xi_{15B}) \stackrel{?}{=} 12r(\Xi_{15A}), \quad r(\Xi_{15D}) \stackrel{?}{=} 22r(\Xi_{15A})$$

où $\Xi_{15A} = \{X,Y\}_1$, $\Xi_{15B} = \{X,Y\}_6$, $\Xi_{15D} = \{X,Y\}_{16}$ correspondent aux courbes $Y^2 + kXY = X(X-1)^2$ pour $k=1$, $k=6$, $k=16$ respectivement.

Ces relations sont d'ailleurs des cas particuliers de relations plus générales

$$m_k = \tfrac{1}{2}r(\{x,y\}_k), \quad \pi r(\{x,y\}_k) = 8D^E(P)$$

pour les courbes $y^2 + y(x^2+kx+1) + x^2 = 0$ et le point de 4-torsion $P=(1,0)$ de la courbe E_k.

Références

1. A. Beilinson, *Higher regulators of modular curves*, Application of Algebraic K-Theory to Algebraic Geometry and Number Theory, Part I, II (Boulder, 1983), Contemp. Math., vol. 55, Amer. Math. Soc., Providence, RI, 1986, pp. 1–34.
2. M. J. Bertin, *Mesure de Mahler d'une famille de polynômes*, Crelle's J. (à paraître).
3. S. Bloch, *Higher regulators, algebraic K-theory and zeta functions of elliptic curves*, (Irvine Lecture, 1977) CRM-Monograph Series, vol. 11, Amer. Math. Soc., Providence, RI, 2000.
4. D. W. Boyd, *Kronecker's theorem and Lehmer's problem for polynomials in several variables*, J. Number Theory **13** (1981), 116–121.
5. ———, *Speculations concerning the Range of Mahler's measure*, Canad. Math. Bull. **24** (1981), 453–469.
6. ———, *Mahler's measure and special values of L-functions*, Experiment. Math. **7** (1998), 37–82.
7. S. Bloch and D. Grayson, *K_2 and L-functions of elliptic curves computer calculations*. I, Contemp. Math. **55** (1986), 79–88.
8. J. E. Cremona, *Algorithms for modular elliptic curves*, Cambridge University Press, 1992; Tables à `ftp ://euclid.ex.ac.uk/pub/cremona/data`.
9. C. Deninger, *Deligne periods of mixed motives, K-theory and the entropy of certain $\mathbb{Z}^n$-actions*, J. Amer. Math. Soc. **10** (1997), 259–281.
10. D. H. Lehmer, *Factorization of certain cyclotomic functions*, Ann. of Math. (2) **34** (1933), 461–479.
11. F. Rodriguez–Villegas, *Modular Mahler measures*, Preprint, 1996.
12. ———, *Modular Mahler measures*. I, Topics in Number Theory (S. D. Ahlgren, G. E. Andrews, and K. Ono, eds.), Kluwer, Dordrecht, 1999, pp. 17–48.

13. ———, *Identities between Mahler measures*, Preprint, 2000; http ://www.ma.utexas.edu/users/villegas/research.html.

14. D. Zagier and H. Gangl, *Classical and elliptic polylogarithms and special values of L-series*, The Arithmetic and Geometry of Algebraic Cycles, Nato Adv. Sci. Inst. Ser. C Math. Phys. Sci., vol. 548, Kluwer, Dordrecht, 2000, pp. 561–615.

Institut de Mathématiques de Jussieu, Université Pierre et Marie Curie, 175, rue du Chevaleret, 75013 Paris, France.

E-mail address: bertin@math.jussieu.fr

Centre de Recherches Mathématiques
CRM Proceedings and Lecture Notes
Volume **36**, 2004

The p-adic Height Pairings of Coleman–Gross and of Nekovář

Amnon Besser

Abstract. We prove that the p-adic height pairing of Nekovář, considered for algebraic curves, gives the p-adic height pairing of Coleman and Gross, defined using Coleman integration.

1. Introduction

In [**7**], Coleman and Gross proposed a definition of a p-adic height pairing on curves over number fields with good reduction at primes above p. The pairing was defined as a sum of local terms and the most interesting terms are the ones corresponding to primes above p where the definition depends on Coleman's theory of p-adic integration. Later, Nekovář (following a construction of Zarhin [**18**] for abelian varieties) constructed in [**12**] a general p-adic height pairing for Galois representations which are crystalline (or even semi-stable) at primes above p, and satisfying certain additional technical assumptions.

Wintenberger raised the question of the equality of the two height pairings for curves. In other words: does the Nekovář height pairing applied to $H^1_{\text{ét}}$ of a curve recover the Coleman–Gross pairing. The purpose of this small note is to answer this in the affirmative. More precisely, let F be a number field and let X be a smooth and proper curve over F, with good reduction at all places v above a fixed prime p. To define either the height of Coleman and Gross or the height of Nekovář the following choices must be made.

- A "global log"—a continuous idele class character
$$\ell\colon \mathbb{A}_F^\times/F^\times \to \mathbb{Q}_p$$
- for each $v \mid p$ a choice of a subspace $W_v \in H^1_{\mathrm{dR}}(X\otimes F_v/F_v)$ complementary to the space of holomorphic forms.

For the definition of the Coleman–Gross height we must insist that the local characters ℓ_v, for $v \mid p$, are ramified in the sense that they do not vanish on the units in F_v (this seems to be overlooked in [**7**]).

2000 *Mathematics Subject Classification.* Primary: 11G30; Secondary: 14F30, 11G25.
This is the final form of the paper.

With these choices the Coleman–Gross height pairing is a pairing

$$h_{\mathrm{CG}}\colon \operatorname{Div}_0(X) \times \operatorname{Div}_0(X) \to \mathbb{Q}_p,$$

where $\operatorname{Div}_0(X)$ denotes the group of zero divisors on X (defined over F), while the Nekovář height pairing is a pairing

$$h_{\mathrm{Nek}}\colon H^1_{\mathrm{f}}(F,V) \times H^1_{\mathrm{f}}(F,V) \to \mathbb{Q}_p,$$

where $V = V_p(J_X) \cong H^1_{\text{ét}}(\overline{X}, \mathbb{Q}_p(1))$, with $\overline{X} = X \otimes_F \overline{F}$, is the Tate-module of the Jacobian of X and H_{f} is finite cohomology in the sense of Bloch and Kato. The relation between the two pairings is provided by the étale Abel–Jacobi map

$$\alpha_X\colon \operatorname{Div}_0(X) \to H^1_{\mathrm{f}}(F,V).$$

Our main result is the following.

THEOREM 1.1. *With the same choices of* (ℓ, W_v) *we have*

$$h_{\mathrm{Nek}}\big(\alpha_X(y), \alpha_X(z)\big) = h_{\mathrm{CG}}(y,z).$$

The proof goes as follows. We first recall the construction of the Coleman–Gross height in Section 2. This is expressed in terms of local heights. Likewise, the Nekovář height can be expressed as a sum of local heights, although it also has a global description which we will not need here. The decomposition into local heights depends on a choice of a "mixed extension." We recall this in Section 3 as well as the particular mixed extension we will use. With this mixed extension, it is known that the local heights at places not dividing p are equal to the local heights of Coleman–Gross. This leaves the comparison of the local heights above p. This is done in the rest of the paper.

We would like to thank Wintenberger and Nekovář for suggesting this problem to us on various occasions. We would also like to thank the referee for several useful suggestions.

2. The Coleman–Gross height pairing

We first recall the theory of p-adic height pairings due to Coleman and Gross [7].

Recall that we have the character $\ell\colon \mathbb{A}_F^\times/F^\times \to \mathbb{Q}_p$. One deduces from ℓ the following data:

- For any place $v \nmid p$ we have $\ell_v(\mathcal{O}_{F_v}^\times) = 0$ for continuity reasons, which implies that ℓ_v is completely determined by the number $\ell_v(\pi_v)$, where π_v is any uniformizer in F_v.
- For any place $v \mid p$ we can decompose

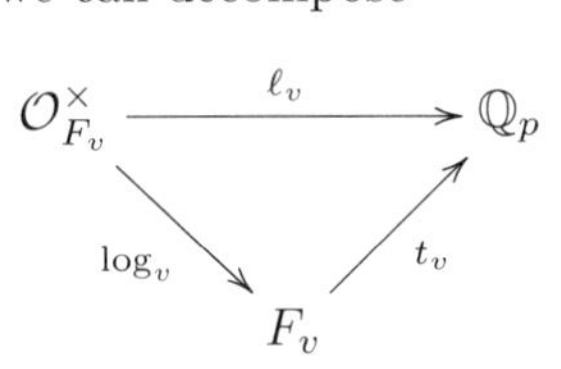

where t_v is a $\mathbb{Q}_p$-linear map. Since we assume that ℓ_v is ramified it is then possible to extend $\log_v$ to $F_v^\times$ in such a way that the diagram remain commutative.

The height pairing is a sum of local terms, $h_{\mathrm{CG}}(y,z) = \sum_v h^v_{\mathrm{CG}}(y,z)$ over all finite places v. When $v \nmid p$ the local term is given [**7**, (1.3)] by

$$h^v_{\mathrm{CG}}(y,z) = \ell_v(\pi_v) \cdot (y,z) \tag{2.1}$$

where (y,z) denotes the intersection multiplicity of the extension of y and z to a regular model of X over $\mathcal{O}_{F_v}$.

We now describe the local contribution at a place $v \mid p$. We modify the definitions of Coleman and Gross to work with integration theory over $\overline{\mathbb{Q}}_p$ instead of over $\mathbb{C}_p$. Let X be a curve over $K = F_v$ with good reduction. Let $\Omega^1_{\log}(X)$ be the space of one forms on X (defined over K) with at most logarithmic singularities. For $\omega \in \Omega^1_{\log}(X)$ we let $\mathrm{Res}(\omega)$ be the residue divisor. This is a divisor of degree 0 defined over K such that over $\overline{K}$ it becomes $\sum_{P\in X} \mathrm{Res}_P(\omega)P$. Let $\Omega^1(X)$ be the space of global holomorphic forms on X. We note that if ω_1 and $\omega_2 \in \Omega^1_{\log}(X)$ and $\mathrm{Res}(\omega_1) = \mathrm{Res}(\omega_2)$ then $\omega_1 - \omega_2 \in \Omega^1(X)$. There is a canonical projection

$$\Psi\colon \Omega^1_{\log}(X) \to H^1_{\mathrm{dR}}(X). \tag{2.2}$$

In [**7**] it is given by certain requirements. Following [**4**] we can describe it as follows:

PROPOSITION 2.1. *Let $\omega \in \Omega^1_{\log}(X)$ and let $U \subset X$ be a wide open subspace on which ω is holomorphic. The class $\Psi(\omega)$ is the image of $\omega|_U \in H^1_{\mathrm{dR}}(U)$ under the unique Frobenius equivariant section to the map $H^1_{\mathrm{dR}}(X) \to H^1_{\mathrm{dR}}(U)$.*

By Frobenius equivariant we mean the following: Both $H^1_{\mathrm{dR}}(X)$ and $H^1_{\mathrm{dR}}(U)$ carry an action of a linear Frobenius. This is because these two cohomology groups are isomorphic to the rigid cohomology groups of the reduction of X and of an open affine subset of it respectively, and rigid cohomology has a natural action of Frobenius. The map $H^1_{\mathrm{dR}}(X) \to H^1_{\mathrm{dR}}(U)$ commutes with Frobenius, and, as is explained in loc. cit., because of weight considerations there is a unique section which commutes with Frobenius as well.

Now recall that we have at our disposal the complementary subspace $W = W_v$.

DEFINITION 2.2. For any divisor y of degree 0 on X we let $\omega_W(y) \in \Omega^1_{\log}(X)$ be the unique form satisfying $\mathrm{Res}(\omega_W(y)) = y$ and $\Psi(\omega_W(y)) \in W$.

Now we need to use the theory of p-adic integration. In [**7**] the theory of Coleman [**6**] is used. Several other p-adic integration theories have been developed in recent years [**3, 4, 8, 17, 19**]. In most of these theories one can at least integrate algebraic differentials on all smooth algebraic varieties over a finite extension of $\mathbb{Q}_p$ and in this domain of integration they are conjectured to be identical. In [**4**] it is shown that the theory developed there using [**17**], as well as the theory of [**8**], coincide for curves with Coleman's integration theory. For our needs these theories are equivalent.

The integration theory depends on a branch of the p-adic log. For this we take the branch $\log_v$ defined by ℓ_v and we can extend it in a unique way to $\overline{K}$. The properties of p-adic integration, which are valid for all versions except [**3, 6**] (which work in the rigid setting but require good or close to good reduction), can be summarized as follows.

PROPOSITION 2.3. *For any smooth algebraic variety X/K and a holomorphic differential ω on X defined over K the theory associates a function $\int \omega\colon X(\overline{K}) \to \overline{K}$ unique up to an additive constant, such that the following properties are satisfied:*

- *The function* $\int \omega$ *is locally analytic and satisfies* $d \int \omega = \omega$.
- *The integration is functorial with respect to arbitrary morphisms in the sense that if* $f\colon Y \to X$ *is defined over* K*, then we have* $\int f^*\omega = f^* \int \omega$ *up to a constant.*
- *If* $\sigma \in \mathrm{Gal}(\overline{K}/K)$ *and* $x, y \in X(\overline{K})$*, then* $\int_{\sigma(x)}^{\sigma(y)} \omega = \sigma\left(\int_x^y \omega\right)$.

The following easy corollary of the above properties is well-known (and is in fact used by Zarhin and Colmez to define the integral).

COROLLARY 2.4. *Let* J *be a commutative algebraic group over* K*, let* ω *be an invariant differential on* J *and let* ω_0 *be its value at the identity element. Let* $\log\colon J(K) \to T_0(J)$ *be the logarithm for the* p*-adic Lie group* J*, depending on the choice of a branch of the logarithm, taking values in the tangent space at* 0 *to* J*. Then we have* $\int_0^P \omega = \langle \omega_0, \log(P) \rangle$*, where* $\langle \cdot, \cdot \rangle$ *is the duality between the cotangent and tangent space at the identity.*

Suppose again that X is a curve as before and let $\omega \in \Omega^1_{\log}(X)$. Then we obtain an integral $\int \omega$ defined on $X(\overline{K})$ minus the singular points of the form. The sum of the values of $\int \omega$ on $z \in \mathrm{Div}_0(X)$, disjoint from the singular points, is well-defined independent of the constant of integration and denoted $\int_z \omega$. Furthermore, since z and ω are defined over K we have $\int_z \omega \in K$.

DEFINITION 2.5. The local height pairing at $v \mid p$ of Coleman and Gross is defined as follows: Let y and z be two divisors of degree 0 on X with disjoint supports. Then their pairing is given by $h^v_{\mathrm{CG}}(y,z) := t_v\left(\int_z \omega_{W_v}(y)\right)$.

3. The Nekovář height in terms of mixed extensions

Nekovář gives two expressions for the height pairing. The one which is relevant for us is given in terms of mixed extensions and is expressed as a sum of local terms. As we saw in the previous section the Coleman–Gross height pairing is also given as a sum of local terms. That the terms at places not dividing p are the same is well-known and we recall this here.

The height pairing of Nekovář is a map

$$h_{\mathrm{Nek}}\colon H^1_{\mathrm{f}}(F, V) \times H^1_{\mathrm{f}}(F, V^*(1)) \to \mathbb{Q}_p,$$

where V is a continuous $G = \mathrm{Gal}(\overline{F}/F)$-representation satisfying certain conditions (see [**12**, 2.1.2]) and H_{f} is finite cohomology in the sense of Bloch and Kato [**5**]. To describe the height pairing we assume that we are given extensions of continuous G-representations,

$$0 \to V \to E_1 \to A \to 0 \tag{3.1}$$

and

$$0 \to B(1) \to E_2 \to V \to 0. \tag{3.2}$$

Here, A and B are finite dimensional $\mathbb{Q}_p$-representations of G which are trivialized by a finite extension (this is more general then in [**12**] and is inspired by [**15**]). Suppose for now that G acts trivially on A and B. Then, such sequences yield (the second by first dualizing) elements $[E_1] \in \mathrm{Hom}(A, H^1(F, V))$ and $[E_2] \in B \otimes H^1(F, V^*(1))$. We assume that these in fact belong to $\mathrm{Hom}(A, H^1_{\mathrm{f}}(F, V))$ and $B \otimes H^1_{\mathrm{f}}(F, V^*(1))$ respectively. The height pairing produces out of these two extensions an element $h(E_1, E_2) \in \mathrm{Hom}(A, B)$. The construction is functorial with respect to maps $A' \to A$ and $B \to B'$. To get the height pairing itself one restricts to the case $A = B = \mathbb{Q}_p$. The more general setup is convenient for example in geometric situations, as we will see.

To describe the height pairing one chooses a mixed extension. This is an embedding of the sequences (3.1) and (3.2) inside a commutative diagram with exact rows and columns as follows [**12**, p. 159]:

$$
\begin{array}{ccccccccc}
 & & & & 0 & & 0 & & \\
 & & & & \downarrow & & \downarrow & & \\
0 & \longrightarrow & B(1) & \longrightarrow & E_2 & \xrightarrow{\pi} & V & \longrightarrow & 0 \\
 & & \| & & \downarrow & & \downarrow & & \\
0 & \longrightarrow & B(1) & \longrightarrow & E & \longrightarrow & E_1 & \longrightarrow & 0 \\
 & & & & \downarrow & & \downarrow & & \\
 & & & & A & = & A & & \\
 & & & & \downarrow & & \downarrow & & \\
 & & & & 0 & & 0 & &
\end{array}
\tag{3.3}
$$

Having fixed such a mixed extension E we obtain local mixed extensions E_v by restricting to the decomposition groups at the various places v. Each of these local mixed extensions gives a local contribution h_{E_v} and the sum $h_E = \sum_v h_{E_v}$ is shown to depend only on E_1 and E_2 so we can define $h(E_1, E_2) = h_E$. We will only need to describe the local contribution at places above p, which we will do in the next section.

Suppose now that X is a curve as in the statement of the theorem and that y and z are two zero divisors on X with disjoint supports. Let $\mathcal{Y}$ and $\mathcal{Z}$ be the supports of y and z respectively. We will use an overline ($\overline{\mathcal{Y}}$ and $\overline{\mathcal{Z}}$) to denote extension of scalars to the algebraic closure of the corresponding field. Then we obtain our mixed extension as follows (compare [**12**, 5.6]). Let $\tilde{A} = (\mathbb{Q}_p^{\overline{\mathcal{Y}}})_0$ be the subspace of the space of functions from $\overline{\mathcal{Y}}$ to $\mathbb{Q}_p$ where the sum of values is 0, and let $\widetilde{B} = (\mathbb{Q}_p^{\overline{\mathcal{Z}}})^0$ be the quotient of the corresponding space for $\mathcal{Z}$ by the subspace of constant functions.

DEFINITION 3.1. The semi geometric mixed extension associated with the cycles y and z is the mixed extension in (3.3) with

$$V = H^1_{\text{ét}}(\overline{X}, \mathbb{Q}_p(1)), \qquad E_1 = H^1_{\text{ét}}(\overline{X} - \overline{\mathcal{Y}}, \mathbb{Q}_p(1)), \qquad E_2 = H^1_{\text{ét}}(\overline{X}; \overline{\mathcal{Z}}, \mathbb{Q}_p(1))$$

(étale cohomology of $\overline{X}$ relative to $\overline{\mathcal{Z}}$), and finally

$$E = H^1_{\text{ét}}(\overline{X} - \overline{\mathcal{Y}}; \overline{\mathcal{Z}}, \mathbb{Q}_p(1)).$$

The associated geometric mixed extension is the extension obtained from the semi geometric mixed extension by the maps $A = \tilde{A}^G \hookrightarrow \tilde{A}$ and $\widetilde{B} \twoheadrightarrow B = \widetilde{B}_G$.

In this mixed extension, $\tilde{A}$ is identified with the kernel of $H^2_{\overline{\mathcal{Y}},\text{ét}}(\overline{X}, \mathbb{Q}_p(1)) \to H^2_{\text{ét}}(\overline{X}, \mathbb{Q}_p(1))$ and $\widetilde{B}$ with the cokernel of $H^0_{\text{ét}}(\overline{X}, \mathbb{Q}_p(0)) \to H^0_{\text{ét}}(\overline{X}; \overline{\mathcal{Z}}, \mathbb{Q}_p(0))$.

Consider now $y, z \in \mathrm{Div}_0(X)$ as $y \in A$ and $z \in B^*$ ($\mathbb{Q}_p$-dual) in the obvious manner. Then it is well-known that $[E_1](y) = \alpha_X(y)$ and $((z \otimes 1) \circ [E_2])(1) = \alpha_X(z)$.

We thus find

$$h_{\mathrm{Nek}}(y,z) = z\big(h_E(y)\big) = \sum_v z\big(h_{E_v}(y)\big).$$

PROPOSITION 3.2. *For $v \nmid p$ we have $z\big(h_{E_v}(y)\big) = h^v_{\mathrm{CG}}(y,z)$, where E is the mixed extension described above and h^v_{CG} is the local height pairing of Coleman and Gross described in* (2.1).

PROOF. This is well-known. See for example [**13**, Proposition 2.16] where a more general result is proved. ☐

To prove Theorem 1.1 it thus remains to prove the following result.

THEOREM 3.3. *For $v \mid p$ we have, with the same choice of the space W_v,*

$$z\big(h_{E_v}(y)\big) = h^v_{\mathrm{CG}}(y,z).$$

The proof of this theorem will occupy the rest of this paper. Before proving this we want to make the following comment: Since we will dualize several times with respect to cup products, one has to check the signs very carefully here. We did not do this, hence, a priori, Theorem 3.3 is only true up to a global sign. This sign can not be -1, given that all the other local terms match up without a sign, however, as can be seen from the fact that both the Nekovář height pairing and the Coleman–Gross height pairing have the property that they vanish if either y or z are principal divisors.

There is also the question of a global sign. I would like to thank the referee for the following observation: The sign used in this paper is minus that used in [**12**] and this sign should in fact be preferable as is discussed in [**13**, 6.4].

4. The local height at a place above p

We now begin to compute the local contribution to the height pairing at a place $v \mid p$ coming from the mixed extension described in the previous section. In this section we do this using a certain splitting (the map w below), which will be described in Section 6.

We begin by recalling the general construction of Nekovář in [**12**, 4.7]. We make several modifications to bring it to a form suitable for comparison with the Coleman-Gross construction.

Let $K = F_v$. The Kummer map,

$$K^\times \otimes \mathbb{Q}_p \xrightarrow{\alpha_K} H^1\big(K, \mathbb{Q}_p(1)\big), \tag{4.1}$$

is an isomorphism and identifies $H^1_{\mathrm{f}}\big(K,\mathbb{Q}_p(1)\big)$ with $\mathcal{O}_K^\times \hat{\otimes} \mathbb{Q}_p$, $\hat{\otimes}$ denoting the completed tensor product, so that there exists a short exact sequence

$$0 \to H^1_{\mathrm{f}}\big(K,\mathbb{Q}_p(1)\big) \to H^1\big(K,\mathbb{Q}_p(1)\big) \xrightarrow{\mathrm{ord}} \mathbb{Q}_p \to 0. \tag{4.2}$$

The p-adic logarithm $\log\colon \mathcal{O}_K^\times \to K$ defines an isomorphism $H^1_{\mathrm{f}}\big(K,\mathbb{Q}_p(1)\big) \cong K$. The branches of the p-adic logarithms, i.e., extensions of log to $K^\times$, are therefore in one to one correspondence with splittings of (4.2). We will denote the splitting induced by a branch $\log_\varpi$ by $\log'_\varpi\colon H^1\big(K,\mathbb{Q}_p(1)\big) \to H^1_{\mathrm{f}}\big(K,\mathbb{Q}_p(1)\big)$. We can view the local component of the idele class character ℓ_v as a map $\ell_v\colon H^1\big(K,\mathbb{Q}_p(1)\big) \to \mathbb{Q}_p$. Then, the character ℓ_v factors as $t_v \circ \log_\varpi$ if and only if we have $\ell_v = \ell_v \circ \log'_\varpi$. Note that in this case we have

$$\big(K \to H^1_{\mathrm{f}}\big(K,\mathbb{Q}_p(1)\big) \xrightarrow{\ell_v} \mathbb{Q}_p\big) = t_v, \tag{4.3}$$

where the first map is the inverse to the map induced the logarithm. It is in fact just the Bloch–Kato exponential map to be discussed later. From now onward we assume that such a choice of $\log_\varpi$ has been made.

We now consider the diagram (3.3) as a diagram of representations of $\mathrm{Gal}(\overline{K}/K)$ and assume that A and B are trivial representations. Recall that $\pi\colon E_2 \to V$. We then have the following diagram

(4.4)

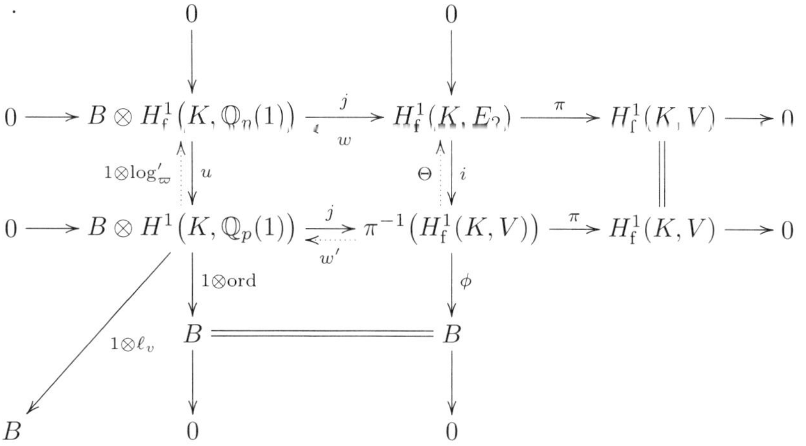

Here, by abuse of notation, we denote by π also the map induced by π on cohomology. This diagram is mostly copied from [**12**, 4.8]. The dotted lines in the diagram denote various splittings. We have already discussed $\log'_\varpi$. The map w depends on the choice of the complementary subspace W_v from the introduction (it does not depend on a choice of $\log_\varpi$) and will be discussed later. The two other splittings are induced, w' by w and Θ by $\log'_\varpi$ in the following formal way.

DEFINITION 4.1. Let $x \in \pi^{-1}\big(H^1_{\mathrm{f}}(K,V)\big)$ and choose $y \in H^1_{\mathrm{f}}(K, E_2)$ such that $\pi(y) = \pi(x)$. Then $\pi\big(x - i(y)\big) = 0$ so $x - i(y) = j(z)$ with $z \in B \otimes H^1\big(K, \mathbb{Q}_p(1)\big)$. Then we define

$$w'(x) = u\big(w(y)\big) + z$$

and

$$\Theta(x) = y + j \circ (1 \otimes \log'_\varpi)(z).$$

The following is an easy diagram chase away.

LEMMA 4.2. *The maps w' and Θ are well-defined and constitute splittings of the respective diagrams. Furthermore, in diagram (4.4) the two possible ways of connecting $H^1_{\mathrm{f}}(K, E_2)$ with $B \otimes H^1\big(K, \mathbb{Q}_p(1)\big)$ (in both directions) give the same map.*

In [**12**, 4.8] Nekovář describes the local height pairing separately in the two cases ℓ_v ramified or not. However it is easy to see that his two definitions coincide with the following.

DEFINITION 4.3. The local height pairing corresponding to the mixed extension E is the element of $h_E \in \mathrm{Hom}(A, B)$ obtained as follows: We have $[E] \in \mathrm{Hom}\big(A, H^1(K, E_2)\big)$, but in fact lies in $\mathrm{Hom}\big(A, \pi^{-1}\big(H^1_{\mathrm{f}}(K,V)\big)\big)$. Then h_E is the composition

$$A \xrightarrow{[E]} \pi^{-1}\big(H^1_{\mathrm{f}}(K,V)\big) \xrightarrow{w'} B \otimes H^1\big(K, \mathbb{Q}_p(1)\big) \xrightarrow{1\otimes\ell_v} B.$$

Since we are assuming the existence of a $\log_{\varpi}$ through which ℓ_v factors, we can write this in a different way.

LEMMA 4.4. *We have* $(1\otimes \ell_v)\circ w' = (1\otimes \ell_v)\circ u\circ w\circ \Theta$.

PROOF. With x, y and z as in Definition 4.1 we have

$$u\circ w\circ \Theta(x) = u\Big(w\Big(y + j((1\otimes \log'_{\varpi})(z))\Big)\Big) = u(w(y) + (1\otimes \log'_{\varpi})(z))$$

so that $u\circ w\circ\Theta(x) - w'(x) = u((1\otimes \log'_{\varpi})(z)) - z$. But the compatibility of $\log_{\varpi}$ with ℓ_v precisely means that ℓ_v vanishes on the image of $u\circ \log'_{\varpi} - \mathrm{Id}$. $\square$

As a consequence we have obtained the description of the local height pairing as the composition

$$h_E = (1\otimes \ell_v)\circ u\circ w\circ \Theta\circ [E]. \tag{4.5}$$

5. The connection with p-adic integration

In this section we go back to the geometric mixed extension of Definition 3.1 and we show that the composition $\Theta\circ [E]\colon A\to H^1_{\mathrm{f}}(K,E_2)$ can be described in terms of Coleman integration.

Assume that we are given a curve X over K, zero divisors y and z with disjoint supports $\mathcal{Y}$ and $\mathcal{Z}$ respectively. We assume that y and z are sums of K-rational points. This implies that $A = \tilde{A}$ in the notation of Section 3. This restriction will be eliminated in Section 6.

In the situation we are considering, we have a natural isomorphism, given by the exponential map of Bloch and Kato,

$$D_{\mathrm{dR}}(V)/F^0 \xrightarrow{\exp} H^1_{\mathrm{f}}(K,V), \tag{5.1}$$

where D_{dR} is the de Rham functor of Fontaine, and similarly with V replaced by E_2. Recall that $D_{\mathrm{dR}}(V)$ has a filtration, which we used above, and is also an extension of scalars from K_0, the maximal unramified extension of $\mathbb{Q}_p$ in K, of $D_{\mathrm{cr}}(V)$, which comes equipped with a semi-linear Frobenius. Thus, $D_{\mathrm{dR}}(V)$ comes with a natural linear Frobenius (raising the semi-linear one to the power $[K_0{:}\mathbb{Q}_p]$ and extending scalars) and the same is true for $D_{\mathrm{dR}}(E_2)$.

By the de Rham conjecture proved by Faltings [**9**], we know that the de Rham functors applied to the local representations are given by $D_{\mathrm{dR}}(V) = H^1_{\mathrm{dR}}(X/K)$ and $D_{\mathrm{dR}}(E_2) = H^1_{\mathrm{dR}}(X;\mathcal{Z}/K)$, where in both cases the cohomology is twisted by 1 so that $F^0D_{\mathrm{dR}} = F^1H_{\mathrm{dR}}$.

An element $\beta\in H^1_{\mathrm{f}}(K,E_2)$ defines a K-valued functional on $F^1H^1_{\mathrm{dR}}(X-\mathcal{Z}/K)$ as follows: By the exponential map (5.1) we may view β as an element of $H^1_{\mathrm{dR}}(X;\mathcal{Z}/K)/F^1$, which is Poincaré dual to $F^1H^1_{\mathrm{dR}}(X-\mathcal{Z}/K)$.

If $D\in A$, then $\Theta\circ[E](D)\in H^1_{\mathrm{f}}(K,E_2)$ gives such a functional which we denote $[E]_D$. To describe this explicitly, note that $F^1H^1_{\mathrm{dR}}(X-\mathcal{Z}/K)$ is the space of one-forms ω on X with logarithmic singularities along $\mathcal{Z}$. Then we have

$$[E]_D(\omega) = \omega\cup \exp^{-1}\big(\Theta\circ[E](D)\big). \tag{5.2}$$

The following result is an extension of [**1**, Proposition 9.2].

PROPOSITION 5.1. *For* $D\in A$ *we have* $[E]_D(\omega) = \int_D \omega$.

To prove this result, we first recall the generalized Jacobian J of X with respect to $\mathcal{Z}$. This can be described in two different ways as either the Picard group of line bundles of degree 0 on X with trivialization along $\mathcal{Z}$, or as divisors of X with support disjoint from $\mathcal{Z}$ modulo divisors of functions that take the value 1 on $\mathcal{Z}$. The maps between the two descriptions takes a divisor D to $\mathcal{O}(D)$ together with the trivialization furnished by the canonical Section 1 and going the other way take a line bundle trivialized along $\mathcal{Z}$ to the divisor of a section which takes the value 1 at all points of $\mathcal{Z}$ under the trivialization. Since we are assuming that $\mathcal{Z}$ splits over K there exists a short exact sequence

$$0 \to \Big(\bigoplus_{x\in\mathcal{Z}} \mathbb{G}_{\mathrm{m}}\Big)\Big/\mathbb{G}_{\mathrm{m}} \to J \to J_X \to 0, \tag{5.3}$$

where J_X is the Jacobian of X. This is given in the first description as follows: The left hand map sends a collection $(\alpha_x) \in \mathbb{G}_{\mathrm{m}}$ to the trivial line bundle with the trivialization being the obvious one skewed by α_x at x. The right hand map forgets the trivializations. In the second description the right hand maps takes the class of the divisor D to the class of the same divisor (but under a stronger relation) and the left hand map sends the collection (α_x) to the divisor of a function that takes the value α_x at x.

The following is known:

LEMMA 5.2. *The Galois representation $E_2 = H^1_{\text{ét}}(\overline{X}; \overline{\mathcal{Z}}, \mathbb{Q}_p(1))$ is isomorphic to $V_p(J)$, the p-adic Tate-module of J tensored with $\mathbb{Q}_p$. Furthermore, the exact sequence* (3.2) *is induced by the short exact sequence* (5.3).

PROPOSITION 5.3. *Let D be a zero-divisor on X. Then we have $[E](D) = \alpha_J([D]) \in H^1(K, V_p(J))$ where α_J is the Kummer map of J and $[D]$ is the class of D in $J(K)$.*

PROOF. This follows by the same argument as the one used in the proof of [**14**, Appendix, Lemma] □

The following is an easy extension of the well-known identification of the tangent space to the Picard scheme (probably, also well-known).

LEMMA 5.4. *There is a canonical isomorphism $T_0(J) \to H^1_{\mathrm{dR}}(X; \mathcal{Z}/K)/F^1$.*

By [**5**, Example 3.10.1] there exists a commutative diagram

$$\begin{array}{ccc} H^1_{\mathrm{dR}}(X;\mathcal{Z}/K)/F^1 = T_0(J) & \xrightarrow{\exp_J} & J(K)\hat{\otimes}\mathbb{Q}_p \\ & \searrow^{\exp} & \downarrow{\scriptstyle \alpha_J} \\ & & H^1_{\mathrm{f}}(K, E_2) \end{array} \tag{5.4}$$

Where $\exp_J$ is the exponential map of the p-adic Lie group $J(K)$ while exp is the Bloch–Kato exponential.

PROPOSITION 5.5. *Assume that X has a K-rational point P_0 and let $g\colon X - \mathcal{Z} \to J$ be the map $P \mapsto P - P_0$. Then the map $g^*\colon \Omega^1(J) \to F^1H^1_{\mathrm{dR}}(X - \mathcal{Z}/K)$ is an isomorphism (recall that the space on the right is the space of forms with logarithmic singularities along $\mathcal{Z}$). Furthermore, this isomorphism fits in a commutative* *diagram*

$$\begin{array}{ccc} \Omega^1(J) & \xrightarrow{g^*} & F^1 H^1_{\mathrm{dR}}(X - \mathcal{Z}/K) \\ {\scriptstyle \mathrm{ev}_0}\downarrow & & \downarrow{\scriptstyle \text{Poincaré duality}} \\ T_0(J)^* & \longrightarrow & (H^1_{\mathrm{dR}}(X;\mathcal{Z}/K)/F^1)^* \end{array}$$

where ev_0 *is evaluation at* 0 *and the bottom map is the dual of the isomorphism of Lemma* 5.4.

PROOF. The first statement is [**16**, Proposition V.5]. The second statement is again well known for the usual Jacobian (see, e.g., the proof of Proposition 2.2 in [**11**]) and the proof is the same. □

PROOF OF PROPOSITION 5.1. The short exact sequence (5.3) induces a short exact sequence of tangent spaces compatible with exponential maps. In addition, the exponential map for J_X is onto. Even more precisely, an integer multiple of every class in $J_X(K)$ is in the image of the exponential map, without tensoring with $\mathbb{Q}_p$. This implies that any divisor D on X, can be, after multiplying by a sufficiently large integer, written as a sum of a divisor, whose class, $[D] \in J(K)$, is in the image of $\exp_J$, and a principal divisor, both disjoint from $\mathcal{Z}$. It is thus sufficient to check the result for these two types of divisors.

We now identify the space $T_0(J)$ with $H^1_{\mathrm{dR}}(X,\mathcal{Z}/K)/F^1$ and the space $\Omega^1(J)$ with $F^1H^1_{\mathrm{dR}}(X - \mathcal{Z}/K)$. The duality between these spaces is given by the cup product.

For the first type the proof is essentially the same as the proof of [**1**, Proposition 9.2]: By assumption there exists $\eta \in H^1_{\mathrm{dR}}(X;\mathcal{Z}/K)/F^1$ with $[D] = \exp_J(\eta)$. We have

$$\begin{aligned} [E](D) &= \alpha_J([D]) && \text{by Proposition 5.3} \\ &= \alpha_J\big(\exp_J(\eta)\big) \\ &= \exp(\eta) && \text{by (5.4).} \end{aligned}$$

In particular we have $[E](D) \in H^1_{\mathrm{f}}(K, E_2)$ so that $\Theta \circ [E](D) = [E](D) = \exp(\eta)$ as well. It follows that $[E]_D(\omega) = \omega \cup \eta$. By Proposition 5.5 there exists a holomorphic invariant differential ω' on J such that $g^*\omega' = \omega$. By functoriality of the p-adic integral (Proposition 2.3) we have $\int_D \omega = \int_0^{[D]} \omega'$, where $[D]$ is the image of D in $J(K)$. Let $\log_J \colon J(K) \to T_0(J) = H^1_{\mathrm{dR}}(X;\mathcal{Z}/K)/F^1$ be the logarithm for the group $J(K)$ (depending on the choice of $\log_\varpi$). By Corollary 2.4, $\int_0^{[D]} \omega' = \langle \omega', \log_J([D])\rangle = \omega \cup \log_J([D])$, where $\langle \cdot, \cdot \rangle$ is the duality between the tangent space and holomorphic forms on the Jacobian. But $\log_J([D]) = \eta$ so the proof in this case is complete.

Consider now the case where D is the divisor of a rational function f. If f take values in $\mathcal{O}_K$ on $\mathcal{Z}$, then D belongs to the image of $\exp_J$. Since $\log(\mathcal{O}_K^\times)$ contains a lattice in K, we may assume, after raising f to a sufficiently high power, that $\log(f)$ takes values on $\mathcal{Z}$ which are in $\log(\mathcal{O}_K^\times)$. By finding a rational function g with $\log(f) = \log(g)$ on $\mathcal{Z}$ and such that g takes values in $\mathcal{O}_K^\times$ on $\mathcal{Z}$. For the divisor of g the result is already proved. Thus, we may assume that $\log(f) = 0$ on $\mathcal{Z}$. From the description of the maps in (5.3) and Lemma 5.2 it follows that $[E](D) = j\big(\bigoplus_{x\in\mathcal{Z}} \alpha_K\big(f(x)\big)\big)$. The definition of $\log'_\varpi$ and the fact that $\log(f) = 0$

on $\mathcal{Z}$ implies that $1 \otimes \log'_{\varpi}\big(\bigoplus_{x\in\mathcal{Z}} \alpha_K\big(f(x)\big)\big) = 0$. It follows from Lemma 4.2 that $\Theta \circ [E](D) = 0$. Now we apply the theory of the double index in the algebraic version of [**4**] (in this case it can also be deduced with a bit of effort from the rigid theory of [**2**]). We take the sum of the double indices $\langle \int \omega, \log(f)\rangle_x$ of the two p-adic integrals $\log(f)$ and $\int(\omega)$ over all points x. A general result (in the rigid case see [**2**, Corollary 4.11]) says that the sum of these indices for two functions, one of which is a log, is always 0. By definition, the sum of the local indices is $\int_{(f)} \omega - \log f\big(\mathrm{Res}(\omega)\big)$ so this last expression is 0. But by our assumptions $\log f\big(\mathrm{Res}(\omega)\big) = 0$ so also $\int_{(f)} \omega = 0$ and the proof is complete. □

6. End of the proof

To finish the description of the height pairing and complete the proof we must describe the map $w\colon H^1_{\mathrm{f}}(K, E_2) \to H^1_{\mathrm{f}}\big(K, B(1)\big)$, giving a splitting of the first line of (4.4). To describe this splitting is equivalent to describing a splitting $H^1_{\mathrm{f}}(K, V) \to H^1_{\mathrm{f}}(K, E_2)$. By the fact that the exponential map (5.1) is an isomorphism for V and E_2 this is the same as describing a section $D_{\mathrm{dR}}(V)/F^0 \to D_{\mathrm{dR}}(E_2)/F^0$ to the natural map in the other direction.

Under the assumptions in [**12**], which are satisfied in our case, there is a unique Frobenius equivariant splitting (in the sense described after Proposition 2.1) of $D_{\mathrm{dR}}(E_2) \to D_{\mathrm{dR}}(V)$. We now throw in the space $W = W_v \in D_{\mathrm{dR}}(V)$, complementary to F^0. This gives the required splitting,

$$D_{\mathrm{dR}}(V)/F^0 \xrightarrow{\text{along } W} D_{\mathrm{dR}}(V) \xrightarrow{\text{Frob equivariant}} D_{\mathrm{dR}}(E_2) \to D_{\mathrm{dR}}(E_2)/F^0 \,. \tag{6.1}$$

This completes the description of the local height pairing.

In our situation we must describe the map

$$\begin{aligned} H^1_{\mathrm{dR}}(X/K)/F^1 &\xrightarrow{\text{along } W} H^1_{\mathrm{dR}}(X/K) \\ &\xrightarrow{\text{Frob equivariant}} H^1_{\mathrm{dR}}(X;\mathcal{Z}/K) \to H^1_{\mathrm{dR}}(X;\mathcal{Z}/K)/F^1 \,. \end{aligned} \tag{6.2}$$

Dualizing, we have the splitting

$$\begin{aligned} F^1H^1_{\mathrm{dR}}(X-\mathcal{Z}/K) &\to H^1_{\mathrm{dR}}(X-\mathcal{Z}/K) \\ &\xrightarrow{\text{Frob equivariant}} H^1_{\mathrm{dR}}(X/K) \to F^1H^1_{\mathrm{dR}}(X/K). \end{aligned} \tag{6.3}$$

Let W' be the annihilator of W with respect to the cup product. It is easy to check that the rightmost map is the projection with respect to the direct sum decomposition $H^1_{\mathrm{dR}}(X/K) = W' \oplus F^1H^1_{\mathrm{dR}}(X/K)$.

PROPOSITION 6.1. *The composition of the two leftmost maps in* (6.3) *is induced by the map* Ψ *of* (2.2)

PROOF. From the description of Ψ in Proposition 2.1 this is clear from the following lemma. □

LEMMA 6.2. *Let X_s be the closed fiber of $\mathcal{X}$. Let $S = \{s_1, \dots, s_n\}$ be a finite set of closed points of X_s such that every point of $\mathcal{Z}$ reduces to a point of S. Then, the map $H^1_{\mathrm{dR}}(X-\mathcal{Z}/K) \to H^1_{\mathrm{rig}}(X_s - S/K)$ is Frobenius equivariant.*

PROOF. Note that the Frobenius structure on $H^1_{\mathrm{dR}}(X-\mathcal{Z}/K)$ is defined via its identification with D_{dR} of étale cohomology and is not directly related to any cohomology on which Frobenius acts naturally. However, as both de Rham and rigid cohomology have Poincaré duality (but not cristalline cohomology, which is badly behaved for open schemes), we can dualize and must show that the natural map $H^1_{\mathrm{rig}}(X_s;S/K) \to H^1_{\mathrm{dR}}(X;\mathcal{Z}/K)$ is Frobenius equivariant. Lift each $z_i \in \mathcal{Z}$ to a section $z_i\colon \operatorname{Spec}(\mathcal{O}_K) \to \mathcal{X}$, and let $z_i^s\colon \operatorname{Spec}(\kappa) \to X_s$ be the reduction. Then, the comparison isomorphisms in a relative situation (see for example [**10**]) imply that the the Frobenius structure on $H^1_{\mathrm{dR}}(X;\mathcal{Z}/K)$, which is coming from its identification with $D_{\mathrm{dR}}(E_2)$, is induced via the isomorphism $H^1_{\mathrm{dR}}(X;\mathcal{Z}/K) \cong H^1_{\mathrm{cr}}(X_s;\bigcup_i \operatorname{Spec}(\kappa)/K_0) \otimes K$, where $H^1_{\mathrm{cr}}(X_s;\bigcup_i \operatorname{Spec}(\kappa)/K_0)$ is the cohomology relative to the maps z_i^s. As both X_s and $\operatorname{Spec}(\kappa)$ are proper the relative crystalline cohomology is the same as the relative rigid cohomology and the lemma follows easily. □

COROLLARY 6.3. *The induced splitting $B^* \otimes K \to F^1 H^1_{\mathrm{dR}}(X - \mathcal{Z}/K)$ is given by $z \to \omega_{W'}(z)$, where $\omega_{W'}(z)$ is the form defined by Definition* 2.2.

PROOF. The induced splitting is given by the following recipe: for $z \in B^*$ take $\omega \in F^1 H^1_{\mathrm{dR}}(X - \mathcal{Z}/K)$ whose residue divisor is z, apply the map (6.3) and subtract the result from ω. The corollary follows immediately. □

PROOF OF THEOREM 3.3. We are still assuming that $\mathcal{Y}$ is split over K. Let $y \in A$. By (4.5) We need to apply $z \in B^*$ to $h_E(y) = (1\otimes \ell_v)\circ u \circ w \circ \Theta \circ [E](y)$. By Proposition 5.1, $\Theta \circ [E](y)$ is a functional on $H^1_{\mathrm{dR}}(X - \mathcal{Z}/K)$ defined by integrating a form on y. According to Corollary 6.3 this is mapped by w to the functional on $B^* \otimes H^1_{\mathrm{f}}(K, \mathbb{Q}_p(1)) \cong B^* \otimes K$ obtained by applying this functional to the form $\omega_{W'}(z)$. Finally, we apply to the result $(1 \otimes \ell_v) \circ u$, which by (4.3) is just $1 \otimes t_v$. This gives $z(h_E(y)) = t_v(\int_y \omega_{W_v'}(z))$. By the proof of Proposition 5.2 in [**7**] we have $\int_y \omega_{W_v'}(z) = \int_z \omega_{W_v}(y)$, so $z(h_E(y)) = h^v_{\mathrm{CG}}(y,z)$ as required.

To finish the proof we must remove the assumption that $\mathcal{Y}$ is split over K. For this it is sufficient to observe that the two local height pairings behave in the same way with respect to finite extensions. Indeed, suppose that K_1 is a finite extension of K. Let ℓ_1 be an extension of ℓ_v to $K_1^\times$. Such an extension can be given by $t_1 \circ \log_v$ where $\log_v$ extends uniquely to $K_1^\times$ and $t_1\colon K_1 \to \mathbb{Q}_p$ extends t_v. Let $W_1 = W_v \cdot K_1$. Let E be a mixed extension over K. Write h_{E/K_1} for the height pairing corresponding to the mixed extension E restricted to K_1, taken with respect to ℓ_1 and W_1. Then it is straightforward to see that $h_{E/K_1} = h_E$. Thus, the local height pairing is unchanged when we extend the field to K_1, provided that we choose the data (ℓ_1, W_1) as above. If we now compare the Coleman–Gross height pairing we see that the form ω_z is unchanged. Since the log is also unchanged the integral $\int_y \omega_z$ is unchanged. Finally, since this integral is in fact in K and t_1 coincides with t_v on K we see that the entire local height pairing is unchanged by our extension of scalars. This completes the proof. □

References

1. A. Besser, *A generalization of Coleman's p-adic integration theory*, Invent. Math. **142** (2000), no. 2, 397–434.
2. ______, *Syntomic regulators and p-adic integration.* II. *K_2 of curves*, Israel J. Math. **120** (2000), part B, 335–360.
3. ______, *Coleman integration using the Tannakian formalism*, Math. Ann. **322** (2002), no. 1, 19–48.
4. ______, *p-adic Arakelov theory*, math. NT/0301029, 2003.

5. S. Bloch and K. Kato, *L-functions and Tamagawa numbers of motives*, The Grothendieck Festschrift, Vol. I, Prog. Math., vol. 86, Birkhäuser, Boston, MA, 1990, pp. 333–400.
6. R. Coleman and E. de Shalit, *p-adic regulators on curves and special values of p-adic L-functions*, Invent. Math. **93** (1988), no. 2, 239–266.
7. R. Coleman and B. Gross, *p-adic heights on curves*, Algebraic Number Theory, Adv. Stud. Pure Math., vol. 17, Academic Press, Boston, MA, 1989, pp. 73–81.
8. P. Colmez, *Intégration sur les variétés p-adiques*, Astérisque, vol. 248, Soc. Math. France, Paris, 1998.
9. G. Faltings, *Crystalline cohomology and p-adic Galois representations*, Algebraic Analysis, Geometry and Number Theory (Baltimore, MD, 1988), Johns Hopkins Univ. Press, Baltimore, MD, 1989, pp. 25–80.
10. M. Kisin, *Potential semi-stability of p-adic étale cohomology*, Israel J. Math. **129** (2002), 157–173.
11. J. S. Milne, *Jacobian varieties*, Arithmetic Geometry (Storrs, CT, 1984), Springer, New York, 1986, pp. 167–212.
12. J. Nekovář, *On p-adic height pairings*, Séminaire de théorie des nombres, Paris, 1990–91, Progr. Math., vol. 108, Birkhäuser, Boston, MA, 1993, pp. 127–202.
13. ———, *On the p-adic height of Heegner cycles*, Math. Ann. **302** (1995), no. 4, 609–686.
14. W. Raskind, *Higher l-adic Abel–Jacobi mappings and filtrations on Chow groups*, Duke Math. J. **78** (1995), no. 1, 33–57.
15. A. J. Scholl, *Height pairings and special values of L-functions*, Motives (Seattle, WA, 1991), Proc. Sympos. Pure Math., vol. 55, Part 1, Amer. Math. Soc., Providence, RI, 1994, pp. 571–598.
16. J.-P. Serre, *Groupes algébriques et corps de classes*, Publ. Inst. Math. Univ. Nancago, vol. 7, Hermann, Paris, 1959.
17. V. Vologodsky, *Hodge structure on the fundamental group and its application to p-adic integration*, Mosc. Math. J. **3** (2003), no. 1, 205–247.
18. Yu. G. Zarhin, *p-adic heights on abelian varieties*, Séminaire de théorie des nombres, Paris 1987–88, Progr. Math., vol. 81, Birkhäuser, Boston, MA, 1990, pp. 317–341.
19. ———, *p-adic abelian integrals and commutative Lie groups*, J. Math. Sci., 81 (1996), no. 3, 2744–2750.

Department of Mathematics, Ben-Gurion University of the Negev, P.O. Box 653, Be'er-Sheva 84105, Israel.

E-mail address: bessera@math.bgu.ac.il

Centre de Recherches Mathématiques
CRM Proceedings and Lecture Notes
Volume **36**, 2004

Dirichlet Series for Squares of Sums of Squares: A Summary

Jonathan Michael Borwein and Kwok-Kwong Stephen Choi

This note is a summary version of the results that the authors presented at CNTA VII in Montreal and detailed proofs may be found in [**2**].

Let σ_k denote the sum of kth powers of the divisors of n. There is a beautiful formula for the generating functions of $\sigma_k(n)$ (see [**8**, Theorem 291, §17])

$$\sum_{n=1}^{\infty} \frac{\sigma_k(n)}{n^s} = \zeta(s)\zeta(s-k), \quad \mathrm{Re}(s) > \max\{1, k+1\} \tag{1}$$

which is in terms of only the Riemann Zeta function $\zeta(s)$. Following Hardy and Wright, by standard techniques, one can prove the following remarkable identity due to Ramanujan (also see [**8**, Theorem 305, §17] or [**13**])

$$\sum_{n=1}^{\infty} \frac{\sigma_a(n)\sigma_b(n)}{n^s} = \frac{\zeta(s)\zeta(s-a)\zeta(s-b)\zeta(s-a-b)}{\zeta(2s-a-b)} \tag{2}$$

for $\mathrm{Re}(s) > \max\{1, a+1, b+1, a+b+1\}$. In this paper, we identify other arithmetical functions enjoying similarly explicit representations. We begin by generalizing the above result and proving that

THEOREM 1. *If f_i and g_i are completely multiplicative arithmetical functions, then we have*

$$\sum_{n=1}^{\infty} \frac{(f_1 * g_1)(n) \cdot (f_2 * g_2)(n)}{n^s} = \frac{L_{f_1 f_2}(s) L_{g_1 g_2}(s) L_{f_1 g_2}(s) L_{g_1 f_2}(s)}{L_{f_1 f_2 g_1 g_2}(2s)}$$

where $L_f(s) := \sum_{n=1}^{\infty} f(n) n^{-s}$ is the Dirichlet series corresponding to f and the convolution $f * g$

$$f * g(n) := \sum_{d \mid n} f(d) g(n/d),$$

provided that $\mathrm{Re}(s)$ *is greater than all the abscissa of absolute convergence of $L_{f_i}(s)$ and $L_{g_i}(s)$.*

2000 *Mathematics Subject Classification.* Primary: 11M41, 11E25.

Research supported by NSERC and by the Canada Research Chair Programme.

The authors wish to thank Ram Murty for pointing out Rankin and Selberg's result and Werner Nowak for Kühleitner's result.

This is the final form of the paper.

This result recovers Hardy and Wright's formulae (1) and (2) immediately. In fact, Rankin and Selberg discovered the formula of Theorem 1 when they considered the convolution L-functions attached to automorphic forms (e.g. see [**9**, §13] or [**11**]), as was pointed out by Ram Murty to the authors.

More generally, for certain classes of Dirichlet series, $\sum_{n=1}^{\infty} A(n)n^{-s}$, our Theorem 1 can be applied to obtain closed forms for the series $\sum_{n=1}^{\infty} A^2(n)n^{-s}$. In particular, if the generating function $L_f(s)$ of an arithmetic function f is expressible as a sum of products of two L-functions:

$$L_f(s) = \sum_{\chi_1,\chi_2} a(\chi_1,\chi_2)L_{\chi_1}(s)L_{\chi_2}(s)$$

for appropriate coefficients $a(\chi_1,\chi_2)$ and Dirichlet characters χ_i, then we are able to find a simple closed form (in term of L-functions) for the generating function $L_f^2(s) := \sum_{n=1}^{\infty} f^2(n)n^{-s}$.

One of our central applications is to the study of the number of representations as a sum of squares. Let $r_N(n)$ be the number of solutions to $x_1^2+x_2^2+\cdots+x_N^2 = n$ (counting permutations and signs). Hardy and Wright record a classical closed form, due to Lorenz, of the generating function for $r_2(n)$ in the terms of $\zeta(s)$ and a Dirichlet L-function, namely,

$$\sum_{n=1}^{\infty} \frac{r_2(n)}{n^s} = 4\zeta(s)L_{-4}(s)$$

where $L_\mu(s) = \sum_{n=1}^{\infty} \left(\frac{\mu}{n}\right)n^{-s}$ is the *primitive L-function* corresponding to the *Kronecker symbol* $\left(\frac{\mu}{n}\right)$. Define

$$\mathcal{L}_N(s) := \sum_{n=1}^{\infty} \frac{r_N(n)}{n^s} \quad \text{and} \quad \mathcal{R}_N(s) := \sum_{n=1}^{\infty} \frac{r_N^2(n)}{n^s}.$$

Simple closed forms for $\mathcal{L}_N(s)$ are known for $N = 2, 4, 6$ and 8, due to the known explicit formula of $r_N(n)$ for these N (e.g. Ref. [**12**, §91]); indeed the corresponding q–series were known to Jacobi,

$$\begin{aligned}
\mathcal{L}_2(s) &= 4\zeta(s)L_{-4}(s),\\
\mathcal{L}_4(s) &= 8(1-4^{1-s})\zeta(s)\zeta(s-1),\\
\mathcal{L}_6(s) &= 16\zeta(s-2)L_{-4}(s) - 4\zeta(s)L_{-4}(s-2),\\
\mathcal{L}_8(s) &= 16(1-2^{1-s}+4^{2-s})\zeta(s)\zeta(s-3).
\end{aligned}$$

Now applying an extension of Theorem 1 to these $\mathcal{L}_N$, we obtain the following closed forms for $\mathcal{R}_N(s)$:

THEOREM 2. *We have*

$$\mathcal{R}_2(s) = \frac{(4\zeta(s)L_{-4}(s))^2}{(1+2^{-s})\zeta(2s)}, \quad \operatorname{Re}(s) > 1,$$

$$\mathcal{R}_4(s) = 64\frac{(8\cdot 2^{3-3s} - 10\cdot 2^{2-2s} + 2^{1-s} + 1)\zeta(s-2)\zeta^2(s-1)\zeta(s)}{(1+2^{1-s})\zeta(2s-2)}, \quad \operatorname{Re}(s) > 3,$$

$$\begin{aligned}
\mathcal{R}_6(s) = 16&\frac{(17-32\cdot 2^{-s})}{(1-16\cdot 2^{-2s})}\frac{\zeta(s-4)L_{-4}^2(s-2)\zeta(s)}{\zeta(2s-4)}\\
&- \frac{128}{(1+4\cdot 2^{-s})}\frac{L_{-4}(s-4)\zeta^2(s-2)L_{-4}(s)}{\zeta(2s-4)}, \quad \operatorname{Re}(s) > 5,
\end{aligned}$$

and

$$\mathcal{R}_8(s) = 256\frac{(32 \cdot 2^{6-2s} - 3 \cdot 2^{3-s} + 1)\zeta(s-6)\zeta^2(s-3)\zeta(s)}{(1+2^{3-s})\zeta(2s-6)} \qquad \mathrm{Re}(s) > 7.$$

An immediate application of Theorem 2 is to the estimation of the average order of $r_N^2(n)$. Using the usual hyperbola method of Dirichlet and a convolution argument, we can prove that

THEOREM 3. *We have*

$$\sum_{n\leq x} r_2^2(n) = 4x\log x + 4\alpha x + O(x^{2/3}), \tag{3}$$

$$\sum_{n\leq x} r_4^2(n) = W_4 x^3 + O(x^2\log^5 x),$$

$$\sum_{n\leq x} r_6^2(n) = W_6 x^5 + O(x^4),$$

and

$$\sum_{n\leq x} r_8^2(n) = W_8 x^7 + O(x^6)$$

where $\alpha := 2\gamma + 8\pi L'_{-4}(1) - 12/\pi^2\zeta'(2) + \frac{1}{3}\log 2 - 1 = 2.0166216\ldots$ *and*

$$W_N := \frac{1}{(N-1)(1-2^{-N})}\frac{\pi^N}{\Gamma^2(N/2)}\frac{\zeta(N-1)}{\zeta(N)}, \qquad (N \geq 3).$$

Here γ *is the Euler's constant.*

This technique can be adjusted to handle all integers $N \geq 2$ except $N = 3$ as below and so we establish all but the most difficult case of the following general conjecture due to Wagon (see [**4**]):

Wagon's conjecture. For integer $N \geq 3$, $\sum_{n\leq x} r_N^2(n) \sim W_N x^{N-1}$ as $x \to \infty$.

Despite lacking the closed form for $\mathcal{R}_N(s)$ in general for $N \geq 5$, we can use the singular series formula for $r_N(n)$ given by Hardy (see [**7**, p. 342] or [**6**, p. 155]), which may be written as

$$r_N(n)\frac{\Gamma(N/2)}{\pi^{N/2}}n^{1-N/2} = \sum_{k=1}^{\infty}\sum_{\substack{1\leq h\leq k\\(h,k)=1}}\left(\frac{G(h,k)}{k}\right)^N e^{-2\pi i h n/k} + O(n^{1-N/4}),$$

where $G(h,k) = \sum_{j=1}^{k} e^{2\pi i h j^2/k}$ is the standard quadratic Gauss sum, to obtain

$$\sum_{n\leq x} r_N^2(n) = W_N x^{N-1} + O(x^{N-2} + x^{3N/4}) \quad (N \geq 5). \tag{4}$$

The error term in (3) is not as good as the best known result. Indeed M. Kühleitner proved in [**10**] that

$$\sum_{n\leq x} r_2^2(n) = 4x\log x + 4\alpha x + O\big(x^{1/2}(\log x)^{11/3}(\log\log x)^{1/3}\big).$$

However, for $N \geq 5$, the estimate $O(x^{N-2})$ in (4) in fact is the best possible.

We similarly study the number of representations by a binary quadratic forms. Let $r_{2,P}(n)$ be the number of solutions of the binary quadratic form $x^2 + Py^2 = n$. Define

$$\mathcal{L}_{2,P}(s) := \sum_{n=1}^{\infty} \frac{r_{2,P}(n)}{n^s} \quad \text{and} \quad \mathcal{R}_{2,P}(s) := \sum_{n=1}^{\infty} \frac{r_{2,P}(n)^2}{n^s}.$$

The closed forms of $\mathcal{L}_{2,P}(s)$ have been studied by a number of people, particular by Glasser, Zucker and Robertson (see [**5, 14**]). In finding the exact evaluation of lattice sums, they are interested in expressing a multiple sum, such as the generating functions of $r_{2,P}(n)$, as a product of simple sums. As a result, plenty of closed forms of Dirichlet series $\sum_{(n,m)\neq(0,0)}(am^2 + bmn + cn^2)^{-s}$ in terms of L-functions have been found. One of the most interesting cases is when the binary quadratic forms have *disjoint discriminants*, i.e., have only one form per genus. Then there are simple closed forms for the corresponding $\mathcal{L}_{2,P}(s)$ (see Ref. [**1**, (9.2.8)])

$$\mathcal{L}_{2,P}(s) = 2^{1-t} \sum_{\mu|P} L_{\varepsilon\mu}(s) L_{-4P\varepsilon/\mu}(s) \tag{5}$$

where P is an odd square-free number, t is the number of distinct factors of P and $\varepsilon := \left(\frac{-1}{\mu}\right)$. Explicitly, (5) holds for all *type one* numbers. These include and may comprise:

$$P = 5, 13, 21, 33, 37, 57, 85, 93, 105, 133, 165, 177, 253, 273, 345, 357, 385, 1365.$$

It is known that there are only finitely many such disjoint discriminants. We call such P **solvable**. We have similar closed forms of L-functions for the quadratic form $x^2 + 2Py^2$ with discriminant $-8P$ (see [**1**, (9.2.9)]):

$$\mathcal{L}_{2,2P}(s) = 2^{1-t} \sum_{\mu|P} L_{\varepsilon\mu}(s) L_{-8P\varepsilon/\mu}(s)$$

for the *type two* integers

$$P = 1, 3, 5, 11, 15, 21, 29, 35, 39, 51, 65, 95, 105, 165, 231.$$

Again applying extensions of Theorem 1, we obtain closed forms for $\mathcal{R}_{2,P}(s)$ and $\mathcal{R}_{2,2P}(s)$.

THEOREM 4. *Let P be a solvable square-free integer and let t be the number of distinct factors of P. We have for P respectively of type one and type two:*

$$\mathcal{R}_{2,P}(s) = 2^{2(1-t)} \sum_{\mu_1,\mu_2|P} L^2_{\varepsilon_{\mu_1^*\mu_2^*}\mu_1^*\mu_2^*}(s) L^2_{-4P\varepsilon_{\mu_1^*\mu_2^*}/\mu_1^*\mu_2^*}(s)\zeta(2s)^{-1}$$
$$\times \prod_{p|2P} \left\{1 + \left[\left(\frac{\varepsilon_{\mu_1^*\mu_2^*}\mu_1^*\mu_2^*}{p}\right) + \left(\frac{-4P\varepsilon_{\mu_1^*\mu_2^*}/\mu_1^*\mu_2^*}{p}\right)\right] p^{-s}\right\}^{-1},$$

and

$$\mathcal{R}_{2,2P}(s) = 2^{2(1-t)} \sum_{\mu_1,\mu_2|P} L^2_{\varepsilon_{\mu_1^*\mu_2^*}\mu_1^*\mu_2^*}(s) L^2_{-8P\varepsilon_{\mu_1^*\mu_2^*}/\mu_1^*\mu_2^*}(s)\zeta(2s)^{-1}$$
$$\times \prod_{p|2P} \left\{1 + \left[\left(\frac{\varepsilon_{\mu_1^*\mu_2^*}\mu_1^*\mu_2^*}{p}\right) + \left(\frac{-8P\varepsilon_{\mu_1^*\mu_2^*}/\mu_1^*\mu_2^*}{p}\right)\right] p^{-s}\right\}^{-1}$$

where $\varepsilon_\mu = \left(\frac{-1}{\mu}\right)$ and $\mu_i^* = \mu_i/(\mu_1, \mu_2)$.

In particular, the prime cases provide:

COROLLARY 5. *We have*

$$\mathcal{R}_{2,p}(s) = \frac{2\zeta^2(s)L_{-4p}^2(s)}{(1+2^{-s})(1+p^{-s})\zeta(2s)} + \frac{2L_p^2(s)L_{-4}^2(s)}{(1-2^{-s})(1+p^{-s})\zeta(2s)}$$

for $p = 5, 13, 37,$ *while*

$$\mathcal{R}_{2,2}(s) = \frac{4\zeta^2(s)L_{-8}^2(s)}{(1+2^{-s})\zeta(2s)}.$$

Similarly,

$$\mathcal{R}_{2,2p}(s) = \frac{2\zeta^2(s)L_{-8p}^2(s)}{(1+2^{-s})(1+p^{-s})\zeta(2s)} + \frac{2L_{-p}^2(s)L_8^2(s)}{(1-2^{-s})(1-p^{-s})\zeta(2s)},$$

for $p = 3, 11$ *while*

$$\mathcal{R}_{2,2p}(s) = \frac{2\zeta^2(s)L_{-8p}^2(s)}{(1+2^{-s})(1+p^{-s})\zeta(2s)} + \frac{2L_p^2(s)L_{-8}^2(s)}{(1-2^{-s})(1-p^{-s})\zeta(2s)}$$

for $p = 5, 29.$

Closed forms for $\mathcal{L}_{2,P}(s)$ are also accessible for some P other than those of *type one* or *type two*. For example, (see [[illegible] Table 6]) one has

$$\mathcal{L}_{2,3}(s) = (2 + 4^{1-s})\zeta(s)L_{-3}(s)$$

and hence we obtain

$$\mathcal{R}_{2,3}(s) = 4\frac{1+2^{3-2s}}{1+3^{-s}}\frac{(\zeta(s)L_{-3}(s))^2}{\zeta(2s)}. \tag{6}$$

We may also derive many formulae for non-square free integers via modular transformations [**1**]. We contain ourselves with the simplest example which is

$$\mathcal{R}_{2,4}(s) = \frac{4 - 2^{2-s} + 2^{4-2s}}{1+2^{-s}}\frac{(\zeta(s)\,L_{-4}(s))^2}{\zeta(2s)}$$

as a consequence of a quadratic transformation leading to

$$\mathcal{L}_{2,4}(s) = (2^{-1} - 2^{-1-s} + 4^{-s})\mathcal{L}_2(s).$$

There are some simple closed forms of the generating functions for more general binary quadratic forms found in [**5**]. Let

$$\mathcal{L}_{(a,b,c)}(s) := \sum_{(n,m)\neq(0,0)} \frac{1}{(am^2+bmn+cn^2)^s} = \sum_{n=1}^{\infty} \frac{r_{(a,b,c)}(n)}{n^s}$$

and $\mathcal{R}_{(a,b,c)}(s) := \sum_{n=1}^{\infty} r_{(a,b,c)}(n)^2/n^s$ where $r_{(a,b,c)}(n)$ is the number of representations of n by the quadratic form $ax^2 + bxy + cy^2$. Then, we have (e.g. [**15**, (26)])

$$\sum_{h(D)} \mathcal{L}_{(a,b,c)}(s) = \omega(D)\zeta(s)L_D(s)$$

where the sum is taken over the $h(D)$ inequivalent reduced quadratic forms of discriminant $D := b^2 - 4ac$ and $\omega(-3) = 6, \omega(-4) = 4$ and $\omega(D) = 2$ for $D < -4$. In particular, for $c = 2, 3, 5, 11, 17, 41$, $h(D) = 1$ and the result is especially simple:

$$\mathcal{L}_{(1,1,c)}(s) = 2\zeta(s)L_D(s).$$

Hence from Theorem 1, we have

$$\mathcal{R}_{(1,1,c)}(s) = \frac{4(\zeta(s)L_D(s))^2}{(1+|D|^{-s})\zeta(2s)},$$

with similar formulae for $(a,b,c) = (1,1,1)$ and $(1,0,1)$. Thanks to the *On-Line Encyclopedia of Integer Sequences*: `http://www.research.att.com/~njas/sequences/` we discover that the sequence $2,3,5,11,17,41$ is exactly the so-called Euler "lucky" numbers which are the numbers n such that $m \to m^2 - m + n$ has prime values for $m = 0,\ldots,n-1$.

Applying standard complex integration methods to $\mathcal{R}_{2,P}(s)$ and $\mathcal{R}_{2,2P}(s)$, we derive the asymptotic formula

THEOREM 6. *Let P be a solvable square-free integer. Let $x > 1$ and $\varepsilon > 0$. We have for either $N = P$ of type one or $N = 2P$ of type two:*

$$\sum_{n\le x} r_{2,N}(n)^2 = \frac{3}{N}\left(\prod_{p|2N}\frac{2p}{p+1}\right)(x\log x + \alpha(N)x) + O(N^{1/4+\varepsilon}x^{3/4+\varepsilon})$$

where the implicit constants are independent of both x and P and

$$\alpha(N) := 2\gamma + \sum_{p|2N}\frac{\log p}{p+1} + 2\frac{L'_{-4N}(1)}{L_{-4N}(1)} - \frac{12}{\pi^2}\zeta'(2) - 1$$

and $\sum_{p|n}$ is the summation over all prime factors of n.

Akin to Wagon's conjecture, we make the following conjecture.

Quadratic conjecture. For any square-free P,

$$\sum_{n\le x} r_{2,P}(n)^2 \sim \frac{3}{P}\left(\prod_{p|2P}\frac{2p}{p+1}\right)x\log x$$

as $x \to \infty$.

In view of Theorem 6, (3) and (6), our conjecture is true for solvable P and for $P = 1,3$. We have also confirmed it for $P = 7$ and 15 from the representations of

$$\mathcal{L}_{2,7}(s) = 2(1 - 2^{1-s} + 2^{1-2s})\zeta(s)L_{-7}(s)$$

and

$$\mathcal{L}_{2,15}(s) = (1 - 2^{1-s} + 2^{1-2s})\zeta(s)L_{15}(s) + (1 + 2^{1-s} + 2^{1-2s})L_{-3}(s)L_5(s)$$

again given in [**5**], which leads to

$$\mathcal{R}_{2,7}(s) = 4\frac{(1 - 3\cdot 2^{-s} + 2^{2-2s})}{(1+2^{-s})(1+7^{-s})}\frac{(\zeta(s)L_{-7}(s))^2}{\zeta(2s)}$$

and

$$\mathcal{R}_{2,15}(s) = \frac{2(1 - 3\cdot 2^{-s} + 2^{2-2s})}{(1+2^{-s})(1+3^{-s})(1+5^{-s})}\frac{(\zeta(s)L_{-15}(s))^2}{\zeta(2s)} + \frac{2(1 + 3\cdot 2^{-s} + 2^{2-2s})}{(1-2^{-s})(1-3^{-s})(1-5^{-s})}\frac{(L_{-3}(s)L_5(s))^2}{\zeta(2s)},$$

and may be analyzed by the methods above.

For the negative P, we have studied only the case $P = -1$. By the elementary formula due to Sierpinski [**16**]

$$r_{2,-1}(n) = 2d(n) - 4d\left(\frac{n}{2}\right) + 4d\left(\frac{n}{4}\right)$$

we obtain

$$\mathcal{L}_{2,-1}(s) = 2(1 - 2^{1-s} + 2^{1-2s})\zeta(s)^2$$

and

$$\mathcal{R}_{2,-1}(s) = 4\frac{(1 - 3 \cdot 2^{-s} + 4 \cdot 2^{-2s})\zeta(s)^4}{(1 + 2^{-s})\zeta(2s)}$$

where $d(n)$ is the divisor function and $d(x) = 0$ if x is not an integer.

We also studied $\mathcal{L}_N(s)$, for all $N \geq 2$, and obtained a Bessel-series in $r_{N-1}(m)$:

$$(7)\quad \begin{aligned} \mathcal{L}_N(s) &= \sum_{n>0} \frac{r_N(n)}{n^s} \\ &= \frac{2N\,\Gamma(s - N - 32)}{\Gamma(s+1)}\pi^{N-1/2}\zeta(2s - N + 1) \\ &\quad + \frac{4N\pi^{s+1}}{\Gamma(s+1)} \sum_{m>0} \frac{m^{1/2s}\, r_{N-1}(m)}{m^{N-3}/4} \sum_{n>0} \frac{n^{N+1}/2}{n^s} K_{s-N-3/2}(2n\pi\sqrt{m}). \end{aligned}$$

Here K_s is the *modified Bessel function of the second kind.* This is especially attractive for $N = 3, 5, 7, 9$ since in these cases $r_{N-1}(n)$ is an explicit divisor function.

Most pleasantly, for "jellium", which is the Wigner sum analogue of Madelung's constant, we have

$$\mathcal{L}_3(1/2) = -\pi + 3\pi \sum_{m>0} \frac{r_2(m)}{\sinh^2(\pi\sqrt{m})},$$

in which the exponential convergence is entirely apparent.

Some similar identities are

$$\mathcal{L}_5(3/2) = -\frac{10}{9}\pi^2 \left(1 + 3\sum_{m>0} \frac{r_4(m)}{\sinh^2(\pi\sqrt{m})}\right),$$

$$\mathcal{L}_7(5/2) = -\frac{28}{45}\pi^3 \left(1 + 3\sum_{m>0} \frac{r_6(m)}{\sinh^2(\pi\sqrt{m})}\right),$$

$$\mathcal{L}_9(7/2) = -\frac{8}{35}\pi^4 \left(1 + 3\sum_{m>0} \frac{r_8(m)}{\sinh^2(\pi\sqrt{m})}\right),$$

and

$$R_{2,4}(2) = 8\mathrm{G}^2$$

where $\mathrm{G} := \sum_{n\geq 0}(-1)^n(2n+1)^{-2}$ is *Catalan's constant.*

For a survey of other rapidly convergent lattice sums of this type see [**1, 3**].

Unfortunately we have not been able to extend this analysis to $R_N(s)$.

References

1. J. M. Borwein and P. B. Borwein, *Pi and the AGM*, A Study in Analytic Number Theory and Computational Complexity, CMS, Monogr. Advanced Texts, vol. 4, John Wiley & Sons, New York, 1987, Paperback, 1998.
2. J. M. Borwein and K. K. S. Choi, *On Dirichlet series for sums of squares*, Ramanujan J., special issue for Robert Rankin, (2002) (to appear).
3. R. E. Crandall, *New representations for the Madelung constant*, Experiment. Math. **8** (1999), no. 4, 367–379.
4. R. Crandall and S. Wagon, *Sums of squares: Computational aspects*, Preprint, 2001.
5. M. Glasser and I. Zucker, *Lattice sums*, Theoretical Chemistry (H. Eyring, ed.), Adv. and Perspectives, vol. 5, Academic Press, New York, 1980, pp. 67–139.
6. E. Grosswald, *Representations of integers as sums of squares*, Springer-Verlag, New York, 1985.
7. G. H. Hardy, *Collected papers*, vol. I, London Math. Society Clarendon Press, Oxford 1966.
8. G. H. Hardy and E. M. Wright, *An introduction to the theory of numbers*, 5th ed., The Clarendon Press, Oxford University Press, New York, 1979.
9. H. Iwaniec, *Topics in classical automorphic forms*, Grad. Stud. Math., vol. 17, Amer. Math. Soc., Providence, RI, 1997.
10. M. Kühleitner, *On a question of A. Schinzel concerning the sum* $\sum_{n\le x}(r(n))^2$, Österreichisch-Ungarisch-Slowakisches Kolloquium Über Zahlentheorie (Maria Trost, 1992), Grazer Math. Ber., vol. 318, Karl-Franzens-Univ. Graz, Graz, 1993, pp. 63–67.
11. M. R. Murty, *Problems in analytic number theory*, Grad. Texts in Math., vol. 206, Readings in Mathematics, Springer-Verlag, New York, 2001.
12. H. Rademacher, *Topics in analytic number theory*, (E. Grosswald, J. Lehner and M. Newman, eds.), Die Grundlehren der mathematischen Wissenschaften, Band 169, Springer-Verlag, New York-Heidelberg, 1973.
13. ———, *Some formulae in the analytic theory of numbers*, Messenger of Math. **45** (1916), 81–84.
14. M. M. Robertson and I. J. Zucker, *Exact values for some two-dimensional lattice sums*, J. Phys. A **8** (1975), 874–881.
15. D. Shanks, *Calculation and applications of Epstein zeta functions*, Math. Comp. **29** (1975), 271–287.
16. W. Sierpinski, *Über die Darstellungen ganzer Zahlen als Differenz von zwei Quadraten*, Wiadom. Mat. **11** (1907), 89–110.

CECM, Department of Mathematics, Simon Fraser University, Burnaby, B.C. V5A 1S6, Canada.

E-mail address, J. Borwein: `jborwein@cecm.sfu.ca`

E-mail address, K.-K. S. Choi: `kkchoi@cecm.sfu.ca`

Centre de Recherches Mathématiques
CRM Proceedings and Lecture Notes
Volume **36**, 2004

Equivariant Whitehead Torsion and Refined Euler Characteristics

David Burns

ABSTRACT. We discuss a natural refinement of the classical construction of the Euler characteristic of a perfect complex of modules.

Introduction

In this article we discuss a natural refinement of the classical notion of Euler characteristic. This refined construction combines the standard algebraic construction of the Euler characteristic of a perfect complex of modules with an equivariant version of aspects of the classical construction of Whitehead torsion [**25, 29**]. It can be applied to any perfect complex of modules that is endowed with certain additional structure of a cohomological nature and takes values in a suitable relative algebraic K-group (which is itself the target of a natural homomorphism from an appropriate Whitehead group).

The central construction of this article was first introduced in order to formulate Tamagawa number conjectures for motives over number fields with respect to coefficient rings which are neither regular or commutative [**8**]. Subsequently, it has also been used to refine and reinterpret all of the central conjectures and results of classical Galois module theory (cf. [**4–9, 26**]), to refine well known isogeny invariance results in Galois cohomology [**21**] and to formulate and study special value conjectures for equivariant Zeta functions associated to varieties over finite fields [**10**]. More recently, it has been observed that our construction also provides an explicit approach to certain aspects of the notion of 'categories of virtual objects' which was introduced by Deligne in [**20**] (in this regard see [**13**, Remark 4]), that it can be used to refine and reinterpret the constructions of 'Euler characteristics of nearly perfect complexes' due to Chinburg, Kolster, Pappas and Snaith (cf. [**15**]) and of 'arithmetic classes of metrised complexes' and 'Hermitian de Rham discriminants' which are introduced by Chinburg, Pappas and Taylor in [**16, 17**] respectively, that it sheds new light on Lichtenbaum's theory of Weil-étale cohomology (cf. [**11**]) and that it contributes towards the formulation of a workable 'non-abelian Iwasawa theory'. Details of all of these applications will appear elsewhere. However, despite

2000 *Mathematics Subject Classification.* Primary: 19A31; Secondary: 13D25, 14G15.
This is the final form of the paper.

such applications, detailed verifications of the most important features of the underlying algebraic constructions have not hitherto appeared in print and the main aim of the present article is to rectify this oversight in as explicit and concrete a manner as possible.

In a little more detail, the main contents of this article is as follows. In §1 we describe Deligne's notion of 'reduced determinant' and then introduce a related notion of 'trivialization'. We construct such trivializations in the context of bounded complexes of finitely generated modules over semisimple algebras (in Lemma 1.3) by means of an equivariant version of a well known construction which is used to define torsion invariants of chain complexes and of Hermitian chain complexes. In §2 we describe a natural algebraic refined Euler characteristic construction. This construction takes values in a relative algebraic K-group and can be applied to any 'trivialized perfect complex'. We prove that it is additive on certain distinguished triangles of trivialized perfect complexes. In §3 we restrict to the case of commutative algebras and interpret the constructions of §1 and §2 in terms of the determinantal formalism of Grothendieck, Knudsen and Mumford that is described in [**23**]. Finally, in §4 we prove a useful description of the torsion subgroups of certain relative algebraic K-groups.

All modules that arise in this article are to be understood as left modules. For any ring R we shall write $\mathfrak{D}(R)$ for the derived category of bounded complexes of finitely generated R-modules. We say that a complex of R-modules is 'perfect' if it is isomorphic in $\mathfrak{D}(R)$ to a bounded complex of finitely generated projective R-modules, and we write $\mathfrak{D}^{\mathrm{perf}}(R)$ for the full triangulated subcategory of $\mathfrak{D}(R)$ which has as objects those complexes which are perfect.

The author is very grateful to Manuel Breuning and Matthias Flach for a number of helpful remarks.

1. Reduced determinants and trivializations

In this section we begin by reviewing the notion of 'reduced determinant' mentioned by Deligne in [**19**, Remarks 2.12 and 2.13]. We then introduce an associated notion of 'trivialization'.

For any field F and any finitely generated F-module M, of dimension d say, we define an F-module of dimension 1 by setting $\det_F(M) := \bigwedge_F^d M$.

We now fix a field E and a finite dimensional central simple E-algebra A, and we use the theory of Galois descent (cf. [**28**, §17, in particular Example 6]) to associate to each pair of isomorphic finitely generated A-modules V and W a canonical E-module of dimension one. To do this we let $\mathrm{Is}_A(V, W)$ denote the set of A-equivariant isomorphisms from V to W, and for any extension field E' of E and any A-module V we set $V' := E' \otimes_E V$. If now E' splits A, then for any finitely generated A-modules V and W, and any indecomposable idempotent e' of the E'-algebra A' we obtain a dimension one E'-module by setting

$$\delta_{A'}(V', W') := \mathrm{Hom}_{E'}\big(\det_{E'}\big(\mathrm{Hom}_{A'}(A'e', V')\big), \det_{E'}\big(\mathrm{Hom}_{A'}(A'e', W')\big)\big).$$

This module is independent (modulo canonical isomorphisms) of the choice of e'. Further, if E'/E is Galois, then the natural action of $\mathrm{Gal}(E'/E)$ on A', V' and W' extends to give an action on $\delta_{A'}(V', W')$ in the following way. For each $\omega \in \mathrm{Gal}(E'/E)$ one obtains an ω-semi-linear isomorphism $\mathrm{Hom}_{A'}(A'e', V') \to$

$\mathrm{Hom}_{A'}(A'\omega(e'), V')$ by sending each homomorphism f to the function $\omega(f)$ defined by setting $\omega(f)(a') := \omega(f(\omega^{-1}(a')))$ for all $a' \in A'\omega(e')$. For each A-module V this gives rise to an ω-semi-linear isomorphism

$$\rho_V^\omega \colon \det_{E'}(\mathrm{Hom}_{A'}(A'e', V')) \xrightarrow{\sim} \det_{E'}(\mathrm{Hom}_{A'}(A'\omega(e'), V'))$$

such that $\rho_V^\omega(f_1\wedge\cdots\wedge f_d) = \omega(f_1)\wedge\cdots\wedge\omega(f_d)$ where $d = \dim_{E'}(\mathrm{Hom}_{A'}(A'e', V')) = \dim_{E'}(\mathrm{Hom}_{A'}(A'\omega(e'), V'))$ and each f_i belongs to $\mathrm{Hom}_{A'}(A'e', V')$. Now, since $\delta_{A'}(V', W')$ can also be computed by using $\omega(e')$ in place of e', the association $(\omega, \theta) \mapsto \rho_W^\omega \circ \theta \circ (\rho_V^\omega)^{-1}$ induces upon $\delta_{A'}(V', W')$ the structure of $\mathrm{Gal}(E'/E)$-module. Further, the sub-module of fixed points

$$\delta_A(V, W) := H^0(\mathrm{Gal}(E'/E), \delta_{A'}(V', W'))$$

is an E-module of dimension one that is independent of the choice of E'.

Now each homomorphism $\phi \in \mathrm{Hom}_A(V, W)$ gives rise to an element

$$\theta(\phi, e') \in \mathrm{Hom}_{E'}(\mathrm{Hom}_{A'}(A'e', V'), \mathrm{Hom}_{A'}(A'e', W'))$$

which is induced by mapping a homomorphism f to $(E'\otimes_E \phi)\circ f$ and the associated 'reduced determinant'

$$\det_A(\phi) := \det_{E'}(\theta(\phi, e'))$$

is an element of $\delta_A(V, W)$ which is independent of the choices of both E' and e'. We set

$$\delta_A^+(V, W) := \{\det_A(\phi) \colon \phi \in \mathrm{Is}_A(V, W)\} \subset \delta_A(V, W).$$

We observe that if $V \neq 0$, then $\delta_A(V, V)$ identifies naturally with E and we write $\delta_A^+(V)$ in place of $\delta_A^+(V, V) \subseteq E$.

If Λ is any semisimple algebra, then we let $\zeta(\Lambda)$ denote its centre and write $\zeta(\Lambda)^{\times +}$ for the image of the reduced norm map $\mathrm{nr}_\Lambda \colon \Lambda^\times \to \zeta(\Lambda)^\times$. If E is a number field, then we write $S_\infty(E)$, $S_\mathbb{R}(E)$ and $S_\mathbb{C}(E)$ for the set of its archimedean, real archimedean and complex archimedean places respectively. For each $v \in S_\infty(E)$ we fix an embedding $\sigma_v \colon E \to \mathbb{C}$ which induces v and we write E_v for the completion of E at v. We recall that a finite dimensional central simple $\mathbb{R}$-algebra is said to be 'ramified' if it is a matrix ring over the skewfield of real quaternions.

LEMMA 1.1. *Let A be a finite dimensional central simple E-algebra, and let V and W be isomorphic finitely generated A-modules:*

(1) *If $\theta \in \mathrm{End}_A(V)$, then $\det_A(\theta) = \mathrm{nr}_{\mathrm{End}_A(V)}(\theta)$.*
(2) *If $V \neq 0$, then $\delta_A^+(V) = \zeta(A)^{\times +}$.*
(3) *If $V \neq 0$, then $\delta_A^+(V, W) = \zeta(A)^{\times +} \cdot \det_A(\theta) \subset \delta_A(V, W)$ for any $\theta \in \mathrm{Is}_A(V, W)$.*
(4) *If E is a number field, then $\zeta(A\otimes_\mathbb{Q}\mathbb{R})^{\times +}$ is equal to*

$$\left\{ (e_v)_v \in \prod_{v\in S_\infty(E)} E_v^\times : e_v > 0 \text{ if } v \in S_\mathbb{R}(E) \text{ and } A\otimes_{E,\sigma_v}\mathbb{R} \text{ is ramified} \right\}.$$

(5) *If E is a number field, then one has*

$$\delta_A^+(V, W) = \delta_A(V, W) \cap \delta_{A\otimes_\mathbb{Q}\mathbb{R}}^+(V\otimes_\mathbb{Q}\mathbb{R}, W\otimes_\mathbb{Q}\mathbb{R}).$$

PROOF. To be explicit we assume that $A = M_n(D)$ with D a division algebra of index f over E.

(1) If e is any indecomposable idempotent of A, then Ae is a simple A-module and $V \cong (Ae)^m$ for some integer m so there exists an isomorphism $\mathrm{End}_A(V) \cong \mathrm{M}_m(\mathrm{End}_A(Ae))$. We let E' be a splitting field for A over E, and write A' for $E' \otimes_E A$. If e' is any indecomposable idempotent of A', then $A'e \cong (A'e')^f$ and so $\mathrm{End}_{A'}(A'e) \cong \mathrm{M}_f(\mathrm{End}_{A'}(A'e'))$. Recalling that $\mathrm{End}_{A'}(A'e') \cong E'$ ([**18**, Lemma 3.19]) the homomorphism $E' \otimes_E \theta \in \mathrm{End}_{A'}(V')$ may be regarded as an element $M(\theta)$ of $\mathrm{M}_m(M_f(\mathrm{End}_{A'}(A'e'))) \cong \mathrm{M}_{mf}(E')$, and then $\mathrm{nr}_{\mathrm{End}_A(V)}(\theta) = \det(M(\theta))$.

On the other hand, regarding θ as an element of $\mathrm{End}_A((Ae)^m)$ one computes $\det_A(\theta)$ as the determinant over E' of the element of $\mathrm{End}_{E'}((e'A'e)^m)$ which is obtained by restricting $E' \otimes_E \theta$ to $(e'A'e)^m$. However, $A'e \cong (A'e')^f$ and restriction to $e'A'e'$ identifies $\mathrm{End}_{A'}(A'e')$ with $\mathrm{End}_{E'}(e'A'e') = E'$. Restricting $E' \otimes_E \theta \in \mathrm{M}_m(\mathrm{End}_{A'}(A'e))$ to $(e'A'e)^m$ thus gives an element of $\mathrm{M}_m(\mathrm{End}_{E'}((e'A'e')^f)) \cong \mathrm{M}_{mf}(E')$. The matrix so obtained is $M(\theta)$ and hence $\det_A(\theta) = \mathrm{nr}_{\mathrm{End}_A(V)}(\theta)$, as required.

(2) The proof of claim (1) shows that if $V \cong (Ae)^m$, then there exists an isomorphism of E-algebras $\lambda \mathrm{End}_A(V) \cong \mathrm{M}_m(\mathrm{End}_A(Ae))$ such that $\det_A(\theta) = \mathrm{nr}_{\mathrm{M}_m(\mathrm{End}_A(Ae))}(\lambda(\theta))$ for each $\theta \in \mathrm{End}_A(V)$. This implies that $\delta_A^+(V)$ is equal to $\zeta(\mathrm{M}_m(\mathrm{End}_A(Ae)))^{\times +}$ and this gives the claimed result since $\mathrm{End}_A(Ae)$ is isomorphic to the opposite ring of D.

(3) If θ and θ' both belong to $\mathrm{Is}_A(V,W)$, then $\theta' = (\theta' \circ \theta^{-1}) \circ \theta$ and $\theta' \circ \theta^{-1} \in \mathrm{Aut}_A(W)$. In addition, for any elements $\phi \in \mathrm{Aut}_A(W)$ and $\theta \in \mathrm{Is}_A(V,W)$ one has $\det_A(\phi \circ \theta) = \det_A(\phi) \cdot \det_A(\theta)$ in $\delta_A(V,W)$ and so the stated result is a direct consequence of claim (2).

(4) The natural decomposition of algebras

$$A \otimes_{\mathbb{Q}} \mathbb{R} \cong \left(\prod_{v \in S_{\mathbb{R}}(E)} A \otimes_{E,\sigma_v} \mathbb{R} \right) \times \left(\prod_{v \in S_{\mathbb{C}}(E)} A \otimes_{E,\sigma_v} \mathbb{C} \right)$$

induces a decomposition

$$\zeta(A \otimes_{\mathbb{Q}} \mathbb{R})^{\times +} = \left(\prod_{v \in S_{\mathbb{R}}(E)} \zeta(A \otimes_{E,\sigma_v} \mathbb{R})^{\times +} \right) \times \left(\prod_{v \in S_{\mathbb{C}}(E)} \zeta(A \otimes_{E,\sigma_v} \mathbb{C})^{\times +} \right).$$

This implies (4) since $\zeta(A \otimes_{E,\sigma_v} E_v)^{\times +}$ is equal to $E_v^{\times}$ unless $E_v = \mathbb{R}$ and $A \otimes_{E,\sigma_v} \mathbb{R}$ is ramified in which case it is equal to the multiplicative group of strictly positive real numbers.

(5) If $\theta \in \mathrm{Is}_A(V,W)$, then claim (3) implies $\delta_A^+(V,W) = \zeta(A)^{\times +} \cdot \det_A(\theta)$, $\delta_A(V,W) = E \cdot \det_A(\theta)$ and $\delta^+_{A \otimes_{\mathbb{Q}} \mathbb{R}}(V \otimes_{\mathbb{Q}} \mathbb{R}, W \otimes_{\mathbb{Q}} \mathbb{R}) = \zeta(A \otimes_{\mathbb{Q}} \mathbb{R})^{\times +} \cdot \det_A(\theta)$, and so one need only prove that $E \cap \zeta(A \otimes_{\mathbb{Q}} \mathbb{R})^{\times +} = \zeta(A)^{\times +}$. However, claim (4) implies that

$$E \cap \zeta(A \otimes_{\mathbb{Q}} \mathbb{R})^{\times +} = \{e \in E^{\times} : \sigma_v(e) > 0 \text{ if } v \in S_{\mathbb{R}}(E) \text{ and } A \otimes_{E,\sigma_v} \mathbb{R} \text{ is ramified}\},$$

and this is equal to $\zeta(A)^{\times +}$ by the Hasse–Schilling–Maass Norm Theorem [**18**, (7.48)]. □

If V and W are isomorphic finitely generated A-modules, then for each element τ of $\delta_A^+(V, W)$ we write $\Phi(\tau)$ for its fibre under the canonical map

$$\det_A \colon \mathrm{Is}_A(V, W) \to \delta_A^+(V, W).$$

For any such τ we also write τ^{-1} for the element of $\delta_A^+(W, V)$ which is equal to $\det_A(\phi^{-1})$ for any (and therefore every) $\phi \in \Phi(\tau)$.

Let V_1, V_2 and V_3 be pairwise isomorphic finitely generated A-modules. If ϕ_1' and ϕ_2' are E'-linear maps from $\mathrm{Hom}_{A'}(A'e', V_1')$ to $\mathrm{Hom}_{A'}(A'e', V_2')$ and from $\mathrm{Hom}_{A'}(A'e', V_2')$ to $\mathrm{Hom}_{A'}(A'e', V_3')$ respectively, then the association

$$\det_{E'}(\phi_2') \otimes_{E'} \det_{E'}(\phi_1') \mapsto \det_{E'}(\phi_2' \circ \phi_1')$$

induces an isomorphism

$$\delta_A(V_2, V_3) \otimes_E \delta_A(V_1, V_2) \cong \delta_A(V_1, V_3). \tag{1.1}$$

For any elements $\tau_1 \in \delta_A^+(V_1, V_2)$ and $\tau_2 \in \delta_A^+(V_2, V_3)$ we shall write $\tau_2 \circ \tau_1$ for the element of $\delta_A^+(V_1, V_3)$ which is equal to $\det_A(\phi_2 \circ \phi_1)$ for any choice of isomorphisms $\phi_1 \in \Phi(\tau_1)$ and $\phi_2 \in \Phi(\tau_2)$. Regarding (1.1) as an identification one therefore has $\tau_2 \otimes_E \tau_1 = \tau_2 \circ \tau_1$.

We now use Wedderburn decompositions to relax the restriction that A is simple. To be specific, we suppose that there exists a finite index set I such that $A = \prod_{i \in I} A_i$ with each A_i a finite dimensional central simple E_i-algebra for a suitable extension field E_i of E. Then $\zeta(A) = \prod_{i \in I} E_i$ and $\zeta(A)^{\times +} = \prod_{i \in I} \zeta(A_i)^{\times +}$. For any finitely generated A-modules V and W there are corresponding decompositions $V = \bigoplus_{i \in I} V_i$ and $W = \bigoplus_{i \in I} W_i$ and hence also $\mathrm{Is}_A(V, W) \cong \bigoplus_{i \in I} \mathrm{Is}_{A_i}(V_i, W_i)$. For any isomorphism $\phi \in \mathrm{Is}_A(V, W)$ we let ϕ_i denote its component in $\mathrm{Is}_{A_i}(V_i, W_i)$ and then set $\det_A(\phi) := \prod_{i \in I} \det_{A_i}(\phi_i)$. We let $\delta_A^+(V, W)$ denote the image of $\mathrm{Is}_A(V, W)$ under $\det_A$, and we write $\delta_A(V, W)$ for the associated $\zeta(A)$-module. A 'trivialization', resp. 'trivialization fibre', of $\mathrm{Is}_A(V, W)$ is an element of $\delta_A^+(V, W)$, resp. a fibre of the map $\det_A \colon \mathrm{Is}_A(V, W) \to \delta_A^+(V, W)$. For each $\tau \in \delta_A^+(V, W)$ we write $\Phi(\tau)$ for the corresponding trivialization fibre of $\mathrm{Is}_A(V, W)$, and trivialization inverses τ^{-1}, compositions $\tau_1 \circ \tau_2$ and tensor products $\tau_1 \otimes_{\zeta(A)} \tau_2$ are defined component-wise. We remark that the terminology of 'trivializations' is suggested naturally by the following example.

EXAMPLE 1.2. Let A be a finite dimensional semisimple E-algebra which is commutative. Then (in terms of the above notation) one has $A = \prod_{i \in I} E_i$ and if $V = \bigoplus_{i \in I} V_i$ is any finitely generated A-module, then one obtains a rank one A-module by setting $\det_A(V) := \bigoplus_{i \in I} \det_{E_i}(V_i)$. If now W is any A-module which is isomorphic to V, then there are canonical identifications

$$\delta_A(V, W) \cong \mathrm{Hom}_A\big(\det_A(V), \det_A(W)\big), \quad \delta_A^+(V, W) \cong \mathrm{Is}_A\big(\det_A(V), \det_A(W)\big).$$

For each finitely generated A-module X we write $\det_A(X)^{-1}$ for the A-linear dual $\mathrm{Hom}_A(\det_A(X), A)$. Taken in conjunction with the canonical evaluation pairing $\det_A(W) \otimes_A \det_A(W)^{-1} \cong A$ each element τ of $\delta_A^+(V, W)$ induces a 'trivialization'

$$\tau_{\mathrm{triv}} \in \mathrm{Is}_A\big(\det_A(V) \otimes_A \det_A(W)^{-1}, A\big)$$

of the A-line $\det_A(V) \otimes_A \det_A(W)^{-1}$. For each $\phi \in \mathrm{Is}_A(V, W)$ we shall often write ϕ_{triv} in place of $\big(\det_A(\phi)\big)_{\mathrm{triv}}$.

In the remainder of this section we discuss trivializations in the context of bounded complexes of finitely generated A-modules. We remark that the basic algebraic construction which we shall first describe (in Lemma 1.3) is simply an equivariant version of a well known construction that is used to define torsion invariants of chain complexes and of Hermitian chain complexes (see, for example, [**25, 29**] or [**3**, §1a]).

For any bounded complex of finitely generated A-modules

$$C^{\cdot} := [0 \to \cdots \to C^i \to C^{i+1} \to \cdots \to 0] \tag{1.2}$$

we let $B^i(C^{\cdot})$, $Z^i(C^{\cdot})$ and $H^i(C^{\cdot})$ denote the A-modules of coboundaries, cocycles and cohomology in degree i, and we write $d^i(C^{\cdot})\colon C^i \to B^{i+1}(C^{\cdot})$ and $\pi^i(C^{\cdot})\colon Z^i(C^{\cdot}) \to H^i(C^{\cdot})$ for the differential and the natural surjective map in degree i. Then for each i there are short exact sequences

$$0 \to Z^i(C^{\cdot}) \xrightarrow{\subseteq} C^i \xrightarrow{d^i(C^{\cdot})} B^{i+1}(C^{\cdot}) \to 0$$

and

$$0 \to B^i(C^{\cdot}) \xrightarrow{\subseteq} Z^i(C^{\cdot}) \xrightarrow{\pi^i(C^{\cdot})} H^i(C^{\cdot}) \to 0.$$

Since A is semisimple, we can choose A-equivariant sections σ^i and μ^i to $d^i(C^{\cdot})$ and $\pi^i(C^{\cdot})$ respectively, and so obtain an isomorphism of A-modules

$$\theta(\sigma^i,\mu^i)\colon B^{i+1}(C^{\cdot})\oplus H^i(C^{\cdot})\oplus B^i(C^{\cdot}) \xrightarrow{\sim} C^i, \quad (x,y,z) \mapsto \sigma^i(x)+\mu^i(y)+z. \tag{1.3}$$

For any $\mathbb{Z}$-graded module $X^{\cdot}$ we shall write X^- and X^+ for the direct sum (in the natural 'ascending' order) of the modules X^i as i runs over all odd and all even integers respectively. We use similar notation in respect of any morphism of $\mathbb{Z}$-graded modules.

For each $\phi \in \mathrm{Is}_A\big(H^-(C^{\cdot}), H^+(C^{\cdot})\big)$ we let

$$\phi(\sigma^{\cdot},\mu^{\cdot})\colon C^- \xrightarrow{\sim} C^+$$

denote the A-module isomorphism which is obtained as the composite

$$\begin{aligned} C^- &\xrightarrow{\sim} \oplus\big(B^{2i+2}(C^{\cdot}) \oplus H^{2i+1}(C^{\cdot}) \oplus B^{2i+1}(C^{\cdot})\big)\\ &\xrightarrow{\sim} \oplus\big(B^i(C^{\cdot})\big) \oplus H^-(C^{\cdot})\\ &\xrightarrow{\sim} \oplus\big(B^i(C^{\cdot})\big) \oplus H^+(C^{\cdot})\\ &\xrightarrow{\sim} \oplus\big(B^{2i+1}(C^{\cdot}) \oplus H^{2i}(C^{\cdot}) \oplus B^{2i}(C^{\cdot})\big)\\ &\xrightarrow{\sim} C^+, \end{aligned}$$

where $\bigoplus$ indicates direct sum over all integers i, the first map is $\bigoplus \theta(\sigma^{2i+1},\mu^{2i+1})^{-1}$, the second map is clear, the third map is $\mathrm{id}\oplus\phi$, the fourth map is clear and the last map is $\bigoplus \theta(\sigma^{2i},\mu^{2i})$.

LEMMA 1.3. *For any element τ of $\delta_A^+\big(H^-(C^{\cdot}), H^+(C^{\cdot})\big)$ there exists a element $\tau(C^{\cdot})$ of $\delta_A^+(C^-,C^+)$ which is equal to $\det_A\big(\phi(\sigma^{\cdot},\mu^{\cdot})\big)$ for any $\phi \in \Phi(\tau)$ and any choice of sections $\{\sigma^i,\mu^j\}_{i,j}$ as described above. If $C^{\cdot}$ is acyclic, then we write $\iota(C^{\cdot})$ for the unique element of $\delta_A^+(C^-,C^+)$ which is obtained in this manner.*

PROOF. We must show that if ϕ and ϕ' both belong to $\Phi(\tau)$ and $\{\widehat{\sigma}^i,\widehat{\mu}^j\}_{i,j}$ is any other choice of sections as above, then one has

$$\det_A\big(\phi(\sigma^{\cdot},\mu^{\cdot})\big) = \det_A\big(\phi'(\widehat{\sigma}^{\cdot},\widehat{\mu}^{\cdot})\big) \in \delta_A^+(C^-,C^+). \tag{1.4}$$

To do this we note first that $\phi'(\widehat{\sigma}^{\cdot}, \widehat{\mu}^{\cdot}) = \kappa_1 \circ \phi'(\sigma^{\cdot}, \mu^{\cdot}) \circ \kappa_2$ with

$$\kappa_1 = \bigoplus_{i\in\mathbb{Z}} \big(\theta(\widehat{\sigma}^{2i}, \widehat{\mu}^{2i}) \circ \theta(\sigma^{2i}, \mu^{2i})^{-1}\big) \in \mathrm{Aut}_A(C^+)$$

and

$$\kappa_2 = \bigoplus_{i\in\mathbb{Z}} \big(\theta(\sigma^{2i+1}, \mu^{2i+1}) \circ \theta(\widehat{\sigma}^{2i+1}, \widehat{\mu}^{2i+1})^{-1}\big) \in \mathrm{Aut}_A(C^-).$$

We claim that $\det_A(\kappa_1) = 1$ and $\det_A(\kappa_2) = 1$. We verify this equality for κ_1, leaving the argument for κ_2 (which is entirely similar) to the reader. It is enough to show that for each splitting field E' of A over E and each indecomposable idempotent e' of $A' := E' \otimes_E A$ the restriction $(E' \otimes_E \kappa_1)e'$ of $E' \otimes_E \kappa_1$ to $e'(E' \otimes_E C^+)$ has determinant equal to 1.
We set $C_i' := E' \otimes_E C^i$, $B_i' := E' \otimes_E B^i(C^{\cdot})$, $Z_i' := E' \otimes_E Z^i(C^{\cdot})$ and $H_i' := E' \otimes_E H^i(C^{\cdot})$ for each integer i. Then one has

$$(E' \otimes_E \kappa_1)e' = \bigoplus_{j\in\mathbb{Z}} \kappa_{e',2j}$$

where

$$\kappa_{e',2j} := \big(E' \otimes_E \theta(\widehat{\sigma}^{2j}, \widehat{\mu}^{2j}) \circ \theta(\sigma^{2j}, \mu^{2j})^{-1}\big)e' \in \mathrm{Aut}_{E'}(e'C_{2j}'),$$

and so it suffices to prove that $\det_{E'}(\kappa_{e',2j}) = 1$ for each integer j. The key point here is that the maps $(E' \otimes_E \widehat{\sigma}^{2j})e'$ and $(E' \otimes_E \sigma^{2j})e'$ are both sections to $\big(E' \otimes_E d^{2j}(C^{\cdot})\big)e'$ and the maps $(E' \otimes_E \widehat{\mu}^{2j})e'$ and $(E' \otimes_E \mu^{2j})e'$ are both sections to $\big(E' \otimes_E \pi^{2j}(C^{\cdot})\big)e'$. It follows that

$$(E' \otimes_E \widehat{\sigma}^{2j})e' - (E' \otimes_E \sigma^{2j})e' \in \mathrm{Hom}_{E'}(e'B_{2j+1}', \mu^{2j}e'H_{2j}' \oplus e'B_{2j}')$$

and that

$$(E' \otimes_E \widehat{\mu}^{2j})e' - (E' \otimes_E \mu^{2j})e' \in \mathrm{Hom}_{E'}(e'H_{2j}', e'B_{2j}').$$

Upon choosing an ordered E'-basis of $e'C_{2j}'$ as the union of E'-bases of $\sigma^{2j}e'B_{2j+1}' \oplus \{0\} \oplus \{0\}$, $\{0\} \oplus \mu^{2j}e'H_{2j}' \oplus \{0\}$ and $\{0\} \oplus \{0\} \oplus e'B_{2j}'$ the matrix of $\kappa_{e',2j}$ is therefore a block matrix of shape

$$\begin{pmatrix} I & * & * \\ 0 & I & * \\ 0 & 0 & I \end{pmatrix}$$

and the determinant of this matrix is 1.

We have now shown that $\det_A\big(\phi'(\sigma^{\cdot}, \mu^{\cdot})\big) = \det_A\big(\phi'(\widehat{\sigma}^{\cdot}, \widehat{\mu}^{\cdot})\big)$ and so to prove (1.4) it only remains to prove that $\det_A\big(\phi(\sigma^{\cdot}, \mu^{\cdot})\big) = \det_A\big(\phi'(\sigma^{\cdot}, \mu^{\cdot})\big)$. To do this we set $W := \bigoplus_{j\in\mathbb{Z}} \sigma^{2j}B^{2j+1}(C^{\cdot}) \oplus \{0\} \oplus B^{2j}(C^{\cdot})$. Then the sections μ^{2i} together induce an isomorphism $C^+ \xrightarrow{\sim} W \oplus H^+(C^{\cdot})$ and, with respect to this decomposition, one has $\phi'(\sigma^{\cdot}, \mu^{\cdot}) \circ \phi(\sigma^{\cdot}, \mu^{\cdot})^{-1} = \mathrm{id}_W \oplus (\phi' \circ \phi^{-1})$ so that $\det_A\big(\phi'(\sigma^{\cdot}, \mu^{\cdot})\big) \otimes_{\zeta(A)} \det_A\big(\phi(\sigma^{\cdot}, \mu^{\cdot})\big)^{-1} = \det_A(\phi' \circ \phi^{-1}) = \tau \otimes_{\zeta(A)} \tau^{-1} = 1$, as required. □

In practice it can be useful to carefully choose the sections $\sigma^{\cdot}$ and $\mu^{\cdot}$ when computing $\tau(C^{\cdot})$ via the recipe of Lemma 1.3. The following lemma can be very helpful in this regard.

LEMMA 1.4. *Suppose given a commutative diagram of short exact sequences of finitely generated A-modules*

$$\begin{array}{ccccccccc} & & 0 & & 0 & & 0 & & \\ & & \downarrow & & \downarrow & & \downarrow & & \\ 0 & \to & K' & \to & V' & \xrightarrow{d'} & W' & \to & 0 \\ & & \downarrow & & \downarrow & & \downarrow{\scriptstyle \phi_1} & & \\ 0 & \to & K & \to & V & \xrightarrow{d} & W & \to & 0 \\ & & \downarrow & & \downarrow{\scriptstyle \phi_2} & & \downarrow & & \\ 0 & \to & K'' & \to & V'' & \xrightarrow{d''} & W'' & \to & 0 \\ & & \downarrow & & \downarrow & & \downarrow & & \\ & & 0 & & 0 & & 0. & & \end{array}$$

Then one may choose an A-equivariant section $\sigma\colon W \to V$ to d such that $\sigma|_{W'}$ is a section to d' (and hence σ induces a section to d'').

PROOF. We first choose any A-equivariant section σ to d. If θ denotes the composite map $W' \xrightarrow{\phi_1} W \xrightarrow{\sigma} V \xrightarrow{\phi_2} V''$, then the commutativity of the above diagram implies that θ belongs to $\operatorname{Hom}_A(W', K'')$. Since W' is a projective A-module θ lifts through $\phi_2|_K$ to an element of $\operatorname{Hom}_A(W', K)$ and since K is injective over A the latter element can in turn be extended through ϕ_1 to an element θ' of $\operatorname{Hom}_A(W, K)$. We claim that the homomorphism $\widehat{\sigma} := \sigma - \theta' \in \operatorname{Hom}_A(W, V)$ is a section to d which restricts to give a section to d'. Indeed, one has $d \circ \widehat{\sigma} = d \circ \sigma - d \circ \theta' = d \circ \sigma = \mathrm{id}_W$ and for each $w' \in W'$ one has $\phi_2(\widehat{\sigma}(\phi_1(w'))) = \phi_2(\sigma(\phi_1(w'))) - \phi_2(\theta'(\phi_1(w'))) = \theta(w') - \theta(w') = 0$, so that $\widehat{\sigma}|_{W'}$ belongs to $\operatorname{Hom}_A(W', V')$, as required. □

To conclude this section we discuss the behaviour of trivializations with respect to short exact sequences of bounded complexes of A-modules.

For any ring R, a 'short exact sequence of complexes of R-modules'

$$0 \to C_1^{\cdot} \xrightarrow{\alpha} C_2^{\cdot} \xrightarrow{\beta} C_3^{\cdot} \to 0$$

is a pair of morphisms of complexes of R-modules α and β such that in each degree i there are short exact sequences

$$0 \to C_1^i \xrightarrow{\alpha^i} C_2^i \xrightarrow{\beta^i} C_3^i \to 0.$$

In particular, if $R = A$ and each $C_i^{\cdot}$ is a bounded complex of finitely generated A-modules, then these short exact sequences together induce a canonical isomorphism

$$i_{\alpha,\beta}\colon \delta_A(C_1^-, C_1^+) \otimes_{\zeta(A)} \delta_A(C_3^-, C_3^+) \xrightarrow{\sim} \delta_A(C_2^-, C_2^+).$$

(Explicitly, if $\phi_i \in \operatorname{Is}_A(C_i^-, C_i^+)$ for $i = 1, 3$, then $i_{\alpha,\beta}(\det_A(\phi_1) \otimes \det_A(\phi_3))$ is equal to the reduced determinant of the composite

$$C_2^- \to C_1^- \oplus C_3^- \xrightarrow{(\phi_1, \phi_3)} C_1^+ \oplus C_3^+ \to C_2^+$$

where the first and third maps are the isomorphisms induced by any choice of sections to the morphisms β^i—this construction is independent of such choices.)

If now $\beta\colon C^{\cdot} \to D^{\cdot}$ is any morphism of bounded complexes of finitely generated R-modules which is injective, resp. surjective, in each degree, then we let $K(\beta)^{\cdot}$

denote the complex which in each degree i has $K(\beta)^i = \operatorname{cok}(\beta^i)$, resp. $K(\beta)^i = \ker(\beta^i)$, and for which the differential $d^i(K(\beta)^\cdot)$ is equal to the morphism induced by, resp. to the restriction of, $d^i(C^\cdot)$. Then there is an associated short exact sequence of complexes

$$0 \to C^\cdot \xrightarrow{\beta} D^\cdot \xrightarrow{\pi_\beta} K(\beta)^\cdot \to 0$$

with π_β denoting the natural projection map, resp.

$$0 \to K(\beta)^\cdot \xrightarrow{\iota_\beta} C^\cdot \xrightarrow{\beta} D^\cdot \to 0$$

with ι_β denoting the natural inclusion map.

For any complex of A-modules $C^\cdot$ we write $H^\cdot(C^\cdot)$ for the complex which in each degree i is equal to $H^i(C^\cdot)$ and for which all differentials are 0.

In the statement of the next result all tensor products are to be considered as taken over $\zeta(A)$.

LEMMA 1.5. *Let $C^\cdot$, $\{\mu^j\}_j$ and $\{\sigma^i\}_i$ be as in Lemma 1.3.*

(1) *The sections μ^j combine to induce a morphism of complexes $\mu\colon H^\cdot(C^\cdot) \to C^\cdot$ which is injective in each degree. The associated complex $K(\mu)^\cdot$ is acyclic, and for each trivialization $\tau \in \delta_A^+(H^-(C^\cdot), H^+(C^\cdot))$ one has*

$$\tau(C^\cdot) = i_{\mu,\pi_\mu}\big(\tau \otimes \iota(K(\mu)^\cdot)\big).$$

(2) *The isomorphisms of (1.3) together induce a morphism of complexes $\sigma\colon C^\cdot \to H^\cdot(C^\cdot)$ which is surjective in each degree. The associated complex $K(\sigma)^\cdot$ is acyclic, and for each trivialization $\tau \in \delta_A^+(H^-(C^\cdot), H^+(C^\cdot))$ one has*

$$\tau(C^\cdot) = i_{\iota_\sigma,\sigma}\big(\iota(K(\sigma)^\cdot) \otimes \tau\big).$$

(3) *If $0 \to C_1^\cdot \xrightarrow{\alpha} C_2^\cdot \xrightarrow{\beta} C_3^\cdot \to 0$ is a short exact sequence of bounded acyclic complexes of finitely generated A-modules, then one has*

$$\iota(C_2^\cdot) = i_{\alpha,\beta}\big(\iota(C_1^\cdot) \otimes \iota(C_3^\cdot)\big).$$

PROOF. Everything but the last parts of (1) and (2) are clear, and both last parts are a straightforward consequence of the explicit description of $\tau(C^\cdot)$ given in Lemma 1.3. As far as (3) is concerned, the fact that each complex $C_j^\cdot$ is acyclic implies that in each degree i one has a commutative diagram of short exact sequences

$$\begin{array}{ccccccccc}
 & & 0 & & 0 & & 0 & & \\
 & & \downarrow & & \downarrow & & \downarrow & & \\
0 & \to & B^i(C_1^\cdot) & \to & C_1^i & \xrightarrow{d^i(C_1^\cdot)} & B^{i+1}(C_1^\cdot) & \to & 0 \\
 & & \downarrow & & \downarrow{\scriptstyle \alpha^i} & & \downarrow & & \\
0 & \to & B^i(C_2^\cdot) & \to & C_2^i & \xrightarrow{d^i(C_2^\cdot)} & B^{i+1}(C_2^\cdot) & \to & 0 \\
 & & \downarrow & & \downarrow{\scriptstyle \beta^i} & & \downarrow & & \\
0 & \to & B^i(C_3^\cdot) & \to & C_3^i & \xrightarrow{d^i(C_3^\cdot)} & B^{i+1}(C_3^\cdot) & \to & 0 \\
 & & \downarrow & & \downarrow & & \downarrow & & \\
 & & 0 & & 0 & & 0. & &
\end{array}$$

By using Lemma 1.4 we choose a section $\sigma_2^i\colon B^{i+1}(C_2^\cdot) \to C_2^i$ to $d^i(C_2^\cdot)$ which induces under restriction, resp. passage to quotient, a section σ_1^i, resp. σ_3^i, to $d^i(C_1^\cdot)$, resp. $d^i(C_3^\cdot)$. Claim (3) now follows upon computing each trivialization $\iota(C_j^\cdot)$ by using this choice of sections $\sigma_j^\cdot$. □

2. Refined Euler characteristics

In this section R is a Dedekind domain with field of fractions F and E is an extension field of F. We let $\mathfrak{M}$ denote the set of maximal ideals of R. For any R-module X and R-algebra Λ we set $X_\Lambda := \Lambda \otimes_R X$. If Λ is equal to the completion (or localization) of R at $v \in \mathfrak{M}$, then we write X_v in place of X_Λ.

Let $\mathcal{A}$ be an R-order in a finite dimensional semisimple F-algebra A. We write $\mathfrak{B}(\mathcal{A})$ and $\mathfrak{B}(A_E)$ for the categories of finitely generated projective $\mathcal{A}$-modules and of finitely generated A_E-modules respectively. We let $K_0(\mathcal{A})$ and $K_0(A_E)$ denote the Grothendieck groups of the categories $\mathfrak{B}(\mathcal{A})$ and $\mathfrak{B}(A_E)$ respectively, and we write $K_1(\mathcal{A})$ and $K_1(A_E)$ for the associated Whitehead groups. We also let $SK_0(\mathcal{A})$ denote the kernel of the morphism from $K_0(\mathcal{A})$ to $K_0(A_E)$ which is induced by the scalar extension functor $E \otimes_R -$ from $\mathfrak{B}(\mathcal{A})$ to $\mathfrak{B}(A_E)$ (this kernel is indeed independent of the choice of E), and we write $K_0(\mathcal{A}, E)$ for the associated relative algebraic K-group, as described explicitly in terms of generators and relations [**27**, p. 215]. (The group $K_0(\mathcal{A}, E)$ is also described by Bass in [**1**, IX, §1] where it is denoted by $K_0'(\mathcal{A} \hookrightarrow A_E)$). In addition, we recall from [**27**] that there is a natural commutative diagram of the form

$$
\begin{array}{ccccccc}
\mathrm{im}(\mathrm{nr}_{\mathcal{A}}) & \subseteq & \zeta(A_E)^{\times+} & & & & \\
\uparrow{\scriptstyle \mathrm{nr}_{\mathcal{A}}} & & \uparrow{\scriptstyle \mathrm{nr}_{A_E}} & & & & \\
K_1(\mathcal{A}) & \xrightarrow{\partial_{2,\mathcal{A}}} & K_1(A_E) & \xrightarrow{\partial_{1,\mathcal{A}}} & K_0(\mathcal{A},E) & \xrightarrow{\partial_{0,\mathcal{A}}} & SK_0(\mathcal{A}) \\
\uparrow{\scriptstyle =} & & \uparrow{\scriptstyle \iota_1} & & \uparrow{\scriptstyle \iota_2} & & \uparrow{\scriptstyle =} \\
K_1(\mathcal{A}) & \to & K_1(A) & \to & K_0(\mathcal{A},F) & \to & SK_0(\mathcal{A}) \\
 & & \downarrow & & \downarrow{\scriptstyle \gamma} & & \\
 & & \bigoplus_{v\in\mathfrak{M}} K_1(A_v)/K_1(\mathcal{A}_v) & \xrightarrow{\partial} & \bigoplus_{v\in\mathfrak{M}} K_0(\mathcal{A}_v,F_v). & &
\end{array} \tag{2.1}
$$

In this diagram nr_{A_E} is the reduced norm map on $K_1(A_E)$ (cf. [**18**, §45A]), $\mathrm{nr}_{\mathcal{A}} := \mathrm{nr}_{A_E} \circ \partial_{2,\mathcal{A}}$ where $\partial_{2,\mathcal{A}}$ is the natural scalar extension morphism, $\partial_{1,\mathcal{A}}\big((A_E^n, \phi)\big) = [\mathcal{A}^n, \phi, \mathcal{A}^n]$ for each n and $\phi \in \mathrm{Aut}_{A_E}(A_E^n)$, $\partial_{0,\mathcal{A}}([P, g, Q]) = (P) - (Q)$ for each object P and Q of $\mathfrak{B}(\mathcal{A})$ and isomorphism $g \in \mathrm{Is}_{A_E}(P_E, Q_E)$, the morphisms in the third row are completely analogous to those in the second, ι_1 and ι_2 are the natural scalar extension morphisms, γ is equal to the product $\prod_{v\in\mathfrak{M}} \gamma_v$ with $\gamma_v([P, g, Q]) = [P_v, F_v \otimes_F g, Q_v]$ and ∂ is induced by the maps $K_1(A_v) \to K_0(\mathcal{A}_v, F_v)$ which are analogous to $\partial_{1,\mathcal{A}}$. We remark that the second and third rows of the diagram are exact [**27**, Theorem 15.5], that ι_1 and ι_2 are injective (indeed, in the sequel we shall usually regard these maps as inclusions), that nr_{A_E} and γ are bijective (concerning γ see, for example, the discussion following [**18**, (49.12)]) and ∂ is injective, and that ∂ is bijective precisely when $SK_0(\mathcal{A})$ coincides with the locally-free class group of $\mathcal{A}$ [**18**, (49.15)]. We further recall that this last condition is satisfied if, for example, $\mathcal{A}$ is commutative or $\mathcal{A} = R[G]$ for any finite group G (loc. cit., Remark 49.11(iv)) or $\mathcal{A}$ is a maximal R-order in A (loc. cit., Theorem 49.32).

We write $\mathrm{P\,Mod}(\mathcal{A})^{\cdot}$ for the category of bounded complexes of finitely generated projective $\mathcal{A}$-modules. Then any object $P^{\cdot}$ of $\mathrm{P\,Mod}(\mathcal{A})^{\cdot}$ has a natural Euler characteristic class $\chi_{\mathcal{A}}P^{\cdot}$ in $K_0(\mathcal{A})$ that we normalize in the following way

$$\chi_{\mathcal{A}}P^{\cdot} := \sum_{i\in\mathbb{Z}}(-1)^{i+1}(P^i) = (P^-) - (P^+) \in K_0(\mathcal{A}).$$

We recall that the mapping cone of any quasi-isomorphism is acyclic and hence that the Euler characteristic construction $\chi_{\mathcal{A}}(-)$ extends to give a well-defined map from $\mathfrak{D}^{\mathrm{perf}}(\mathcal{A})$ to $K_0(\mathcal{A})$.

Our first aim in this section is to show that if $P^{\cdot} \in \mathrm{Ob}\big(\mathrm{P\,Mod}(\mathcal{A})^{\cdot}\big)$ and $\tau \in \delta^+_{A_E}\big(H^-(P^{\cdot}_E), H^+(P^{\cdot}_E)\big)$, then one can associate to the pair $(P^{\cdot},\tau)$ a canonical pre-image $\chi_{\mathcal{A}}\big(P^{\cdot},\tau(P^{\cdot}_E)\big)$ of $\chi_{\mathcal{A}}P^{\cdot}$ under the map $\partial_{0,\mathcal{A}}\colon K_0(\mathcal{A},E) \to K_0(\mathcal{A})$. We shall then show that this construction naturally extends to the derived category and is additive on suitable distinguished triangles.

THEOREM 2.1. *Fix an object $P^{\cdot}$ of $\mathrm{P\,Mod}(\mathcal{A})^{\cdot}$ with the property that P^-_E and P^+_E are isomorphic A_E-modules. Then for each $\kappa \in \delta^+_{A_E}(P^-_E, P^+_E)$ there exists a canonical element $\chi_{\mathcal{A}}(P^{\cdot},\kappa)$ of $K_0(\mathcal{A},E)$ which satisfies each of the following conditions:*

(1) $\partial_{0,\mathcal{A}}\big(\chi_{\mathcal{A}}(P^{\cdot},\kappa)\big) = \chi_{\mathcal{A}}P^{\cdot}$ *in* $SK_0(\mathcal{A})$.

(2) *If τ and τ' are any elements of $\delta^+_{A_E}\big(H^-(P^{\cdot})_E, H^+(P^{\cdot})_E\big)$, then $\tau^{-1}\circ\tau' \in \zeta(A_E)^{\times +}$ and in $K_0(\mathcal{A},E)$ one has $\chi_{\mathcal{A}}\big(P^{\cdot},\tau'(P^{\cdot}_E)\big) - \chi_{\mathcal{A}}\big(P^{\cdot},\tau(P^{\cdot}_E)\big) = \partial_{1,\mathcal{A}}\big(\mathrm{nr}^{-1}_{A_E}(\tau^{-1}\circ\tau')\big)$.*

(3) *Let τ be an element of $\delta^+_{A_E}\big(H^-(P^{\cdot})_E, H^+(P^{\cdot})_E\big)$ and write $-\tau$ for the element of $\delta^+_{A_E}(H^-(P^{\cdot})_E, H^+(P^{\cdot})_E)$ that is equal to $\det_A(-\phi)$ for any $\phi\in\tau$. Then $(-\tau)^{-1} \in \delta^+_{A_E}\big(H^-(P^{\cdot}[1])_E, H^+(P^{\cdot}[1])_E\big)$, and in $K_0(\mathcal{A},E)$ one has $\chi_{\mathcal{A}}\big(P^{\cdot}[1],(-\tau)^{-1}(P^{\cdot}[1]_E)\big) = -\chi_{\mathcal{A}}\big(P^{\cdot},\tau(P^{\cdot}_E)\big)$.*

(4) *Let τ be an element of $\delta^+_{A_E}\big(H^-(P^{\cdot})_E, H^+(P^{\cdot})_E\big)$. Let $Q^{\cdot}$ be an object of $\mathrm{P\,Mod}(\mathcal{A})^{\cdot}$ and $\rho\colon P^{\cdot} \to Q^{\cdot}$ an $\mathcal{A}$-equivariant quasi-isomorphism. If τ_ρ denotes the unique element of $\delta^+_{A_E}\big(H^-(Q^{\cdot})_E, H^+(Q^{\cdot})_E\big)$ which is equal to*

$$\det_{A_E}\big(H^+(E\otimes_R\rho)\circ\phi\circ H^-(E\otimes_R\rho)^{-1}\big)$$

for any $\phi\in\Phi(\tau)$, then in $K_0(\mathcal{A},E)$ one has

$$\chi_{\mathcal{A}}\big(P^{\cdot},\tau(P^{\cdot}_E)\big) = \chi_{\mathcal{A}}\big(Q^{\cdot},\tau_\rho(Q^{\cdot}_E)\big).$$

In particular, if $P^{\cdot}$ is acyclic, then one has $\chi_{\mathcal{A}}\big(P^{\cdot},\iota(P^{\cdot}_E)\big) = 0$.

(5) *Assume that $R=\mathbb{Z}$ and that $E\subseteq\mathbb{R}$. If $\tau\in\delta^+_{A_E}\big(H^-(P^{\cdot})_E, H^+(P^{\cdot})_E\big)$, then $\chi_{\mathcal{A}}\big(P^{\cdot},\tau(P^{\cdot}_E)\big)$ belongs to the subgroup $K_0(\mathcal{A},\mathbb{Q})$ of $K_0(\mathcal{A},E)$ if and only if $\tau\in\delta_A\big(H^-(P^{\cdot})_{\mathbb{Q}}, H^+(P^{\cdot})_{\mathbb{Q}}\big)$.*

PROOF. Our proof of this result will proceed via a series of lemmas.

LEMMA 2.2. *Let P_1, P_2 be finitely generated projective $\mathcal{A}$-modules and ϕ, ϕ' elements of $\mathrm{Is}_{A_E}(P_{1,E},P_{2,E})$. If $\det_{A_E}(\phi) = \det_{A_E}(\phi')$ in $\delta_{A_E}(P_{1,E},P_{2,E})$, then one has $[P_1,\phi,P_2] = [P_1,\phi',P_2]$ in $K_0(\mathcal{A},E)$.*

PROOF. The defining relations of $K_0(\mathcal{A},E)$ imply that

$$[P_1,\phi',P_2] = [P_1,\phi,P_2] + [P_2,\phi'\circ\phi^{-1},P_2].$$

In addition, one has $[P_2, \phi' \circ \phi^{-1}, P_2] = \partial_{1,\mathcal{A}}((P_{2,E}, \phi' \circ \phi^{-1}))$, and Lemma 1.1(1) and diagram (2.1) combine to imply that

$$\begin{aligned}(P_{2,E}, \phi' \circ \phi^{-1}) \in \ker(\partial_{1,\mathcal{A}}) &\Leftrightarrow (P_{2,E}, \phi' \circ \phi^{-1}) \in \mathrm{im}(\partial_{2,\mathcal{A}}) \\ &\Leftrightarrow \det_{A_E}(\phi' \circ \phi^{-1}) \in \mathrm{im}(\mathrm{nr}_{\mathcal{A}}) \subseteq \zeta(A_E)^{\times +} \\ &\Leftrightarrow \det_{A_E}(\phi') \otimes_{\zeta(A_E)} \det_{A_E}(\phi)^{-1} \in \mathrm{im}(\mathrm{nr}_{\mathcal{A}}).\end{aligned}$$

Given these observations, the claimed equality is clear. □

Following this lemma we may set

$$\chi_{\mathcal{A}}(P^{\cdot}, \kappa) := [P^-, \phi, P^+] \in K_0(\mathcal{A}, E)$$

for any $\phi \in \Phi(\kappa)$. With this definition it is at once clear that

$$\chi_{\mathcal{A}} P^{\cdot} = \partial_{0,\mathcal{A}}(\chi_{\mathcal{A}}(P^{\cdot}, \kappa)) \in SK_0(\mathcal{A}),$$

and so this construction verifies claim (1) of the theorem.

Concerning claim (2), it is clear that $\tau^{-1} \circ \tau' \in \delta^+_{A_E}(H^-(P^{\cdot})_E, H^-(P^{\cdot})_E)$ and that Lemma 1.1(2) identifies the latter group with $\zeta(A_E)^{\times +}$. In addition, we observe that the equality of claim (2) follows directly from [**9**, Proposition 1.2.1(ii)].

Before dealing with claim (3) we recall that in each degree i one has $P^{\cdot}[1]^i = P^{i+1}, d^i(P^{\cdot}[1]) = -d^{i+1}(P^{\cdot})$ and $\pi^i(P^{\cdot}[1]) = \pi^{i+1}(P^{\cdot})$. In particular, in the notation of Lemma 1.3, having chosen sections σ^i and μ^j to $d^i(P^{\cdot})$ and $\pi^i(P^{\cdot})$ respectively we obtain sections to $d^i(P^{\cdot}[1])$ and $\pi^i(P^{\cdot}[1])$ by setting $\sigma[1]^i := -\sigma^{i+1}$ and $\mu[1]^i = \mu^{i+1}$. Hence, if $\phi \in \tau$, then one has

$$\begin{aligned}&\chi_{\mathcal{A}}(P^{\cdot}, \tau(P^{\cdot}_E)) + \chi_{\mathcal{A}}(P^{\cdot}[1], (-\tau)^{-1}(P^{\cdot}[1]_E)) \\ &\quad = [P_1^-, \phi(\sigma^{\cdot}, \mu^{\cdot}), P_1^+] + [P^{\cdot}[1]^-, (-\phi)^{-1}(\sigma[1]^{\cdot}, \mu[1]^{\cdot}), P_1^{\cdot}[1]^+] \\ &\quad = [P_1^-, \phi(\sigma^{\cdot}, \mu^{\cdot}), P_1^+] + [P_1^+, (-\phi)^{-1}(\sigma[1]^{\cdot}, \mu[1]^{\cdot}), P_1^-] \\ &\quad = [P_1^-, \phi(\sigma^{\cdot}, \mu^{\cdot}) \circ (-\phi)^{-1}(\sigma[1]^{\cdot}, \mu[1]^{\cdot}), P_1^-] \\ &\quad = \partial_{1,\mathcal{A}}((P^-_{1,E}, \phi(\sigma^{\cdot}, \mu^{\cdot}) \circ (-\phi)^{-1}(\sigma[1]^{\cdot}, \mu[1]^{\cdot}))) \\ &\quad = \partial_{1,\mathcal{A}}((P^-_{1,E}, -\mathrm{id}_{P^-_{1,E}}))\end{aligned}$$

where the third equality follows from the defining relations of $K_0(\mathcal{A}, E)$, and the last is valid because $\phi(\sigma^{\cdot}, \mu^{\cdot}) \circ (-\phi)^{-1}(\sigma[1]^{\cdot}, \mu[1]^{\cdot}) = -\mathrm{id}_{P^-_{1,E}}$ (as one verifies by explicit computation). Now $(P^-_{1,E}, -\mathrm{id}_{P^-_{1,E}}) = \partial_{2,\mathcal{A}}((P_1^-, -\mathrm{id}_{P_1^-}))$ and so the exactness of the second row of (2.1) implies that the last expression in the above displayed formula is zero. This implies the equality of claim (3).

We next remark that claim (5) follows as a consequence of Lemma 1.1(5), as can be shown directly by means of the argument used to prove [**9**, Proposition 1.2.1(iv)]. It therefore now only remains for us to prove claim (4). To do this we first deal with a special case and then check that this special case is in fact sufficient to deal with the general situation.

LEMMA 2.3. *Let $\rho\colon P^{\cdot} \to Q^{\cdot}$ be a quasi-isomorphism of objects of* $\mathrm{P\,Mod}(\mathcal{A})^{\cdot}$ *such that in each degree i all of the induced maps $\rho^i\colon P^i \to Q^i$, $\rho^i_Z\colon Z^i(P^{\cdot}) \to Z^i(Q^{\cdot})$ and $\rho^i_B\colon B^i(P^{\cdot}) \to B^i(Q^{\cdot})$ are surjective. Then in $K_0(\mathcal{A}, E)$ one has $\chi_{\mathcal{A}}(P^{\cdot}, \tau(P^{\cdot}_E)) = \chi_{\mathcal{A}}(Q^{\cdot}, \tau_\rho(Q^{\cdot}_E))$.*

PROOF. For each integer i we define $Z(\rho)^i := \ker(\rho^i_Z)$ and $B(\rho)^i := \ker(\rho^i_B)$ and we recall that $K(\rho)^i$ denotes $\ker(\rho^i)$. Since both P^i and Q^i are projective $\mathcal{A}$-modules and ρ^i is surjective it follows that $K(\rho)^i$ is a projective $\mathcal{A}$-module. In addition, one has $Z(\rho)^i = Z^i(K(\rho)^{\cdot})$ and since ρ^i_Z is surjective also $B(\rho)^i = B^i(K(\rho)^{\cdot})$. In addition, $B(\rho)^i = Z(\rho)^i$ since $H^i(\rho)$ is bijective and so $K(\rho)^{\cdot}$ is a bounded acyclic complex of finitely generated projective $\mathcal{A}$-modules.

In each degree i we can use Lemma 1.4 to choose sections $\sigma^i\colon B^{i+1}(P^{\cdot})_E \to P^i_E$, $\widetilde{\sigma}^i\colon B^{i+1}(Q^{\cdot})_E \to Q^i_E, \mu^i\colon H^i(P^{\cdot})_E \to Z^i(P^{\cdot})_E$ and $\widetilde{\mu}^i\colon H^i(Q^{\cdot})_E \to Z^i(Q^{\cdot})_E$ to $E\otimes_R d^i(P^{\cdot})$, $E\otimes_R d^i(Q^{\cdot})$, $E\otimes_R \pi^i(P^{\cdot})$ and $E\otimes_R \pi^i(Q^{\cdot})$ respectively, in such a way that both of the following diagrams commute

$$\begin{array}{ccc} B^{i+1}(P^{\cdot})_E & \xrightarrow{\sigma^i} & P^i_E \\ \big\downarrow{\scriptstyle \rho^{i+1}_B} & & \big\downarrow{\scriptstyle \rho^i} \\ B^{i+1}(Q^{\cdot})_E & \xrightarrow{\widetilde{\sigma}^i} & Q^i_E \end{array} \qquad \begin{array}{ccc} H^i(P^{\cdot})_E & \xrightarrow{\mu^i} & Z^i(P^{\cdot})_E \\ \big\downarrow{\scriptstyle H^i(\rho)} & & \big\downarrow{\scriptstyle \rho^i_Z} \\ H^i(Q^{\cdot})_E & \xrightarrow{\widetilde{\mu}^i} & Z^i(Q^{\cdot})_E. \end{array}$$

For any $\phi\in\Phi(\tau)$ we let ϕ_ρ denote the composite isomorphism

$$H^-(Q^{\cdot})_E \xrightarrow{H^-(E\otimes_R\rho)^{-1}} H^-(P^{\cdot})_E \xrightarrow{\phi} H^+(P^{\cdot})_E \xrightarrow{H^+(E\otimes_R\rho)} H^+(Q^{\cdot})_E.$$

Then, letting $\phi(\sigma^{\cdot})$ denote the restriction of $\phi(\sigma^{\cdot},\mu^{\cdot})$ to $K(\rho)^-_E$, one has an isomorphism of short exact sequences of A_E-modules

$$\begin{array}{ccccccccc} 0 & \to & K(\rho)^-_E & \xrightarrow{\subseteq} & P^-_E & \xrightarrow{\rho^-} & Q^-_E & \to & 0 \\ & & \big\downarrow{\scriptstyle \phi(\sigma^{\cdot})} & & \big\downarrow{\scriptstyle \phi(\sigma^{\cdot},\mu^{\cdot})} & & \big\downarrow{\scriptstyle \phi_\rho(\widetilde{\sigma}^{\cdot},\widetilde{\mu}^{\cdot})} & & \\ 0 & \to & K(\rho)^+_E & \xrightarrow{\subseteq} & P^+_E & \xrightarrow{\rho^+} & Q^+_E & \to & 0. \end{array}$$

Taken in conjunction with the short exact sequences of projective $\mathcal{A}$-modules

$$0\to K(\rho)^- \xrightarrow{\subseteq} P^- \xrightarrow{\rho^-} Q^- \to 0, \quad 0\to K(\rho)^+ \xrightarrow{\subseteq} P^+ \xrightarrow{\rho^+} Q^+ \to 0$$

this diagram gives an equality in $K_0(\mathcal{A},E)$

$$[P^-,\phi(\sigma^{\cdot},\mu^{\cdot}),P^+] = [K(\rho)^-,\phi(\sigma^{\cdot}),K(\rho)^+] + [Q^-,\phi_\rho(\widetilde{\sigma}^{\cdot},\widetilde{\mu}^{\cdot}),Q^+],$$

and so Lemma 2.3 will follow if we can show that $[K(\rho)^-,\phi(\sigma^{\cdot}),K(\rho)^+] = 0$ in $K_0(\mathcal{A},E)$.

LEMMA 2.4. *Each exact sequence* $0\to B(\rho)^i \xrightarrow{\subseteq} K(\rho)^i \xrightarrow{d^i(K(\rho)^{\cdot})} B(\rho)^{i+1}\to 0$ *is split.*

PROOF. It is clearly enough to prove that each $\mathcal{A}$-module $B(\rho)^{i+1}$ is projective. This can be proved by a decreasing induction on i using the fact that $K(\rho)^{\cdot}$ is a bounded acyclic complex of projective $\mathcal{A}$-modules. □

Following Lemma 2.4 we choose a section $\lambda_i\colon B(\rho)^{i+1}\to K(\rho)^i$ to $d^i(K(\rho)^{\cdot})$ in each degree i, and then let γ denote the composite isomorphism

$$K(\rho)^- \xrightarrow{\lambda_-^{-1}} \bigoplus_{i\in\mathbb{Z}} B(\rho)^i \xrightarrow{\lambda_+} K(\rho)^+$$

where λ_- and λ_+ denote the maps $\oplus_{i\in\mathbb{Z}}(\lambda_{2i+1}\oplus \mathrm{id}_{B(\rho)^{2i+1}})$ and $\oplus_{i\in\mathbb{Z}}(\lambda_{2i}\oplus\mathrm{id}_{B(\rho)^{2i}})$ respectively. Both isomorphisms $\phi(\sigma^{\cdot})$ and $E\otimes_R\gamma$ belong to the trivialization fibre associated to $\iota\big(K(\rho)^{\cdot}_E\big)$. Lemma 2.2 therefore implies that

$$[K(\rho)^-, \phi(\sigma^{\cdot}), K(\rho)^+] = [K(\rho)^-, E\otimes_R \gamma, K(\rho)^+]$$

and the latter element is 0 because $\gamma\big(K(\rho)^-\big) = K(\rho)^+$. This completes our proof of Lemma 2.3. $\square$

To complete the proof of Theorem 2.1(4) we now need to show that the special case dealt with in Lemma 2.3 is actually sufficient to cover all cases.

LEMMA 2.5. *Let $P^{\cdot}$, $Q^{\cdot}$ and ρ be as in Theorem 2.1(4). Then there exists an object $R^{\cdot}$ of* P Mod$(\mathcal{A})^{\cdot}$ *and $\mathcal{A}$-equivariant quasi-isomorphisms $\lambda\colon R^{\cdot}\to P^{\cdot}$ and $\theta\colon R^{\cdot}\to Q^{\cdot}$ such that in each degree i one has $H^i(\rho\circ\lambda) = H^i(\theta)$ and in addition all morphisms $\lambda^i\colon R^i\to P^i$, $\lambda^i|_{Z^i(R^{\cdot})}\colon Z^i(R^{\cdot})\to Z^i(P^{\cdot})$, $\lambda^i|_{B^i(R^{\cdot})}\colon B^i(R^{\cdot})\to B^i(P^{\cdot})$, $\theta^i\colon R^i\to Q^i$, $\theta^i|_{Z^i(R^{\cdot})}Z^i(R^{\cdot})\to Z^i(Q^{\cdot})$ and $\theta^i|_{B^i(R^{\cdot})}\colon B^i(R^{\cdot})\to B^i(Q^{\cdot})$ are surjective.*

PROOF. In each degree i we choose a finitely generated free $\mathcal{A}$-module X^i for which there is a surjection $\kappa^i\colon X^i\to B^i(Q^{\cdot})$. (We take $X^i=0$ if $B^i(Q^{\cdot})=0$.) We consider the following two diagrams

$$\begin{array}{ccccc}
X^i\oplus B^i(P^{\cdot}) & \overset{\subseteq}{\longrightarrow} & X^i\oplus Z^i(P^{\cdot}) & \xrightarrow{(0,\pi^i(P^{\cdot}))} & H^i(P^{\cdot})\\
\big\downarrow{\scriptstyle(\kappa^i,\rho^i)} & & \big\downarrow{\scriptstyle(\kappa^i,\rho^i)} & & \big\downarrow{\scriptstyle H^i(\rho)}\\
B^i(Q^{\cdot}) & \overset{\subseteq}{\longrightarrow} & Z^i(Q^{\cdot}) & \xrightarrow{d^{i-1}(Q^{\cdot})} & H^i(Q^{\cdot}),
\end{array}$$

$$\begin{array}{ccccc}
X^{i-1}\oplus Z^{i-1}(P^{\cdot}) & \xrightarrow{((\mathrm{id},0),\subseteq)} & X^{i-1}\oplus X^i\oplus P^{i-1} & \xrightarrow{(0,(\mathrm{id},0),(0,d^{i-1}(P^{\cdot})))} & X^i\oplus B^i(P^{\cdot})\\
\big\downarrow{\scriptstyle(\kappa^{i-1},\rho^{i-1})} & & \big\downarrow{\scriptstyle\nu^{i-1}} & & \big\downarrow{\scriptstyle(\kappa^i,\rho^i)}\\
Z^{i-1}(Q^{\cdot}) & \overset{\subseteq}{\longrightarrow} & Q^{i-1} & \xrightarrow{d^{i-1}(Q^{\cdot})} & B^i(Q^{\cdot}),
\end{array}$$

where $\nu^{i-1}\colon X^{i-1}\oplus X^i\oplus P^{i-1}\to Q^{i-1}$ is the homomorphism $(\kappa^{i-1},\widehat{\kappa}^i,\rho^{i-1})$ for any lift $\widehat{\kappa}^i\in \mathrm{Hom}_{\mathcal{A}}(X^i,Q^{i-1})$ of κ^i through the surjection $d^{i-1}(Q^{\cdot})\colon Q^{i-1}\to B^i(Q^{\cdot})$. It is easily verified that both diagrams commute, that all rows are short exact sequences and that all vertical maps are surjective.

We now let $R^{\cdot}$ denote the complex which in degree i is equal to $R^i := X^i\oplus X^{i+1}\oplus P^i$ and with differential $d^i(R^{\cdot})$ given by $(x^i,x^{i+1},p^i)\mapsto \big(x^{i+1},0,d^i(P^{\cdot})(p^i)\big)$ for each $x^i\in X^i$, $x^{i+1}\in X^{i+1}$ and $p^i\in P^i$. Then in each degree i one has

$$\begin{aligned}
\ker\big(d^i(R^{\cdot})\big) &= \big(X^i,0,\ker(d^i(P^{\cdot}))\big) = \big(X^i,0,Z^i(P^{\cdot})\big),\\
\mathrm{im}\big(d^{i-1}(R^{\cdot})\big) &= \big(X^i,0,\mathrm{im}(d^{i-1}(P^{\cdot}))\big) = \big(X^i,0,B^i(P^{\cdot})\big)
\end{aligned}$$

and hence $H^i(R^{\cdot})$ is canonically isomorphic to $H^i(P^{\cdot})$. In addition, there are morphisms $\theta\colon R^{\cdot}\to Q^{\cdot}$ and $\lambda\colon R^{\cdot}\to P^{\cdot}$ which are given in each degree i by setting $\theta^i := \nu^i$ and $\lambda^i(x^i,x^{i+1},p^i) := p^i$ for each $x^i\in X^i$, $x^{i+1}\in X^{i+1}$ and $p^i\in P^i$. It is not difficult to check that θ and λ are morphisms of complexes which satisfy all of the properties claimed in Lemma 2.5. $\square$

Using the notation of Lemma 2.5 we let ${}_{\lambda}\tau$ denote the unique element of $\delta^+_{A_E}\big(H^-(R^{\cdot})_E, H^+(R^{\cdot})_E\big)$ that satisfies $({}_{\lambda}\tau)_{\lambda} = \tau$. Then the equalities $H^i(\rho\circ\lambda) = H^i(\theta)$ in each degree i together imply that $({}_{\lambda}\tau)_{\theta} = \tau_{\rho}$. From (two applications of)

Lemma 2.3 we therefore deduce that

$$\begin{aligned}\chi_{\mathcal{A}}(P^{\cdot}, \tau(P^{\cdot}_E)) &= \chi_{\mathcal{A}}(R^{\cdot}, {}_{\lambda}\tau(R^{\cdot}_E)) \\ &= \chi_{\mathcal{A}}(Q^{\cdot}, \tau_{\rho}(Q^{\cdot}_E)),\end{aligned}$$

and this proves Theorem 2.1(4). □

DEFINITION 2.6. An 'E-trivialized perfect complex of $\mathcal{A}$-modules' is a pair (X, τ) with $X \in \mathrm{Ob}(\mathfrak{D}^{\mathrm{perf}}(\mathcal{A}))$ and $\tau \in \delta^{+}_{A_E}(H^{-}(X)_E, H^{+}(X)_E)$. The 'refined Euler characteristic' $\chi_{\mathcal{A}}(X, \tau)$ of (X, τ) is the element of $K_0(\mathcal{A}, E)$ that can be computed as $\chi_{\mathcal{A}}(P^{\cdot}, {}_{\alpha}\tau(P^{\cdot}_E))$ where $\alpha\colon P^{\cdot} \to X$ is an $\mathcal{A}$-equivariant quasi-isomorphism with $P^{\cdot} \in \mathrm{Ob}(\mathrm{P\,Mod}(\mathcal{A})^{\cdot})$ and ${}_{\alpha}\tau$ is the unique element of $\delta^{+}_{A_E}(H^{-}(P^{\cdot})_E, H^{+}(P^{\cdot})_E)$ which satisfies $({}_{\alpha}\tau)_{\alpha} = \tau$.

REMARK 2.7. If $\beta\colon Q^{\cdot} \to X$ is any other $\mathcal{A}$-equivariant quasi-isomorphism with $Q^{\cdot} \in \mathrm{Ob}(\mathrm{P\,Mod}(\mathcal{A})^{\cdot})$, then there exists an $\mathcal{A}$-equivariant quasi-isomorphism $\gamma\colon Q^{\cdot} \to P^{\cdot}$ such that $\beta = \alpha \circ \gamma$ in $\mathfrak{D}(\mathcal{A})$. Upon applying Theorem 2.1(4) with $\rho = \gamma$ one verifies that the definition of $\chi_{\mathcal{A}}(X, \tau)$ is indeed independent of the choice of quasi-isomorphism α which is used to compute it.

We recall that if

$$X_1 \xrightarrow{u} X_2 \xrightarrow{v} X_3 \xrightarrow{w} X_1[1] \tag{2.2}$$

is any distinguished triangle in $\mathfrak{D}^{\mathrm{perf}}(\mathcal{A})$, then in $K_0(\mathcal{A})$ one has

$$\chi_{\mathcal{A}} X_1 - \chi_{\mathcal{A}} X_2 + \chi_{\mathcal{A}} X_3 = 0.$$

In the remainder of this section we discuss corresponding results for refined Euler characteristics. In this regard we first deal with an important special case.

For any homomorphism of $\mathcal{A}$-modules θ we let θ_E denote the associated homomorphism $E \otimes_R \theta$ of A_E-modules.

THEOREM 2.8. *Assume that the triangle (2.2) is such that $H^i(w)_E = 0$ for all integers i. Then the long exact cohomology sequence associated to (2.2) gives rise to short exact sequences*

$$0 \to H^{-}(X_1)_E \xrightarrow{H^{-}(u)_E} H^{-}(X_2)_E \xrightarrow{H^{-}(v)_E} H^{-}(X_3)_E \to 0$$

$$0 \to H^{+}(X_1)_E \xrightarrow{H^{+}(u)_E} H^{+}(X_2)_E \xrightarrow{H^{+}(v)_E} H^{+}(X_3)_E \to 0$$

which together induce an isomorphism

$$\begin{aligned} i_{u,v}\colon \delta_{A_E}(H^{-}(X_1)_E, H^{+}(X_1)_E) \otimes_{\zeta(A_E)} \delta_{A_E}(H^{-}(X_3)_E, H^{+}(X_3)_E) \\ \xrightarrow{\sim} \delta_{A_E}(H^{-}(X_2)_E, H^{+}(X_2)_E).\end{aligned}$$

If $\tau_j \in \delta_{A_E}(H^{-}(X_j)_E, H^{+}(X_j)_E)$ for $j = 1, 2, 3$ are any trivializations which satisfy

$$i_{u,v}(\tau_1 \otimes \tau_3) = \tau_2,$$

then in $K_0(\mathcal{A}, E)$ one has

$$\chi_{\mathcal{A}}(X_1, \tau_1) - \chi_{\mathcal{A}}(X_2, \tau_2) + \chi_{\mathcal{A}}(X_3, \tau_3) = 0.$$

PROOF. We first choose objects $P^{\cdot}_1$ and $P^{\cdot}_3$ of $\mathrm{P\,Mod}(\mathcal{A})^{\cdot}$ together with quasi-isomorphisms $\alpha_j\colon P^{\cdot}_j \to X_j$ for $j = 1, 3$. We then choose a morphism $\mu\colon P^{\cdot}_3[-1] \to P^{\cdot}_1$ which is such that $\alpha_1 \circ \mu = -w[-1] \circ \alpha_3[-1]$ in $\mathfrak{D}(\mathcal{A})$ and we let $P^{\cdot}_2$ denote the mapping cone of μ. Then, by the axioms of a triangulated category, there exists a quasi-isomorphism $\alpha_2\colon P^{\cdot}_2 \to X_2$ which makes the following diagram commute in $\mathfrak{D}(\mathcal{A})$

$$\begin{array}{ccccccc}
P^{\cdot}_3[-1] & \xrightarrow{\mu} & P^{\cdot}_1 & \xrightarrow{\theta} & P^{\cdot}_2 & \xrightarrow{\phi} & P^{\cdot}_3 \\
\big\downarrow{\scriptstyle \alpha_3[-1]} & & \big\downarrow{\scriptstyle \alpha_1} & & \big\downarrow{\scriptstyle \alpha_2} & & \big\downarrow{\scriptstyle \alpha_3} \\
X_3[-1] & \xrightarrow{-w[-1]} & X_1 & \xrightarrow{u} & X_2 & \xrightarrow{v} & X_3
\end{array}$$

where θ and ϕ are the natural morphisms (coming from the definition of $P^{\cdot}_2$ as the mapping cone of μ) and the lower row is the distinguished triangle induced by (2.2).

Recalling Definition 2.6, it is therefore enough for us to prove that in $K_0(\mathcal{A}, E)$ one has

$$\chi_{\mathcal{A}}\big(P^{\cdot}_1, {}_{\alpha_1}\tau_1(P^{\cdot}_{1,E})\big) - \chi_{\mathcal{A}}\big(P^{\cdot}_2, {}_{\alpha_2}\tau_2(P^{\cdot}_{2,E})\big) + \chi_{\mathcal{A}}\big(P^{\cdot}_3, {}_{\alpha_3}\tau_3(P^{\cdot}_{3,E})\big) = 0. \tag{2.3}$$

To prove this the key point is that the above commutative diagram induces an isomorphism between the images under $E\otimes_R -$ of the long exact cohomology sequences associated to both (2.2) and the canonical short exact sequence of complexes

$$0 \to P^{\cdot}_1 \xrightarrow{\theta} P^{\cdot}_2 \xrightarrow{\phi} P^{\cdot}_3 \to 0,$$

and hence that the equality $i_{u,v}(\tau_1 \otimes \tau_3) = \tau_2$ implies the existence of isomorphisms $t_j \in {}_{\alpha_j}\tau_j$ for each $j \in \{1, 2, 3\}$ which together lie in a commutative diagram of the form

$$\begin{array}{ccccccccc}
0 & \to & H^-(P^{\cdot}_1)_E & \xrightarrow{H^-(\theta)_E} & H^-(P^{\cdot}_2)_E & \xrightarrow{H^-(\phi)_E} & H^-(P^{\cdot}_3)_E & \to & 0 \\
 & & \big\downarrow{\scriptstyle t_1} & & \big\downarrow{\scriptstyle t_2} & & \big\downarrow{\scriptstyle t_3} & & \\
0 & \to & H^+(P^{\cdot}_1)_E & \xrightarrow{H^+(\theta)_E} & H^+(P^{\cdot}_2)_E & \xrightarrow{H^+(\phi)_E} & H^+(P^{\cdot}_3)_E & \to & 0.
\end{array} \tag{2.4}$$

Indeed, the required equality (2.3) will follow if we can extend the latter diagram to a commutative diagram of the form

$$\begin{array}{ccccccccc}
 & & 0 & & 0 & & 0 & & \\
 & & \big\downarrow & & \big\downarrow & & \big\downarrow & & \\
0 & \to & P^-_{1,E} & \xrightarrow{\theta^-_E} & P^-_{2,E} & \xrightarrow{\phi^-_E} & P^-_{3,E} & \to & 0 \\
 & & \big\downarrow & & \big\downarrow & & \big\downarrow & & \\
 & & B_1 \oplus H^-(P^{\cdot}_1)_E & \xrightarrow{\gamma_1} & B_2 \oplus H^-(P^{\cdot}_2)_E & \xrightarrow{\gamma_2} & B_3 \oplus H^-(P^{\cdot}_3)_E & & \\
 & & \big\downarrow{\scriptstyle \mathrm{id}\oplus t_1} & & \big\downarrow{\scriptstyle \mathrm{id}\oplus t_2} & & \big\downarrow{\scriptstyle \mathrm{id}\oplus t_3} & & \\
 & & B_1 \oplus H^+(P^{\cdot}_1)_E & \xrightarrow{\gamma_3} & B_2 \oplus H^+(P^{\cdot}_2)_E & \xrightarrow{\gamma_4} & B_3 \oplus H^+(P^{\cdot}_3)_E & & \\
 & & \big\downarrow & & \big\downarrow & & \big\downarrow & & \\
0 & \to & P^+_{1,E} & \xrightarrow{\theta^+_E} & P^+_{2,E} & \xrightarrow{\phi^+_E} & P^+_{3,E} & \to & 0 \\
 & & \big\downarrow & & \big\downarrow & & \big\downarrow & & \\
 & & 0 & & 0 & & 0 & &
\end{array} \tag{2.5}$$

with $B_i = \bigoplus_{j\in\mathbb{Z}} B^j(P^{\cdot}_{i,E})$ for $i \in \{1, 2, 3\}$, $\gamma_1 = \big(\theta', H^-(\theta)_E\big)$, $\gamma_2 = \big(\phi', H^-(\phi)_E\big)$, $\gamma_3 = \big(\theta', H^+(\theta)_E\big)$, $\gamma_4 = \big(\phi', H^+(\phi)_E\big)$ where $\theta'\colon B_1 \to B_2$ and $\phi'\colon B_2 \to B_3$ are the natural homomorphisms that are induced by θ and ϕ respectively and all unlabelled vertical maps the isomorphisms that are induced by a choice of sections to each of the natural homomorphisms $P^i_{j,E} \to B^{i+1}(P^\cdot_{j,E})$ and $Z^i(P^\cdot_{j,E}) \to H^i(P^\cdot_{j,E})$. To show that the existence of such a diagram is sufficient, we note that the vertical composite homomorphisms belong to ${}_{\alpha_i}\tau_i(P^\cdot_{i,E})$ respectively, and so the commutativity of the outer diagram would combine with the short exact sequences

$$0 \to P_1^- \xrightarrow{\theta^-} P_2^- \xrightarrow{\phi^-} P_3^- \to 0, \quad 0 \to P_1^+ \xrightarrow{\theta^+} P_2^+ \xrightarrow{\phi^+} P_3^+ \to 0$$

and the defining relations of $K_0(\mathcal{A}, E)$ to imply (2.3). On the other hand, constructing such a diagram (2.5) from (2.4) amounts to choosing sections to each of the homomorphisms $P^i_{j,E} \to B^{i+1}(P^\cdot_{j,E})$ and $Z^i(P^\cdot_{j,E}) \to H^i(P^\cdot_{j,E})$ in such a way that the top and bottom two squares of the given diagram commute. To do this we first observe that by using the exactness of the rows of (2.4) it can be shown (by a descending inductive argument) that in each degree i one has commutative diagrams of short exact sequences of the form

$$\begin{array}{ccccccccc}
 & & 0 & & 0 & & 0 & & \\
 & & \downarrow & & \downarrow & & \downarrow & & \\
0 & \to & Z^i(P^\cdot_{1,E}) & \to & P^i_{1,E} & \xrightarrow{d^i(P^\cdot_{1,E})} & B^{i+1}(P^\cdot_{1,E}) & \to & 0 \\
 & & \downarrow \theta^i_E & & \downarrow \theta^i_E & & \downarrow \theta^{i+1}_E & & \\
0 & \to & Z^i(P^\cdot_{2,E}) & \to & P^i_{2,E} & \xrightarrow{d^i(P^\cdot_{2,E})} & B^{i+1}(P^\cdot_{2,E}) & \to & 0 \\
 & & \downarrow \phi^i_E & & \downarrow \phi^i_E & & \downarrow \phi^{i+1}_E & & \\
0 & \to & Z^i(P^\cdot_{3,E}) & \to & P^i_{3,E} & \xrightarrow{d^i(P^\cdot_{3,E})} & B^{i+1}(P^\cdot_{3,E}) & \to & 0 \\
 & & \downarrow & & \downarrow & & \downarrow & & \\
 & & 0 & & 0 & & 0 & &
\end{array}$$

and

$$\begin{array}{ccccccccc}
 & & 0 & & 0 & & 0 & & \\
 & & \downarrow & & \downarrow & & \downarrow & & \\
0 & \to & B^i(P^\cdot_{1,E}) & \to & Z^i(P^\cdot_{1,E}) & \xrightarrow{\pi^i(P^\cdot_{1,E})} & H^i(P^\cdot_{1,E}) & \to & 0 \\
 & & \downarrow \theta^i_E & & \downarrow \theta^i_E & & \downarrow H^i(\theta)_E & & \\
0 & \to & B^i(P^\cdot_{2,E}) & \to & Z^i(P^\cdot_{2,E}) & \xrightarrow{\pi^i(P^\cdot_{2,E})} & H^i(P^\cdot_{2,E}) & \to & 0 \\
 & & \downarrow \phi^i_E & & \downarrow \phi^i_E & & \downarrow H^i(\phi)_E & & \\
0 & \to & B^i(P^\cdot_{3,E}) & \to & Z^i(P^\cdot_{3,E}) & \xrightarrow{\pi^i(P^\cdot_{3,E})} & H^i(P^\cdot_{3,E}) & \to & 0 \\
 & & \downarrow & & \downarrow & & \downarrow & & \\
 & & 0 & & 0 & & 0. & &
\end{array}$$

By means of Lemma 1.4 we may now choose sections to the homomorphisms $d^i(P^\cdot_{j,E})$ and $\pi^i(P^\cdot_{j,E})$ in a manner that is compatible with these diagrams, and by using any such choice of sections one can construct a commutative diagram of the required form (2.5). □

REMARK 2.9. The triangle (2.2) gives rise to distinguished triangles of the form

$$X_2 \xrightarrow{v} X_3 \xrightarrow{w} X_1[1] \xrightarrow{-u[1]} X_2[1]$$

and

$$X_3 \xrightarrow{w} X_1[1] \xrightarrow{-u[1]} X_2[1] \xrightarrow{-v[1]} X_3[1].$$

Taken in conjunction with Theorem 2.1(3) this observation leads to criteria analogous to that of Theorem 2.8 for triangles (2.2) in which either $H^i(u)_E = 0$ for all integers i or $H^i(v)_E = 0$ for all integers i.

REMARK 2.10. Each $\mathcal{A}$-module M may be viewed as a complex $M[0]$ which is concentrated in degree 0. In particular, if M is finite and $M[0] \in \mathrm{Ob}(\mathfrak{D}^{\mathrm{perf}}(\mathcal{A}))$, then it gives rise to a natural class $\chi^t_{\mathcal{A}}(M)$ in $K_0(\mathcal{A}, F)$. Indeed, if $P^{\cdot}$ is any object of $\mathrm{P\,Mod}(\mathcal{A})^{\cdot}$ that is isomorphic in $\mathfrak{D}^{\mathrm{perf}}(\mathcal{A})$ to $M[0]$, then $P^{\cdot}_F$ is acyclic and one sets $\chi^t_{\mathcal{A}}(M) := \chi_{\mathcal{A}}\big(P^{\cdot}, \iota(P^{\cdot}_F)\big) \in K_0(\mathcal{A}, F)$. Thus, more explicitly, for any such module M there is a short exact sequence $0 \to P^{-1} \xrightarrow{\phi} P^0 \to M \to 0$, in which P^{-1} and P^0 are finitely generated projective $\mathcal{A}$-modules and $\chi^t_{\mathcal{A}}(M) = [P^{-1}, F\otimes_R \phi, P^0]$ in $K_0(\mathcal{A}, F)$. Now for any $\mathcal{A}$-module M we let M_{tors} and M_{tf} denote the $\mathcal{A}$-submodule of R-torsion elements and the quotient M/M_{tors} respectively. We assume that $\mathfrak{D}(\mathcal{A}) = \mathfrak{D}^{\mathrm{perf}}(\mathcal{A})$ and that M_{tf} is $\mathcal{A}$-projective for each finitely generated $\mathcal{A}$-module M. For any object X of $\mathfrak{D}(\mathcal{A})$ we let $H^{\cdot}(X)$, $H^{\cdot}(X)_{\mathrm{tors}}$ and $H^{\cdot}(X)_{\mathrm{tf}}$ denote the complexes which in degree i are equal to $H^i(X)$, $H^i(X)_{\mathrm{tors}}$ and $H^i(X)_{\mathrm{tf}}$ respectively, and in which all differentials are 0. Then there exists a natural exact sequence of complexes

$$0 \to H^{\cdot}(X)_{\mathrm{tors}} \to H^{\cdot}(X) \to H^{\cdot}(X)_{\mathrm{tf}} \to 0$$

and an isomorphism $X \xrightarrow{\sim} H^{\cdot}(X)$ in $\mathfrak{D}(\mathcal{A})$. As a consequence, there is a distinguished triangle in $\mathfrak{D}(\mathcal{A})$ of the form

$$H^{\cdot}(X)_{\mathrm{tors}} \to X \to H^{\cdot}(X)_{\mathrm{tf}} \to H^{\cdot}(X)_{\mathrm{tors}}[1].$$

Further, if $\tau \in \delta^+_{A_E}\big(H^-(X)_E, H^+(X)_E\big)$ then, with respect to this triangle, the trivializations $\tau_1 = 1$, $\tau_2 = \tau$ and $\tau_3 = \tau$ satisfy the hypotheses of Theorem 2.8 and so for any $\phi \in \Phi(\tau)$ one has in $K_0(\mathcal{A}, E)$

$$\begin{aligned}\chi_{\mathcal{A}}(X, \tau) &= \chi_{\mathcal{A}}(H^{\cdot}(X)_{\mathrm{tf}}, \tau) + \chi_{\mathcal{A}}\big(H^{\cdot}(X)_{\mathrm{tors}}, 1\big),\\ &= [H^-(X)_{\mathrm{tf}}, \phi, H^+(X)_{\mathrm{tf}}] + \sum_{i\in\mathbb{Z}}(-1)^i\chi^t_{\mathcal{A}}\big(H^i(X)_{\mathrm{tors}}\big).\end{aligned}$$

To end this section we briefly discuss extensions of Theorem 2.8 in the context of general triangles (2.2). In this regard we let $\Delta^{\cdot}$ denote the long exact sequence of cohomology associated to (2.2) which we view as an acyclic complex with (to fix conventions) $H^0(X_1)$ placed in degree 0. Then there are natural isomorphisms of A_E-modules

$$\begin{aligned}\Delta^-_E &\cong H^-(X_1)_E \oplus H^+(X_2)_E \oplus H^-(X_3)_E,\\ \Delta^+_E &\cong H^+(X_1)_E \oplus H^-(X_2)_E \oplus H^+(X_3)_E,\end{aligned}$$

which together induce an isomorphism

$$i_{\Delta^{\cdot}} : \delta_{A_E}(H^-_{1,E}, H^+_{1,E}) \otimes \delta_{A_E}(H^+_{2,E}, H^-_{2,E}) \otimes \delta_{A_E}(H^-_{3,E}, H^+_{3,E}) \xrightarrow{\sim} \delta_{A_E}(\Delta^-_E, \Delta^+_E)$$

where $H^-_{j,E} := H^-(X_j)_E$, $H^+_{j,E} := H^+(X_j)_E$ and all unadorned tensor products are to be considered as taken with respect to $\zeta(A_E)$. By a further development of the explicit method used to prove Theorem 2.8 (which in particular makes heavy use of Lemmas 1.4 and 1.5) it can be shown that there exists an element $u_{\Delta^\cdot}$ of $\zeta(A)^\times$ which satisfies $(u_{\Delta^\cdot})^2 = 1$ and is such that if $\tau_i \in \delta_{A_E}\big(H^-(X_i)_E, H^+(X_i)_E\big)$ for $i \in \{1,2,3\}$ are any trivializations satisfying

$$i_{\Delta^\cdot}(\tau_1 \otimes \tau_2^{-1} \otimes \tau_3) = u_{\Delta^\cdot} \iota(\Delta^\cdot_E) \in \delta_{A_E}(\Delta^-_E, \Delta^+_E),$$

then in $K_0(\mathcal{A}, E)$ one has $\chi_{\mathcal{A}}(X_1, \tau_1) - \chi_{\mathcal{A}}(X_2, \tau_2) + \chi_{\mathcal{A}}(X_3, \tau_3) = 0$.

EXAMPLE 2.11. If in (2.2) each complex $X_{j,E}$ is acyclic outside degrees 0 and 1, then Theorem 2.8 can be used to show that one can take $u_{\Delta^\cdot}$ to be equal to $\mathrm{nr}_A\big((H^0(X_2)_F, -\mathrm{id}_{H^0(X_2)_F})\big)$.

REMARK 2.12. Example 2.11 shows that, in general, [**9**, Proposition 1.2.2] is incorrect as stated (since it assumes that one can take $u_{\Delta^\cdot} = 1$). However, fortunately we may reassure the reader by observing that Theorem 2.8 implies that the existence of a commutative diagram as in (loc. cit., Remark 1.2.3) does indeed ensure that in $K_0(\mathcal{A}, F)$ one has $\chi_{\mathcal{A},F}(C^\cdot_1, \tau_1) - \chi_{\mathcal{A},F}(C^\cdot_2, \tau_2) + \chi_{\mathcal{A},F}(C^\cdot_3, \tau_3) = 0$ (in the notation of loc. cit.) and it is this observation rather than the general statement of (loc. cit., Proposition 1.2.2) which underlies applications both in the remainder of loc. cit. as well as in [**4**–**6**].

In general, the explicit computation of the 'locally constant sign function' $u_{\Delta^\cdot}$ as above is rather involved and far from illuminating. Indeed, it seems that the only sensible approach to describing conditions on trivializations which are sufficient to ensure additivity of refined Euler characteristics along general triangles of the form (2.2) is to suppress any desire to be explicit and make use of Deligne's theory of virtual objects as in [**13**]. However, whilst such an approach does indeed allow one to formulate a precise generalization of Theorem 2.8 (and hence to correct the statement of [**9**, Proposition 1.2.2)]), it is sufficiently different from the explicit methods used here that we prefer to avoid any further discussion in the present manuscript and simply refer the reader to [**7**] for more details in this regard.

3. The commutative case

In this section we assume that the algebra A is commutative and we interpret the constructions of §1 and §2 in terms of the determinant formalism that is described by Knudsen and Mumford in [**23**].

We first give a very brief review of the formalism of graded invertible modules (and refer the reader to [**23**] for a more comprehensive discussion). For any commutative noetherian ring R a graded invertible R-module is a pair (L, α) consisting of an invertible (that is, rank one projective) R-module L and a locally-constant function $\alpha\colon \mathrm{Spec}\, R \to \mathbb{Z}$. The category $\mathcal{P}(R)$ of graded invertible R-modules and isomorphisms of such is a symmetric monoidal category in the sense of [**24**] with tensor product $(L,\alpha)\otimes(M,\beta) = (L\otimes_R M, \alpha+\beta)$ and unit object $(R, 0)$. It is endowed with the natural associativity constraint and a commutativity constraint determined by the 'Koszul rule of signs': if $\wp$ is any prime ideal of R, $\ell \in L_\wp$ and $m \in M_\wp$, then the $\wp$-localization of the isomorphism $(L,\alpha)\otimes(M,\beta) \cong (M,\beta)\otimes(L,\alpha)$ sends $\ell\otimes m$ to $(-1)^{\alpha(\wp)\beta(\wp)} \cdot m \otimes \ell$.

For each finitely generated projective R-module P one obtains an associated graded invertible R-module by setting

$$\mathrm{Det}_R(P) := \big(\mathrm{Det}_R(P), \mathrm{rk}_R(P)\big)$$

where $\mathrm{Det}_R(P)$ denotes the (locally defined) highest exterior power of P and $\mathrm{rk}_R(P)$ denotes the R-rank of P, and for a bounded complex $P^\cdot$ of such modules one sets

$$\mathrm{Det}_R\, P^\cdot := \bigotimes_{i\in\mathbb{Z}} \mathrm{Det}_R(P^i)^{(-1)^{i+1}}$$

where, for each $(L,\alpha) \in \mathrm{Ob}\big(\mathcal{P}(R)\big)$ one sets $(L,\alpha)^{-1} := \big(\mathrm{Hom}_R(L,R), -\alpha\big) \in \mathrm{Ob}\big(\mathcal{P}(R)\big)$, and the tensor product is taken in $\mathcal{P}(R)$.

We write

$$j_R(P^\cdot)\colon \mathrm{Det}_R P^\cdot \xrightarrow{\sim} \mathrm{Det}_R(P^-) \otimes \mathrm{Det}_R(P^+)^{-1}$$

for the canonical morphism in $\mathcal{P}(R)$.

The determinant Det_R is well defined as a functor from the subcategory of $\mathfrak{D}^{\mathrm{perf}}(R)$ where morphisms are restricted to isomorphisms to the category $\mathcal{P}(R)$. In addition, if $X \in \mathrm{Ob}\big(\mathfrak{D}^{\mathrm{perf}}(R)\big)$ is such that each module $H^i(X)$ has finite projective dimension over R (which is automatically the case if, for example, R is a regular ring), then there is also a canonical morphism in $\mathcal{P}(R)$

$$i_R(X)\colon \mathrm{Det}_R X \xrightarrow{\sim} \mathrm{Det}_R\big(H^-(X)\big) \otimes \mathrm{Det}_R\big(H^+(X)\big)^{-1}$$

[**23**, Remark following Theorem 2].

If now R is equal to a semisimple algebra A and V and W are isomorphic finitely generated A-modules, then each element τ of $\delta_A^+(V,W) \cong \mathrm{Is}_A\big(\det_A(V), \det_A(W)\big)$ (cf. Example 1.2) induces a canonical morphism in $\mathcal{P}(A)$

$$\tau_{\mathrm{Triv}}\colon \mathrm{Det}_A(V) \otimes_A \mathrm{Det}_A(W)^{-1} \xrightarrow{\sim} (A,0).$$

In particular, if $C^\cdot$ is the complex (1.2) and $\tau \in \delta_A^+\big(H^-(C^\cdot), H^+(C^\cdot)\big)$, then by comparing the construction of $\tau(C^\cdot)$ in Lemma 1.3 to the explicit description of $i_A(C^\cdot)$ one finds that

$$\tau_{\mathrm{Triv}} \circ i_A(C^\cdot) = \tau(C^\cdot)_{\mathrm{Triv}} \circ j_A(C^\cdot) \in \mathrm{Hom}_{\mathcal{P}(A)}\big(\mathrm{Det}_A C^\cdot, (A,0)\big). \tag{3.1}$$

We now let R be a Dedekind domain with fraction field F and E an extension field of F. We assume that $\mathcal{A}$ is an R-order in a finite dimensional semisimple F-algebra A.

THEOREM 3.1. *Assume $X \in \mathrm{Ob}\big(\mathfrak{D}^{\mathrm{perf}}(\mathcal{A})\big)$ and $\tau \in \delta_{A_E}^+\big(H^-(X)_E, H^+(X)_E\big)$. Then $\chi_{\mathcal{A}}(X,\tau) = 0$ in $K_0(\mathcal{A},E)$ if and only if in $\mathcal{P}(\mathcal{A})$ one has*

$$\tau_{\mathrm{Triv}} \circ i_{A_E}(X_E)(\mathrm{Det}_{\mathcal{A}} X) = (\mathcal{A}, 0).$$

PROOF. Upon recalling the explicit description of $\chi_{\mathcal{A}}(X,\tau)$ given in Definition 2.6, and then taking into account the equality (3.1), one finds that it is enough to prove the following result: if P_1 and P_2 are finitely generated projective $\mathcal{A}$-modules and $\phi\colon P_{1,E} \xrightarrow{\sim} P_{2,E}$ is an isomorphism of A_E-modules, then $[P_1,\phi,P_2] = 0$ in $K_0(\mathcal{A},E)$ if and only if $\det_{A_E}(\phi)\big(\det_{\mathcal{A}}(P_1)\big) = \det_{\mathcal{A}}(P_2)$.

LEMMA 3.2. $[P_1,\phi,P_2] = [\det_{\mathcal{A}}(P_1), \det_{A_E}(\phi), \det_{\mathcal{A}}(P_2)]$ *in* $K_0(\mathcal{A},E)$.

PROOF. We first deal with the case that $E = F$. In this case we recall the isomorphism γ from diagram (2.1) and we observe that this isomorphism commutes with taking determinants. It is therefore enough in this case to prove the claimed equality in each group $K_0(\mathcal{A}_v, F_v)$. Further, to do this we may (and will) suppose that the (commutative) ring $\mathcal{A}_v$ is indecomposable, and hence that $P_{1,v}$ and $P_{2,v}$ are both free $\mathcal{A}_v$-modules of the same rank. Now after choosing an $\mathcal{A}_v$-equivariant isomorphism $\kappa\colon P_{2,v} \xrightarrow{\sim} P_{1,v}$ and setting $\kappa_{F_v} := F_v \otimes_{R_v} \kappa$ one has $[P_{2,v}, \kappa_{F_v}, P_{1,v}] = 0$ in $K_0(\mathcal{A}_v, F_v)$ and hence

$$[P_{1,v}, \phi_v, P_{2,v}] = [P_{1,v}, \phi_v, P_{2,v}] + [P_{2,v}, \kappa_{F_v}, P_{1,v}] = [P_{1,v}, f, P_{1,v}],$$

where $f = \kappa_{F_v} \circ \phi_v$. The element $[P_{1,v}, f, P_{1,v}]$ lies in the image of $K_1(A_v)$. Further, thinking of f as a matrix one has $f = \det_{A_v}(f)$ in $K_1(A_v)$ and hence in $K_0(\mathcal{A}_v, F_v)$ one has

$$\begin{aligned}[P_{1,v}, f, P_{1,v}] &= [\det_{\mathcal{A}_v}(P_{1,v}), \det_{A_v}(f), \det_{\mathcal{A}_v}(P_{1,v})] \\ &= [\det_{\mathcal{A}_v}(P_{1,v}), \det_{A_v}(\phi_v), \det_{\mathcal{A}_v}(P_{2,v})] \\ &\qquad + [\det_{\mathcal{A}_v}(P_{2,v}), \det_{A_v}(\kappa_{F_v}), \det_{\mathcal{A}_v}(P_{1,v})] \\ &= [\det_{\mathcal{A}_v}(P_{1,v}), \det_{A_v}(\phi_v), \det_{\mathcal{A}_v}(P_{2,v})],\end{aligned}$$

where the last equality follows because $\det_{A_v}(\kappa_{F_v})\big(\det_{\mathcal{A}_v}(P_{2,v})\big) = \det_{\mathcal{A}_v}(P_{1,v})$. The last two displayed formulas now combine to prove the result of the lemma in the case $E = F$.

To deal with the general case we proceed as follows. Any two projective $\mathcal{A}$-modules which become isomorphic after tensoring with E are already isomorphic after tensoring with F, and in $K_0(\mathcal{A}, E)$ one has

$$[P_1, \phi, P_2] = [P_1, \psi, P_2] + [P_2, \phi_\psi, P_2]$$

where ψ is any choice of isomorphism $P_{1,F} \xrightarrow{\sim} P_{2,F}$ of A-modules and $\phi_\psi := \phi \circ \psi^{-1}$. Now, identifying ϕ_ψ with the associated element of $K_1(A_E)$, one has $[P_2, \phi_\psi, P_2] = \partial_{1,\mathcal{A}}(\phi_\psi)$, $\phi_\psi = \det_{A_E}(\phi_\psi)$ and

$$\partial_{1,\mathcal{A}}\big(\det_{A_E}(\phi_\psi)\big) = [\det_{\mathcal{A}}(P_2), \det_{A_E}(\phi_\psi), \det_{\mathcal{A}}(P_2)].$$

From the case $E = F$ considered above one also knows that

$$[P_1, \psi, P_2] = [\det_{\mathcal{A}}(P_1), \det_A(\psi), \det_{\mathcal{A}}(P_2)]$$

in $K_0(\mathcal{A}, F)$, and hence in $K_0(\mathcal{A}, E)$ one has

$$\begin{aligned}[P_1, \phi, P_2] &= [\det_{\mathcal{A}}(P_1), \det_A(\psi), \det_{\mathcal{A}}(P_2)] + [\det_{\mathcal{A}}(P_2), \det_{A_E}(\phi_\psi), \det_{\mathcal{A}}(P_2)] \\ &= [\det_{\mathcal{A}}(P_1), \det_{A_E}(\phi), \det_{\mathcal{A}}(P_2)],\end{aligned}$$

as required. □

To complete the proof of Proposition 3.1 it is now sufficient to prove that $[\det_{\mathcal{A}}(P_1), \det_{A_E}(\phi), \det_{\mathcal{A}}(P_2)] = 0$ in $K_0(\mathcal{A}, E)$ if and only if

$$\det_{A_E}(\phi)\big(\det_{\mathcal{A}}(P_1)\big) = \det_{\mathcal{A}}(P_2),$$

and this equivalence follows directly from the argument used to prove [**14**, Lemma 1(ii)]. □

4. Rational torsion elements

Let G be a finite group. In this section we describe the torsion subgroup $K_0(\mathbb{Z}[G],\mathbb{Q})_{\mathrm{tors}}$ of $K_0(\mathbb{Z}[G],\mathbb{Q})$ in a manner that can be very useful in concrete applications. This description was first obtained in an unpublished preprint of Flach and the present author [**12**] and has since been used in several arithmetical contexts (see, for example, [**9**, Theorem 2.2.4, Lemma 2.2.6] and [**5**, Corollary 6.3(i)]. In addition, the result of [**13**, Proposition 4.1] shows that the description we present here is of interest in the context of the equivariant Tamagawa number conjecture for a large class of motives.

Upon recalling the isomorphism γ from diagram (2.1) it is clear that we only need to describe the group $K_0(\mathbb{Z}_p[G],\mathbb{Q}_p)_{\mathrm{tors}}$ for each prime p. To this end we now fix a prime p. For each subgroup H of G we let ρ^G_H denote the restriction functor from $\mathfrak{B}(\mathbb{Z}_p[G])$ to $\mathfrak{B}(\mathbb{Z}_p[H])$ and $\rho^{G,*}_H\colon K_0(\mathbb{Z}_p[G],\mathbb{Q}_p)\to K_0(\mathbb{Z}_p[H],\mathbb{Q}_p)$ the morphism induced by the assignment $[P,\phi,Q]\mapsto[\rho^G_HP,\rho^G_H\phi,\rho^G_HQ]$. If J is a normal subgroup of G and X is any $\mathbb{Z}_p[G]$-module, then we write X^J for the $\mathbb{Z}_p[G/J]$-module which is given by $\{x\in X\colon j(x)=x \text{ for all } j\in J\}$. This construction induces a functor from $\mathfrak{B}(\mathbb{Z}_p[G])$ to $\mathfrak{B}(\mathbb{Z}_p[G/J])$ and we let $q^G_{G/J,*}\colon K_0(\mathbb{Z}_p[G],\mathbb{Q}_p)\to K_0(\mathbb{Z}_p[G/J],\mathbb{Q}_p)$ denote the associated morphism $[P,\phi,Q]\mapsto[P^J,\phi^J,Q^J]$.

THEOREM 4.1. *One has*

$$K_0(\mathbb{Z}_p[G],\mathbb{Q}_p)_{\mathrm{tors}}=\cap\ker(q^H_{Q,*}\circ\rho^{G,*}_H)$$

where the intersection runs over all cyclic subgroups H of G and over all quotients Q of H which are of order prime to p.

PROOF. The fact that $K_0(\mathbb{Z}_p[G],\mathbb{Q}_p)_{\mathrm{tors}}$ is contained in the displayed intersection is a direct consequence of the fact that each group $K_0(\mathbb{Z}_p[Q],\mathbb{Q}_p)$ is torsion-free. Indeed, the latter is true because $\mathbb{Z}_p[Q]$ is the (unique) maximal $\mathbb{Z}_p$-order in $\mathbb{Q}_p[Q]$ (since p does not divide $\#Q$). However, to prove the reverse inclusion it is convenient to combine Artin's Induction Theorem with a 'Hom-description' of relative algebraic K-groups in the spirit of Fröhlich. To do this we fix an algebraic closure $\mathbb{Q}^c_p$ of $\mathbb{Q}_p$ and write $\mathbb{Z}^c_p$ for its valuation ring and $G_{\mathbb{Q}_p}$ for the Galois group $\mathrm{Gal}(\mathbb{Q}^c_p/\mathbb{Q}_p)$. For any finite group Γ we let $R_{\Gamma,p}$ denote the ring of $\mathbb{Q}^c_p$-valued characters of Γ, and we recall from [**22**, Chapter II] that there is a canonical surjective homomorphism

$$h_{\Gamma,p}\colon \mathrm{Hom}_{G_{\mathbb{Q}_p}}(R_{\Gamma,p},\mathbb{Q}^{c\times}_p)\to K_0(\mathbb{Z}_p[\Gamma],\mathbb{Q}_p)$$

for which

$$\mathrm{Hom}_{G_{\mathbb{Q}_p}}(R_{\Gamma,p},\mathbb{Z}^{c\times}_p)=h^{-1}_{\Gamma,p}\big(K_0(\mathbb{Z}_p[\Gamma],\mathbb{Q}_p)_{\mathrm{tors}}\big) \tag{4.1}$$

(that is, $\mathrm{Hom}_{G_{\mathbb{Q}_p}}(R_{\Gamma,p},\mathbb{Z}^{c\times}_p)$ is the full pre-image of $K_0(\mathbb{Z}_p[\Gamma],\mathbb{Q}_p)_{\mathrm{tors}}$ under $h_{\Gamma,p}$). For any subgroup H of G we let

$$\mathrm{coind}^H_G\colon \mathrm{Hom}_{G_{\mathbb{Q}_p}}(R_{G,p},\mathbb{Q}^{c\times}_p)\to\mathrm{Hom}_{G_{\mathbb{Q}_p}}(R_{H,p},\mathbb{Q}^{c\times}_p)$$

denote the map induced by the induction map $\mathrm{ind}_G^H\colon R_{H,p} \to R_{G,p}$, and we recall from loc. cit. that there is a commutative diagram

$$\begin{array}{ccc} \mathrm{Hom}_{G_{\mathbb{Q}_p}}(R_{G,p}, \mathbb{Q}_p^{c\times}) & \xrightarrow{h_{G,p}} & K_0(\mathbb{Z}_p[G], \mathbb{Q}_p) \\ \Big\downarrow \mathrm{coind}_G^H & & \Big\downarrow \rho_H^{G,*} \\ \mathrm{Hom}_{G_{\mathbb{Q}_p}}(R_{H,p}, \mathbb{Q}_p^{c\times}) & \xrightarrow{h_{H,p}} & K_0(\mathbb{Z}_p[H], \mathbb{Q}_p). \end{array} \tag{4.2}$$

Letting $\mathcal{C}(G)$ denote the set of cyclic subgroups of G, Artin's Induction Theorem implies that there exists an integer N such that, for each $\chi \in R_{G,p}$, there are characters $\{\chi_C \in R_{C,p}\colon C \in \mathcal{C}(G)\}$ and integers $\{n_C\colon C \in \mathcal{C}(G)\}$ such that $N \cdot \chi = \sum_{C\in\mathcal{C}(G)} n_C \cdot \mathrm{ind}_C^G \chi_C$. For each homomorphism $f \in \mathrm{Hom}_{G_{\mathbb{Q}_p}}(R_{G,p}, \mathbb{Q}_p^{c\times})$ and character $\chi \in R_{G,p}$ one therefore has

$$\begin{aligned} f(\chi)^N &= f(N\cdot\chi) \\ &= f\left(\sum_{C\in\mathcal{C}(G)} n_C \cdot \mathrm{ind}_C^G \chi_C\right) \\ &= \prod_{C\in\mathcal{C}(G)} \left(\mathrm{coind}_G^C f(\chi_C)\right)^{n_C}. \end{aligned}$$

This implies that $f(\chi) \in \mathbb{Z}_p^{c\times}$ for all $\chi \in R_{G,p}$ if and only if $\mathrm{coind}_G^C f(\phi) \in \mathbb{Z}_p^{c\times}$ for all $C \in \mathcal{C}(G)$ and $\phi \in R_{C,p}$. From (4.1) (with Γ taken to be both G and C) we deduce that $h_{G,p}(f) \in K_0(\mathbb{Z}_p[G], \mathbb{Q}_p)_{\mathrm{tors}}$ if and only if $h_{C,p}\big(\mathrm{coind}_G^C(f)\big) \in K_0(\mathbb{Z}_p[C], \mathbb{Q}_p)_{\mathrm{tors}}$ for each $C \in \mathcal{C}(G)$. From diagram (4.2) we also know that $h_{C,p}\big(\mathrm{coind}_G^C(f)\big) = \rho_C^{G,*}\big(h_{G,p}(f)\big)$ for each $C \in \mathcal{C}(G)$.

We now fix a subgroup $H \in \mathcal{C}(G)$ and let $\#H = p^n r$ with $n \geq 0$ and r coprime to p. We set $\Sigma := \{0, 1, \ldots, n\}$ and for each $i \in \Sigma$ we let P_i denote the subgroup of H which has order p^i, let e_i denote the idempotent $p^{-i} \cdot \sum_{g\in P_i} g$ of $\mathbb{Q}_p[P_i]$ and set $\widehat{e}_i := e_i - e_{i+1}$ (where, for convenience, we have set $e_{n+1} := 0$). For each $i \in \Sigma$ we set $\mathcal{M}_p(H)_i := \widehat{e}_i\mathbb{Z}_p[H]$ and we write $m_{H,i}\colon K_0(\mathbb{Z}_p[H], \mathbb{Q}_p) \to K_0\big(\mathcal{M}_p(H)_i, \mathbb{Q}_p\big)$ for the homomorphism induced by the scalar extension functor $\mathcal{M}_p(H)_i \otimes_{\mathbb{Z}_p[H]} -$. We observe that $\oplus_{i\in\Sigma}\mathcal{M}_p(H)_i$ is the maximal $\mathbb{Z}_p$-order in $\mathbb{Q}_p[H]$ and hence that $K_0(\mathbb{Z}_p[H], \mathbb{Q}_p)_{\mathrm{tors}} = \cap_{i\in\Sigma}\ker(m_{H,i})$.

We now fix an integer $i \in \Sigma$ and compute $\ker(m_{H,i})$. Writing $\mathbb{Q}_p(p^i)$ for the extension of $\mathbb{Q}_p$ which is generated by the set of roots of unity of order p^i, the order $\mathcal{M}_p(H)_i$ identifies with $\mathbb{Z}_p[p^i] \otimes_{\mathbb{Z}_p} \mathbb{Z}_p[C]$ where $\mathbb{Z}[p^i]$ denotes the valuation ring of $\mathbb{Q}_p(p^i)$ and C denotes the subgroup of H of order r. We set $H_i := P_i \times C$ and $\Gamma := H_i/P_i$. The following diagram is commutative

$$\begin{array}{ccccc} K_0(\mathbb{Z}_p[H], \mathbb{Q}_p) & \xrightarrow{m_{H,i}} & K_0\big(\mathcal{M}_p(H)_i, \mathbb{Q}_p\big) & \xrightarrow{\sim} & \dfrac{(\mathbb{Q}_p(p^i) \otimes_{\mathbb{Q}_p} \mathbb{Q}_p[C])^\times}{(\mathbb{Z}_p[p^i] \otimes_{\mathbb{Z}_p} \mathbb{Z}_p[C])^\times} \\ \Big\downarrow \rho_{H_i}^{H,*} & & \Big\downarrow \rho_{\widehat{e}_i\mathbb{Z}_p[C]}^{\mathcal{M}_p(H)_i} & & \Big\downarrow \mathcal{N}_i \\ K_0(\mathbb{Z}_p[H_i], \mathbb{Q}_p) & \xrightarrow{\widehat{m}_{H_i,i}} & K_0(\widehat{e}_i\mathbb{Z}_p[C], \mathbb{Q}_p) & \xrightarrow{\sim} & \dfrac{\mathbb{Q}_p[C]^\times}{\mathbb{Z}_p[C]^\times} \\ \Big\downarrow q_{\Gamma,*}^{H_i} & & \Big\downarrow q_{\mathbb{Z}_p[\Gamma],*}^{\widehat{e}_i\mathbb{Z}_p[C]} & & \Big\downarrow q \\ K_0(\mathbb{Z}_p[\Gamma], \mathbb{Z}_p) & \xrightarrow{=} & K_0(\mathbb{Z}_p[\Gamma], \mathbb{Q}_p) & \xrightarrow{\sim} & \dfrac{\mathbb{Q}_p[\Gamma]^\times}{\mathbb{Z}_p[\Gamma]^\times}. \end{array}$$

In this diagram the second morphism in each row comes from the relevant localization sequence (as in (2.1)), $\mathcal{N}_i$ is induced by $\mathrm{Norm}_{\mathbb{Q}_p(p^i)/\mathbb{Q}_p} \otimes_{\mathbb{Q}_p} \mathrm{id}_{\mathbb{Q}_p[C]}$ and is bijective (since $\mathbb{Q}_p(p^i)/\mathbb{Q}_p$ is totally ramified), q is the bijection induced by the isomorphism $C \xrightarrow{\sim} \Gamma$ which comes from restricting the natural quotient map $H_i \to \Gamma$, and $\widehat{m}_{H_i,i}$ is induced by scalar extension through the morphism $\mathbb{Z}_p[H_i] \to \widehat{e}_i\mathbb{Z}_p[C]$ induced by multiplication by $\widehat{e}_i$. From the commutativity of the above diagram we deduce that $\ker(m_{H,i}) = \ker(q^{H_i}_{\Gamma,*} \circ \rho^{H,*}_{H_i})$ and taken in conjunction with the earlier reductions this equality (for each H and i) implies the stated proposition. □

REMARK 4.2. If F is any subfield of $\mathbb{C}$, then a 'Hom-description' of the group $K_0(\mathbb{Z}[G], F)$ which is precisely in the spirit of [**22**, Chapter II] is given by Agboola and the present author in [**1**, Theorem 3.4].

References

1. A. Agboola and D. Burns, *Twisted forms and relative K-theory*, (submitted for publication).
2. H. Bass, *Algebraic K-theory*, W. A. Benjamin, Inc., New York-Amsterdam, 1968.
3. J.-M. Bismut, H. Gillet, and C. Soulé, *Analytic torsion and holomorphic determinant bundles.* I, Commun. Math. Phys. **115** (1988), 49–78.
4. W. Bley and D. Burns, *Equivariant Tamagawa numbers, fitting ideals and Iwasawa theory*, Compositio Math. **126** (2001), 213–247.
5. ———, *Equivariant epsilon constants, discriminants and étale cohomology*, Proc. London Math. Soc. **87** (2003), 545–590.
6. M. Breuning, *On equivariant global epsilon constants for certain dihedral extensions*, Math. Comp. (to appear).
7. M. Breuning and D. Burns, *Equivariant Tamagawa numbers and classical Galois module theory*, King's College London, Preprint, 2003.
8. ———, *Iwasawa theory and p-adic Hodge theory over non-commutative algebras.* I, King's College London, Preprint, 1997.
9. ———, *Equivariant Tamagawa numbers and Galois module theory*, Compositio Math. **129** (2001), 203–237.
10. ———, *On the values of equivariant zeta functions of curves over finite fields*, (submitted for publication).
11. ———, *On nearly perfect complexes and Weil-étale cohomology*, manuscript in preparation.
12. D. Burns and M. Flach, *Equivariant Tamagawa numbers of motives*, Caltech, Preprint, 1998.
13. ———, *Tamagawa numbers for motives with (non-commutative) coefficients*, Documenta Math. **6** (2001), 501–570.
14. ———, *Tamagawa numbers for motives with (non-commutative) coefficients.* II, Amer. J. Math. **125** (2003), 475–512.
15. D. Burns, B. Köck, and V. P. Snaith, *Refined and l-adic Euler characteristics of nearly perfect complexes*, J. Algebra **272** (2004), 247–272.
16. T. Chinburg, G. Pappas, and M. J. Taylor, *Arithmetic classes of metrised complexes and epsilon constants*, Ann. Sci. École Norm. Sup. (4) **35** (2002), 307–352.
17. ———, *Duality and Hermitian Galois module structure*, Proc. London Math. Soc. **87** (2003), 54–108.
18. C. W. Curtis and I. Reiner, *Methods of representation theory.* I–II, Wiley Classics Library, 1994.
19. P. Deligne, *Valeurs de fonctions L et périods d'intégrales*, Proc. Sympos. Pure Math., vol. 33, Automorphic Forms, Representations and L-Functions (Oregon State Univ., Corvallis, 1977), Part 2, Amer. Math. Soc., Providence, R.I., 1979, pp. 313–346.
20. ———, *Le déterminant de la cohomologie*, Contemp. Math. **67**, Amer. Math. Soc., Providence, RI, 1987, 313–346.
21. M. Flach, *Euler characteristics in relative K-groups*, Bull. London Math. Soc. **32** (2000), 272–284.
22. A. Fröhlich, *Galois module structure of algebraic integers*, Results in Mathematics and Related Areas (3), vol. 1. Springer-Verlag, Berlin, 1983.

23. F. Knudsen and D. Mumford, *The projectivity of the moduli space of stable curves* I. *Preliminaries on 'det' and 'Div'*, Math. Scand. **39** (1976), 19–55.
24. S. MacLane, *Categories for the working mathematician*, Grad. Texts in Math., vol. 5, Springer-Verlag, New York-Berlin, 1971.
25. J. Milnor, *Whitehead torsion*, Bull. Amer. Math. Soc. **72** (1966), 358–426.
26. V. P. Snaith, *Burns' equivariant Tamagawa invariant* $T\Omega^{\mathrm{loc}}(N/\mathbb{Q}, 1)$ *for some quaternion fields*, J. London Math. Soc. **68** (2003), 599–614.
27. R. G. Swan, *Algebraic K-theory*, Lecture Notes in Math., vol. 76, Springer, Berlin, 1968.
28. W. C. Waterhouse, *Introduction to affine group schemes*, Grad. Texts in Math., vol. 66, Springer-Verlag, New York-Berlin, 1979.
29. J. H. C. Whitehead, *Simple homotopy types*, Amer. J. Math. **72** (1950) 1–57.

Department of Mathematics, King's College London, London WC2R 2LS, United Kingdom.

E-mail address: david.burns@kcl.ac.uk

Centre de Recherches Mathématiques
CRM Proceedings and Lecture Notes
Volume **36**, 2004

Questions about the Reductions Modulo Primes of an Elliptic Curve

Alina Carmen Cojocaru

1. Introduction

This is largely a survey paper in which we discuss new and old problems about the reductions E_p modulo primes p of a fixed elliptic curve E defined over the field of rational numbers. We investigate, in particular, how the "noncyclic" part of the group of points of E_p is distributed, thus making progress toward a conjecture of R. Takeuchi. The new result is Theorem 2 of Section 3.

Many interesting questions that resemble classical problems in number theory, such as Artin's primitive root conjecture, the twin prime conjecture, the Buniakowski–Schinzel hypothesis, can be formulated using the group of points of the reduction modulo a prime of a fixed elliptic curve defined over a global field, say over $\mathbb{Q}$. Before discussing these questions in detail, let us recall the basic definitions and properties of elliptic curves. For precise references or more detailed facts, the reader is referred to [**33, 34**].

An *elliptic curve E defined over* $\mathbb{Q}$ is the locus of an equation of the form

$$E : y^2 = x^3 + ax + b \tag{1.1}$$

(where $a, b \in \mathbb{Z}$ are such that the discriminant $-16(4a^3 + 27b^2)$ is nonzero), together with a point at infinity $\mathcal{O}$, given in projective coordinates by $[0{:}1{:}0]$.

When doing arithmetic over E we are usually concerned with rational solutions of (1.1) or solutions of congruences modulo primes defined by (1.1). We will briefly consider both these aspects.

Let $E(\mathbb{Q})$ be the set of rational points on E together with the point at infinity. Remarkably, we can define an addition law on $E(\mathbb{Q})$ by the simple geometric rule: the sum of three points equals zero (the point at infinity) if and only if the points

---

2000 *Mathematics Subject Classification.* 14H52, 11R45, 11N36.

This research was supported by an NSERC postdoctoral fellowship.

I am indebted to M. Ram Murty and Manjul Bhargava for their useful comments on previous versions of the paper. I am also indebted to the audiences of my talks on results discussed in this paper, which I gave at CNTA VII and at various universities (Laval, McGill, Carleton, Queen's, Toronto, McMaster, Waterloo, British Columbia, UIUC, and UCLA) during 2002. I am thankful to the organizers of CNTA VII for providing me with financial support for participating in this conference.

This is the final form of the paper.

lie on the same line. With respect to this law, $E(\mathbb{Q})$ becomes an abelian group and, moreover, by a famous result of Mordell published in 1922, a finitely generated abelian group. Therefore $E(\mathbb{Q}) \simeq \mathbb{Z}^r \oplus E(\mathbb{Q})_{\text{tors}}$, where r is some nonnegative integer called *the (arithmetic) rank of E over $\mathbb{Q}$*, and $E(\mathbb{Q})_{\text{tors}}$ is the torsion part of $E(\mathbb{Q})$.

We remark that we can talk about an elliptic curve E defined over any field K (for the purpose of this paper, K will have characteristic $\neq 2, 3$). Again, we can define an abelian group structure on the set $E(K)$ of K-rational points of E (N.B. $E(K)$ is a finitely generated abelian group if K is a global field). In particular, if E is defined over $\mathbb{Q}$, we can also talk about the group $E(\overline{\mathbb{Q}})$, where $\overline{\mathbb{Q}}$ is the algebraic closure of $\mathbb{Q}$.

From now on let E be a fixed elliptic curve defined over $\mathbb{Q}$. There are some natural maps that we can define on $E(\overline{\mathbb{Q}})$. For each $n \in \mathbb{Z}$, let us consider the multiplication by n map defined on $E(\overline{\mathbb{Q}})$. These are group homomorphisms and also morphisms of curves, called isogenies of E over $\overline{\mathbb{Q}}$. If we let $\text{End}_{\overline{\mathbb{Q}}}(E)$ denote the ring of all isogenies of E over $\overline{\mathbb{Q}}$, then we obtain a canonical embedding $\mathbb{Z} \leq \text{End}_{\overline{\mathbb{Q}}}(E)$, which, in "most of the cases," is in fact a ring isomorphism. If it is an isomorphism, then we say that E is *without complex multiplication* (non-CM). If the embedding is strict, then we obtain that $\text{End}_{\overline{\mathbb{Q}}}(E)$ is an order in an imaginary quadratic field $\mathbb{Q}(\sqrt{-D})$ of class number 1 (hence $D \in \{1, 2, 3, 7, 11, 19, 43, 67, 163\}$, by a famous 1965's result of Heegner–Stark–Baker). In this situation we say that E is *with complex multiplication* (CM) and that $\mathbb{Q}(\sqrt{-D})$ is its CM field.

The CM and non-CM situations are very different, as can be seen, for example, from the following. For an arbitrary positive integer k we let $E[k]$ be *the group of k-division points of E*, that is, the points of $E(\overline{\mathbb{Q}})$ annihilated by k. We know that, as $\mathbb{Z}$-modules, $E[k] \simeq \mathbb{Z}/k\mathbb{Z} \oplus \mathbb{Z}/k\mathbb{Z}$. If we adjoin to $\mathbb{Q}$ the x- and y-coordinates of the k-division points of E, then we obtain a finite Galois extension of $\mathbb{Q}$ which is unramified outside kN, with N denoting the conductor of E in the sense explained below, and which contains the cyclotomic field $\mathbb{Q}(\zeta_k)$. We denote this extension by $\mathbb{Q}(E[k])$ and call it *the k-division field of E*. We can then define a natural representation

$$\phi_k \colon \text{Gal}\big(\mathbb{Q}(E[k])/\mathbb{Q}\big) \to \text{GL}_2(\mathbb{Z}/k\mathbb{Z}),$$

called *the Galois representation associated to $E[k]$*, which is easily seen to be injective. Is this representation surjective? In 1941, M. Deuring showed that in the case of a CM elliptic curve E, ϕ_k is not surjective for any integer $k > 2$. Moreover, in this situation we have that

$$\phi(k)^2 \leq [\mathbb{Q}(E[k]){:}\mathbb{Q}] \leq k^2 \tag{1.2}$$

for any $k > 2$, where $\phi(\cdot)$ is the Euler function. By contrast, in 1972 J.-P. Serre showed that in the case of a non-CM elliptic curve E, there exists a positive integer $A(E)$, depending on E, such that ϕ_k is surjective for any positive integer k coprime to $A(E)$. Therefore for such E and k we have

$$[\mathbb{Q}(E[k]){:}\mathbb{Q}] = \#\,\text{GL}_2(\mathbb{Z}/k\mathbb{Z}) = k^4 \prod_{q|k}\left(1 - \frac{1}{q}\right)\left(1 - \frac{1}{q^2}\right), \tag{1.3}$$

where the product is over prime divisors q of k.

Now let us consider the reduction E_p of E modulo a prime $p > 3$:

$$E_p : y^2 \equiv x^3 + ax + b \pmod{p}.$$

This is a (not necessarily smooth) curve defined over $\mathbb{F}_p$. There is an arithmetic invariant $N = N_E \in \mathbb{Z}$ of the elliptic curve E, called *the conductor of* E, which gives us precise information about the reduction of E modulo each prime p. One of the properties of this invariant is that E_p is smooth if and only if $p \nmid N$. Clearly, N is a divisor of the discriminant of the polynomial in x defining E.

From now on we will consider only primes $p \nmid N$, called *primes of good reduction for* E. For such a p, E_p is an elliptic curve defined over $\mathbb{F}_p$. This paper is concerned with *questions about the group* $E_p(\mathbb{F}_p)$, *as* p *varies*. The most basic questions about this group are to find its cardinality and to determine its structure.

The Riemann hypothesis for E_p, proved by H. Hasse in 1933, asserts that if we write

$$\#E_p(\mathbb{F}_p) = p + 1 - a_p$$

for some $a_p \in \mathbb{Z}$, then

$$|a_p| \leq 2\sqrt{p}.$$

We say that p is of *supersingular reduction* if $a_p = 0$ and of *ordinary reduction* otherwise.

Hasse's bound implies that the polynomial $X^2 - a_p X + p$ has two complex conjugate roots π_p, $\overline{\pi}_p$ such that $|\pi_p| = p^{1/2}$. As it turns out, the complex number π_p plays the role of the Frobenius map $(x, y) \mapsto (x^p, y^p)$ on E_p, after identifying the endomorphism ring $\mathrm{End}_{\overline{\mathbb{F}}_p}(E_p)$ with an order in an imaginary quadratic field (ordinary case) or with an order in a quaternion algebra (supersingular case), where $\overline{\mathbb{F}}_p$ denotes an algebraic closure of $\mathbb{F}_p$. Moreover, we have that $\mathbb{Z}[\pi_p] \leq \mathrm{End}_{\mathbb{F}_p}(E_p)$. Now $\mathrm{End}_{\mathbb{F}_p}(E_p)$ is always an order in an imaginary quadratic field. Thus if we denote by Δ_p the discriminant of this order and by b_p the index of $\mathbb{Z}[\pi_p]$ in $\mathrm{End}_{\mathbb{F}_p}(E_p)$, we obtain by the formula

$$a_p^2 - 4p = b_p^2 \Delta_p. \tag{1.4}$$

The classical theory of elliptic curves provides us also with information about the structure of $E_p(\mathbb{F}_p)$. If we let k be a nonzero integer, and if $E_p(\overline{\mathbb{F}}_p)[k]$ denotes the group of $\overline{\mathbb{F}}_p$-points of E_p annihilated by k, then $E_p(\overline{\mathbb{F}}_p)[k]$ is isomorphic to $\mathbb{Z}/k\mathbb{Z} \oplus \mathbb{Z}/k\mathbb{Z}$ for any $(k, p) = 1$, and to $\mathbb{Z}/k\mathbb{Z}$ or $\{\mathcal{O}\}$ for any $(k, p) \neq 1$. The group $E_p(\mathbb{F}_p)$ can be viewed as a subgroup of $E_p(\overline{\mathbb{F}}_p)[k]$ for some k such that $\#E_p(\mathbb{F}_p) \mid k$, and so can be written as

$$E_p(\mathbb{F}_p) \simeq \mathbb{Z}/d_p\mathbb{Z} \oplus \mathbb{Z}/d_p e_p\mathbb{Z}$$

for uniquely determined positive integers d_p, e_p. More precisely, d_p and e_p are given by the formulas

$$d_p = \gcd\left\{b_p, \frac{a_p + b_p\delta_p - 2}{2}\right\}, \quad e_p = \frac{p + 1 - a_p}{d_p^2},$$

where δ_p is 0 if $\Delta_p \equiv 0 \pmod{4}$ and 1 if $\Delta_p \equiv 1 \pmod{4}$. We refer the reader to [**8**] for an explanation of how to deduce these formulas.

There is a rich family of questions regarding the group $E_p(\mathbb{F}_p)$ as p varies over primes of good reduction for E: what is the percentage of the primes p for which $E_p(\mathbb{F}_p)$ has order divisible by a fixed integer?; what can we say about the asymptotic distribution of the d_p's?; what about the e_p's, or the $\mathbb{Q}(\pi_p)$'s, or the a_p's, or even the b_p's? We will discuss some of these questions in what follows.

Throughout the paper, we will use the notation introduced above, as well as classical notation. More precisely, p will denote a prime of good reduction for a given elliptic curve E defined over $\mathbb{Q}$; l, q will denote arbitrary rational primes; k will denote a positive integer; x will denote a positive real number; for a positive integer n, $\phi(n)$ will denote the Euler function of n, $\nu(n)$ the number of distinct prime divisors of n, and $\mu(n)$ the Möbius function of n; for a positive real number x, $\pi(x)$ will denote the number of primes $\leq x$ and $\operatorname{li} x$ the logarithmic integral $\int_2^x (\log t)^{-1}\, dt$; det and tr will denote the determinant and trace of a given matrix. Whenever we write O_A, $\ll_A$, $\gg_A$, we will mean that the implied constants depend on A; whenever we write O, $\ll$, $\gg$, we will mean that the implied constants are absolute.

2. Divisibility of $\#E_p(\mathbb{F}_p)$ by fixed integers

We start our discussion with the simplest of the questions posed above.

QUESTION 1. Let E be an elliptic curve defined over $\mathbb{Q}$ and of conductor N. Let k be a fixed positive integer. What is the asymptotic behavior of the function

$$\#\{p \leq x : p \nmid kN, k \mid \#E_p(\mathbb{F}_p)\},$$

as $x \to \infty$?

The key fact which we will use in answering this question is the following classical property of the representation ϕ_k:

LEMMA 1. *Let E be an elliptic curve defined over $\mathbb{Q}$ and of conductor N. Let p be a prime and k be a positive integer such that $p \nmid kN$. If σ_p denotes the Frobenius at p in $\mathbb{Q}(E[k])/\mathbb{Q}$, then*

$$\operatorname{tr} \phi_k(\sigma_p) \equiv a_p \pmod{k}, \quad \det \phi_k(\sigma_p) \equiv p \pmod{k}.$$

Therefore,

$$k \mid \#E_p(\mathbb{F}_p) \iff \sigma_p \subseteq C_k, \tag{2.1}$$

where

$$C_k := \{g \in \operatorname{Gal}(\mathbb{Q}(E[k])/\mathbb{Q}) : \det \phi_k(g) + 1 - \operatorname{tr} \phi_k(g) \equiv 0 \pmod{k}\}. \tag{2.2}$$

By using the classical Chebotarev density theorem we then obtain:

THEOREM 1. *Let E be an elliptic curve defined over $\mathbb{Q}$ and of conductor N. Let k be a fixed positive integer. Then, as $x \to \infty$,*

$$\#\{p \leq x : p \nmid kN, k \mid \#E_p(\mathbb{F}_p)\} \sim \frac{\#C_k}{[\mathbb{Q}(E[k]){:}\mathbb{Q}]} \operatorname{li} x.$$

We remark that it can be shown that $\#C_k$ is $\approx k$ in the CM case and $\approx k^3$ in the non-CM case. Thus, from (1.2) and (1.3), we can deduce that the density above should be $\approx 1/k$.

3. Distribution of the d_p's and the cyclicity of $E_p(\mathbb{F}_p)$

Let us be a little more courageous and ask:

QUESTION 2. Let E be an elliptic curve defined over $\mathbb{Q}$ and of conductor N. Let d be a fixed positive integer. What is the asymptotic behavior of the function

$$\#\{p \leq x : p \nmid dN, d_p = d\},$$

as $x \to \infty$?

This is already a more difficult question: not only do we want $d | \# E_p(\mathbb{F}_p)$, but also we ask for the stronger conditions that $E_p(\mathbb{F}_p)$ contain a subgroup isomorphic to $\mathbb{Z}/d\mathbb{Z} \oplus \mathbb{Z}/d\mathbb{Z}$ and that this be the biggest two-copy of a cyclic group that occurs among the subgroups of $E_p(\mathbb{F}_p)$.

The case $d = 1$ has been extensively studied in the past 25 years. In 1975 [**1**], I. Borosh, C. J. Moreno, and H. Porta made several computations which led them to expect that for many elliptic curves E we have $d_p = 1$ (i.e., the group $E_p(\mathbb{F}_p)$ is cyclic) for infinitely many primes p. In 1976 [**31**], J.-P. Serre confirmed this expectation by proving, under the assumption of the generalized Riemann hypothesis (GRH) for the Dedekind zeta functions of the division fields $\mathbb{Q}(E[k])$ of E, that

$$\#\{p \leq x : p \nmid N, E_p(\mathbb{F}_p) \text{ cyclic}\} \sim \sum_{k \geq 1} \frac{\mu(k)}{[\mathbb{Q}(E[k]):\mathbb{Q}]} \operatorname{li} x. \tag{3.1}$$

As explained for example in [**10**], the density above is positive if and only if $\mathbb{Q}(E[2]) \neq \mathbb{Q}$.

There are many aspects of the cyclicity problem that we can consider. Can we remove GRH in Serre's result? Can we obtain effective error terms? What is the size of the smallest prime p for which $E_p(\mathbb{F}_p)$ is cyclic? In 1979 [**26**], M. R. Murty removed GRH in (3.1) in the case of a CM elliptic curve. His unconditional proof made use of class field theoretical (CFT) properties of the division fields of CM elliptic curves (not available in the non-CM case!) and of the large sieve for number fields in the form of a Bombieri–Vinogradov theorem for number fields (due to R. J. Wilson). In 1987 [**27**], M. R. Murty also showed, unconditionally, that for certain non-CM elliptic curves E there exist infinitely many primes p for which $E_p(\mathbb{F}_p)$ is cyclic. In 1990 [**19**], R. Gupta and M. R. Murty used sieve methods to show, unconditionally and for any elliptic curve E such that $\mathbb{Q}(E[2]) \neq \mathbb{Q}$, that

$$\#\{p \leq x : p \nmid N, E_p(\mathbb{F}_p) \text{ cyclic}\} \gg_N \frac{x}{(\log x)^2}.$$

Hence, unconditionally, there are infinitely many primes p for which $E_p(\mathbb{F}_p)$ is cyclic if $\mathbb{Q}(E[2]) \neq \mathbb{Q}$. In [**2, 3**], the author managed to replace GRH in (3.1) with a $\frac{3}{4}$-quasi GRH (that is, a zero-free region of $\operatorname{Re}(s) > \frac{3}{4}$) in the case of a non-CM elliptic curve E. In [**3, 5**] she also obtained a new simpler unconditional proof of (3.1) in the case of a CM elliptic curve E by removing the use of CFT and of the large sieve for number fields, and by using instead the sieve of Eratosthenes. We remark that the proofs of all these results will lead to error terms of the form $\mathrm{O}_N(x \log\log x/(\log x)^2)$ (the conditional formulas for non-CM and CM curves) and $\mathrm{O}_N(x/(\log x)^B)$ or $\mathrm{O}_N(x/[(\log x)(\log\log\log x)])$ (the unconditional formulas for CM curves), where B is any positive integer. However, it is shown in [**3, 10**] that if we assume the full strength of GRH, then these error terms can be improved to $\mathrm{O}_N\big(x^{5/6}(\log x)^{2/3}\big)$ in the non-CM case and $\mathrm{O}_N(x^{3/4}(\log x)^{1/2})$ in the CM case. The dependence of the O_N-constant on N can also be made explicit. Then, by comparing the main term with the error term, we obtain that, under GRH, the smallest prime p for which $E_p(\mathbb{F}_p)$ is cyclic is $\mathrm{O}_\varepsilon\big((\log N)^{4+\varepsilon}\big)$ in the non-CM case and $\mathrm{O}_\varepsilon\big((\log N)^{2+\varepsilon}\big)$ in the CM case, for any $0 < \varepsilon < 1$.

Now let us return to the general situation stated in Question 2. The same technique as for $d = 1$ can be applied for an arbitrary d and will lead to very similar results. More precisely, we obtain the following:

THEOREM 2. *Let E be an elliptic curve defined over $\mathbb{Q}$ and of conductor N. Let d be a positive integer:*

(1) *Assume that E is without* CM:

(a) *If a $\frac{3}{4}$-quasi* GRH *holds for the Dedekind zeta functions of the division fields of E, then*

$$\#\{p \le x : p \nmid dN, d_p = d\} = \sum_{k \ge 1} \frac{\mu(k)}{[\mathbb{Q}(E[dk]):\mathbb{Q}]} \operatorname{li} x + \mathrm{O}_{d,N}\left(\frac{x \log\log x}{(\log x)^2}\right).$$

(b) *If* GRH *holds for the Dedekind zeta functions of the division fields of E, then*

$$\#\{p \le x : p \nmid dN, d_p = d\} = \sum_{k \ge 1} \frac{\mu(k)}{[\mathbb{Q}(E[dk]):\mathbb{Q}]} \operatorname{li} x + \mathrm{O}_{d,N}(x^{5/6}(\log x)^{2/3}).$$

(2) *Assume that E is with* CM:

(a) *Unconditionally,*

$$\#\{p \le x : p \nmid dN, d_p = d\} = \sum_{k \ge 1} \frac{\mu(k)}{[\mathbb{Q}(E[dk]):\mathbb{Q}]} \operatorname{li} x + \mathrm{O}_{d,N}\left(\frac{x}{(\log x)(\log\log\log x)}\right).$$

(b) *If* GRH *holds for the Dedekind zeta functions of the division fields of E, then*

$$\#\{p \le x : p \nmid dN, d_p = d\} = \sum_{k \ge 1} \frac{\mu(k)}{[\mathbb{Q}(E[dk]):\mathbb{Q}]} \operatorname{li} x + \mathrm{O}_{d,N}\big(x^{3/4}(\log x)^{1/2}\big).$$

As already mentioned, the proof of this theorem is based on the same arguments as the ones used in the case $d = 1$ (see [**2, 3, 5, 10**]). For clarity, we will give a summary of the proof of part 1(b). The complete proof of the theorem will appear in an upcoming paper.

Before discussing the ideas of the proof, let us point out that, in this generality, Theorem 2 provides answers to a conjecture posed in [**35**]. To be more precise, it was predicted in [**35**, Conjectures 2 and 2'] that, for an elliptic curve E defined over $\mathbb{Q}$ and for an integer d,

$$\#\{p \le x : p \nmid dN, d_p = d\} \sim \sum_{k \ge 1} \frac{\mu(k)}{[\mathbb{Q}(E[dk]):\mathbb{Q}]} \operatorname{li} x,$$

as $x \to \infty$. Thus, by part 1(a) of Theorem 2, this conjectural law is true in the non-CM case under a $\frac{3}{4}$-quasi GRH, and by part 2(a), it is true in the CM case under no hypothesis.

The first key ingredient in the proof of Theorem 2 is the following simple, but important, result (for a proof, see [**26**, pp. 153–154]).

LEMMA 2. *Let E be an elliptic curve defined over $\mathbb{Q}$ and of conductor N. Let p and d be such that $p \nmid dN$. Then $E_p(\mathbb{F}_p)$ contains a subgroup isomorphic to $\mathbb{Z}/d\mathbb{Z} \oplus \mathbb{Z}/d\mathbb{Z}$ if and only if p splits completely in $\mathbb{Q}(E[d])/\mathbb{Q}$.*

Another key ingredient is an effective version of the Chebotarev density theorem, due to J. Lagarias and A. Odlyzko [**22**].

LEMMA 3 (Effective Chebotarev Density Theorem). *Let $L/\mathbb{Q}$ be a finite Galois extension, of Galois group G, degree n_L and discriminant* $\operatorname{disc}(L/\mathbb{Q})$. *Let C be a conjugacy set (that is, a finite union of conjugacy classes) of G. We set*

$$\pi_C(x, L/\mathbb{Q}) := \#\{p \leq x : \sigma_p \subseteq C\},$$

where σ_p is the Frobenius of p in $L/\mathbb{Q}$.

(1) *There exist positive constants c_1 and c_2, with c_1 effective and c_2 absolute, such that, if*

$$\sqrt{\frac{\log x}{n_L}} \geq c_2 \max(\log |d_L|, |d_L|^{1/n_L}),$$

then

$$\pi_C(x, L/\mathbb{Q}) = \frac{\#C}{\#G} \operatorname{li} x + \mathrm{O}\left((\#\widetilde{C}) x \exp\left(-c_1 \sqrt{\frac{\log x}{n_L}} \right) \right),$$

where $\widetilde{C}$ denotes the set of conjugacy classes contained in C.

(2) *Assume that a δ-quasi* GRH *holds for the Dedekind zeta function of L (that is, we have a zero-free region of* $\operatorname{Re}(s) > \delta$ *for the Dedekind zeta function of L). Then*

$$\pi_C(x, L/\mathbb{Q}) = \frac{\#C}{[L{:}\mathbb{Q}]} \operatorname{li} x + \mathrm{O}\left((\#C) x^{\delta} \left(\frac{\log |\operatorname{disc}(L/\mathbb{Q})|}{[L : \mathbb{Q}]} + \log x \right) \right).$$

SKETCH OF PROOF OF PART 1(B) OF THEOREM 2. By Lemma 2 and Hasse's bound, we can write

$$\#\{p \leq x : p \nmid dN, d_p = d\} = \sum_{k \leq 2\sqrt{x}} \mu(k) \pi_1(x, \mathbb{Q}(E[dk])/\mathbb{Q}).$$

We split this sum into two parts, according to whether $k < y$ or $y < k \leq 2\sqrt{x}$ for some parameter $y = y(x)$ to be chosen later. We use part 2 of Lemma 3 with $\delta = \frac{1}{2}$ to estimate the first subsum, and obtain

$$\sum_{k \leq y} \mu(k) \pi_1(x, \mathbb{Q}(E[dk])/\mathbb{Q}) = \sum_{k \leq y} \frac{\mu(k)}{[\mathbb{Q}(E[dk]){:}\mathbb{Q}]} \operatorname{li} x + \mathrm{O}_{d,N}(y x^{1/2} \log x).$$

To estimate the sum running over $y < k \leq 2\sqrt{x}$ we note that if a prime p splits completely in $\mathbb{Q}(E[dk])$, then it also does in $\mathbb{Q}(\zeta_{dk})$, since $\mathbb{Q}(\zeta_{dk}) \subseteq \mathbb{Q}(E[dk])$. Hence, for such a prime p, we have $p \equiv 1 \pmod{dk}$. On the other hand, by

Lemma 4, p must also satisfy $p+1-a_p \equiv 0 \pmod{d^2k^2}$. Therefore we can write

$$\begin{aligned}
&\sum_{y<k\leq 2\sqrt{x}} \mu(k)\pi_1\big(x, \mathbb{Q}(E[dk])/\mathbb{Q}\big) \\
&\quad\leq \sum_{y<k\leq 2\sqrt{x}} \#\{p \leq x : p \nmid dN, p \equiv 1 \pmod{dk}, p+1-a_p \equiv 0 \pmod{d^2k^2}\} \\
&\quad\leq \sum_{\substack{|a|\leq 2\sqrt{x} \\ a\neq 2}} \sum_{\substack{y<k\leq 2\sqrt{x} \\ k|a-2}} \sum_{\substack{p\leq x \\ a_p=a \\ k^2|p+1-a}} 1 + \sum_{y<k\leq 2\sqrt{x}} \sum_{\substack{p\leq x \\ a_p=2 \\ k^2|p-1}} 1 \\
&\quad=: \sum_1 + \sum_2 .
\end{aligned}$$

Now we use elementary estimates to get

$$\sum_1 \ll \sum_{y<k\leq 2\sqrt{x}} \left(\frac{x}{k^2}+1\right)\frac{\sqrt{x}}{k} \ll \frac{x\sqrt{x}}{y^2} + \sqrt{x}\log x,$$

$$\sum_2 \ll \sum_{y<k\leq 2\sqrt{x}} \left(\frac{x}{k^2}+1\right) \ll \frac{x}{y} + \sqrt{x}.$$

We choose

$$y := \left(\frac{x}{\log x}\right)^{1/3}$$

and, by putting together all of the above, we obtain

$$\#\{p \leq x : p \nmid dN, d_p = d\} = \sum_{k\leq y} \frac{\mu(k)}{[\mathbb{Q}(E[dk]){:}\mathbb{Q}]} \operatorname{li} x + \mathrm{O}_{d,N}\big(x^{5/6}(\log x)^{2/3}\big). \tag{3.2}$$

By using properties (1.3) of $[\mathbb{Q}(E[dk]){:}\mathbb{Q}]$ we can also estimate the tail

$$\sum_{k>y} \frac{\mu(k)}{[\mathbb{Q}(E[dk]){:}\mathbb{Q}]} \operatorname{li} x,$$

and, with a little more care, we can obtain the dependence on N of the occurring $\mathrm{O}_{d,N}$-constants. This requires knowledge of upper bounds in terms of N for the constant $A(E)$ mentioned in Section 1. Such bounds can be found in [**6, 32**]. We can finally show that

$$\begin{aligned}
\#\{p \leq x : p \nmid dN, d_p = d\} = \sum_{k\geq 1} \frac{\mu(k)}{[\mathbb{Q}(E[dk]){:}\mathbb{Q}]} \operatorname{li} x + \mathrm{O}_d\big(x^{5/6}(\log Nx)^{2/3}\big) \\
+ \mathrm{O}_d\left(\frac{(\log\log x)(\log Nx)}{\log x}(\log N)^3(\log\log 2N)^6\right),
\end{aligned}$$

from which we can deduce (under the assumption that the density above is positive) that the size of the smallest prime p such that $d_p = d$ is $\mathrm{O}_{d,\varepsilon}\big((\log N)^{4+\varepsilon}\big)$ for any $0 < \varepsilon < 1$. □

Let us remark that in the above proof of (3.2) we did not use that E was a non-CM elliptic curve. However, the proof of the improved error term occurring in part 2(b) of Theorem 2 does use that in that situation we have a CM elliptic curve. Also, we remark that our proof of (3.2) can be generalized to elliptic curves defined over any number field. This is an advantage that we have over the proofs of parts 1(a) and 2(b) of the theorem, in which we invoke the Brun–Titchmarsh theorem whose number field version is not straightforward. Finally, we point out that it is an easy exercise to obtain the explicit dependence of the occurring O-constants on the integer d.

4. Primitive points on $E_p(\mathbb{F}_p)$

Having obtained satisfactory answers to the question of how often $d_p = 1$, hence $E_p(\mathbb{F}_p)$ is cyclic, we can proceed to asking a stronger question:

QUESTION 3. Let E be an elliptic curve defined over $\mathbb{Q}$, of conductor N and of arithmetic rank ≥ 1. Let $a \in E(\mathbb{Q})$ be a point of infinite order. What is the asymptotic behavior of

$$\#\{p \leq x : p \nmid N, E_p(\mathbb{F}_p) = \langle a \pmod{p} \rangle\},$$

as $x \to \infty$ (N.B. we may need to exclude the primes p dividing the denominators of the coordinates of a)?

This question was formulated by S. Lang and H. Trotter in 1976 [**24**]. They conjectured the existence of the density of the primes p for which $E_p(\mathbb{F}_p) = \langle a \pmod{p} \rangle$, a prediction that could be viewed as an elliptic curve analogue of Artin's primitive root conjecture. We should observe, though, that while in the classical situation of E. Artin the group $\mathbb{F}_p^*$ is always cyclic, in the elliptic curve situation of Lang–Trotter the group $E_p(\mathbb{F}_p)$ may not be cyclic. Therefore it was only natural that we studied the cyclicity of $E_p(\mathbb{F}_p)$ first.

In 1986 [**18**], R. Gupta and M. R. Murty showed that, under GRH, the above conjecture of Lang–Trotter is true for CM elliptic curves. More precisely, they used Chebotarev density arguments and sieve methods for the fields $\mathbb{Q}(E[k], k^{-1}a)$ to prove:

THEOREM 3 (R. Gupta, M. R. Murty, 1986). *Let E be a* CM *elliptic curve defined over $\mathbb{Q}$, of conductor N and of arithmetic rank ≥ 1. Let K be the* CM *field of E. Let $a \in E(\mathbb{Q})$ be a point of infinite order. If* GRH *holds for the Dedekind zeta functions of the fields $\mathbb{Q}(E[k], k^{-1}a)$, then there exists a constant $C(E, a)$, depending on E and a (and defined in terms of $\mathbb{Q}(E[k], k^{-1}a)$), such that*

$$\begin{aligned}\#\{p \leq x : p \nmid N, p \text{ splits in } K, E_p(\mathbb{F}_p) = \langle a \pmod{p} \rangle\}\\ = C(E, a)\, \mathrm{li}\, x + \mathrm{O}_N\left(\frac{x \log\log x}{(\log x)^2}\right).\end{aligned}$$

If K is one of $\mathbb{Q}(\sqrt{-11})$, $\mathbb{Q}(\sqrt{-19})$, $\mathbb{Q}(\sqrt{-43})$, $\mathbb{Q}(\sqrt{-67})$, $\mathbb{Q}(\sqrt{-163})$, then the density $C(E, a)$ is positive.

Currently, nothing is known about the primes p which are inert in the CM field K of E. In addition, nothing is known about Question 3 in the case of a non-CM curve.

In 1986 and 1990, R. Gupta and M. R. Murty considered higher rank analogues of Question 3, which were also formulated, with conjectural answers, by Lang and Trotter in 1976 (see [**24**]). They provided conditional and unconditional answers to these questions in both the CM and non-CM cases. For a detailed discussion and precise formulations of their results, we refer the reader to [**18, 19, 29**].

5. Square-free orders for $E_p(\mathbb{F}_p)$

We turn our attention to the integers e_p appearing in the description of the group $E_p(\mathbb{F}_p)$. For example, by observing that the cyclicity of $E_p(\mathbb{F}_p)$ is ensured if the order of the group is square-free, we can ask:

QUESTION 4. Let E be an elliptic curve defined over $\mathbb{Q}$ and of conductor N. What is the asymptotic behavior of the function

$$\#\{p \leq x : p \nmid N, \#E_p(\mathbb{F}_p) \text{ square-free}\},$$

as $x \to \infty$?

In other words, we are interested in the primes p for which both the conditions that $d_p = 1$ and e_p is square-free are satisfied.

In trying to answer this question, we start with the basic observation that

$$\#E_p(\mathbb{F}_p) \text{ is square-free} \iff q^2 \nmid \#E_p(\mathbb{F}_p) \text{ for any prime } q.$$

Thus we want to estimate the number of primes p for which $q \nmid d_p$ for any q and $q^2 \nmid e_p$ for any q. We have seen in Section 4 that the behavior of the d_p's is well understood. The study of the behavior of the e_p's is more challenging, one of the difficulties that we encounter being that the distribution of the e_p's is related to the distribution of primes p (not) splitting completely in "too small" field extensions. Luckily, however, in the CM case we have the following beautiful characterization of the squarefreeness of e_p (for a proof, see [**4**]):

LEMMA 4. *Let E be a* CM *elliptic curve defined over $\mathbb{Q}$, of conductor N and with complex multiplication by the full ring of integers $\mathcal{O}_K$ of an imaginary quadratic field K. Let $p \nmid N$ be a prime of ordinary good reduction for E. Then $\#E_p(\mathbb{F}_p)$ is square-free if and only if p splits completely in K, as $(p) = (\pi_p)(\overline{\pi}_p)$, and π_p does not split completely in $K(E[\mathfrak{q}])$ for any inert prime ideal $\mathfrak{q}$ of K, in $K(E[\mathfrak{q}^2])$ for any ramified prime ideal $\mathfrak{q}$ of K, and in $K(E[\mathfrak{q}\overline{\mathfrak{q}}])$, $K(E[\mathfrak{q}^2])$, and $K(E[\overline{\mathfrak{q}}^2])$ for any split prime ideal $\mathfrak{q}$ of K. Here, for an ideal (α) of K, we denote by $E[(\alpha)]$ the kernel of the multiplication by α map.*

Therefore, by Remark (2.1) made in Section 2 and by the above lemma, in order to answer Question 4 for CM elliptic curves with CM by K, we need to estimate the number of primes that satisfy Chebotarev conditions in division fields of the form $K(E[q])$ and $K(E[\mathfrak{q}^2])$. These extensions are "sufficiently large," and then, by using Chebotarev density theorems and sieve methods, we can prove:

THEOREM 4. *Let E be a* CM *elliptic curve defined over $\mathbb{Q}$ and of conductor N, with complex multiplication by the full ring of integers of an imaginary quadratic field.*

(1) *Unconditionally, we have that*

$$\#\{p \leq x : p \nmid N, \#E_p(\mathbb{F}_p) \textit{ square-free}\} = \sum_{k \geq 1} \frac{\mu(k)\#C_{k^2}}{[\mathbb{Q}(E[k^2]):\mathbb{Q}]} \operatorname{li} x + \mathrm{O}_N\left(\frac{x}{(\log x)(\log\log\log x)}\right).$$

(2) *Under* GRH *for the Dedekind zeta functions of the division fields of E, we have that*

$$\#\{p \leq x : p \nmid N, \#E_p(\mathbb{F}_p) \textit{ square-free}\} = \sum_{k \geq 1} \frac{\mu(k)\#C_{k^2}}{[\mathbb{Q}(E[k^2]):\mathbb{Q}]} \operatorname{li} x + \mathrm{O}_N\left(x^{7/8}(\log x)^7\right).$$

Here, the conjugacy sets C_{k^2} *are as in* (2.2).

Similar, but conditional, asymptotic formulas can also be obtained in the non-CM case, but by using a different and more elaborate approach. For proofs of all these results, we refer the reader to [**3, 4**].

6. Prime orders for $E_p(\mathbb{F}_p)$

What if the order of $E_p(\mathbb{F}_p)$ is not only square-free, but also a prime (which means, $d_p = 1$ and e_p is prime)? In other words:

QUESTION 5. Let E be an elliptic curve defined over $\mathbb{Q}$ and of conductor N. What is the asymptotic behavior of the function

$$\#\{p \leq x : p \nmid N, \#E_p(\mathbb{F}_p) \text{ prime}\},$$

as $x \to \infty$?

In 1988 [**21**], N. Koblitz showed the relevance of this question to the theoretical study of elliptic curve cryptography. We recall that certain public-key cryptosystems (based on the intractability of the discrete logarithm problem) can be implemented using the group of points of an elliptic curve defined over a finite field. A desirable property of this curve is that the cyclic group generated by a fixed point of the curve have order divisible by a large prime. By choosing the elliptic curve such that its group of points has (large) prime order, the desired property is clearly satisfied for any point.

The basic observation in approaching Question 5 is that

$$\#E_p(\mathbb{F}_p) \text{ is a prime } \text{“}\Longleftrightarrow\text{”}\ q \nmid \#E_p(\mathbb{F}_p) \text{ for any prime } q \leq \sqrt{p}+1,$$

where by "$\Leftrightarrow$" we mean that care should be taken with the prime values of $\#E_p(\mathbb{F}_p)$ which may lie between $\sqrt{p}-1$ and $\sqrt{p}+1$. Then, by using remark (2.1), we can write

$$\#\{p \leq x : \#E_p(\mathbb{F}_p) \text{ prime}\} \text{ “}=\text{” } \sum_{k \leq \sqrt{x}+1} \mu(k)\#\{p \leq x : \sigma_p \subseteq C_k\}. \tag{6.1}$$

This formula suggests that the number of primes $p \leq x$ for which $\#E_p(\mathbb{F}_p)$ is prime should be

$$\sim C(E)\frac{x}{(\log x)^2}$$

for some (positive) constant $C(E)$ depending on E, as already conjectured by Koblitz in [**21**].

Now let us remark further that Question 5 could be viewed as an elliptic curve analogue of the classical twin prime conjecture: we want to count primes p for which $p + (1 - a_p)$ is also a prime. As in the classical case, the densities $\approx 1/k$ of the primes p for which $k \mid \#E_p(\mathbb{F}_p)$ will direct us to serious difficulties in trying to estimate (6.1). Nevertheless, we recall that an ingenious use of the lower bound sieve allowed J. Chen to prove in 1973 that there are infinitely many primes p such that $p + 2$ has at most 2 prime divisors. In 2001 [**25**], S. A. Miri and V. K. Murty carried out an elliptic curve version of Chen's method and showed:

THEOREM 5 (Miri and Murty [**25**]). *Let E be a non-CM elliptic curve defined over $\mathbb{Q}$ and of conductor N. Assume that GRH holds for the Dedekind zeta functions of the division fields of E. Then there exist*

$$\gg_N \frac{x}{(\log x)^2}$$

primes $p \leq x, p \nmid N$, for which $\#E_p(\mathbb{F}_p)$ has at most 16 prime divisors (counted with multiplicities).

A similar result holds in the CM case, *unconditionally*, as will be explained in [**7**].

7. Distribution of the Frobenius fields

In our discussions of Questions 2–5 we were more successful in the CM case than in the non-CM case. One of the features of the CM curves which contributed to our success concerns the imaginary quadratic fields $\mathbb{Q}(\pi_p)$ that were introduced in Section 1. More precisely, this special feature is related to the following question:

QUESTION 6. Let E be an elliptic curve defined over $\mathbb{Q}$ and of conductor N. Let K be an imaginary quadratic field. What is the asymptotic behavior of

$$\#\{p \leq x : p \nmid N, a_p \neq 0, \mathbb{Q}(\pi_p) = K\},$$

as $x \to \infty$?

It is not difficult to answer this question in the CM case. First we observe that $\mathbb{Q}(\pi_p) \subseteq \mathrm{End}_{\overline{\mathbb{F}}_p}(E_p) \otimes_{\mathbb{Z}} \mathbb{Q}$. Then, if E has CM by some field K and if p is an ordinary prime for E, we have $K = \mathrm{End}_{\overline{\mathbb{F}}_p}(E_p) \otimes_{\mathbb{Z}} \mathbb{Q}$. Thus we obtain:

THEOREM 6. *Let E be a CM elliptic curve defined over $\mathbb{Q}$ and of conductor N. Let K be the CM field of E. Then $\mathbb{Q}(\pi_p) = K$ for any prime $p \nmid N$ of ordinary reduction for E.*

To answer Question 6 in the non-CM case is far more difficult. It was conjectured by S. Lang and H. Trotter in 1976 [**23**] that if E is a non-CM curve, then, as $x \to \infty$,

$$\#\{p \leq x : p \nmid N, \mathbb{Q}(\pi_p) = K\} \sim C(E, K) \frac{x^{1/2}}{\log x}$$

for some positive constant $C(E, K)$ depending on E and K.

In 1981 [**32**], J.-P. Serre asserted that one could obtain nontrivial upper bounds for the above quantity, by using Chebotarev type arguments and Selberg's sieve. Recently, the author, É. Fouvry and M. R. Murty completed the proof of such results by using the much simpler square sieve (see [**9**]). To be more precise, if p is a prime such that $\mathbb{Q}(\pi_p) = K$ for some imaginary quadratic field K of discriminant Δ, then $\Delta(a_p^2 - 4p)$ must be equal to a square. The square sieve is a device that suggests how to obtain upper bounds for the number of squares in a finite (multi)set of positive integers. By applying this sieve to the sequence $\{\Delta(a_p^2 - 4p)\}_{p \leq x}$, and by invoking effective versions of the Chebotarev density theorem, we can prove:

THEOREM 7 (Cojocaru, Fouvry, and Murty [**9**]). *Let E be a non-CM elliptic curve defined over $\mathbb{Q}$ and of conductor N. Let $K = \mathbb{Q}(\sqrt{-D})$ be an imaginary quadratic field, where D is positive and square-free. Let $x \geq 3$.*

(1) *Unconditionally, we have that*

$$\#\{p \le x : p \nmid N, \mathbb{Q}(\pi_p) = K\} \ll_N \frac{x(\log\log x)^{13/12}}{(\log x)^{25/24}}(1+\nu(D)).$$

If $\mathcal{D}_E(x)$ *denotes the set of (distinct) square-free parts of* $4p - a_p^2$ *for primes* $p \le x$ *of ordinary reduction for* E, *then there exists* $x_0 = x_0(N)$ *such that, for any* $x \ge x_0$,

$$\#\big(\mathcal{D}_E(\infty) \cap [1,x]\big) \gg_N \log\log\log x.$$

In particular, there are infinitely many distinct fields $\mathbb{Q}(\pi_p)$.

(2) *If we assume* GRH *for the Dedekind zeta functions of the division fields of* E, *then*

$$\#\{p \le x : p \nmid N, \mathbb{Q}(\pi_p) = K\} \ll_N x^{17/18}\log x$$

and

$$\#\mathcal{D}_E(x) \gg_N \frac{x^{1/18}}{(\log x)^2}.$$

We note that additional improvements of the above bounds can be obtained by assuming the validity of Artin's holomorphy conjecture and of a pair correlation conjecture (see [**9**]).

8. Distribution of the a_p's

Now let us investigate the integers a_p. More precisely, let us consider:

QUESTION 7. Let E be an elliptic curve defined over $\mathbb{Q}$ and of conductor N. Let a be a fixed integer. What is the asymptotic behavior of

$$\#\{p \le x : p \nmid N, a_p = a\},$$

as $x \to \infty$?

For example, consider a CM elliptic curve E with CM by $\mathbb{Q}(i)$ and try to count the primes p for which $a_p = 2$. By Theorem 6 and (1.4)), this implies that $p = n^2 + 1$ for some integer n. Thus our question can be viewed as an elliptic curve analogue of the classical Buniakowski–Schinzel hypothesis about the prime values of the polynomial $n^2 + 1$.

In 1976 [**23**], S. Lang and H. Trotter conjectured that there exists a positive constant $C(E, a)$, depending on E and a, such that, as $x \to \infty$,

$$\#\{p \le x : p \nmid N, a_p = a\} \sim C(E,a)\frac{x^{1/2}}{\log x}.$$

The only exception to this conjectural rule occurs in the case $a = 0$ and E a CM curve, about which we know:

THEOREM 8 (Deuring [**11**]). *Let* E *be a* CM *elliptic curve defined over* $\mathbb{Q}$ *and of conductor* N. *Then, with finitely many exceptions, a prime* $p \nmid N$ *is supersingular if and only if it is inert in the* CM *field of* E, *and so, as* $x \to \infty$,

$$\#\{p \le x : p \nmid N, a_p = 0\} \sim \frac{1}{2} \cdot \frac{x}{\log x}.$$

If $a \neq 0$ and E is CM, then the Lang–Trotter conjecture is consistent with classical conjectures of Hardy–Littlewood about the distribution of primes in quadratic extensions. If E is non-CM, the Lang–Trotter conjecture seems at least as difficult.

It is easy to obtain upper bounds for the number of primes p for which $a_p = a$ if E is a CM curve and $a \neq 0$. Indeed, in this case we can use (1.4) with $\Delta_p = \Delta$ for some Δ independent of p to deduce that

$$\#\{p \leq x : p \nmid N, a_p = a\} = \#\left\{p \leq x : p \nmid N, p = \left(\frac{a}{2}\right)^2 - \left(\frac{b_p}{2}\right)^2 \Delta\right\} \ll x^{1/2}.$$

By using sieve theory (e.g., Brun's sieve), one could actually improve this bound to obtain:

THEOREM 9. *Let E be a* CM *elliptic curve defined over $\mathbb{Q}$ and of conductor N. Let a be a nonzero arbitrary integer. Then*

$$\#\{p \leq x : p \nmid N, a_p = a\} \ll \frac{x^{1/2}}{\log x}.$$

The non-CM situation is not that easy anymore. The first results in this case were obtained by Serre [**32**] (by invoking Chebotarev type arguments) and by V. K. Murty [**30**] (by appealing to different arguments from Serre's). More precisely, we have:

THEOREM 10. *Let E be a non-*CM *elliptic curve defined over $\mathbb{Q}$ and of conductor N. Let a be an arbitrary integer:*

(1) (Serre [**32**]). *If $a \neq 0, \pm 2$, then*

$$\#\{p \leq x : p \nmid N, a_p = a\} \ll_N \frac{x(\log\log x)^{2/3}(\log\log\log x)^{1/3}}{(\log x)^{4/3}}.$$

If $a = \pm 2$, then

$$\#\{p \leq x : p \nmid N, a_p = a\} \ll_N \frac{x(\log\log x)^{1/2}(\log\log\log x)^{1/4}}{(\log x)^{5/4}}.$$

If $a = 0$, then

$$\#\{p \leq x : p \nmid N, a_p = 0\} \ll_N \frac{x(\log\log x)(\log\log\log x)^{1/2}}{(\log x)^{3/2}}.$$

(2) (Serre, [**32**]). *Assume* GRH *for the Dedekind zeta functions of the division fields of E. If $a \neq 0, \pm 2$, then*

$$\#\{p \leq x : p \nmid N, a_p = a\} \ll_N \frac{x^{5/6}}{(\log x)^{1/3}}.$$

If $a = \pm 2$, then

$$\#\{p \leq x : p \nmid N, a_p = a\} \ll_N \frac{x^{7/8}}{(\log x)^{1/2}}.$$

If $a = 0$, then

$$\#\{p \leq x : p \nmid N, a_p = 0\} \ll_N x^{3/4}.$$

(3) (Murty [**30**]). *Assume that we have analytic continuation, functional equation and a Riemann Hypothesis for the L-series attached to the symmetric powers of the Galois representations of E (for definitions, see* [**30**]). *Then, as* $x \to \infty$,

$$\#\{p \leq x : p \nmid N, a_p = a\} \ll_N x^{3/4}(\log x)^{1/2}.$$

It was also noted by N. Elkies and M. R. Murty (see [**14**, pp. 25–26] and [**16**]) that the upper bound $O_N(x^{3/4})$ for the number of supersingular primes $\leq x$ can be obtained *unconditionally* by using a result of M. Kaneko [**20**].

The question of finding lower bounds is even more difficult. The only results known at the moment concern supersingular primes.

THEOREM 11. *Let E be a non-*CM *elliptic curve defined over* $\mathbb{Q}$ *and of conductor N:*

(1) (Elkies [**15**]). *There are infinitely many primes p such that* $a_p = 0$. *Moreover, if* GRH *holds for the Dedekind zeta function of any quadratic extension of* $\mathbb{Q}$, *then, as* $x \to \infty$,

$$\#\{p \leq x : p \nmid N, a_p = 0\} \gg \log\log x.$$

(2) (Fouvry and Murty [**17**]). *For any* $\delta > 0$, *there exists a positive real number* $x_0(E, \delta)$ *such that*

$$\#\{p \leq x : p \nmid N, a_p = 0\} \geq \frac{\log\log\log x}{(\log\log\log\log x)^{1+\delta}}$$

for any $x > x_0(E, \delta)$.

9. Distribution of the b_p's

Finally, let us turn our attention to the integers b_p appearing in (1.4). It is clear from the definition that they are natural companions of the integers a_p. It is natural to ask:

QUESTION 8. Let E be an elliptic curve defined over $\mathbb{Q}$ and of conductor N. Let b be a fixed positive integer. What is the asymptotic behavior of

$$\#\{p \leq x : p \nmid bN, b_p = b\},$$

as $x \to \infty$?

The case $b = 1$ of this question was investigated by the author and W. Duke in [**8**]. The starting point in these investigations is to put the integers a_p and b_p in a more convenient context. This is accomplished by realizing that the matrix

$$\begin{pmatrix} (a_p + b_p\delta_p)/2 & b_p \\ (b_p(\Delta_p - \delta_p))/4 & (a_p - b_p\delta_p)/2 \end{pmatrix},$$

reduced modulo an integer k such that $p \nmid kN$, represents the Frobenius at p in the k-division field $\mathbb{Q}(E[k])$ (see [**13**]). Consequently, we have:

LEMMA 5. *Let E be an elliptic curve defined over* $\mathbb{Q}$ *and of conductor N. Let p be a prime and k a positive integer such that* $p \nmid kN$. *Then* $k \mid b_p$ *if and only if p splits completely in the subfield* J_k *of* $\mathbb{Q}(E[k])$ *fixed by the scalars of* $\mathrm{Gal}(\mathbb{Q}(E[k])/\mathbb{Q})$.

Using this result we can then write

$$(9.1)\quad \#\{p \le x : p \nmid bN, b_p = b\} = \sum_{k \le 2\sqrt{x}} \mu(k) \#\{p \le x : p \nmid bN, p \text{ splits completely in } J_{bk}\}.$$

As in our discussions about the distributions of the d_p's and e_p's, we can apply the Chebotarev density theorem to estimate (9.1), but, again, this will be possible only in a short range of the indices k. The challenging part will be to estimate the remaining sum. We point out that, unlike the full k-division field, J_k does not contain the cyclotomic field $\mathbb{Q}(\zeta_k)$. Hence the Brun–Titchmarsh theorem that we usually rely on cannot be invoked anymore.

Surprisingly, if E is a CM elliptic curve, then (9.1) can be handled fairly simply, as follows. First we distinguish between ordinary and supersingular primes p. Then we note that the ordinary primes contribute very little to our estimates (and this does not happen anymore if E is non-CM!). Indeed, by Lemma 5, we get that

$$\#\{p \le x : p \nmid bN, a_p \ne 0, b_p = b\} \le \#\left\{p \le x : a_p \ne 0, p = \left(\frac{a_p}{2}\right)^2 - \frac{b^2\Delta}{4}\right\} \ll x^{1/2},$$

where we recall that $\Delta = \Delta_p$ is independent of p in the ordinary case. The main contribution to our final estimates will be given by the supersingular primes of E. However, we note that for supersingular primes p such that $b_p = b$ we have $-4p = b^2\Delta_p$, hence b can only be 1 or 2. Thus, if $b \ge 3$, then

$$\#\{p \le x : p \nmid bN, a_p = 0, b_p = b\} = 0,$$

and if $b = 1$ or 2, then, by Theorem 8 and Lemma 5,

$$\#\{p \le x : p \nmid bN, a_p = 0, b_p = 1\} = \#\{p \le x : p \nmid N, p \text{ is inert in } K \text{ and } J_2\},$$
$$\#\{p \le x : p \nmid bN, a_p = 0, b_p = 2\} = \#\{p \le x : p \nmid 2N, p \text{ is inert in } K \text{ and splits in } J_2\}.$$

We finish off by invoking the unconditional effective Chebotarev density theorem for the fields K and J_2 (see part 1 of Lemma 3), and we obtain:

THEOREM 12 (Cojocaru and Duke [**8**]). *Let E be a* CM *elliptic curve defined over $\mathbb{Q}$ and of conductor N. Let b be a positive integer. Then*

$$\#\{p \le x : p \nmid bN, b_p = b\} = C(E)\,\mathrm{li}\, x + \mathrm{error}(x),$$

where

$$C(E) = \begin{cases} \frac{1}{2} - 1/[J_2{:}\mathbb{Q}] + 1/[J_2K{:}\mathbb{Q}] & \text{if } b = 1 \\ 1/[J_2{:}\mathbb{Q}] - 1/[J_2K{:}\mathbb{Q}] & \text{if } b = 2 \\ 0 & \text{if } b \ge 3 \end{cases}$$

and

$$\mathrm{error}(x) = \begin{cases} \mathrm{O}_N(x/(\log x)^B) & \text{if } b = 1 \text{ or } b = 2 \\ \mathrm{O}(x^{1/2}) & \text{if } b \ge 3, \end{cases}$$

for any $B > 0$.

Let us note that, as with Theorem 9, by using sieve theory the error term $O(x^{1/2})$ can be improved to $O(x^{1/2}/\log x)$.

Clearly, the above proof cannot be emulated if E is a non-CM elliptic curve. In this situation, the key observation for handling the ordinary primes p that split completely in J_{bk} for (large) indices k is that $k^2 \mid 4p - a_p^2$, and so $m(4p - a_p^2)$ is a square for some (small) positive integer m. We can now invoke the square sieve that we already appealed to in our approach on Question 6. We obtain:

THEOREM 13 (Cojocaru and Duke [**8**]). *Let E be a non-CM elliptic curve defined over $\mathbb{Q}$ and of conductor N. Let b be a positive integer. Assume that GRH holds for the division fields of E. Then*

$$\#\{p \leq x : p \nmid bN, b_p = b\} = \sum_{k \geq 1} \frac{\mu(k)}{[J_{bk}:\mathbb{Q}]} \operatorname{li} x + O_{b,N,\varepsilon}(x^{53/54+\varepsilon})$$

for any $0 < \varepsilon < 1$.

For the details of the proof in the case $b = 1$, we refer the reader to [**8**]. The general case is dealt with the same, mutatis mutandis.

We remark that the integers b_p^2 can be interpreted as the orders of the Tate–Shafarevich groups of the curves E_p viewed as constant curves defined over their own function fields (see the explanations in [**8**]). This may lead to new interesting meanings of the above two results. We also remark that if $b_p = 1$, then $d_p = 1$ (i.e., the group $E_p(\mathbb{F}_p)$ is cyclic), and so Question 8 can be viewed as a refinement of Question 2 which we discussed at the beginning of the paper.

10. Concluding remarks

As already suggested in our discussions above, there are many questions about $E_p(\mathbb{F}_p)$ which remain unanswered. Among them, the most notable ones are the Lang–Trotter conjectures and the Koblitz conjecture. Other questions of interest concern the positivity of the asymptotic constants that occur in some of our results (e.g., Theorems 2, 4, and 13), and upper bounds for the smallest prime p for which $E_p(\mathbb{F}_p)$ has a given property. These questions will be discussed in a forthcoming paper.

Most of the results that we discussed have (certain) generalizations to elliptic curves defined over number fields. A natural direction of research is to study if similar results can be obtained for elliptic curves defined over function fields. Even more generally, one can study analogous questions about reductions of abelian varieties and Drinfeld modules, even though the formulation of these questions may not be straightforward.

Finally, we remark that there is a vast literature concerning reductions modulo primes of an elliptic curve. In our present survey, many important questions, such as the Sato–Tate conjecture, had to be left out. The only criterion used in selecting the aspects of $E_p(\mathbb{F}_p)$ that we discussed was the author's own current research.

References

1. I. Borosh, C. J. Moreno, and H. Porta, *Elliptic curves over finite fields*. II, Math. Comput. **29** (1975), 951–964.
2. A. C. Cojocaru, *On the cyclicity of the group of $\mathbb{F}_p$-rational points of non-CM elliptic curves*, J. Number Theory **96** (2002), 335–350.

3. ———, *Cyclicity of elliptic curves modulo p*, Ph.D. thesis, Queen's University, Kingston, 2002.
4. ———, *Square-free orders for* CM *elliptic curves modulo p*, Preprint, 2002.
5. ———, *Cyclicity of* CM *elliptic curves modulo p*, Trans. Amer. Math. Soc. **355** (2003), no. 7, 2651–2662.
6. ———, *On the surjectivity of the Galois representations associated to non-*CM *elliptic curves*, Canad. Math. Bull. (to appear).
7. ———, *Bounded number of prime factors for the orders of the reductions of a* CM *elliptic curve*, Preprint, 2004.
8. A. C. Cojocaru and W. Duke, *Reductions of an elliptic curve and their Tate–Shafarevich groups*, Math. Ann. (to appear).
9. A. C. Cojocaru, É. Fouvry, and M. R. Murty, *The square sieve and the Lang–Trotter conjecture*, Preprint, 2001.
10. A. C. Cojocaru and M. R. Murty, *Cyclicity of elliptic curves modulo p and elliptic curve analogues of Linnik's problem*, Canad. J. Math. (to appear).
11. M. Deuring, *Die Typen der Multiplikatorenringe elliptischer Functionenkörper*, Abh. Math. Sem. Hansischen Univ. **14** (1941), 197–272.
12. ———, *Teilbarkeitseigenschatten der singulären Moduln der elliptischen Funktionen und die Diskriminante der Klassengleichung*, Comment. Math. Helv. **19** (1946), 74–82.
13. W. Duke and Á. Tóth, *The splitting of primes in division fields of elliptic curves*, Experiment. Math. **11** (2003), no. 4, 555–565.
14. N. D. Elkies, *Supersingular primes of a given elliptic curve over a number field*, Ph.D. thesis, Harvard University, Cambridge, MA, 1987.
15. ———, *The existence of infinitely many supersingular primes for every elliptic curve over* $\mathbb{Q}$, Invent. Math. **89** (1987), no. 3, 561–567.
16. ———, *Distribution of supersingular primes*, Journées Arithmétiques, 1989 (Luminy, 1989), Astérisque, vol. 198–200, Soc. Math. France, Paris, 1991, pp. 127–132.
17. É. Fouvry and M. R. Murty, *On the distribution of supersingular primes*, Canad. J. Math. **48** (1996), no. 1, 81–104.
18. R. Gupta and M. R. Murty, *Primitive points on elliptic curves*, Compositio Math. **58** (1986), no. 1, 13–44.
19. ———, *Cyclicity and generation of points modulo p on elliptic curves*, Invent. Math. **101** (1990), no. 1, 225–235.
20. M. Kaneko, *Supersingular j-invariants* mod p, Osaka J. Math. **26** (1989), no. 4, 849–855.
21. N. Koblitz, *Primality of the number of points on an elliptic curve over a finite field*, Pacific J. Math. **131** (1988), no. 1, 157–165.
22. J. Lagarias and A. Odlyzko, *Effective versions of the Chebotarev density theorem*, Algebraic Number Fields: L-functions and Galois Properties (Durham, 1975), Academic Press, London, 1977, pp. 409–464.
23. S. Lang and H. Trotter, *Frobenius distributions in* GL_2*-extensions*, Lecture Notes in Math., vol. 504, Springer-Verlag, Berlin–New York, 1976.
24. ———, *Primitive points on elliptic curves*, Bull. Amer. Math. Soc. **83** (1977), no. 2, 289–292.
25. S. A. Miri and V. K. Murty, *An application of sieve methods to elliptic curves*, Progress in Cryptology—INDOCRYPT 2001 (Chennai, 2001), Lecture Notes in Comput. Sci., vol. 2247, Springer, Berlin, 2001, pp. 91–98.
26. M. R. Murty, *On Artin's conjecture*, J. Number Theory **16** (1983), no. 2, 147–168.
27. ———, *On the supersingular reduction of elliptic curves*, Proc. Indian Acad. Sci. Math. Sci. **97** (1987), no. 1-3, 247–250.
28. ———, *Recent developments in elliptic curves*, Proc. of the Ramanujan Centennial International Conference (Annamalainagar, 1987), RMS Publ., vol. 1, Ramanujan Math. Soc., Annamalainagar, 1988, pp. 45–53.
29. ———, *Artin's conjecture and elliptic analogues*, Sieve Methods, Exponential Sums and their Applications in Number Theory (Cardiff, 1995), London Math. Soc. Lecture Note Ser., vol. 237, Cambridge Univ. Press, Cambridge, 1997, pp. 26–344.
30. V. K. Murty, *Explicit formulae and the Lang–Trotter conjecture*, Rocky Mountain J. Math. **15** (1985), no. 2, 535–551.
31. J.-P. Serre, *Résumé des cours de* 1977–1978, Ann. Collège France, Collège de France, Paris, 1978, pp. 67–70.

32. ———, *Quelques applications du théorème de densité de Chebotarev*, Inst. Hautes Études Sci. Publ. Math. **54** (1981), pp. 123–201.
33. J. H. Silverman, *The arithmetic of elliptic curves*, Grad. Texts in Math., vol. 106, Springer-Verlag, New York, 1986.
34. ———, *Advanced topics in the arithmetic of elliptic curves*, Grad. Texts in Math., vol. 151, Springer-Verlag, New York, 1994.
35. R. Takeuchi, *On the distribution of the group of rational points of reductions of an elliptic curve*, Preprint, 2003.

DEPARTMENT OF MATHEMATICS, PRINCETON UNIVERSITY, FINE HALL, WASHINGTON ROAD, PRINCETON, NJ 08544, USA.
E-mail address: cojocaru@math.princeton.edu

Centre de Recherches Mathématiques
CRM Proceedings and Lecture Notes
Volume **36**, 2004

New Perspectives in Arakelov Geometry

Caterina Consani and Matilde Marcolli

Abstract. In this paper we give a unified description of the Archimedean and the totally split degenerate fibers of an arithmetic surface, using operator algebras and Connes' theory of spectral triples in noncommutative geometry.

1. Introduction

The aim of this article is to report on some of the recent results obtained by the two authors on a noncommutative interpretation of the totally split degenerate fibers of an arithmetic surface. The results stated in the first part of the paper were announced in a talk given by the first author at the VII meeting of the CNTA in May 2002 at Montreal. The material presented in this note is based on the papers [**11, 12**].

Let $\mathcal{X}$ be an arithmetic surface defined over $\operatorname{Spec}(\mathbb{Z})$ (or over $\operatorname{Spec}(\mathcal{O}_{\mathbb{K}})$, for a number field $\mathbb{K}$), having the smooth algebraic curve $X_{/\mathbb{Q}}$ as its generic fiber. It is well known that, as a Riemann surface, $X_{/\mathbb{C}}$ admits always a uniformization by means of a Schottky group Γ. In fact, the presence of this uniformization plays a fundamental role in the theory of the "fiber at infinity" of $\mathcal{X}$ described in [**11, 22**]. In analogy to Mumford's p-adic uniformization of algebraic curves (cf. [**26**]), the Riemann surface $X_{/\mathbb{C}}$ can be interpreted as the boundary at infinity of a 3-manifold $\mathfrak{X}_\Gamma$ defined as the quotient of the real hyperbolic 3-space $\mathbb{H}^3$ by the action of the Schottky group Γ. The space $\mathfrak{X}_\Gamma$ contains in its interior an infinite link of bounded geodesics. Manin gave an expression for the Arakelov Green function on $X_{/\mathbb{C}}$ in terms of configurations of geodesics in $\mathfrak{X}_\Gamma$, thus interpreting this tangle as the dual graph $\mathcal{G}$ of the "closed fiber at infinity" of $\mathcal{X}$.

In the first part of this paper we concentrate on Manin's description of such dual graph and we exhibit the suspension flow $\mathcal{S}_T$ of a dynamical system T on the limit set of the Schottky group Γ as a geometric model for the dual graph $\mathcal{G}$.

2000 *Mathematics Subject Classification.* 58B34, 14G40, 46L55, 11G30.

This paper was partly written during visits of the second author to Florida State University and University of Toronto. The authors thank these institutions for the hospitality.

The first author is partially supported by NSERC grant 72016789.

The second author is partially supported by the Humboldt Foundation and the German Government (Sofja Kovalevskaya Award).

The authors thank Alain Connes for many extremely helpful discussions.

This is the final form of the paper.

In particular, the first cohomology group of $\mathcal{S}_T$ determines a model of the first cohomology of the dual graph of the "fiber at infinity". Furthermore, the first (co)homology group of $\mathcal{S}_T$ carries a natural filtration.

A crucial feature of this construction is the fact (proved in [**11**, Section 5]) that this dynamical cohomology contains a subspace isomorphic to the Archimedean cohomology of [**10**]: the group of invariants for the action of the local monodromy "at infinity". This space has also the correct geometric properties, in order to be interpreted in terms of a cohomology theory associated to a degeneration in a "neighborhood of arithmetic infinity". Our result identifies such space with a distinguished subspace of the cohomology of a topological space constructed in terms of geodesics in $\mathfrak{X}_\Gamma$, our geometric model of the dual graph. Under this identification, the graded structure associated to the filtration on the (co)homology of $\mathcal{S}_T$ corresponds to the graded structure given by Tate twists on the Archimedean cohomology of [**10**].

The Cuntz–Krieger algebra $\mathcal{O}_A$ associated to the shift T describes the "ring of functions" on a noncommutative space, which represents the quotient of the limit set Λ_Γ of the Schottky group, by the action of Γ. In terms of the geometry of the fiber at arithmetic infinity, this space can be thought of as the set of components of the special fiber, or equivalently the vertices of the dual graph $\mathcal{G}$, whereas the quotient $\Lambda_\Gamma \times_\Gamma \Lambda_\Gamma$ gives the edges of $\mathcal{G}$. The algebra $\mathcal{O}_A$ carries a refined information on the action of the Schottky group Γ on its limit set. In particular, we construct a *spectral triple* for this algebra.

In noncommutative geometry, the notion of a spectral triple provides the correct generalization of the classical structure of a Riemannian manifold. The two notions agree on a commutative space. In the usual context of Riemannian geometry, the definition of the infinitesimal element ds on a smooth spin manifold can be expressed in terms of the inverse of the classical Dirac operator D. This is the key remark that motivates the theory of spectral triples. In particular, the geodesic distance between two points on the manifold is defined in terms of D^{-1} (cf. [**8**, Section VI]). The spectral triple that describes a classical Riemannian spin manifold is (A, H, D), where A is the algebra of complex valued smooth functions on the manifold, H is the Hilbert space of square integrable spinor sections, and D is the classical Dirac operator (a square root of the Laplacian). These data determine completely and uniquely the Riemannian geometry on the manifold. It turns out that, when expressed in this form, the notion of spectral triple extends to more general noncommutative spaces, where the data (A, H, D) consist of a C^*-algebra A (or more generally of a smooth subalgebra of a C^*-algebra) with a representation as bounded operators on a Hilbert space H, and an operator D on H that verifies the main properties of a Dirac operator. The notion of smoothness is determined by D: the smooth elements of A are defined by the intersection of domains of powers of the derivation given by commutator with $|D|$. The basic geometric structure encoded by the theory of spectral triples is Riemannian geometry, but in more refined cases, such as Kähler geometry, the additional structure can be easily encoded as additional symmetries.

In our constructions, the Dirac operator D is obtained from the grading operator associated to a filtration on the cochains of the complex that computes the dynamical cohomology. The induced operator on the subspace identified with the Archimedean cohomology agrees with the "logarithm of Frobenius" of [**10, 15**].

This structure further enriches the geometric interpretation of the Archimedean cohomology, giving it the meaning of spinors on a noncommutative manifold, with the logarithm of Frobenius introduced in [**15**] in the role of the Dirac operator.

An advantage of this construction is that a completely analogous formulation exists in the case of Mumford curves. This provides a unified description of the Archimedean and totally split degenerate fibers of an arithmetic surface.

Let p be a finite prime where $\mathcal{X}$ has totally split degenerate reduction. Then, the completion $\widehat{X}_p$ at p of the generic fiber of $\mathcal{X}$ is a split-degenerate stable curve over $\mathbb{Q}_p$ (also called a Mumford curve) uniformized by the action of a p-adic Schottky group Γ. The dual graph of the reduction of $\widehat{X}_p$ coincides with a finite graph obtained as the quotient of a tree Δ_Γ by the action of Γ.

The curve $\widehat{X}_p$ is holomorphically isomorphic to a quotient of a subset of the ends of the Bruhat–Tits tree Δ of $\mathbb{Q}_p$ by the action of Γ. Thus, in this setting, the Bruhat–Tits tree at p replaces the hyperbolic space $\mathbb{H}^3$ "at infinity", and the analog of the tangle of bounded geodesics in $\mathfrak{X}_\Gamma$ is played by doubly infinite walks in Δ_Γ/Γ.

In analogy with the Archimedean construction, we define the system

$$(\mathcal{W}(\Delta_\Gamma/\Gamma), T)$$

where T is an invertible shift map on the set $\mathcal{W}(\Delta_\Gamma/\Gamma)$ of doubly-infinite walks on the graph Δ_Γ/Γ. The first cohomology group $H^1(\mathcal{W}(\Delta_\Gamma/\Gamma)_T, \mathbb{Z})$ of the mapping torus $\mathcal{W}(\Delta_\Gamma/\Gamma)_T$ of T inherits a natural filtration using which we introduce a dynamical cohomology group. We again have a Cuntz–Krieger graph algebra $C^*(\Delta_\Gamma/\Gamma)$ and we can construct a spectral triple as in the case at infinity, where again the Dirac operator is related to the grading operator Φ that computes the local factor as a regularized determinant, as in [**16, 17**]. In [**12**], we also suggested a possible way of extending such construction to places that are not of split degenerate reduction, inspired by the "foam space" construction of [**5, 21**].

2. Notation

Throughout this paper we will denote by K one among the following fields:

(a) the complex numbers $\mathbb{C}$,

(b) a finite extension of $\mathbb{Q}_p$.

When (b) occurs, we write $\mathcal{O}_K$ for the ring of integers of K, $\mathfrak{m} \subset \mathcal{O}_K$ for the maximal ideal and $\pi \in \mathfrak{m}$ for a uniformizer (i.e., $\mathfrak{m} = (\pi)$). We also denote by k the residue classes field $k = \mathcal{O}/\mathfrak{m}$.

We denote by $\mathbb{H}^3$ the three-dimensional real hyperbolic space i.e., the quotient

$$\mathbb{H}^3 = \mathrm{SU}(2)\backslash \mathrm{PGL}(2, \mathbb{C}).$$

This space can also be described as the upper half space $\mathbb{H}^3 \simeq \mathbb{C} \times \mathbb{R}^+$ endowed with the hyperbolic metric. The group $\mathrm{PSL}(2, \mathbb{C})$ acts on $\mathbb{H}^3$ by isometries. The complex projective line $\mathbb{P}^1(\mathbb{C})$ is identified with the conformal boundary at infinity of $\mathbb{H}^3$ and the action of $\mathrm{PSL}(2, \mathbb{C})$ on $\mathbb{H}^3$ extends to an action on $\overline{\mathbb{H}}^3 := \mathbb{H}^3 \cup \mathbb{P}^1(\mathbb{C})$. The group $\mathrm{PSL}(2, \mathbb{C})$ acts on $\mathbb{P}^1(\mathbb{C})$ by fractional linear transformations.

For an integer $g \geq 1$, a Schottky group of rank g is a discrete subgroup $\Gamma \subset \mathrm{PSL}(2, K)$, which is purely loxodromic and isomorphic to a free group of rank g. We denote by Λ_Γ the limit set of the action of Γ. One sees that Λ_Γ is contained in $\mathbb{P}^1(K)$. This set can also be described as the closure of the set of the attractive and repelling fixed points $z^{\pm}(g)$ of the loxodromic elements $g \in \Gamma$. In the case $g = 1$ the limit set consists of two points, but for $g \geq 2$ the limit set is usually a fractal of some Hausdorff dimension $0 \leq \delta = \dim_H(\Lambda_\Gamma) < 2$. We denote by $\Omega_\Gamma = \Omega_\Gamma(K)$ the domain of discontinuity of Γ, that is, the complement of Λ_Γ in $\mathbb{P}^1(K)$.

When $K = \mathbb{C}$, the quotient space $\mathfrak{X}_\Gamma := \mathbb{H}^3/\Gamma$ is topologically a handlebody of genus g, and the quotient $X_{/\mathbb{C}} = \Omega_\Gamma/\Gamma$ is a Riemann surface of genus g. The covering $\Omega_\Gamma \to X_{/\mathbb{C}}$ is called a Schottky uniformization of $X_{/\mathbb{C}}$. Every complex Riemann surface $X_{/\mathbb{C}}$ admits a Schottky uniformization. The handlebody $\mathfrak{X}_\Gamma$ can be compactified by adding the conformal boundary at infinity $X_{/\mathbb{C}}$ to obtain $\overline{\mathfrak{X}}_\Gamma := \mathfrak{X}_\Gamma \cup X_{/\mathbb{C}} = (\mathbb{H}^3 \cup \Omega_\Gamma)/\Gamma$.

A directed graph E consists of data $E = (E^0, E^1, E^1_+, r, s, \iota)$, where E^0 is the set of vertices, E^1 is the set of oriented edges $w = \{e, \epsilon\}$, where e is an edge of the graph and $\epsilon = \pm 1$ is a choice of orientation. The set E^1_+ consists of a choice of orientation for each edge, namely one element in each pair $\{e, \pm 1\}$. The maps $r, s\colon E^1 \to E^0$ are the range and source maps, and ι is the involution on E^1 defined by $\iota(w) = \{e, -\epsilon\}$.

A directed graph is finite if E^0 and E^1 are finite sets. It is locally finite if each vertex emits and receives at most finitely many oriented edges in E^1. A vertex v in a directed graph is a sink if there is no edge in E^1_+ with source v. A juxtaposition of oriented edges $w_1 w_2$ is said to be admissible if $w_2 \neq \iota(w_1)$ and $r(w_1) = s(w_2)$. A (finite, infinite, doubly infinite) walk in a directed graph E is an admissible (finite, infinite, doubly infinite) sequence of elements in E^1.

We denote by $\mathcal{W}^n(E)$ the set of walks of length n, by $\mathcal{W}^*(E) = \bigcup_n \mathcal{W}^n(E)$, by $\mathcal{W}^+(E)$ the set of infinite walks, and by $\mathcal{W}(E)$ the set of doubly infinite walks. A directed graph is a directed tree if, for any two vertices, there exists a unique walk in $\mathcal{W}^*(E)$ connecting them.

The edge matrix A_+ of a locally finite (or row finite) directed graph is an $(\#E^1_+) \times (\#E^1_+)$ (possibly infinite) matrix. The entries satisfy $A_+(w_i, w_j) = 1$ if $w_i w_j$ is an admissible path, and $A_+(w_i, w_j) = 0$ otherwise. The directed edge matrix of E is a $\#E^1 \times \#E^1$ (possibly infinite) matrix with entries $A(w_i, w_j) = 1$ if $w_i w_j$ is an admissible walk and $A(w_i, w_j) = 0$ otherwise.

Even when not explicitly stated, all Hilbert spaces and algebras of operators we consider will be separable, i.e., they admit a dense (in the norm topology) countable subset.

3. A dynamical theory at infinity

Let $\Gamma \subset \mathrm{PSL}(2, \mathbb{C})$ be a Schottky group of rank $g \geq 2$. Given a choice of a set of generators $\{g_i\}_{i=1}^g$ for Γ, there is a bijection between the elements of Γ and the set of all admissible walks in the Cayley graph of Γ, namely reduced words in the $\{g_i\}_{i=1}^{2g}$, where we use the notation $g_{i+g} := g_i^{-1}$, for $i = 1, \ldots, g$.

In the following we consider the sets $\mathcal{S}^+$ and $\mathcal{S}$ of resp. right-infinite, doubly infinite admissible sequences in the $\{g_i\}_{i=1}^{2g}$:

$$\mathcal{S}^+ = \{a_0 a_1 \ldots a_\ell \ldots \mid a_i \in \{g_i\}_{i=1}^{2g}, a_{i+1} \neq a_i^{-1}, \forall i \in \mathbb{N}\}, \tag{3.1}$$

$$\mathcal{S} = \{\ldots a_{-m} \ldots a_{-1} a_0 a_1 \ldots a_\ell \ldots \mid a_i \in \{g_i\}_{i=1}^{2g}, a_{i+1} \neq a_i^{-1}, \forall i \in \mathbb{Z}\}. \tag{3.2}$$

The admissibility condition simply means that we only allow "reduced" words in the generators, without cancellations.

On the space $\mathcal{S}$ we consider the topology generated by the sets $W^s(x,\ell) = \{y \in \mathcal{S} \mid x_k = y_k, k \geq \ell\}$, and $W^u(x,\ell) = \{y \in \mathcal{S} \mid x_k = y_k, k \leq \ell\}$ for $x \in \mathcal{S}$ and $\ell \in \mathbb{Z}$.

There is a two-sided shift operator T acting on $\mathcal{S}$ as the map

$$T(\ldots a_{-m}\ldots a_{-1}a_0a_1\ldots a_\ell\ldots) = \ldots a_{-m+1}\ldots a_0a_1a_2\ldots a_{\ell+1}\ldots . \tag{3.3}$$

DEFINITION 3.1. A subshift of finite type $(\mathcal{S}_A, T)$ consists of all doubly infinite sequences in the elements of a given finite set W (alphabet) with the admissibility condition specified by a $\#W \times \#W$ elementary matrix,

$$\mathcal{S}_A = \{(w_k)_{k\in\mathbb{Z}} : w_k \in W, A(w_k, w_{k+1}) = 1\},$$

and with the action of the invertible shift $(Tw)_k = w_{k+1}$.

LEMMA 3.2. *The space $\mathcal{S}$ can be identified with the subshift of finite type $\mathcal{S}_A$ with the symmetric $2g\times 2g$ matrix A given by the directed edge matrix of the Cayley graph of Γ.*

The two-sided shift operator T on $\mathcal{S}$ of (3.3) decomposes $\mathcal{S}$ in a product of expanding and contracting directions, so that $(\mathcal{S}, T)$ is a Smale space.

The following topological space is defined in terms of the Smale space $(\mathcal{S}, T)$ and will be considered as a geometric realization of the "dual graph" associated to the fiber at arithmetic infinity of the arithmetic surface $\mathcal{X}$.

DEFINITION 3.3. The mapping torus (suspension flow) of the dynamical system $(\mathcal{S}, T)$ is defined as

$$\mathcal{S}_T := \mathcal{S} \times [0,1]/(x,0) \sim (Tx, 1). \tag{3.4}$$

The space $\mathcal{S}_T$ is a very natural space associated to the noncommutative space

$$\Lambda_\Gamma \times_\Gamma \Lambda_\Gamma \simeq \mathcal{S}/\mathbb{Z}, \tag{3.5}$$

with $\mathbb{Z}$ acting via the invertible shift T of (3.3), namely the *homotopy quotient* (cf. [**3, 6**]),

$$\mathcal{S}_T = \mathcal{S} \times_{\mathbb{Z}} \mathbb{R}. \tag{3.6}$$

Namely, it is a commutative space that provides, up to homotopy, a geometric model for (3.5), where the noncommutative space (3.5) can be identified with the quotient space of a foliation (3.6) whose generic leaf is contractible (a copy of $\mathbb{R}$).

3.1. (Co)homology of $\mathcal{S}_T$. In this paragraph we give an explicit description of the (co)homology $H^1(\mathcal{S}_T, \mathbb{Z})$.

The shift T acting on $\mathcal{S}$ induces an automorphism of the C^*-algebra of continuous functions $C(\mathcal{S})$. With an abuse of notation we still denote it by T. Consider the crossed product C^*-algebra $C(\mathcal{S}) \rtimes_T \mathbb{Z}$. This is a suitable norm completion of $C(\mathcal{S})[T, T^{-1}]$ with product $(V * W)_k = \sum_{r\in\mathbb{Z}} V_k \cdot (T^r W_{r+k})$, for $V = \sum_k V_k T^k$, $W = \sum_k W_k T^k$, and $V * W = \sum_k (V*W)_k T^k$. The K-theory group $K_0(C(\mathcal{S}) \rtimes_T \mathbb{Z})$ is described by the co–invariants of the action of T (cf. [**4, 27**]).

THEOREM 3.4. *The cohomology $H^1(\mathcal{S}_T)$ satisfies the following properties*

(1) *There is an identification of $H^1(\mathcal{S}_T, \mathbb{Z})$ with the K_0-group of the crossed product C^*-algebra for the action of T on $\mathcal{S}$,*

$$H^1(\mathcal{S}_T, \mathbb{Z}) \cong K_0(C(\mathcal{S}) \rtimes_T \mathbb{Z}). \tag{3.7}$$

(2) *The identification* (3.7) *endows* $H^1(\mathcal{S}_T, \mathbb{Z})$ *with a filtration by free abelian groups* $F_0 \hookrightarrow F_1 \hookrightarrow \ldots \hookrightarrow F_n \hookrightarrow \cdots$, *with* $\operatorname{rank} F_0 = 2g$ *and* $\operatorname{rank} F_n = 2g(2g-1)^{n-1}(2g-2)+1$, *for* $n \geq 1$, *so that*

$$H^1(\mathcal{S}_T, \mathbb{Z}) = \varinjlim_n F_n.$$

(3) *This filtration is induced by a filtration* $\mathcal{P}_n$ *on the locally constant functions* $C(\mathcal{S}^+, \mathbb{Z})$ *which depend on "future coordinates", with* $\mathcal{P}_n$ *given by locally constant functions that depend only on the first* $n+1$ *coordinates. The Pimsner–Voiculescu exact sequence for the* K*-theory* (3.7) *is,*

$$0 \to \mathbb{Z} \to C(\mathcal{S}, \mathbb{Z}) \xrightarrow{\delta = 1 - T} C(\mathcal{S}, \mathbb{Z}) \to H^1(\mathcal{S}_T, \mathbb{Z}) \to 0. \tag{3.8}$$

(4) *The cohomology* $H^1(\mathcal{S}_T, \kappa) = H^1(\mathcal{S}_T, \mathbb{Z}) \otimes \kappa$, *for* $\kappa = \mathbb{R}$ *or* $\mathbb{C}$, *is computed by*

$$0 \to \mathbb{C} \to \mathcal{P} \otimes \kappa \xrightarrow{\delta} \mathcal{P} \otimes \kappa \to H^1(\mathcal{S}_T, \mathbb{C}) \to 0. \tag{3.9}$$

The vector space $\mathcal{P} \otimes \kappa$, *for* $\mathcal{P} = C(\mathcal{S}^+, \mathbb{Z}) = \varinjlim_n \mathcal{P}_n$, *admits a Hilbert space completion* $\mathcal{L} = L^2(\Lambda_\Gamma, \mu)$, *where* μ *is the Patterson–Sullivan measure on the limit set* Λ_Γ (*cf.* [**31**]) *satisfying*

$$(\gamma^* \, d\mu)(x) = |\gamma'(x)|^{\delta_H} \, d\mu(x), \quad \forall \gamma \in \Gamma. \tag{3.10}$$

Notice that the adjoint δ^* in the L^2-inner product of the coboundary δ of the cohomology $H^1(\mathcal{S}_T)$ is an important operator associated to the dynamics of the (one sided) shift T on the limit set Λ_Γ, namely, the operator $1 - \mathcal{R}$, where $\mathcal{R}$ is the Perron–Frobenius operator of T. This is the analog of the Gauss–Kuzmin operator studied in [**24, 25**] in the case of modular curves.

For simplicity of notation, in the following we will use the same notation $\mathcal{P}$ and $\mathcal{P}_n$ for the $\mathbb{Z}$-modules and for the κ-vector spaces. It will be clear from the context which one we refer to. We identify the vector spaces $\mathcal{P}_n$ with finite-dimensional subspaces of $L^2(\Lambda_\Gamma, \mu)$, by identifying locally constant functions on $\mathcal{S}^+$ with locally constant functions on Λ_Γ.

The following result computes the first homology of $\mathcal{S}_T$.

PROPOSITION 3.5. *The homology group* $H_1(\mathcal{S}_T, \mathbb{Z})$ *has a filtration by free abelian groups* $\mathcal{K}_N$,

$$H_1(\mathcal{S}_T, \mathbb{Z}) = \varinjlim_N \mathcal{K}_N, \tag{3.11}$$

with

$$K_N = \operatorname{rank}(\mathcal{K}_N) = \begin{cases} (2g-1)^N + 1 & N \text{ even} \\ (2g-1)^N + (2g-1) & N \text{ odd.} \end{cases}$$

The group $H_1(\mathcal{S}_T, \mathbb{Z})$ *can also be written as*

$$H_1(\mathcal{S}_T, \mathbb{Z}) = \bigoplus_{N=0}^{\infty} \mathcal{R}_N$$

where $\mathcal{R}_n$ *is a free abelian group of ranks* $R_1 = 2g$ *and*

$$R_N = \operatorname{rank}(\mathcal{R}_N) = \frac{1}{N} \sum_{d|N} \mu(d) \, (2g-1)^{N/d},$$

for $N > 1$, with μ the Möbius function. This is isomorphic to a free abelian group on countably many generators.

Thus, the $\mathbb{Z}$-modules F_n are obtained as quotients $\mathcal{P}_n/\delta\mathcal{P}_{n-1}$, for $\delta(f) = f - f\circ T$, where $\mathcal{P} = \bigcup_n \mathcal{P}_n$ is the module of continuous $\mathbb{Z}$-valued functions on $\mathcal{S}$ that depend only on "future coordinates". The $\mathbb{Z}$-modules $\mathcal{K}_N$ are generated by all admissible words $a_0\dots a_N$ such that the word $a_N a_0$ is also admissible. Combining Theorem 3.4 with Proposition 3.5, we can compute explicitly the pairing of homology and cohomology for $\mathcal{S}_T$.

PROPOSITION 3.6. *Let F_n and $\mathcal{K}_N$ be the filtrations defined, respectively, in Theorem 3.4 and Proposition 3.5. There is a pairing*

$$\langle\cdot,\cdot\rangle\colon F_n\times\mathcal{K}_N\to\mathbb{Z}\quad \langle[f],x\rangle = N\cdot f(\bar{x}), \tag{3.12}$$

with $x = a_0\dots a_N$. Here the representative $f\in[f]$ is a function that depends on the first $n+1$ terms $a_0\dots a_n$ of sequences in $\mathcal{S}$, and $\bar{x}$ is the truncation of the periodic sequence $\overline{a_0\dots a_N}$ after the first n terms. This pairing descends to the direct limits of the filtrations, where it agrees with the classical cohomology/homology pairing

$$\langle\cdot,\cdot\rangle\colon H^1(\mathcal{S}_T,\mathbb{Z})\times H_1(\mathcal{S}_T,\mathbb{Z})\to\mathbb{Z}. \tag{3.13}$$

3.2. Dynamical (co)homology. We define the dynamical cohomology H^1_{dyn} as the graded vector space given by the sum of the graded pieces of the filtration of $H^1(\mathcal{S}_T)$, introduced in Theorem 3.4. These graded pieces Gr_n are considered with coefficients in the nth Hodge–Tate twist $\mathbb{R}(n)$, for $n\in\mathbb{Z}$. Similarly, we define the dynamical homology H_1^{dyn} as the graded vector space given by the sum of the terms in the filtration of $H_1(\mathcal{S}_T)$, introduced in Proposition 3.5. These vector spaces are again considered with twisted $\mathbb{R}(n)$-coefficients. The pair

$$H^1_{\mathrm{dyn}}\oplus H_1^{\mathrm{dyn}} \tag{3.14}$$

provides a geometric setting, defined in terms of the dynamics of the shift operator T, which contains a copy of the Archimedean cohomology of [**10**] and of its dual.

DEFINITION 3.7. Let $H^1(\mathcal{S}_T,\kappa) = \lim_n F_n$, for a filtration F_n as in Theorem 3.4, with real or complex coefficients. Let $\mathrm{Gr}_n = F_n/F_{n-1}$ be the corresponding graded pieces, with $\mathrm{Gr}_0 = F_0$.

(1) We define a graded linear subspace $\mathcal{V}$ of the Hilbert space $\mathcal{L}$, as the span of the elements
$$\widehat{\Pi}_n\chi_{\mathcal{S}^+(w_{n,k})},$$
with $\chi_{\mathcal{S}^+(w_{n,k})}$ the characteristic function of $\mathcal{S}^+(w_{n,k})\subset\mathcal{S}^+$ with
$$w_{n,k} := a_0a_1\dots a_{n-1} = \underbrace{g_kg_k\dots g_k}_{n-times}.$$
The operator $\widehat{\Pi}_n$ is the projection $\widehat{\Pi}_n = \Pi_n - \Pi_{n-1}$, with Π_n the orthogonal projection of $\mathcal{L}$ onto $\mathcal{P}_n$.

(2) We define the dynamical cohomology as
$$H^1_{\mathrm{dyn}} := \bigoplus_{n\le 0}\mathrm{gr}^{\Gamma}_{2n} H^1_{\mathrm{dyn}}, \tag{3.15}$$

where we set

$$\mathrm{gr}^{\Gamma}_{2n} H^1_{\mathrm{dyn}} := \mathrm{Gr}_{-n} \otimes_{\mathbb{R}} \mathbb{R}(n) \tag{3.16}$$

with $\mathbb{R}(n) = (2\pi\sqrt{-1})^n \mathbb{R}$.

Furthermore, we define the graded subspace of H^1_{dyn}

$$\overline{\mathcal{V}} := \bigoplus_{n\le 0} \mathrm{gr}^{\Gamma}_{2n} \overline{\mathcal{V}},$$

where $\mathrm{gr}^{\Gamma}_{2n} \overline{\mathcal{V}}$ is generated by the elements $(2\pi\sqrt{-1})^n \chi_{-n+1,k}$, for $\chi_{n,k} := [\chi_{\mathcal{S}^+(w_{n,k})}] \in \mathrm{Gr}_{n-1}$.

(3) The dynamical homology H_1^{dyn} is defined as

$$H_1^{\mathrm{dyn}} := \bigoplus_{n\ge 1} \mathrm{gr}^{\Gamma}_{2n} H_1^{\mathrm{dyn}}, \tag{3.17}$$

where we set

$$gr^{\Gamma}_{2n} H_1^{\mathrm{dyn}} := \mathcal{K}_{n-1} \otimes_{\mathbb{R}} \mathbb{R}(n). \tag{3.18}$$

We also define $\mathcal{W} \subset H_1^{\mathrm{dyn}}$ as the graded subspace $\mathcal{W} = \bigoplus_{n\ge 1} \mathrm{gr}^{\Gamma}_{2n} \mathcal{W}$, where $\mathrm{gr}^{\Gamma}_{2n} \mathcal{W}$ is generated by the $2g$ elements $(2\pi\sqrt{-1})^n \underbrace{g_k g_k \cdots g_k}_{n-times}$.

The choice of indexing the grading by gr^{Γ}_{2n} instead of gr^{Γ}_{n} is motivated by comparison to the grading on the cohomological construction of [**10**].

In [**11**], we showed that the subspaces $\mathcal{V}$ and $\overline{\mathcal{V}}$ realize copies of the Archimedean cohomology of [**10**] embedded in the space of cochains $\mathcal{L}$ of the dynamical cohomology and in the dynamical cohomology itself, while the pair $\overline{\mathcal{V}} \oplus \mathcal{W}$ realizes a copy of the cohomology of the cone of the "local monodromy map" N of [**10**] inside the pair of dynamical cohomology and homology $H^1_{\mathrm{dyn}} \oplus H_1^{\mathrm{dyn}}$. The isomorphism between $\mathcal{V}$ and the Archimedean cohomology is realized by a natural choice of a basis of holomorphic differentials for the Archimedean cohomology, constructed from the data of the Schottky uniformization as in [**22**].

An explicit geometric description for the space $\mathcal{V}$ in terms of geodesics in $\mathfrak{X}_\Gamma$ is obtained by interpreting the characteristic function $\chi_{\mathcal{S}^+(w_{n,k})}$ as the "best approximation" within $\mathcal{P}_n$ to a distribution supported on the periodic sequence of period g_k,

$$f(g_k g_k g_k g_k g_k \ldots) = 1 \quad \text{and} \quad f(a_0 a_1 a_2 a_3 \ldots) = 0 \qquad \text{otherwise.}$$

Such periodic sequence $g_k g_k g_k g_k g_k \ldots$ describes the closed geodesic in $\mathfrak{X}_\Gamma$ that is the oriented core of one of the handles in the handlebody. Thus, the subspace $\mathrm{gr}^{\Gamma}_{2n} \mathcal{V} \subset \mathcal{P}_n$ is spanned by the "best approximations" within $\mathcal{P}_n$ to cohomology classes supported on the core handles of the handlebody. In other words, this interpretation views the index $n \le 0$ of the graded structure $\mathcal{V} = \bigoplus_n \mathrm{gr}^{\Gamma}_{2n} \mathcal{V}$ as a measure of "zooming in", with increasing precision for larger $|n|$, on the core handles of the handlebody $\mathfrak{X}_\Gamma$.

3.3. A spectral triple from dynamics. Recall that a spectral triple consists of the following data (cf. [**9**]).

DEFINITION 3.8. A spectral triple $(\mathcal{A}, \mathcal{H}, D)$ consists of a C^*-algebra $\mathcal{A}$ with a representation

$$\rho \colon \mathcal{A} \to \mathcal{B}(\mathcal{H})$$

as bounded operators on a Hilbert space $\mathcal{H}$, and an operator D (called the Dirac operator) on $\mathcal{H}$, which satisfies the following properties:

(1) D is self-adjoint.
(2) For all $\lambda \notin \mathbb{R}$, the resolvent $(D-\lambda)^{-1}$ is a compact operator on $\mathcal{H}$.
(3) The commutator $[D, a]$ is a bounded operator on $\mathcal{H}$, for all $a \in \mathcal{A}_0$, a dense involutive subalgebra of $\mathcal{A}$.

We consider the Cuntz Krieger algebra $\mathcal{O}_A$ (cf. [**13**] [**14**]) defined as the universal C^*-algebra generated by partial isometries $S_1, \ldots, S_{2g}$, satisfying the relations

$$\sum_j S_j S_j^* = I \tag{3.19}$$

$$S_i^* S_i = \sum_j A_{ij}\, S_j S_j^*, \tag{3.20}$$

where $A = (A_{ij})$ is the $2g \times 2g$ transition matrix of the subshift of finite type $(\mathcal{S}, T)$, namely the matrix whose entries are $A_{ij} = 1$ whenever $|i-j| \neq g$, and $A_{ij} = 0$ otherwise.

The algebra $\mathcal{O}_A$ can be also described in terms of the action of the free group Γ on its limit set Λ_Γ (cf. [**28, 30**]), so that we can regard $\mathcal{O}_A$ as a noncommutative space replacing the classical quotient Λ_Γ/Γ. In fact, the action of Γ on $\Lambda_\Gamma \subset \mathbb{P}^1(\mathbb{C})$ determines a unitary representation of $\mathcal{O}_A$ on the Hilbert space $L^2(\Lambda_\Gamma, \mu)$, given by

$$(T_{\gamma^{-1}} f)(x) := |\gamma'(x)|^{\delta_H/2} f(\gamma x), \quad \text{and} \quad (P_\gamma f)(x) := \chi_\gamma(x) f(x), \tag{3.21}$$

where δ_H is the Hausdorff dimension of Λ_Γ and the element $\gamma \in \Gamma$ is identified with a reduced word in the generators $\{g_j\}_{j=1}^g$ and their inverses, and χ_γ is the characteristic function of the cylinder $\Lambda_\Gamma(\gamma)$ of all (right) infinite reduced words that begin with the word γ. This determines an identification of $\mathcal{O}_A$ with the (reduced) crossed product C^*-algebra, $\mathcal{O}_A \cong C(\Lambda_\Gamma) \rtimes \Gamma$.

We then consider on the Hilbert space $\mathcal{L} = L^2(\Lambda_\Gamma, \mu)$, the unbounded linear self adjoint operator $D \colon \mathcal{L} \to \mathcal{L}$ given by the grading operator of the filtration $\mathcal{P}_n$, namely,

$$D = \sum_n n \widehat{\Pi}_n. \tag{3.22}$$

The restriction of this operator to the subspace $\mathcal{V}$ of the dynamical cohomology, isomorphic to the Archimedean cohomology of [**10**], agrees with the "Frobenius" operator Φ considered in [**10**], which computes the local factor as a regularized determinant as in [**15**].

We extend the operator (3.22) to an operator $\mathcal{D}$ on $\mathcal{H} = \mathcal{L} \oplus \mathcal{L}$ as

$$\mathcal{D}|_{\mathcal{L}\oplus 0} = \sum_n (n+1)(\widehat{\Pi}_n \oplus 0) \quad \mathcal{D}|_{0 \oplus \mathcal{L}} = -\sum_n n(0 \oplus \widehat{\Pi}_n). \tag{3.23}$$

The presence of a shift by one in the grading operator reflects the shift by one in the grading that appears in the duality isomorphisms on the cohomology of the cone of the monodromy map N as in Proposition 4.8 of [**10**]. The Dirac operator (3.23) takes into account the presence of this shift.

There is another possible natural choice for the sign of the Dirac operator, instead of the one in (3.23). Instead of being determined by the sign of the operator Φ on the Archimedean cohomology and its dual, this other choice is determined by the duality map that exists on the complex of [**10**], realized by powers of the monodromy map (cf. [**10**, Proposition 4.8]). In this case, the sign would then be given by the operator

$$F = \begin{pmatrix} 0 & 1 \\ 1 & 0 \end{pmatrix},$$

that exchanges the two copies of $\mathcal{L}$ in $\mathcal{H}$.

We assume that the Schottky group Γ has limit set Λ_Γ of Hausdorff dimension $\delta_H < 1$. Then the data above define the "dynamical spectral triple" at arithmetic infinity.

THEOREM 3.9. *The data* $(\mathcal{O}_A, \mathcal{H}, \mathcal{D})$, *where the algebra* $\mathcal{O}_A$ *acts diagonally on* $\mathcal{H} = \mathcal{L} \oplus \mathcal{L}$, *and the operator* $\mathcal{D}$ *is given by* (3.23) *form a spectral triple in the sense of Connes, as in Definition* 3.8.

The bound on the commutators with the generators S_i of $\mathcal{O}_A$ and their adjoints is obtained in [**11**] in terms of the Poincaré series of the Schottky group.

The spectral triple defined this way appears to be related to spectral triples for AF algebras, recently introduced in [**1**]. In such constructions the Dirac operator generalizes the grading operator (3.22), by operators of the form $D = \sum \alpha_n \widehat{\Pi}_n$, where $\widehat{\Pi}_n = \Pi_n - \Pi_{n-1}$ are the projections associated to a filtration of the AF algebra, and the coefficients α_n given by a sequence of positive real numbers, satisfying certain growth conditions.

The C^*-algebra $C(\Lambda_\Gamma)$ is a commutative AF-algebra (approximately finite-dimensional), obtained as the direct limit of the finite-dimensional commutative C^*-algebras generated by characteristic functions of a covering of Λ_Γ. This gives rise to the filtration $\mathcal{P}_n$ in Theorem 3.4, hence our choice of Dirac operator fits into the setting of [**1**] for the AF algebra $C(\Lambda_\Gamma)$. On the other hand, while in the construction of [**1**] the eigenvalues α_n can be chosen sufficiently large, so that the resulting spectral triple for the AF algebra would be finitely summable, when we consider the Cuntz–Krieger algebra $\mathcal{O}_A \cong C(\Lambda_\Gamma) \rtimes \Gamma$, the boundedness of commutators (condition 3 of Definition 3.8) can only be satisfied for the special choice of $\alpha_n = cn$, with c a constant, cf. [**1**, Remark 2.2], which does not yield a finitely summable spectral triple. The reason for this lies in a well known result of Connes [**7**] which shows that non amenable discrete groups (as is the case for the Schottky group Γ) do not admit finitely summable spectral triples. Thus, if the dense subalgebra of $\mathcal{O}_A$ with which $\mathcal{D}$ has bounded commutators contains group elements, then the Dirac operator $\mathcal{D}$ cannot be finitely summable.

In our construction, the choice of sign for $\mathcal{D}$ is prescribed by the graded structure of the cohomology theory of [**10**] and by the identification of the Archimedean cohomology of [**10**] and its dual with subspaces of the dynamical cohomology and homology as in [**11**]. This way, the requirement that the Dirac operator agrees with the operator Φ of [**10**] on these subspaces fixes the choice of the sign of the Dirac operator, which carries the topological information on the noncommutative manifold. In the construction of spectral triples for AF algebras of [**1**] only the metric aspect of the spectral triple is retained, that is, the operator considered is of the form $|D|$, while the sign is not discussed.

A possible way to refine the construction of the spectral triple and deal with the lack of finite summability is through the fact that the Cuntz–Krieger algebra $\mathcal{O}_A$ has a second description as a crossed product algebra. Namely, up to stabilization (i.e., tensoring with compact operators) we have

$$\mathcal{O}_A \simeq \mathcal{F}_A \rtimes_T \mathbb{Z}, \tag{3.24}$$

where $\mathcal{F}_A$ is an approximately finite-dimensional (AF) algebra, cf. [**13, 14**]. This algebra can be described in terms of a groupoid C^*-algebra associated to the "unstable manifold" in the Smale space $(\mathcal{S}, T)$. In fact, consider the algebra $\mathcal{O}_A^{\mathrm{alg}}$ generated algebraically by the S_i and S_i^* subject to the Cuntz–Krieger relations (3.19)–(3.20). Elements in $\mathcal{O}_A^{\mathrm{alg}}$ are linear combinations of monomials $S_\mu S_\nu^*$, for multi-indices μ, ν, cf. [**14**]. The AF algebra $\mathcal{F}_A$ is generated by elements $S_\mu S_\nu^*$ with $|\mu| = |\nu|$, and is filtered by finite-dimensional algebras $\mathcal{F}_{A,n}$ generated by elements of the form $S_\mu P_i S_\nu^*$ with $|\mu| = |\nu| = n$ and $P_i = S_i S_i^*$ the range projections, and embeddings determined by the matrix A. The commutative algebra $C(\Lambda_\Gamma)$ sits as a subalgebra of $\mathcal{F}_A$ generated by all range projections $S_\mu S_\mu^*$. The embedding is compatible with the filtration and with the action of the shift T, which is implemented on $\mathcal{F}_A$ by the transformation $a \mapsto \sum_i S_i a S_i^*$. (cf. [**14**].)

The fact that the algebra can be written in the form (3.24) implies that, by Connes' result on hyperfiniteness [**7**], it may carry a finitely summable spectral triple. It is an interesting problem whether the construction of a finitely summable triple can be carried out in a way that is of arithmetic significance.

3.4. Local factor. For an arithmetic variety $\mathcal{X}$ over $\operatorname{Spec}\mathbb{Z}$, the "Archimedean factor" (local factor at arithmetic infinity) $L_\kappa(H^m, s)$ is a product of Gamma functions, with exponents and arguments that depend on the (pure) Hodge structure $H^m = H^m(X, \mathbb{C}) = \bigoplus_{p+q=m} H^{p,q}$. More precisely, (cf. [**29**])

$$L_\kappa(H^m, s) = \begin{cases} \prod_{p,q} \Gamma_{\mathbb{C}}(s - \min(p,q))^{h^{p,q}} & \kappa = \mathbb{C} \\ \prod_{p<q} \Gamma_{\mathbb{C}}(s-p)^{h^{p,q}} \prod_p \Gamma_{\mathbb{R}}(s-p)^{h^{p+}} \Gamma_{\mathbb{R}}(s-p+1)^{h^{p-}} & \kappa = \mathbb{R}, \end{cases} \tag{3.25}$$

where the $h^{p,q}$, with $p + q = m$, are the Hodge numbers, $h^{p,\pm}$ is the dimension of the $\pm(-1)^p$-eigenspace of de Rham conjugation on $H^{p,p}$, and

$$\Gamma_{\mathbb{C}}(s) := (2\pi)^{-s}\Gamma(s) \quad \Gamma_{\mathbb{R}}(s) := 2^{-1/2}\pi^{-s/2}\Gamma(s/2).$$

Deninger produced a unified description of the factors at arithmetic infinity and at the finite primes, in the form of a Ray–Singer determinant (cf. [**15**]). The factor (3.25) satisfies

$$L_\kappa(H^m, s) = \det_\infty\left(\frac{1}{2\pi}(s - \Phi)|_{\mathcal{V}^m}\right)^{-1}, \tag{3.26}$$

where $\mathcal{V}^m$ is an infinite-dimensional real vector space. The zeta regularized determinant of an unbounded self adjoint operator T is defined as $\det_\infty(s - T) = \exp(-d\zeta_T(s,z)/dz|_{z=0})$.

In [**10**], the graded spaces $\mathcal{V}^\cdot$ are identified with inertia invariants of a double complex of real Tate-twisted differential forms on $X_{/\mathbb{C}}$ with suitable cutoffs. Namely, such complex is endowed with the action of an endomorphism N, which represents a "logarithm of the local monodromy at arithmetic infinity", and the spaces $\mathcal{V}^\cdot$ are identified with the kernel of the map N on the hypercohomology. In particular, in the case of an arithmetic surface, we showed in [**11**] that the Archimedean cohomology group $\mathcal{V}^1$ is identified with the subspace $\mathcal{V}$ of $\mathcal{P}$ and with the corresponding subspace of the dynamical cohomology $H^1_{\rm dyn}$.

The dynamical spectral triple of Theorem 3.9 is not finitely summable. However, it is possible to recover from these data the local factor at arithmetic infinity (3.26) for $m = 1$.

PROPOSITION 3.10. *Consider the zeta functions*

$$\zeta_{\pi(\mathcal{V}),D}(s,z) := \sum_{\lambda \in \mathrm{Spec}(D)} \mathrm{Tr}\big(\pi(\mathcal{V})\Pi(\lambda,D)\big)(s-\lambda)^{-z}, \tag{3.27}$$

for $\pi(\mathcal{V})$ the orthogonal projection on the norm closure of $\mathcal{V}$ in $\mathcal{L}$, and

$$\zeta_{\pi(\mathcal{V},\overline{F}_\infty=\mathrm{id}),D}(s,z) := \sum_{\lambda \in \mathrm{Spec}(D)} \mathrm{Tr}\big(\pi(\mathcal{V},\overline{F}_\infty=\mathrm{id})\Pi(\lambda,D)\big)(s-\lambda)^{-z}, \tag{3.28}$$

for $\pi(\mathcal{V},\overline{F}_\infty=\mathrm{id})$ the orthogonal projection on the norm closure of $\mathcal{V}^{\overline{F}_\infty=\mathrm{id}}$. The corresponding regularized determinants satisfy

$$\exp\Big(-\frac{d}{dz}\zeta_{\pi(\mathcal{V}),D/2\pi}(s/2\pi,z)|_{z=0}\Big)^{-1} = L_{\mathbb{C}}\big(H^1(X),s\big), \tag{3.29}$$

$$\exp\Big(-\frac{d}{dz}\zeta_{\pi(\mathcal{V},\overline{F}_\infty=\mathrm{id}),D/2\pi}(s/2\pi,z)|_{z=0}\Big)^{-1} = L_{\mathbb{R}}\big(H^1(X),s\big). \tag{3.30}$$

Moreover, the operator $\pi(\mathcal{V})$ acts on the range of the spectral projections $\Pi(\lambda,D)$ as certain elements of the algebra $\mathcal{O}_A$.

Here $\overline{F}_\infty$ is the involution induced by the real structure on $X_{/\mathbb{R}}$, which corresponds to the change of orientation on the geodesics in $\mathfrak{X}_\Gamma$ and on $\mathcal{S}_T$.

4. A dynamical theory for Mumford curves

Throughout this chapter K will denote a finite extension of $\mathbb{Q}_p$ and Δ_K the Bruhat–Tits tree associated to $G = \mathrm{PGL}(2,K)$. In the following we recall few results about the action of a Schottky group on a Bruhat–Tits tree and on C^*-algebras of graphs. Detailed explanations are contained in [**2, 18–20, 26, 30**].

Recall that the Bruhat–Tits tree is constructed as follows. One considers the set of free $\mathcal{O}$-modules of rank 2: $M \subset V$. Two such modules are *equivalent* $M_1 \sim M_2$ if there exists an element $\lambda \in K^*$, such that $M_1 = \lambda M_2$. The group $\mathrm{GL}(V)$ of linear automorphisms of V operates on the set of such modules *on the left*: $gM = \{gm \mid m \in M\}$, $g \in \mathrm{GL}(V)$. Notice that the relation $M_1 \sim M_2$ is equivalent to the condition that M_1 and M_2 belong to the same orbit of the center $K^* \subset \mathrm{GL}(V)$. Hence, the group $G = \mathrm{GL}(V)/K^*$ operates (on the left) on the set of classes of equivalent modules.

We denote by Δ_K^0 the set of such classes and by $\{M\}$ the class of the module M. Because $\mathcal{O}$ is a principal ideals domain and every module M has two generators, it follows that

$$\{M_1\},\{M_2\} \in \Delta_K^0, M_1 \supset M_2 \implies M_1/M_2 \simeq \mathcal{O}/\mathfrak{m}^l \oplus \mathcal{O}/\mathfrak{m}^k,\ l,k \in \mathbb{N}.$$

The multiplication of M_1 and M_2 by elements of K preserves the inclusion $M_1 \supset M_2$, hence the natural number

$$d(\{M_1\},\{M_2\}) = |l-k| \tag{4.1}$$

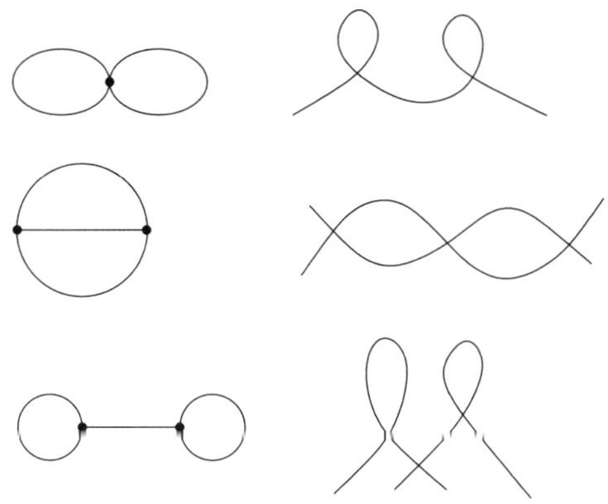

FIGURE 1. The graphs Δ_Γ/Γ for genus $g = 2$, and the corresponding fibers.

is well-defined.

The graph Δ_K of the group $\mathrm{PGL}(2, K)$ is the infinite graph with set of vertices Δ_K^0, in which two vertices $\{M_1\}, \{M_2\}$ are adjacent and hence connected by an edge if and only if $d(\{M_1\}, \{M_2\}) = 1$. (cf. [**20, 26**].)

For a Schottky group $\Gamma \subset \mathrm{PGL}(2, K)$ there is a smallest subtree $\Delta'_\Gamma \subset \Delta_K$ containing the axes of all elements of Γ. The set of ends of Δ'_Γ in $\mathbb{P}^1(K)$ is Λ_Γ, the limit set of Γ. The group Γ carries Δ'_Γ into itself so that the quotient Δ'_Γ/Γ is a finite graph that coincides with the dual graph of the closed fibre of the minimal smooth model of the algebraic curve C/K holomorphically isomorphic to $X_\Gamma := \Omega_\Gamma/\Gamma$ (cf. [**26**, p. 163]). There is a smallest tree Δ_Γ on which Γ acts and such that Δ_Γ/Γ is the (finite) graph of the specialization of C. The curve C is a k-split degenerate, stable curve. When the genus of the fibers is at least 2—i.e., when the Schottky group has at least $g \geq 2$ generators—the curve X_Γ is called a Schottky–Mumford curve.

The possible graphs Δ_Γ/Γ and the corresponding fiber for the case of genus 2 are illustrated in Figure 1.

To a locally finite directed graph one associates a C^*-algebra in the following way.

A Cuntz–Krieger family consists of a collection $\{P_v\}_{v\in E^0}$ of mutually orthogonal projections and a $\{S_w\}_{w\in E^1_+}$ of partial isometries, satisfying the conditions: $S_w^* S_w = P_{r(w)}$ and, for all $v \in s(E^1_+)$, $P_v = \sum_{w:s(w)=v} S_w S_w^*$.

The Cuntz–Krieger elements $\{P_v, S_w\}$ satisfy the relation

$$S_w^* S_w = \sum A_+(w, \tilde{w}) S_{\tilde{w}} S_{\tilde{w}}^*,$$

with A_+ the edge matrix of the graph.

One defines a universal C^*-algebra $C^*(E)$ generated by a Cuntz–Krieger family. If E is a finite graph with no sinks, we have $C^*(E) \simeq \mathcal{O}_{A_+}$, where $\mathcal{O}_{A_+}$ is the Cuntz–Krieger algebra of the edge matrix A_+. If the directed graph is a tree Δ, then $C^*(\Delta)$ is an AF algebra strongly Morita equivalent to the commutative C^*-algebra $C_0(\partial\Delta)$. A monomorphism of directed trees induces an injective $*$-homomorphism of the corresponding C^*-algebras.

If $G \subset \mathrm{Aut}(E)$ is a group acting freely on the directed graph E, with quotient graph E/G, then the crossed product C^*-algebra $C^*(E) \rtimes G$ is strongly Morita equivalent to $C^*(E/G)$. In particular, if Δ is the universal covering tree of a directed graph E and Γ is the fundamental group, then the algebra $C^*(E)$ is strongly Morita equivalent to $C_0(\partial\Delta) \rtimes \Gamma$.

In the following we consider the Bruhat–Tits tree Δ_K for a fixed finite extension K of $\mathbb{Q}_p$, and the corresponding C^*-algebra $C^*(\Delta_K)$, which is strongly Morita equivalent to the abelian C^*-algebra of complex valued functions $C(\mathbb{P}^1(K))$.

Notice that the commutative C^*-algebra $C^*(X_K)$ of complex valued continuous functions on the Mumford curve $X_K = \Omega_\Gamma/\Gamma$ is strongly Morita equivalent to the crossed product $C_0(\Omega_\Gamma) \rtimes \Gamma$, with $C_0(\Omega_\Gamma)$ a Γ-invariant ideal of $C\big(\mathbb{P}^1(K)\big)$ with quotient algebra $C(\Lambda_\Gamma)$. The algebra $C^*(\Delta_\Gamma/\Gamma)$, in turn, is strongly Morita equivalent to $C^*(\Delta_\Gamma) \rtimes \Gamma$ and to $C(\Lambda_\Gamma) \rtimes \Gamma$, where $\partial\Delta_\Gamma = \Lambda_\Gamma$. Similarly, one sees that the algebra $C^*(\Delta_K/\Gamma)$ is strongly Morita equivalent to the crossed product algebra $C^*(\Delta_K) \rtimes \Gamma$, which in turn is strongly Morita equivalent to $C^*\big(\mathbb{P}^1(K)\big) \rtimes \Gamma$.

Thus, up to Morita equivalence, the graph algebra $C^*(\Delta_K/\Gamma)$ can be regarded as a way of extending the commutative C^*-algebra $C^*(X_K)$ (functions on the Mumford curve) by the Cuntz–Krieger algebra $C^*(\Delta_\Gamma/\Gamma)$ associated to the edge matrix of the finite graph Δ_Γ/Γ.

In this paragraph we introduce a dynamical system associated to the space $\mathcal{W}(\Delta/\Gamma)$ of walks on the directed tree Δ on which Γ acts. In particular, we are interested in the cases when $\Delta = \Delta_K$, Δ_Γ.

For $\Delta = \Delta_\Gamma$, we obtain a subshift of finite type associated to the action of the Schottky group Γ on the limit set Λ_Γ, of the type that was considered in [**11**].

Let $\overline{V} \subset \Delta_\Gamma$ be a finite subtree whose set of edges consists of one representative for each Γ-class. This is a fundamental domain for Γ in the weak sense (following the notation of [**20**]), since some vertices may be identified under the action of Γ. Correspondingly, $V \subset \mathbb{P}^1(K)$ is the set of ends of all infinite paths starting at points in $\overline{V}$.

Consider the set $\mathcal{W}(\Delta_\Gamma/\Gamma)$ of doubly infinite walks on the finite graph Δ_Γ/Γ. These are doubly infinite admissible sequences in the finite alphabet given by the edges of $\overline{V}$ with both possible orientations. On $\mathcal{W}(\Delta_\Gamma/\Gamma)$ we consider the topology generated by the sets $\mathcal{W}^s(\omega,\ell) = \{\tilde\omega \in \mathcal{W}(\Delta_\Gamma/\Gamma) : \tilde\omega_k = \omega_k, k \geq \ell\}$ and $\mathcal{W}^u(\omega,\ell) = \{\tilde\omega \in \mathcal{W}(\Delta_\Gamma/\Gamma) : \tilde\omega_k = \omega_k, k \leq \ell\}$, for $\omega \in \mathcal{W}(\Delta_\Gamma/\Gamma)$ and $\ell \in \mathbb{Z}$. With this topology, the space $\mathcal{W}(\Delta_\Gamma/\Gamma)$ is a totally disconnected compact Hausdorff space.

The invertible shift map T, given by $(T\omega)_k = \omega_{k+1}$, is a homeomorphism of $\mathcal{W}(\Delta_\Gamma/\Gamma)$. We can describe again the dynamical system $(\mathcal{W}(\Delta_\Gamma/\Gamma), T)$ in terms of subshifts of finite type.

LEMMA 4.1. *The space $\mathcal{W}(\Delta_\Gamma/\Gamma)$ with the action of the invertible shift T is a subshift of finite type, where $\mathcal{W}(\Delta_\Gamma/\Gamma) = \mathcal{S}_A$ with A the directed edge matrix of the finite graph Δ_Γ/Γ.*

We consider the mapping torus of T:

$$\mathcal{W}(\Delta_\Gamma/\Gamma)_T := \mathcal{W}(\Delta_\Gamma/\Gamma) \times [0,1]/(Tx,0) \sim (x,1). \tag{4.2}$$

4.1. Genus two example. In the example of Mumford–Schottky curves of genus $g = 2$, the tree Δ_Γ is illustrated in Figure 2.

In the first case in the figure, the tree Δ_Γ is just a copy of the Cayley graph of the free group Γ on two generators, hence we can identify doubly infinite walks

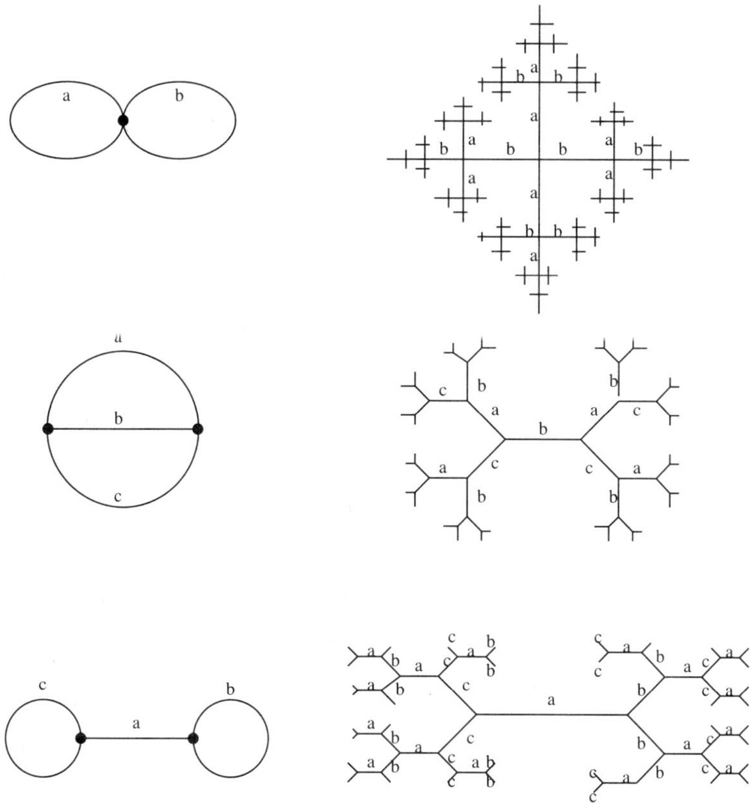

FIGURE 2. The graphs Δ_Γ/Γ for genus $g = 2$, and the corresponding trees Δ_Γ.

in Δ_Γ with doubly infinite reduced words in the generators of Γ and their inverses. The directed edge matrix is given by

$$A = \begin{pmatrix} 1 & 1 & 0 & 1 \\ 1 & 1 & 1 & 0 \\ 0 & 1 & 1 & 1 \\ 1 & 0 & 1 & 1 \end{pmatrix}.$$

In the second case in Figure 2, we label by $a = e_1$, $b = e_2$ and $c = e_3$ the oriented edges in the graph Δ_Γ/Γ, so that we have a corresponding set of labels $E = \{a, b, c, \bar{a}, \bar{b}, \bar{c}\}$ for the edges in the covering Δ_Γ. A choice of generators for the group $\Gamma \simeq \mathbb{Z} * \mathbb{Z}$ acting on Δ_Γ is obtained by identifying the generators g_1 and g_2 of Γ with the chains of edges $a\bar{b}$ and $a\bar{c}$. Doubly infinite walks in the tree Δ_Γ are admissible doubly infinite sequences of such labels, where admissibility is determined by the directed edge matrix

$$A = \begin{pmatrix} 0 & 1 & 0 & 0 & 0 & 1 \\ 1 & 0 & 1 & 0 & 0 & 0 \\ 0 & 1 & 0 & 1 & 0 & 0 \\ 0 & 0 & 1 & 0 & 1 & 0 \\ 0 & 0 & 0 & 1 & 0 & 1 \\ 1 & 0 & 0 & 0 & 1 & 0 \end{pmatrix}.$$

The third case in Figure 2 is analogous. A choice of generators for the group $\Gamma \simeq \mathbb{Z} * \mathbb{Z}$ acting on Δ_Γ is given by $ab\bar{a}$ and c. Doubly infinite walks in the tree Δ_Γ are admissible doubly infinite sequences in the alphabet $E = \{a, b, c, \bar{a}, \bar{b}, \bar{c}\}$, with admissibility determined by the directed edge matrix

$$A = \begin{pmatrix} 0 & 0 & 1 & 0 & 0 & 1 \\ 1 & 1 & 0 & 0 & 0 & 0 \\ 0 & 0 & 1 & 1 & 0 & 0 \\ 0 & 1 & 0 & 0 & 1 & 0 \\ 1 & 0 & 0 & 0 & 1 & 0 \\ 0 & 0 & 0 & 1 & 0 & 1 \end{pmatrix}.$$

The construction is analogous for genus $g > 2$, for the various possible finite graphs Δ_Γ/Γ. The directed edge matrix can then be written in block form as

$$A = \begin{pmatrix} \alpha_{11} & \alpha_{12} \\ \alpha_{21} & \alpha_{22} \end{pmatrix},$$

where each block α_{ij} is a $\#(\Delta_\Gamma/\Gamma)^1_+ \times \#(\Delta_\Gamma/\Gamma)^1_+$-matrix with $\alpha_{12} = \alpha_{12}^t$, $\alpha_{21} = \alpha_{21}^t$, and $\alpha_{11} = \alpha_{22}^t$.

4.2. Cohomology of $\mathcal{W}(\Delta/\Gamma)_T$. Let $\Delta = \Delta_\Gamma$. We identify the first cohomology group $H^1(\mathcal{W}(\Delta_\Gamma/\Gamma)_T, \mathbb{Z})$ with the group of homotopy classes of continuous maps of $\mathcal{W}(\Delta_\Gamma/\Gamma)_T$ to the circle. Let $C(\mathcal{W}(\Delta_\Gamma/\Gamma), \mathbb{Z})$ be the $\mathbb{Z}$-module of integer valued continuous functions on $\mathcal{W}(\Delta_\Gamma/\Gamma)$, and let

$$C(\mathcal{W}(\Delta_\Gamma/\Gamma), \mathbb{Z})_T := \mathrm{Coker}(\delta),$$

for $\delta(f) = f - f \circ T$. The analog of Theorem 3.4 holds:

PROPOSITION 4.2. *The map $f \mapsto \left[\exp\left(2\pi i t f(x)\right)\right]$, which associates to an element $f \in C(\mathcal{W}(\Delta_\Gamma/\Gamma), \mathbb{Z})$ a homotopy class of maps from $\mathcal{W}(\Delta_\Gamma/\Gamma)_T$ to the circle, gives an isomorphism $C(\mathcal{W}(\Delta_\Gamma/\Gamma), \mathbb{Z})_T \simeq H^1(\mathcal{W}(\Delta_\Gamma/\Gamma)_T, \mathbb{Z})$. Moreover, there is a filtration of $C(\mathcal{W}(\Delta_\Gamma/\Gamma), \mathbb{Z})_T$ by free $\mathbb{Z}$-modules $F_0 \subset F_1 \subset \cdots \subset F_n \cdots$, of rank $\theta_n - \theta_{n-1} + 1$, where θ_n is the number of admissible words of length $n+1$ in the alphabet, so that we have*

$$H^1\left(\mathcal{W}(\Delta_\Gamma/\Gamma)_T, \mathbb{Z}\right) = \varinjlim_n F_n.$$

The quotients F_{n+1}/F_n are also torsion-free.

The space $\mathcal{W}(\Delta_\Gamma/\Gamma)_T$ corresponds to a space of "bounded geodesics" on the graph Δ_K/Γ, where geodesics, in this setting, are just doubly infinite walks in Δ_K/Γ. In particular, a closed geodesic is the image under the quotient map $\pi_\Gamma\colon \Delta_K \to \Delta_K/\Gamma$ of a doubly infinite walk in the Bruhat–Tits tree Δ_K with ends given by the pair $z^+(\gamma), z^-(\gamma)$ of fixed points of some element $\gamma \in \Gamma$. Similarly, a bounded geodesic is an element $\omega \in \mathcal{W}(\Delta_K/\Gamma)$ which is the image, under the quotient map, of a doubly infinite walk in Δ_K with both ends on $\Lambda_\Gamma \subset \mathbb{P}^1(K)$. This implies that a bounded geodesic is a walk of the form $\omega = \pi_\Gamma(\tilde{\omega})$, for some $\tilde{\omega} \in \mathcal{W}(\Delta_\Gamma/\Gamma)$. By construction, any such walk is an axis of Δ_Γ.

Orbits of $\mathcal{W}(\Delta_\Gamma/\Gamma)$ under the action of the invertible shift T correspond bijectively to orbits of the complement of the diagonal in $\Lambda_\Gamma \times \Lambda_\Gamma$ under the action of Γ. Thus, we see that $\mathcal{W}(\Delta_\Gamma/\Gamma)_T$ gives a geometric realization of the space of "bounded geodesics" on the graph Δ_K/Γ, much as, in the case of the geometry at arithmetic infinity, we used the mapping torus of the shift T as a model of the tangle of bounded geodesics in a hyperbolic handlebody.

As in the case at infinity, we can consider the Pimsner–Voiculescu exact sequence computing the K-theory groups of the crossed product C^*-algebra

$$C\big(\mathcal{W}(\Delta_\Gamma/\Gamma)\big) \rtimes_T \mathbb{Z},$$

$$(4.3)\quad 0 \to H^0\big(\mathcal{W}(\Delta_\Gamma/\Gamma)_T, \mathbb{Z}\big) \to C\big(\mathcal{W}(\Delta_\Gamma/\Gamma), \mathbb{Z}\big) \xrightarrow{\delta=1-T} C\big(\mathcal{W}(\Delta_\Gamma/\Gamma), \mathbb{Z}\big) \to H^1(\mathcal{W}(\Delta_\Gamma/\Gamma)_T, \mathbb{Z}) \to 0.$$

In the corresponding sequence

$$(4.4)\qquad 0 \to H^0\big(\mathcal{W}(\Delta_\Gamma/\Gamma)_T, \kappa\big) \to \mathcal{P} \xrightarrow{\delta} \mathcal{P} \to H^1(\mathcal{W}(\Delta_\Gamma/\Gamma)_T, \kappa) \to 0,$$

for the cohomology for $H^*(\mathcal{W}(\Delta_\Gamma/\Gamma)_T, \kappa)$, with $\kappa = \mathbb{R}$ or $\mathbb{C}$, we can take the vector space $\mathcal{P}$ obtained, as in the case at infinity, by tensoring with κ the $\mathbb{Z}$-module $\mathcal{P} \subset C(\mathcal{W}(\Delta_\Gamma/\Gamma), \mathbb{Z})$ of functions of future coordinates where $\mathcal{P} \simeq C(\mathcal{W}^+(\Delta_\Gamma/\Gamma), \mathbb{Z})$. This has a filtration $\mathcal{P} = \bigcup_n \mathcal{P}_n$, where $\mathcal{P}_n$ is identified with the submodule of $C(\mathcal{W}^+(\Delta_\Gamma/\Gamma), \mathbb{Z})$ generated by characteristic functions of

$$\mathcal{W}^+(\Delta_\Gamma/\Gamma, \rho) \subset \mathcal{W}^+(\Delta_\Gamma/\Gamma),$$

where $\rho \in \mathcal{W}^*(\Delta_\Gamma/\Gamma)$ is a finite walk $\rho = w_0 \ldots w_n$ of length $n+1$, and $\mathcal{W}^+(\Delta_\Gamma/\Gamma, \rho)$ is the set of infinite paths $\omega \in \mathcal{W}^+(\Delta_\Gamma/\Gamma)$, with $\omega_k = w_k$ for $0 \leq k \leq n+1$. This filtration defines the terms $F_n = \mathcal{P}_n/\delta\mathcal{P}_{n-1}$ in the filtration of the dynamical cohomology of the Mumford curve, as in Proposition 4.2. Again, we will use the same notation in the following for the free $\mathbb{Z}$-module $\mathcal{P}_n$ of functions of at most $n+1$ future coordinates and the vector space obtained by tensoring $\mathcal{P}_n$ by κ.

We obtain a Hilbert space completion of the space $\mathcal{P}$ of cochains in (4.4) by considering $\mathcal{L} = L^2(\Lambda_\Gamma, \mu)$ defined with respect to the measure on $\Lambda_\Gamma = \partial\Delta_\Gamma$ given by assigning its value on the clopen set $V(v)$, given by the ends of all paths in Δ_Γ starting at a vertex v, to be

$$\mu\big(V(v)\big) = q^{-d(v)-1},$$

with $q = \mathrm{card}(\mathcal{O}/\mathfrak{m})$.

In [**11**, Section 4], we showed how the mapping torus $\mathcal{S}_T$ of the subshift of finite type $(\mathcal{S}, T)$, associated to the limit set of the Schottky group, maps surjectively to the tangle of bounded geodesics inside the hyperbolic handlebody, through a map that resolves all the points of intersection of different geodesics. In the case of the Mumford curve, where we replace the real hyperbolic 3-space by the Bruhat–Tits building Δ_K, the analog of the surjective map from $\mathcal{S}_T$ to the tangle of bounded geodesics is a map from $\mathcal{W}(\Delta_\Gamma/\Gamma)_T$ to the dual graph Δ_Γ/Γ. Here is a description of this map.

As before, we write elements of $\mathcal{W}(\Delta_\Gamma/\Gamma)$ as admissible doubly infinite sequences

$$\omega = \ldots w_{i_{-m}} \cdots w_{i_{-1}} w_{i_0} w_{i_1} \cdots w_{i_n} \cdots,$$

with the $w_{i_k} = \{e_{i_k}, \epsilon_{i_k}\}$ oriented edges on the graph Δ_Γ/Γ. We consider each oriented edge w of normalized length one, so that it can be parameterized as $w(t) = \{e(t), \epsilon\}$, for $0 \le t \le 1$, with $\overline{w}(t) = \{e(1-t), -\epsilon\}$. Since $\omega \in \mathcal{S}_A$ is an admissible sequence of oriented edges we have $w_{i_k}(1) = w_{i_{k+1}}(0) \in \Delta_\Gamma^{(0)}$.

We consider a map of the covering space $\mathcal{W}(\Delta_\Gamma/\Gamma) \times \mathbb{R}$ of $\mathcal{W}(\Delta_\Gamma/\Gamma)_T$ to $|\Delta_\Gamma|$ of the form

$$\widetilde{E}(\omega, \tau) = w_{i_{[\tau]}}(\tau - [\tau]). \tag{4.5}$$

Here $|\Delta_\Gamma|$ denotes the geometric realization of the graph. By construction, the map $\widetilde{E}$ satisfies $\widetilde{E}(T\omega, \tau) = \widetilde{E}(\omega, \tau + 1)$, hence it descends to a map E of the quotient

$$E \colon \mathcal{W}(\Delta_\Gamma)_T \to |\Delta_\Gamma|. \tag{4.6}$$

We then obtain a map to $|\Delta_\Gamma/\Gamma|$, by composing with the quotient map of the Γ action, $\pi_\Gamma : \Delta_\Gamma \to \Delta_\Gamma/\Gamma$, that is,

$$\overline{E} := \pi_\Gamma \circ E : \mathcal{W}(\Delta_\Gamma)_T \to |\Delta_\Gamma/\Gamma|. \tag{4.7}$$

Thus, we obtain the following.

PROPOSITION 4.3. *The map $\overline{E}$ of (4.7) is a continuous surjection from the mapping torus $\mathcal{W}(\Delta_\Gamma)_T$ to the geometric realization $|\Delta_\Gamma/\Gamma|$ of the finite graph Δ_Γ/Γ.*

The fibers of the map (4.7) are explicitly described as

$$\overline{E}^{-1}(w_i(t)) = \mathcal{W}(\Delta_\Gamma)(w_i) \times \{t\} \cup \mathcal{W}(\Delta_\Gamma)(\overline{w}_i) \times \{1-t\}, \tag{4.8}$$

where $w_i(t)$, for $t \in [0,1]$ is a parameterized oriented edge in the graph Δ_Γ/Γ, and $\mathcal{W}(\Delta_\Gamma)(w_i) \subset \mathcal{W}(\Delta_\Gamma)$ consists of

$$\mathcal{W}(\Delta_\Gamma)(w_i) = \{\omega \in \mathcal{W}(\Delta_\Gamma) : w_{i_0} = w_i\},$$

for $\omega = \ldots w_{i_{-m}} \cdots w_{i_{-1}} w_{i_0} w_{i_1} \cdots w_{i_n} \cdots$.

We can also consider the construction described above, where, instead of using the tree Δ_Γ, we use larger Γ-invariant trees inside the Bruhat–Tits tree Δ.

The set of doubly infinite walks $\mathcal{W}(\Delta_K/\Gamma)$ can be identified with the set of admissible doubly infinite sequences in the oriented edges of a fundamental domain for the action of Γ. Therefore, we obtain the following identification

$$\begin{aligned} \mathcal{W}(\Delta_K/\Gamma) & \\ \simeq V \times \mathcal{W}^*(\Delta_\Gamma/\Gamma) & \times V \cup V \times \mathcal{W}^+(\Delta_\Gamma/\Gamma) \cup \mathcal{W}^+(\Delta_\Gamma/\Gamma) \times V \cup \mathcal{W}(\Delta_\Gamma/\Gamma), \end{aligned} \tag{4.9}$$

where we distinguish between walks that wander off from the finite graph Δ_Γ/Γ along one of the paths leading to an end in V, in one or in both directions, and those that stay confined within the finite graph Δ_Γ/Γ. Here we include in $\mathcal{W}^*(\Delta_\Gamma/\Gamma) = \cup_n \mathcal{W}^n(\Delta_\Gamma/\Gamma)$ also the case $\mathcal{W}^0(\Delta_\Gamma/\Gamma) = \varnothing$, where the walk in Δ_K/Γ does not intersect the finite graph Δ_Γ/Γ at all.

In the topology induced by the p-adic norm, $\mathbb{P}^1(K)$ is a totally disconnected compact Hausdorff space, and so is $\mathcal{W}(\Delta_K/\Gamma)$ by (4.9). Again, we consider the invertible shift map T, which is a homeomorphism of $\mathcal{W}(\Delta_K/\Gamma)$, and we form the mapping torus

$$\mathcal{W}(\Delta_K/\Gamma)_T := \mathcal{W}(\Delta_K/\Gamma) \times [0,1]/(Tx, 0) \sim (x, 1). \tag{4.10}$$

We obtain the analog of Proposition 4.2, though in the case of Δ_K the $\mathbb{Z}$-modules F_n will not be finitely generated. On the other hand, we can restrict to "neighborhoods" of the tree Δ_Γ inside Δ_K, which correspond to the reduction maps modulo powers of the maximal ideal.

In fact, in the theory of Mumford curves, it is important to consider also the reduction modulo powers $\mathfrak{m}^n$ of the maximal ideal $\mathfrak{m} \subset \mathcal{O}_K$, which provides infinitesimal neighborhoods of order n of the closed fiber.

For each $n \geq 0$, we consider a subgraph $\Delta_{K,n}$ of the Bruhat–Tits tree Δ_K defined by setting

$$\Delta^0_{K,n} := \{v \in \Delta^0_K : d(v, \Delta'_\Gamma) \leq n\},$$

with respect to the distance (4.1), with $d(v, \Delta'_\Gamma) := \inf\{d(v, \tilde{v}) : \tilde{v} \in (\Delta'_\Gamma)^0\}$, and

$$\Delta^1_{K,n} := \{w \in \Delta^1_K : s(w), r(w) \in \Delta^0_{K,n}\}.$$

Thus, we have $\Delta_{K,0} = \Delta'_\Gamma$. We have $\Delta_K = \bigcup_n \Delta_{K,n}$.

For all $n \in \mathbb{N}$, the graph $\Delta_{K,n}$ is invariant under the action of the Schottky group Γ on Δ, and the finite graph $\Delta_{K,n}/\Gamma$ gives the dual graph of the reduction $X_K \otimes \mathcal{O}/\mathfrak{m}^{n+1}$. They form a directed family with inclusions $j_{n,m} \colon \Delta_{K,n} \hookrightarrow \Delta_{K,m}$, for all $m \geq n$, with all the inclusions compatible with the action of Γ.

In [**12**] we introduce the dynamical cohomology and discuss spectral geometry for the C^*-algebras $C^*(\Delta_{K,n}/\Gamma) \simeq C^*(\Delta_{K,n}) \rtimes \Gamma$.

4.3. Spectral triples and Mumford curves. We consider the Hilbert space $\mathcal{H} = \mathcal{L} \oplus \mathcal{L}$ and the operator $\mathcal{D}$ defined as

$$\mathcal{D}|_{\mathcal{L}\oplus 0} = -\frac{2\pi}{R\log q}\sum (n+1)\widehat{\Pi}_n \quad \mathcal{D}|_{0\oplus\mathcal{L}} = \frac{2\pi}{R\log q}\sum n\widehat{\Pi}_n, \tag{4.11}$$

where $\widehat{\Pi}_n = \Pi_n - \Pi_{n-1}$ are the orthogonal projections associated to the filtration $\mathcal{P}_n$, the integer R is the length of all the words representing the generators of Γ (this can be taken to be the same for all generators, possibly after blowing up a finite number of points on the special fiber, as explained in [**12**]), and $q = \mathrm{card}(\mathcal{O}/\mathfrak{m})$.

The same argument used in [**11**] for the case at arithmetic infinity adapts to the case of Mumford curves to prove the following result. (Note: the statement below corrects an unfortunate mistake that occurred in [**12**, Section 5.4].)

THEOREM 4.4. *Consider the tree Δ_Γ of the p-adic Schottky group acting on Δ_K.*

(1) *There is a representation of the algebra $C^*(\Delta_\Gamma/\Gamma)$ by bounded linear operators on the Hilbert space $\mathcal{L}$.*
(2) *The data $(C^*(\Delta_\Gamma/\Gamma), \mathcal{H}, \mathcal{D})$, with the algebra acting diagonally on $\mathcal{H} = \mathcal{L} \oplus \mathcal{L}$, and the Dirac operator $\mathcal{D}$ of* (4.11) *form a spectral triple.*

In [**12**] we recovered arithmetic information such as the local L-factors of [**16**] from the dynamical cohomology and the data of the spectral triple of Theorem 4.4.

Recall that, for a curve X over a global field $\mathbb{K}$, assuming semi-stability at all places of bad reduction, the local Euler factor at a finite place v has the following description [**29**]:

$$L_v\big(H^1(X), s\big) = \det\big(1 - \mathrm{Fr}^*_v\, N(v)^{-s} \mid H^1(\overline{X}, \mathbb{Q}_\ell)^{I_v}\big)^{-1}. \tag{4.12}$$

Here Fr^*_v is the geometric Frobenius acting on ℓ-adic cohomology of $\overline{X} = X \otimes \mathrm{Spec}(\overline{\mathbb{K}})$, with $\overline{\mathbb{K}}$ an algebraic closure and ℓ a prime with $(\ell, q) = 1$, where $N(v)$ is the cardinality of the residue field $k(v)$ at v. The determinant is evaluated on the inertia invariants $H^1(\overline{X}, \mathbb{Q}_\ell)^{I_v}$ at v (all of $H^1(\overline{X}, \mathbb{Q}_\ell)$ when v is a place of good reduction).

Suppose v is a place of $k(v)$-split degenerate reduction. Then the completion of X at v is a Mumford curve X_Γ. In this case, the Euler factor (4.12) takes the following form:

$$L_v\big(H^1(X), s\big) = (1 - q^{-s})^{-g} \tag{4.13}$$

where $q = N(v)$. This factor is computed by the zeta regularized determinant

$$\det{}_{\infty,\pi(\mathcal{V}),i\mathcal{D}}(s) = L_v(H^1(X), s)^{-1}, \tag{4.14}$$

where

$$\det{}_{\infty,a,i\mathcal{D}}(s) := \exp\big(-\zeta'_{a,i\mathcal{D},+}(s,0)\big)\exp\big(-\zeta'_{a,i\mathcal{D},-}(s,0)\big), \tag{4.15}$$

for

$$\begin{aligned} \zeta_{a,i\mathcal{D},+}(s,z) &:= \sum_{\lambda \in \mathrm{Spec}(i\mathcal{D}) \cap i[0,\infty)} \mathrm{Tr}(a\Pi_\lambda)(s+\lambda)^{-z} \\ \zeta_{a,i\mathcal{D},-}(s,z) &:= \sum_{\lambda \in \mathrm{Spec}(i\mathcal{D}) \cap i(-\infty,0)} \mathrm{Tr}(a\Pi_\lambda)(s+\lambda)^{-z}. \end{aligned} \tag{4.16}$$

The element $a = \pi(\mathcal{V})$ is the projection onto a linear subspace $\mathcal{V}$ of $\mathcal{H}$, which is obtained via embeddings of the cohomology of the dual graph Δ_Γ/Γ into the space of cochains of the dynamical cohomology.

The projection $\pi(\mathcal{V})$ acts on the range of the spectral projections $\widehat{\Pi}_n$ of D as elements Q_n in the AF algebra core of the C^*-algebra $C^*(\Delta_\Gamma/\Gamma)$.

Notice how, unlike the local factor at infinity, the factor at the non-Archimedean places involves the full spectrum of D and not just its positive or negative part. It is believed that this difference should correspond to the presence of an underlying geometric space based on loop geometry, which manifests itself as loops at the non-Archimedean places and as "half loops" (holomorphic disks) at arithmetic infinity.

There is another difference between Archimedean and non-Archimedean cases. At the Archimedean prime the local factor is described in terms of zeta functions for a Dirac operator D (cf. [**11, 15**]). On the other hand, at the non-Archimedean places, in order to get the correct normalization as in [**17**], we need to introduce a rotation of the Dirac operator by the imaginary unit, $D \mapsto iD$. This rotation corresponds to the Wick rotation that moves poles on the real line to poles on the imaginary line (zeroes for the local factor) and appears to be a manifestation of a rotation from Minkowskian to Euclidean signature $it \mapsto t$, as already remarked by Manin [**23**, p. 135], who wrote that "*imaginary time motion*" *may be held responsible for the fact that zeroes of* $\Gamma(s)^{-1}$ *are purely real whereas the zeroes of all non-Archimedean Euler factors are purely imaginary.* It is expected, therefore, that a more refined construction would involve a version of spectral triples for Minkowskian signature.

References

1. C. Antonescu and E. Christensen, *Spectral triples for AF C^*-algebras and metrics on the Cantor set*; `math.OA/0309044`.
2. T. Bates, D. Pask, I. Raeburn, and W. Szymański, *The C^*-algebras of row finite graphs*, New York J. Math. **6** (2000), 307–324.
3. P. Baum and A. Connes, *Geometric K-theory for Lie groups and foliations*, Brown University/IHES, 1982, Preprint; Enseign. Math. (2) **46** (2000), no. 1-2, 3–42.

4. M. Boyle and D. Handelman, *Orbit equivalence, flow equivalence, and ordered cohomology*, Israel J. Math. **95** (1996), 169–210.
5. L. O. Chekhov, A. D. Mironov, and A. V. Zabrodin, *Multiloop calculations in p-adic string theory and Bruhat–Tits trees*, Comm. Math. Phys. **125** (1989), no. 4, 675–711.
6. A. Connes, *Cyclic cohomology and the transverse fundamental class of a foliation*, Geometric Methods in Operator Algebras (Kyoto, 1983), Pitman Res. Notes Math. Ser., vol. 123, Longman Sci. Tech., Harlow 1986, pp. 52–144.
7. ———, *Compact metric spaces, Fredholm modules, and hyperfiniteness*, Ergodic Theory Dynam. Systems **9** (1989), 207–220.
8. ———, *Noncommutative geometry*, Academic Press, San Diego, CA, 1994.
9. ———, *Geometry from the spectral point of view*, Lett. Math. Phys. **34** (1995), no. 3, 203–238.
10. C. Consani, *Double complexes and Euler L-factors*, Compositio Math. **111** (1998), no. 3, 323–358.
11. C. Consani and M. Marcolli, *Noncommutative geometry, dynamics and ∞-adic Arakelov geometry*, Selecta Math. (N.S.) (to appear); `math.AG/0205306`.
12. ———, *Spectral triples from Mumford curves*, Int. Math. Res. Not. (2003), no. 36, 1945–1972.
13. J. Cuntz, *A class of C^*-algebras and topological Markov chains. II. Reducible chains and the Ext-functor for C^*-algebras*, Invent. Math. **63** (1981), no. 1, 25–40.
14. J. Cuntz and W. Krieger, *A class of C^*-algebras and topological Markov chains*, Invent. Math. **56** (1980), no. 3, 251–268.
15. C. Deninger, *On the Γ-factors attached to motives*, Invent. Math. **104** (1991), no. 2, 245–261.
16. ———, *Local L-factors of motives and regularized determinants*, Invent. Math. **107** (1992), no. 1, 135–150.
17. ———, *Motivic L-functions and regularized determinants*, Motives (Seattle, WA, 1991), Proc. Sympos. Pure Math., vol. 55, Part 1, Amer. Math. Soc., Providence, RI, 1994, pp. 707–743.
18. A. Kumjian and D. Pask, *C^*-algebras of directed graphs and group actions*, Ergodic Theory Dynam. Systems **19** (1999), no 6, 1503–1519.
19. A. Kumjian, D. Pask, and I. Raeburn, *Cuntz–Krieger algebras of directed graphs*, Pacific J. Math. **184** (1998), no. 1, 161–174.
20. Yu. I. Manin, *p-adic automorphic functions*, J. Soviet Math. **5** (1976), 279-333.
21. ———, *Cubic forms*, North Holland, 1974.
22. ———, *Three-dimensional hyperbolic geometry as ∞-adic Arakelov geometry*, Invent. Math. **104** (1991), no. 2, 223–244.
23. ———, *Lectures on zeta functions and motives*, Columbia University Number Theory Seminar (New York, 1992), Astérisque, vol. 228 , Soc. Math. France, Paris, 1995, pp. 121–163.
24. Yu. I. Manin and M. Marcolli, *Continued fractions, modular symbols, and noncommutative geometry*, Selecta Math. (N.S.) **8** (2002), no. 3, 475–520.
25. M. Marcolli, *Limiting modular symbols and the Lyapunov spectrum*, J. Number Theory **98** (2003), no. 2, 348–376.
26. D. Mumford, *An analytic construction of degenerating curves over complete local rings*, Compositio Math. **24** (1972), 129–174.
27. W. Parry and S. Tuncel, *Classification problems in ergodic theory*, London Math. Soc. Lecture Note Ser., vol. 67, Cambridge Univ. Press, Cambridge–New York, 1982.
28. G. Robertson, *Boundary actions for affine buildings and higher rank Cuntz–Krieger algebras*, C^*-algebras (Münster, 1999), Springer, Berlin, 2000, pp. 182–202.
29. J.-P. Serre, *Facteurs locaux des fonctions zêta des variétés algébriques (définitions et conjectures)*, Séminaire Delange–Pisot–Poitou. 11e année: 1969/70, Secrétariat Mathématique, Paris, 1970, exp. 19.
30. J. Spielberg, *Free-product groups, Cuntz–Krieger algebras, and covariant maps*, Internat. J. Math. **2** (1991), no. 4, 457–476.
31. D. Sullivan, *On the ergodic theory at infinity of an arbitrary discrete group of hyperbolic motions*, Riemann Surfaces and Related Topics: Proceedings of the 1978 Stony Brook Conference (Stony Brook, NY, 1978), Ann. of Math. Stud., vol. 97, Princeton Univ. Press, Princeton, NJ, 1981, pp. 465–496.

Department of Mathematics, University of Toronto, 100 St. George St., Toronto, ON M5S 3G3, Canada.
E-mail address: kc@math.toronto.edu

Max-Planck-Institut für Mathematik, P.O. Box 7280, 53072 Bonn, Germany.
E-mail address: marcolli@mpim-bonn.mpg.de

Centre de Recherches Mathématiques
CRM Proceedings and Lecture Notes
Volume **36**, 2004

A Fourteenth Lecture on Fermat's Last Theorem

Henri Darmon

I informed you earlier of the death of Fermat. He is still alive, and we no longer fear for his health, even though we had counted him among the dead a short time ago . . . .

Letter of Bernard Medon to Nicholas Heinsius, 1652

The title of this lecture alludes to Ribenboim's delightful treatise on Fermat's Last Theorem [**23**]. Fifteen years after the publication of [**23**], Andrew Wiles finally succeeded in solving Fermat's 350-year-old conundrum. That same year, perhaps to console himself of Fermat's demise, Ribenboim published a second book, this time on *Catalan's conjecture* that there are no consecutive perfect powers other than 8 and 9. As we have learned at this Congress, Preda Mihailescu has just disposed of this conjecture as well. His breakthrough comes only 8 years after the publication of Ribenboim's book on Catalan's equation.

Such is the magic of Ribenboim's books: the age-old problems which they treat have invariably been solved, in comparatively short order! So it is with some eagerness that we await the publication of Ribenboim's next tome (hoping it will be devoted to the Riemann Hypothesis, or the Birch and Swinnerton–Dyer conjecture, . . .).

This "fourteenth lecture" is meant as a tribute both to Ribenboim and to the spirit of Fermat: the fascination with concrete Diophantine problems, especially those that draw us, seemingly inexorably, to central topics in the subject (cyclotomic fields, elliptic curves, reciprocity laws, modular forms, . . .).

Since Fermat's and Catalan's equations have surrendered their mysteries, let me begin with a question which combines features of both: the so-called *generalized Fermat equation*

$$x^p + y^q = z^r. \tag{1}$$

This equation, in which the exponents in Fermat's equation are allowed to be different, is discussed in [**4, 8, 16**]; most of this lecture is merely an *update* of the expository article [**4**]. Its main novelty resides in the connection which is drawn,

---

2000 *Mathematics Subject Classification.* Primary: 11D41; Secondary: 11G18 and 14G25.

This is a transcription of the author's Ribenboim Prize lecture given at the CNTA meeting in Montreal in May 2002.

This is the final form of the paper.

following [**6**], between hypergeometric abelian varieties and (1), as well as in the point of view which differs, albeit superficially, from that of [**4**].

It is useful to allow the exponents p, q or r to be equal to ∞. In that case the somewhat ad-hoc convention

$$z^\infty = \begin{cases} 1 & \text{if } z = 1; \\ \infty & \text{otherwise} \end{cases}$$

is adopted, so that the Catalan equation becomes a special case of (1) with $r = \infty$. The usual convention that $1/\infty = 0$ is adopted as well.

In studying (1), it is important to consider only the so-called *primitive* integer solutions, satisfying

$$\gcd(x, y, z) = 1.$$

This is because nonprimitive solutions are often not hard to generate and are therefore less interesting. For example, if a and b are arbitrary integers, the exponent n is odd, and $c = a^n + b^n$, then $(ac, bc, c^{(n+1)/2})$ is a parametrized family of nonprimitive solution to the generalized Fermat equation with exponents $(n, n, 2)$. When considering only primitive solutions, the story is quite different, as the following conjecture suggests.

CONJECTURE 1. *When $1/p + 1/q + 1/r < 1$, the generalized Fermat equation has no nontrivial primitive solutions except for*

$$1^p + 2^3 = 3^2, \qquad 2^5 + 7^2 = 3^4, \qquad 7^3 + 13^2 = 2^9,$$
$$2^7 + 17^3 = 71^2, \qquad 3^5 + 11^4 = 122^2,$$
$$17^7 + 76271^3 = 21063928^2, \qquad 1414^3 + 2213459^2 = 65^7,$$
$$9262^3 + 15312283^2 = 113^7, \qquad 43^8 + 96222^3 = 30042907^2$$
$$33^8 + 1549034^2 = 15613^3.$$

Andrew Granville (a student of Ribenboim, who also holds the first Ribenboim prize) and the author proved the following theorem in the Spring of 1993, shortly before Wiles' momentous announcement in June of that year.

THEOREM 2. *Suppose (p, q, r) are fixed and satisfy $1/p + 1/q + 1/r < 1$. Then the equation $x^p + y^q = z^r$ has finitely many primitive integer solutions.*

PROOF. The proof (which is also presented, in slightly different forms, in [**4, 8, 16**]) relies on the properties of the *Hecke Triangle Group* defined by the abstract presentation:

$$\Gamma_{p,q,r} := \langle \gamma_0, \gamma_1, \gamma_\infty \mid \gamma_0^p = \gamma_1^q = \gamma_\infty^r = \gamma_0\gamma_1\gamma_\infty = 1 \rangle. \tag{2}$$

(If one of the exponents p, q or r is equal to ∞, the corresponding γ_j is taken to be of infinite order.)

The group $\Gamma_{p,q,r}$ is finite if and only if $1/p + 1/q + 1/r > 1$, and is *infinite and nonabelian* if and only if $1/p + 1/q + 1/r < 1$. The reason for this is suggested by the "triangle group" terminology: if τ_0, τ_1, and τ_∞ denote the reflections about the sides of a triangle $\Delta_{p,q,r}$ with opposite angles π/p, π/q and π/r, then the rotations

$$\sigma_0 = \tau_1\tau_\infty, \qquad \sigma_1 = \tau_\infty\tau_0, \qquad \sigma_\infty = \tau_0\tau_1$$

satisfy precisely the same relations as the γ_j in (2). Identifying γ_j with σ_j realizes $\Gamma_{p,q,r}$ as a group of isometries of the underlying plane geometry, with two copies of $\Delta_{p,q,r}$ as fundamental region. The triangle $\Delta_{p,q,r}$ can be drawn on the sphere if $1/p+1/q+1/r>1$, on the usual Euclidean plane if $1/p+1/q+1/r=1$, and on the Poincaré upper half-plane

$$\mathcal{H}=\{z\in\mathbb{C}\mid \operatorname{Im}(z)>0\}$$

if $1/p+1/q+1/r<1$. In the latter case one obtains a model for $\Gamma_{p,q,r}$ as a discrete group acting by isometries on $\mathcal{H}$ with respect to the hyperbolic metric, and hence an embedding $\Gamma_{p,q,r}\subset \mathbf{PSL}_2(\mathbb{R})$. It is an instructive exercise to show that this embedding (by a judicious choice of $\Delta_{p,q,r}$) can be conjugated into an embedding

$$\Gamma_{p,q,r}\to \mathbf{PSL}_2(\mathcal{O}), \tag{3}$$

where $\mathcal{O}$ is the ring of integers of a number field. (More precisely: the real subfield of the field of pqrth roots of unity.) In particular it follows that $\Gamma_{p,q,r}$ has plenty of finite nonabelian quotients (obtained by reducing the embedding of (3) modulo an ideal of $\mathcal{O}$).

By definition, the group $\Gamma_{p,q,r}$ is a quotient of the group $\Gamma_{\infty,\infty,\infty}$, which can itself be identified with the fundamental group of $\mathbb{P}_1-\{0,1,\infty\}$, by assigning to γ_0, γ_1 and γ_∞ certain well-chosen homotopy classes of loops around 0, 1 and ∞. By composing with the natural surjective homomorphism

$$\Gamma_{\infty,\infty,\infty}\to\Gamma_{p,q,r},$$

any surjective homomorphism $\varphi\colon \Gamma_{p,q,r}\to G$ gives a surjection

$$\pi_1(\mathbb{P}_1-\{0,1,\infty\})\to G,$$

which corresponds in the usual way to a ramified Galois covering of topological spaces with Galois group G:

$$\pi\colon X\to\mathbb{P}_1. \tag{4}$$

This covering is of "signature (p,q,r)" in the sense of [**25**, Section 6.4]: it is unramified over $\mathbb{P}_1-\{0,1,\infty\}$, and has ramification indices dividing p, q, and r over 0, 1, and ∞ respectively.

The finite covering X inherits through π a natural structure of a *compact Riemann surface*. By Riemann's existence theorem, X arises from an algebraic curve defined over $\mathbb{C}$. Such a curve can of course be defined over a finitely generated extension of $\mathbb{Q}$, and hence, by specialization, over some number field K. We will now view X in this way, so that π becomes a rational function defined over K.

Let $\Pi_{p,q,r}$ be the set of primitive solutions to $x^p+y^q=z^r$, and let

$$\Sigma_{p,q,r}=\{a^p/c^r \text{ such that } (a,b,c)\in\Pi_{p,q,r}\}\subset\mathbb{P}_1(\mathbb{Q}).$$

To prove Theorem 2 it is enough to show that $\Sigma_{p,q,r}$ is finite. For $t\in\mathbb{P}_1(\mathbb{Q})$, let $L_t=K(\pi^{-1}(t))$ be the field generated over K by the coordinates of $\pi^{-1}(t)$.

The following proposition, an extension of the Chevalley–Weil theorem for ramified coverings, gives some control over the primes of ramification of L_t:

PROPOSITION 3. *There exists a finite set S of primes of K (depending on π, but not on $t\in\Sigma_{p,q,r}$) such that L_t is unramified outside of S, for all $t\in\Sigma_{p,q,r}$.*

SKETCH OF PROOF. Let S be the finite set of primes at which (4) has bad reduction and which divide pqr. For each place v of K which is not in S, write K_v and $\mathcal{O}_v$ for the completion of K at v and its ring of integers respectively. The main point is that, since (a,b,c) is a *primitive* solution,

(1) The numerator of $t = a^p/c^r$ is a pth power in $K_v^\times/\mathcal{O}_v^\times$;
(2) The numerator of $t - 1 = -b^q/c^r$ is a qth power in $K_v^\times/\mathcal{O}_v^\times$;
(3) The denominator of $t = a^p/c^r$ is an rth power in $K_v^\times/\mathcal{O}_v^\times$.

The result follows from the fact that X has ramification indices p, q, r over 0, 1, ∞ by a variant of the Chevalley–Weil theorem [**1**] for ramified coverings. □

To control the possible L_t's that could arise, we invoke the following fundamental result of Hermite–Minkowski.

PROPOSITION 4. *There exists only finitely many extensions of K of degree $\leq d$ which are unramified outside S.*

Hence the fields L_t, as t ranges over $\Sigma_{p,q,r}$, are finite in number. In particular, the compositum:

$$L = \bigcup_{t \in \Sigma_{p,q,r}} L_t$$

is a *finite* extension of K. Note that

$$\Sigma_{p,q,r} \text{ is contained in } \pi\big(X(L)\big), \tag{5}$$

so that Theorem 2 is reduced to a question about the finiteness of algebraic points on the curve X over a single number field L.

PROPOSITION 5. *The curve X has genus strictly greater than* 1.

PROOF. It is a consequence of the Riemann–Hurwitz formula that the Euler characteristic $2 - 2g(X)$ of X is equal to

$$\chi(X) = \deg(\pi)\left(\frac{1}{p} + \frac{1}{q} + \frac{1}{r} - 1\right).$$

Note that it is here that the assumption on (p, q, r) made in the statement of Theorem 2 enters into the proof. □

To complete the proof of Theorem 2, it remains to roll out the big ammunition: Falting's theorem, which asserts that $X(L)$ is finite since X is a curve of genus > 1 and L is a number field. Theorem 2 now follows from (5). □

REMARKS. (1) The reduction of Theorem 2 to Faltings' theorem is similar to an approach of Elkies used to show that the *abc* conjecture implies the Mordell conjecture [**10**]. The proof of Theorem 2 simply reverses the direction of Elkies' argument, using the Mordell conjecture to deduce a very special case of the *abc* conjecture.

(2) The proof sketched above is also explained in [**4, 8**]. Its main virtue lies in the fact that it suggests a general approach to equation (1) (and, in particular, to Fermat's equation). It is summarized in the following loosely stated principle, which is in some sense the main thesis of this lecture:

PRINCIPLE 6. There is a dictionary between the distinct strategies for studying $x^p + y^q = z^r$ and the finite quotients of the Hecke triangle group $\Gamma_{p,q,r}$.

To be more precise, let G be a finite quotient of $\Gamma_{p,q,r}$ and let φ be the natural surjective homomorphism. The attendant strategy for gaining insights into (1) may be described as follows.

(1) The geometric part: Study the corresponding G-covering

$$\pi\colon X \to \mathbb{P}_1.$$

(E.g.: Find an equation—or, better yet, a moduli interpretation—for X, a "minimal" field of definition K, the set S of primes of K where π has bad reduction, etc.).

(2) The arithmetic part: determine a finite extension L of K satisfying (5).

(3) The Diophantine part: understand the collection $X(L)$ as precisely as possible.

As the reader might expect, carrying out this strategy represents a fairly tall order, but this breakdown provides a conceptual template with which to organize the various approaches that have been followed over the years to study equation (1). To illustrate this point, we begin with a discussion of the "classical" Fermat equation

$$x^p + y^p = z^p, \quad p \text{ an odd prime.} \tag{6}$$

In increasing order of sophistication, one might approach (6) by exploiting:

1. Abelian quotients of $\Gamma_{p,p,p}$. Note that $\Gamma_{p,p,p}^{\mathrm{ab}} = \mathbb{Z}/p\mathbb{Z} \times \mathbb{Z}/p\mathbb{Z}$. If we take for φ the natural map from $\Gamma_{p,p,p}$ to its abelianization:

$$\varphi\colon \Gamma_{p,p,p} \to \mathbb{Z}/p\mathbb{Z} \times \mathbb{Z}/p\mathbb{Z},$$

the corresponding curve X is the Fermat curve itself, with equation

$$F_p : u^p + v^p = w^p, \quad \text{where } \pi(u,v,w) = (u/w)^p.$$

Any primitive solution (a,b,c) to (6) gives a rational point on F_p, for which $\pi(a,b,c) = a^p/c^p$ belongs to $\Sigma_{p,q,r}$. Therefore the extension L can be chosen to be $\mathbb{Q}$. The study of (6) has thus been "reduced," tautologically, to the study of rational points on the Fermat curve. Note here that the arithmetic step is trivial; the difficulty is concentrated in the study of the Diophantine equation (6) which has not yet succumbed to a "direct" attack.

2. Solvable quotients of $\Gamma_{p,p,p}$. Just beyond the abelian quotients lie the solvable ones. Suppose for example that $\varphi\colon \Gamma_{p,p,p} \to G$ has solvable image, and that $G/[G,G] = \mathbb{Z}/p\mathbb{Z} \times \mathbb{Z}/p\mathbb{Z}$. In that case X is an unramified covering of the Fermat curve. This probably accounts for the key role played by the study of the Jacobians of Fermat curves in certain approaches to (6). (Cf. for example [**20**].)

If one assumes further that G is a two-step nilpotent p-group, then equations for π can be chosen in such a way that $\mathbb{Q}(\pi^{-1}(a^p/c^p))$ is an extension of $\mathbb{Q}(\zeta_p)$ which is unramified outside p. This approach leads naturally to a careful study of unramified abelian extensions of $\mathbb{Q}(\zeta_p)$. A highlight of this circle of ideas is the celebrated theorem of Kummer on Fermat's equation.

THEOREM 7 (Kummer). *If there is a nontrivial solution to $x^p + y^p = z^p$, then there is an unramified cyclic extension of $\mathbb{Q}(\zeta_p)$ of degree p.*

It can be seen that some of the difficulty has been *transferred* from the Diophantine to the arithmetic part of the argument. Kummer's theorem opens a new universe of deep questions about cyclotomic fields, class numbers, special values of zeta functions, Bernoulli numbers, congruences, ..., all of which are very nicely documented in Ribenboim's "13 lectures."

While extremely fruitful both in proving various cases of (6) and in spurring the further study of cyclotomic fields, Kummer's approach ultimately runs into the following difficulties:

(1) Sometimes, an unramified cyclic extension of $\mathbb{Q}(\zeta_p)$ of degree p does exist. (p is then called an *irregular prime.*) It is known that there are infinitely many irregular primes, and, although Fermat's Last Theorem is now proved, we have not yet exhausted the wealth of deep questions raised by the possible occurrence of such primes.

(2) Even more germane to this lecture, the approach based on solvable quotients of $\Gamma_{p,q,r}$ does not extend to equation (1) in general. For example if the exponents p, q, and r in this equation are pairwise coprime, then $\Gamma_{p,q,r}$ has no abelian, and hence no solvable, quotients! This reflects the elementary observation that equation (1) in this case cannot be attacked by a factorization like the one

$$(x+y)(x+e^{2\pi i/p}y)\cdots(x+\epsilon^{2\pi i(p-1)/p}y) = z^p$$

which is the starting point for Kummer's work on Fermat's Last Theorem.

For these two reasons, it becomes appealing to explore strategies based on

3. Nonsolvable quotients of $\boldsymbol{\Gamma_{p,q,r}}$. Perhaps the most obvious nonsolvable finite quotients are obtained by considering the linear representations of $\Gamma_{p,q,r}$ with coefficients, say, in a finite field F. The first nonabelian examples of such representations are, of course, the two-dimensional ones.

One gains some flexibility by considering *projective* representations

$$\Gamma_{p,q,r} \to \mathbf{PGL}_2(F),$$

which amounts to studying the linear representations of the central extension

$$\widetilde{\Gamma}_{p,q,r} := \langle \gamma_0, \gamma_1, \gamma_\infty \mid \gamma_0^p = \gamma_1^q = \gamma_\infty^r = 1, \gamma_0\gamma_1\gamma_\infty = -1 \rangle,$$

(where -1 denotes a central involution).

Definition 8. A *Frey representation* attached to $x^p + y^q = z^r$ with coefficients in F is a representation

$$\varphi\colon \widetilde{\Gamma}_{p,q,r} \to \mathbf{GL}_2(F).$$

Two Frey representations are identified if one can be obtained from the other by extending the field F of scalars and conjugating. A Frey representation φ is said to be *even* if $\varphi(-1) = 1$, and *odd* if $\varphi(-1) = -1$.

With these concepts in hand, let us now return to discussing the classical Fermat equation of prime exponent p. In this context it is most natural perhaps to study Frey representations with coefficients in a field of characteristic p. The following theorem (attributed, somewhat anachronistically, to Hecke) indicates that in pursuing this line of inquiry one's options become very restricted. This is actually an encouraging sign, since experience indicates that an approach to a problem which is canonical and allows for little "wriggle room" is often more fruitful and deserving of study.

Proposition 9. *There is a unique irreducible Frey representation attached to $x^p + y^p = z^p$ with coefficients in characteristic p. This representation is odd.*

We refer the reader to Theorem 1.5 of [**6**] for a proof of this purely group-theoretic statement.

To construct the Frey representation of Proposition 9, the generators γ_0 and γ_1 can be sent to the matrices

$$\begin{pmatrix} 1 & 1 \\ 0 & 1 \end{pmatrix} \quad \text{and} \quad \begin{pmatrix} 1 & 0 \\ -4 & 1 \end{pmatrix}.$$

It is hardly surprising that the corresponding covering

$$\pi\colon X \longrightarrow \mathbb{P}_1 - \{0, 1, \infty\},$$

being essentially unique, should correspond to a well studied mathematical object. This is indeed the case: the curve X can be realized as the covering of modular curves

$$\pi\colon X(2p) \longrightarrow X(2),$$

where $X(n)$ is the modular curve attached to the full congruence subgroup of level n. An equation for a universal elliptic curve over $X(2)$ is provided by the Legendre family

$$y^2 = x(x-1)(x-t).$$

With this choice of π, it can be seen that the field $\mathbb{Q}(\pi^{-1}(a^p/c^p))$ is closely related to the field of definition of the points of order p of the Frey curve

$$E_{a,b,c}: \quad y^2 = x(x-a^p)(x+b^p).$$

This chain of reasoning has led us to recover Frey's proposed strategy for studying Fermat's Last Theorem.

Let

$$\rho_{a,b,c}\colon G_{\mathbb{Q}} \longrightarrow \mathbf{GL}_2(\mathbb{F}_p)$$

be the Galois representation attached to the p-division points of the elliptic curve $E_{a,b,c}$.

Carrying out the arithmetic step requires a precise understanding of the behavior of $\rho_{a,b,c}$ at the "bad primes"—more precisely, of its restriction to the inertia groups at these primes. Since so much is known about the local properties of Galois representations attached to elliptic curves, this can be done, and the outcome is summarized in the following proposition, which can be viewed as a nonabelian analogue of Theorem 7 of Kummer.

THEOREM 10 (Frey). *If there is a nontrivial solution to $x^p + y^p = z^p$, then there is a Galois representation*

$$\rho_{a,b,c}\colon G_{\mathbb{Q}} \longrightarrow \mathbf{GL}_2(\mathbb{F}_p)$$

(with cyclotomic determinant) which is

(1) *Unramified outside* $2p$.
(2) *"Finite" at* p.
(3) *"Semistable" at* 2.

We will not go into a definition of the technical terms marked in quotes, referring to Section 2.2. of [**7**] for a more detailed discussion.

We are now faced with the problem of understanding or classifying the representations satisfying the conclusion of Theorem 10. This leads to considerations involving *nonabelian* class field theory, which, in its modern formulation, attempts to understand nonabelian extensions of $\mathbb{Q}$ (or of number fields), and more specifically, finite-dimensional representations of $G_{\mathbb{Q}}$. The Langlands philosophy predicts that these representations should be related to *modular forms.* And this is precisely what has been proved by Wiles [**28**] for the representation $\rho_{a,b,c}$. More precisely,

THEOREM 11 (Wiles). *Suppose that $\rho_{a,b,c}$ is irreducible. Then it is attached to a cusp form of level $N = \prod_{\ell | abc} \ell$.*

The following deep result of Ribet, which makes crucial use of the fact that ρ is unramified outside of 2 and p and is finite at p, then leads to the following conclusion:

THEOREM 12 (Ribet). *The representation $\rho_{a,b,c}$ is attached to a cusp form* (mod p) *of level* 2.

But a direct calculation now reveals that there are no such modular forms! (The modular curve $X_0(2)$ has genus 0 and hence there are no cusp forms of weight 2 on $\Gamma_0(2)$.) Hence:

(7) The representation $\rho_{a,b,c}$ must be reducible.

The Diophantine step in the argument now consists in showing that a mod p Galois representation arising from the division points of an elliptic curve cannot be reducible, at least if p is large enough. This amounts to studying the rational points on modular curves of the form $X_0(p)$ (or, eventually, in this situation, $X_0(2p)$). In some sense, the difficult argument used to prove (7) has enabled us to *transfer* a Diophantine question about Fermat curves, to a similar one, of ostensibly similar difficulty, about modular curves. Fortunately, something has been gained in this transfer. Indeed we know, thanks to the fundamental work of Mazur on the Eisenstein ideal [**18, 19**], that the modular curve $X_0(p)$ is *more tractable* than the Fermat curve of exponent p. More precisely, one has

THEOREM 13 (Mazur). *If $p > 163$, then the curve $X_0(p)$ has no rational points other than the cusps, and hence $\rho_{a,b,c}$ is irreducible.*

This contradiction to (7) proves Fermat's Last Theorem, at least for $p > 163$. (Of course, the small exponents had already been handled a long time ago, and in any case many could be disposed of by running through the above argument more carefully.)

Let us now return to the generalized Fermat equation $x^p + y^q = z^r$. As in the case of the Frey representations attached to $x^p + y^p = z^p$, one has the following

RIGIDITY PRINCIPLE. The Frey representations attached to a given triple (p, q, r) of exponents, with coefficients in a field $\mathbb{F}$ of a given characteristic, are *essentially* unique.

For precise statements in this direction see [**6**, Theorems 1.7 and 1.8].

Once again, the Frey representations attached to (1), being essentially unique, should correspond to familiar, previously studied mathematical objects. This is indeed the case. The two-dimensional representations of $\widetilde{\Gamma}_{p,q,r}$, correspond to rank two local systems on $\mathbb{P}_1 - \{0, 1, \infty\}$ with prescribed monodromies (of finite orders, p, q, and r) over the three deleted points. Such systems have been well-studied classically (by mathematicians like Gauss, Riemann, and Poincaré), and are intimately connected to hypergeometric differential equations.

The Frey representations attached to (1) can be realized on the torsion of certain *hypergeometric abelian varieties*, so-called because their periods are expressed in terms of values of classical hypergeometric functions at rational arguments. A fairly general discussion of this point is given in Section 1.3 of [**6**], and we will content ourselves with focusing on an illustrative special case. Consider from now on the equation

$$x^p + y^p = z^r, \tag{8}$$

where p and r are odd primes, and let $K = \mathbb{Q}(\zeta_r)^+$. We take the point of view that r is fixed and that p is allowed to vary, corresponding to the Diophantine problem of understanding whether an rth power can be expressed as a sum of two relatively prime perfect powers.

PROPOSITION 14. *There exists two abelian varieties J_r^+ and J_r^- over $\mathbb{Q}(t)$ having good reduction outside of $t = 0$, 1, and ∞ such that*

(1) $\mathrm{End}_K(J_r^\pm) \otimes \mathbb{Q} \simeq K$;
(2) $\dim(J_r^\pm) = [K{:}\mathbb{Q}]$. (*The abelian variety $J_r^\pm$ is said to be of "$\mathbf{GL}_2$-type" over K.*) *For each prime $\mathfrak{p}$ of K, such an abelian variety gives rise to a two-dimensional* mod p *representation of $G_{K(t)}$,*
$$\rho_{\mathfrak{p}}^\pm : G_{K(t)} \to \mathbf{GL}_2(\mathbb{F}),$$
where $\mathbb{F}$ is the residue field of K above $\mathfrak{p}$.
(3) *For each rational prime p, and $\mathfrak{p}|p$, the representation $\rho_{\mathfrak{p}}^+$ (resp. $\rho_{\mathfrak{p}}^-$) is the* (unique, *up to twist, and automorphisms of $\mathbb{F}$*) *even (resp. odd) Frey representation attached to $x^p + y^p = z^r$.*

The hypergeometric abelian varieties of Proposition 14 play the same role in the study of equation (8) as the elliptic curves in the Legendre family in the study of equation (6).

Here is what happens in carrying out the main steps in the study of (1), following our earlier sketch of the proof of Fermat's Last Theorem. (For more details, see [**6**].)

To carry out the *arithmetic step*, one is led to consider the problem of establishing the modularity (in a suitable sense, that is made precise in [**6**]) of hypergeometric abelian varieties: this amounts to showing that $J_r^\pm(t)$, for $t \in \mathbb{Q}$, say, is modular, in the sense that its ℓ-adic representations correspond to a *Hilbert modular form* over the totally real field K.

Let $\mathfrak{r}$ be the unique prime of K above r. Then the $\mathfrak{r}$-torsion subgroup $J_r^\pm(t)[\mathfrak{r}]$ is a two-dimensional $\mathbb{F}_r$-vector space on which G_K acts linearly. It can be shown that this action *extends* to an action of $G_{\mathbb{Q}}$, and that

THEOREM 15. (1) *The $G_{\mathbb{Q}}$-representation $J_r^+(t)[\mathfrak{r}]$ is reducible.*
(2) *The representation $J_r^-(t)[\mathfrak{r}]$ is isomorphic to (a twist of) the representation $E[r]$, where*
$$E\colon y^2 = x(x-1)(x-t).$$

SKETCH OF PROOF. By Proposition 14, the $\mathbb{F}_r[G_{\mathbb{Q}}]$-modules $J_r^+[\mathfrak{r}]$ and $J_r^-[\mathfrak{r}]$ are even and odd Frey representations attached to the Fermat equation $x^r + y^r = z^r$. Hence Proposition 9 implies that the representation attached to $J_r^+[\mathfrak{r}]$ is reducible (since there are no even irreducible Frey representations attached to the classical

Fermat equation) and that the $G_{\mathbb{Q}(t)}$-module $J_r^-[\mathfrak{r}]$ coincides (up to a twist) with the r-torsion points of the Frey curve. □

Applying the modularity result of [**3**], we then have:

COROLLARY 16. *The* mod r *representation* $J_r^{\pm}(t)[\mathfrak{r}]$ *is modular, i.e., corresponds to a classical modular form over* $\mathbb{Q}$, mod r.

Now applying the technique of base change over an abelian extension of $\mathbb{Q}$, one may conclude that $J_r^{\pm}(t)[\mathfrak{r}]$ corresponds to a Hilbert modular form over K, mod r. Now we are in a good situation to apply the lifting theorems of Wiles, which assert, roughly, that an r-adic representation is modular if the corresponding mod r representation is. (So Wiles' modularity theorem would play the same role, in establishing the modularity of hypergeometric abelian varieties, as the Langlands–Tunnell theorem in Wiles' original argument!) Unfortunately the lifting statement of Wiles and some of its subsequent generalizations (cf. for example [**26, 27**]) come with a number of technical restrictions which preclude us from deducing the modularity of all the hypergeometric abelian varieties involved in the argument. (See [**5**] for a more detailed discussion.) But note how this strategy for proving the modularity of all hypergeometric abelian varieties emerges naturally from the study of the generalized Fermat equation (and is arguably the most valuable insight to have emerged so far from the author's study of this equation).

When $r = 2$ and 3, the hypergeometric abelian varieties $J_r^{\pm}$ correspond to families of elliptic curves (which include the (uni)versal families over $X_0(2)$ and $X_0(3)$ respectively). This fact allowed Merel and the author [**9**] to prove the following theorem, generalizing classical results of Fermat and Euler in the case of $n = 4$ and 3 respectively.

THEOREM 17. *The equations* $x^n + y^n = z^2$ $(n \geq 4)$ *and* $x^n + y^n = z^3$ $(n \geq 3)$ *have no nontrivial primitive solutions.*

REMARK. The proofs for small n are handled by Poonen [**22**] by classical descent techniques.

In trying to extend the analysis to larger values of r, one runs into considerable difficulties of a more than technical nature. Why? The proof of Theorem 17 exploits the isomorphisms

$$\Gamma_{\infty,\infty,2} = \Gamma_0(2); \qquad \Gamma_{\infty,\infty,3} = \Gamma_0(3).$$

When $r \geq 5$, by contrast, the group $\Gamma_{\infty,\infty,r}$ is *nonarithmetic*, and it then seems like a very difficult problem to extend Mazur's Diophantine results on modular curves to the ramified coverings of $\mathbb{P}_1$ obtained by considering the torsion points of hypergeometric abelian varieties over $\mathbb{Q}(\zeta_r)^+(t)$. One of the difficulties of the nonarithmetic case resides in the fact that the corresponding coverings have very few correspondences and there is no analogue of the ring of Hecke operators which plays such an important role in the arguments of [**18, 19**].

The difficulty of the general question has not prevented various mathematicians from proving a number of partial results which would fill a handsome chapter in any sequel to Ribenboim's "13 lectures." For example, Alain Kraus also proved the following result about the equation $x^3 + y^3 = z^p$ by using modular form methods [**15**]:

THEOREM 18 (Kraus, 1998). *The equation*

$$x^3 + y^3 = z^p$$

has no nontrivial primitive solution for $17 \le p \le 10,000$.

He does this by establishing criteria (strikingly reminiscent of the criteria of Sophie Germain and Legendre for disposing the first case of Fermat's Last Theorem) which guarantee that the equation $x^3 + y^3 = z^p$ has no nontrivial primitive solution. These criteria are presumably satisfied for all p but were checked numerically in the range given by the theorem.

There is also the following theorem of Jordan Ellenberg concerning the equation $x^4 + y^2 = z^p$:

THEOREM 19 (Ellenberg, 1999). *The equation*

$$x^4 + y^2 = z^p$$

has no nontrivial primitive solutions, if $p > 211$.

The proof uses some new modularity results on $\mathbb{Q}$-curves obtained in collaboration with Chris Skinner [**12**].

Although it falls somewhat outside the scope of these lectures since it concerns the equation

$$Ax^p + By^p = Cz^p, \tag{9}$$

I cannot resist mentioning the following gem which would have to figure prominently in any sequel to [**23**] concerned with (9).

THEOREM 20 (Halberstadt and Kraus). *Suppose A, B and C are odd and relatively prime. There exists a set Π of primes of positive density (depending, in an effectively computable way, on A, B, C) such that, for all $p \in \Pi$, the equation*

$$Ax^p + By^p = Cz^p$$

has no nontrivial solution.

The proof of Halberstadt and Kraus, an application of their "symplectic method," is based on a careful study of the module of p-division points of the Frey curve endowed with its natural symplectic structure arising from the Weil pairing. Ultimately, it relies crucially on the fact that the Frey curve

$$y^2 = x(x - Au^p)(x + Cw^p)$$

has a discriminant of the form

$$\Delta = 2^{-8}(ABC)^2(uvw)^{2p}$$

in which the exponent of 2 that appears is negative!

Results have also been obtained for the more general equation

$$Ax^p + By^q = Cz^r \tag{10}$$

which blends the features of (1) and (9). In this context let us mention the work of Bennett and Skinner [**2**] and Ivorra [**14**] on the equation $x^n + y^n = Cz^2$, and of Kraus on the equations $x^m - y^m = Rz^n$ [**17**].

In conclusion, the generalized Fermat equation seems to share with its classical counterpart the power of generating important mathematical questions. Aside from two key special cases—Fermat's Last Theorem, and Catalan's conjecture—and the fragmentary results in Theorem 17 [**9**], Theorem 18 [**15**], and Theorem 19 [**11**] mentioned above, almost everything about it remains unknown. Ribenboim should be delighted by this state of affairs!

References

1. S. Beckmann, *Ramified primes in the field of moduli of branched coverings of curves*, J. Algebra **125** (1989), no. 1, 236–255.
2. M. Bennett and C. Skinner, *Ternary Diophantine equations via Galois representations and modular forms*, Canad. J. Math. (to appear).
3. C. Breuil, B. Conrad, F. Diamond, and R. Taylor, *On the modularity of elliptic curves over* $\mathbb{Q}$: *wild* 3*-adic exercises*, J. Amer. Math. Soc. **14** (2001), no. 4, 843–939.
4. H. Darmon, *Faltings plus epsilon, Wiles plus epsilon, and the generalized Fermat equation*, C. R. Math. Rep. Acad. Sci. Canada **19** (1997), no. 1, 3–14.
5. ______, *Modularity of fibres in rigid local systems*, Ann. of Math. (2) **149** (1999), no. 3, 1079–1086.
6. ______, *Rigid local systems, Hilbert modular forms, and Fermat's last theorem*, Duke Math. J. **102** (2000), no. 3, 413–449.
7. H. Darmon, F. Diamond, and R. Taylor, *Fermat's last theorem*, Elliptic Curves, Modular Forms & Fermat's Last Theorem (Hong Kong, 1993), Internat. Press, Cambridge, MA, 1997, pp. 2–140.
8. H. Darmon and A. Granville, *On the equations* $z^m = F(x, y)$ *and* $Ax^p + By^q = Cz^r$, Bull. London Math. Soc. **27** (1995), no. 6, 513–543.
9. H. Darmon and L. Merel, *Winding quotients and some variants of Fermat's last theorem*, J. Reine Angew. Math. **490** (1997), 81–100.
10. N. D. Elkies, *ABC implies Mordell*, Internat. Math. Res. Notices (1991), no. 7, 99–109.
11. J. S. Ellenberg, *Galois representations attached to* $\mathbb{Q}$*-curves and the generalized Fermat equation* $A^4 + B^2 = C^p$, Amer. J. Math. (to appear).
12. J. S. Ellenberg and C. Skinner, *On the modularity of* $\mathbb{Q}$*-curves*, Duke Math. J. **109** (2001), no. 1, 97–122.
13. E. Halberstadt and A. Kraus, *Courbes de Fermat: résultats et problèmes*, J. Reine Angew. Math. **548** (2002), 167–234.
14. W. Ivorra, Ph.D. thesis, Université Paris VI, Paris (in progress).
15. A. Kraus, *Sur l'équation* $a^3 + b^3 = c^p$, Experiment. Math. **7** (1998), no. 1, 1–13.
16. ______, *On the equation* $x^p + y^q = z^r$*: a survey*, Ramanujan J. **3** (1999), no. 3, 315–333.
17. ______, *Une question sur les équations* $x^m - y^m = Rz^n$, Compositio Math. **132** (2002), no. 1, 1–26.
18. B. Mazur, *Modular curves and the Eisenstein ideal*, Inst. Hautes Études Sci. Publ. Math. **47** (1977), 33–186.
19. ______, *Rational isogenies of prime degree*, Invent. Math. **44** (1978), no. 2, 129–162 (with an appendix by D. Goldfeld).
20. W. G. McCallum, *The arithmetic of Fermat curves*, Math. Ann. **294** (1992), no. 3, 503–511.
21. P. Mihailescu, *Primary units and a proof of Catalan's conjecture*, Crelle's Journal (submitted).
22. B. Poonen, *Some Diophantine equations of the form* $x^n + y^n = z^m$, Acta Arith. **86** (1998), no. 3, 193–205.
23. P. Ribenboim, *Thirteen lectures on Fermat's last theorem*, Springer-Verlag, New York–Heidelberg, 1979.
24. ______, *Catalan's conjecture. Are* 8 *and* 9 *the only consecutive powers?* Academic Press, Inc., Boston, MA, 1994.
25. J.-P. Serre, *Topics in Galois theory*, Res. Notes Math., vol. 1, Jones and Bartlett Publishers, Boston, MA, 1992 (Lecture notes prepared by H. Damon. With a foreword by H. Darmon and the author).
26. C. Skinner and A. Wiles, *Residually reducible representations and modular forms*, Inst. Hautes Études Sci. Publ. Math. **89** (1999), 5–126.
27. ______, *Nearly ordinary deformations of irreducible residual representations*, Ann. Fac. Sci. Toulouse Math. (6) **10** (2001), no. 1, 185–215.
28. A. Wiles, *Modular elliptic curves and Fermat's last theorem*, Ann. of Math. (2) **141** (1995), no. 3, 443–551.

Department of Mathematics and Statistics, McGill University, 805 Sherbrooke W., Montréal, QC H3A 2K6, Canada.

E-mail address: darmon@math.mcgill.ca

Centre de Recherches Mathématiques
CRM Proceedings and Lecture Notes
Volume **36**, 2004

Number Fields with Large Class Group

William Duke

Abstract. After a review of the quadratic case, a general problem about the existence of number fields of a fixed degree with extremely large class numbers is formulated. This problem is solved for abelian cubic fields. Then some conditional results proven elsewhere are discussed about totally real number fields of a fixed degree, each of whose normal closure has the symmetric group as Galois group.

1. Introduction

It was Littlewood who first addressed the question of how large the class number h of an imaginary quadratic field $\mathbb{Q}(\sqrt{d})$ can be as a function of $|d|$ as $d \to -\infty$ through fundamental discriminants. In 1927 [**14**] he showed, assuming the generalized Riemann hypothesis (GRH), that for all fundamental $d < 0$

$$h \leq 2\big(c + \mathrm{o}(1)\big)|d|^{1/2} \log\log|d|, \tag{1.1}$$

where $c = e^{\gamma}/\pi$, where γ is Euler's constant. Furthermore he showed, still under GRH, that there are infinitely many such d with

$$h \geq \big(c + \mathrm{o}(1)\big)|d|^{1/2} \log\log|d|. \tag{1.2}$$

He deduced these results from the class number formula of Dirichlet, which for $d < -4$ states that

$$h = \frac{|d|^{1/2}}{\pi} L(1, \chi)$$

for $\chi(\cdot) = (\frac{d}{\cdot})$ the Kronecker symbol, and corresponding results about the Dirichlet L-value $L(1,\chi)$. His analysis is based on the idea that under GRH one can approximate $\log L(s,\chi)$ by a very short sum over primes of the type[1]

$$\log L(1,\chi) = \sum_{p \leq (\log|d|)^{1/2}} \chi(p)p^{-1} + \mathrm{O}(1). \tag{1.3}$$

2000 *Mathematics Subject Classification.* Primary: 11R; Secondary: 11M.

Research supported in part by NSF grant DMS–98–01642. This paper is an expanded version of a lecture given at the CNTA 7 in Montreal in May 2002. I thank the CNTA for providing generous financial support for this conference.

This is the final form of the paper.

[1]In fact, Littlewood used a more elaborate approximation in order to obtain good constants.

This easily gives (1.1) up to the constant, since

$$\sum_{p\leq x} p^{-1} \sim \log\log x$$

as $x \to \infty$. To prove (1.2) using (1.3), again up to the constant, one must produce infinitely many $d < 0$ with $\chi(p) = 1$ for all p with $p \leq (\log|d|)^{1/2}$, which is easily accomplished. Later Chowla [**3, 4**] proved (1.2) unconditionally.

In questions about the size of the class number of real quadratic fields the distribution of the regulator $R = \log\varepsilon$, for ε a fundamental unit for the field, causes new problems in the class number formula for $d > 0$:

$$h = \frac{|d|^{1/2}}{R} L(1,\chi).$$

For example, the conjecture of Gauss that $h = 1$ for infinitely many $d > 0$ is at present insurmountable, even under GRH. However, the obvious bound $R \gg \log d$ and Littlewood's [**14**] GRH result $L(1,\chi) \ll \log\log d$ give, for $d > 0$, the conditional upper bound

$$h \ll d^{1/2}(\log\log d/\log d).$$

In 1977 Montgomery–Weinberger [**16**] proved unconditionally that

$$h \gg d^{1/2}(\log\log d/\log d)$$

for infinitely many $d > 0$, thereby showing that this conditional upper bound is best possible, up to the constant. They make use of the special sequence of d, studied first by Chowla, namely square-free $d = 4t^2 + 1$, for which $\varepsilon = 2t + \sqrt{d}$ and hence $R \leq \log d$. They use zero-density estimates for Dirichlet L-functions to allow short approximations to enough $\log L(1,\chi)$, even in a quadratic sequence of discriminants, to produce large values of $L(1,\chi)$.

2. A general problem

Surprisingly little is known about such extremal class number problems for number fields of degree $n > 2$. First we must define a suitable family of number fields with which to work.

Given a number field K of degree n let G be the Galois group of its normal closure $\widehat{K}$ over $\mathbb{Q}$. If $H \subseteq G$ is the subgroup fixing K pointwise, then G acts on G/H as a permutation group. Given an ordering of the cosets of G/H we thus associate to K an embedding $G \hookrightarrow S_n$. As is well known, the resulting cycle structure of the Frobenius of an unramified prime determines its splitting type in K and that of complex conjugation (the Frobenius at ∞) the signature of K.

Given any subgroup $G \subseteq S_n$ we define $\mathcal{K}(G)$ to be the set of all number fields K of degree n that have an ordering of their cosets so that their associated embedding is G. In order for there to be a uniform upper bound for the class number we need to further restrict the family $\mathcal{K}(G)$. Suppose, for example, that G is *primitive* as a permutation group (see [**24**]) and that $\mathcal{K}^+(G)$ consists of the totally positive fields in $\mathcal{K}(G)$. By the class number formula for $K \in \mathcal{K}^+(G)$ we have

$$h = \frac{d^{1/2}}{2^{n-1}R} L(1,\chi) \tag{2.1}$$

where $d = \mathrm{disc}(K)$ is the discriminant, R is the regulator and

$$L(s,\chi) = \zeta_K(s)/\zeta(s) \tag{2.2}$$

is an Artin L-function, $\zeta_K(s)$ being the Dedekind zeta function of K. In general, it is not even known that $L(s,\chi)$ is entire, although it is conjectured that it is (Artin), and further that GRH holds for these $L(s,\chi)$. If we assume this, then the method of Littlewood shows that

$$L(1,\chi) \ll (\log\log d)^{n-1}.$$

Remak [**17**] proved that if K contains no non-trivial subfields (which is true if G is primitive) then

$$R \gg (\log d)^{n-1}, \tag{2.3}$$

with the implied constants depending only on n. Thus, under GRH, we have the upper bound for the class number of $K \in \mathcal{K}^+(G)$

$$h \ll d^{1/2}(\log\log d/\log d)^{n-1}. \tag{2.4}$$

More generally, by prescribing the location in G of complex conjugation and applying the more general estimates for regulators given by Silverman [**20**], we could formulate such an estimate for an imprimitive G and an arbitrary signature. In any case, it seems to be extremely difficult to prove any result as strong as (2.4) unconditionally.

The problem arises to show that (2.4) is sharp, that is, that there exist $K \in \mathcal{K}^+(G)$ with arbitrarily large discriminant for which

$$h \gg d^{1/2}(\log\log d/\log d)^{n-1}. \tag{2.5}$$

At this level of generality this is clearly difficult since it subsumes the inverse Galois problem. However, for certain G we have enough examples of $K \in \mathcal{K}^+(G)$ to make progress, as in the quadratic case. We remark that in such cases the main difficulty (and interest) of this problem is to show that the conditional upper bound (2.4) cannot be improved upon, apart from the constant.

3. Abelian cubic fields

The simplest case to consider after the quadratic is that of abelian cubic fields, that is $\mathcal{K}^+(G)$ where G is the cyclic group of order 3 in S_3. In this section we will prove the following unconditional result.

THEOREM 1. *There is an absolute constant $c > 0$ such that there exist abelian cubic fields with arbitrarily large discriminant d for which*

$$h > cd^{1/2}(\log\log d/\log d)^2.$$

First we must construct a sequence of abelian cubic fields for which we have a good upper bound for the regulator. Consider the polynomial for $t \in \mathbb{Z}^+$ defined by

$$f_t(x) = x^3 - tx^2 - (t+3)x - 1,$$

easily checked to be irreducible over $\mathbb{Q}$. The discriminant of f_t is given by $\operatorname{disc}(f_t) = g(t)^2$ where

$$g(t) = t^2 + 3t + 9.$$

Thus f_t defines an abelian cubic field $K = K_t$ obtained by adjoining to $\mathbb{Q}$ any root of f_t. These fields have been studied extensively after Shanks' paper [**19**]. It can be checked that exactly one of these roots, say ε, is positive and satisfies

$$t+1 < \varepsilon < t+2. \tag{3.1}$$

The other roots are $-1/(1+\varepsilon)$ and $-1-1/\varepsilon$. As shown in [**22**], ε and $\varepsilon+1$ give a basis for the units of the order $\mathbb{Z}[\varepsilon]$. The regulator R of this order thus satisfies

$$R = \log^2 \varepsilon + \log^2(\varepsilon+1) - \log\varepsilon \log(\varepsilon+1) \sim \log^2 t$$

as $t \longrightarrow \infty$, as is easily seen from (3.1). It is shown in [**23**, Corollary to Proposition 1] that, if $g(t) = t^2+3t+9$ is square-free, then $\mathbb{Z}[\varepsilon]$ is the maximal order in K (see also [**8**]). Thus we can state the following lemma.

LEMMA 1. *For $t \in \mathbb{Z}^+$ let K be the splitting field of $f_t(x)$. Then K is an abelian cubic number field. If $g(t) = t^2+3t+9$ is square-free then $\mathbb{Z}[\varepsilon]$ is the maximal order in K and we have that $d = \operatorname{disc}(K) = g(t)^2$ and*

$$R = \operatorname{reg}(K) = \tfrac{1}{16}\big(1+o(1)\big)\log^2 d.$$

For these K the class number formula (2.1) can be written

$$h = \frac{d^{1/2}}{4R}|L(1,\chi)|^2 \tag{3.2}$$

where now χ denotes a primitive Dirichlet character of order 3 and conductor $m = g(t)$. In order to force $|L(1,\chi)|^2$ to be large we will need to show that enough square-free values of $g(t) = t^2+3t+9$ exist when t is restricted to arithmetic progressions. This will allow us to produce many primes p that split completely in K and hence for which $\chi(p) = 1$.

The following lemma is a modification of Lemma 1 of [**16**].

LEMMA 2. *Suppose that $q, a \in \mathbb{Z}^+$ satisfy $6 \mid q$ and $(g(a), q) = 1$ and let*

$$N(x;q,a) = \#\{1 \le t \le x : t \equiv a \,(\mathrm{mod}\, q) \text{ with } g(t) \text{square-free}\}.$$

Then for $x \ge 2$

$$N(x;q,a) = \frac{x}{q}c_q + O(x^{2/3}\log x),$$

where c_q is given in (3.5) below and satisfies $\frac12 < c_q < 1$. The implied constant is absolute.

PROOF. We have that

$$N(x;q,a) = \sum_{\substack{1\le t\le x\\ t\equiv a\,(\mathrm{mod}\,q)}} \sum_{\substack{r^2 \mid g(t)\\ (r,q)=1}} \mu(r),$$

where we may assume that $(r,q) = 1$ in the second sum since $\big(g(t),q\big) = 1$ for $t \equiv a \,(\mathrm{mod}\, q)$. Thus rearranging gives

$$N(x;q,a) = \sum_{\substack{r \le g(x)\\ (r,q)=1}} \mu(r)\,\#\{1 \le t \le x : t \equiv a\,(\mathrm{mod}\,q) \text{ and } g(t) \equiv 0\,(\mathrm{mod}\,r^2)\}.$$

For a value of y with $1 \le y \le x$ to be chosen later the above sum over $r > y$ is

$$\le \sum_{s \le g(x)y^{-2}} \#\{(t,r) : 1 \le t \le x \text{ and } g(t) = r^2 s\}$$

$$\le \sum_{s \le 14x^2y^{-2}} \#\{(u,v) : 1 \le u \le 2x+3 \text{ and } u^2 - 4sv^2 = 27\},$$

on using that $g(t) = \big((2t+3)^2+27\big)/4$. For $x \geq 2$ this is $\ll (x/y)^2 \log x$ since, by Hilfssatz 2 of [**10**, p. 660], we have that

$$\#\{(u,v) : 1 \leq u \leq 2x+3 \text{ and } u^2 - 4sv^2 = 27\} \ll \log x,$$

with an absolute constant (note that in case s is a square this is trivial). Thus we have for $x \geq 2$

$$N(x;q,a) = \sum_{\substack{r \leq y \\ (r,q)=1}} \mu(r)\#\{1 \leq t \leq x : t \equiv a(q) \text{ and } g(t) \equiv 0(r^2)\} + O(x^2y^{-2}\log x). \tag{3.3}$$

For $(m,6)=1$ the number of solutions to the congruence $g(t) \equiv 0 \,(\mathrm{mod}\, m)$ is given by

$$c(m) = \prod_{p|m}\left(1 + \left(\frac{-3}{p}\right)\right), \tag{3.4}$$

which is multiplicative and satisfies $c(m) = c(m^2)$. Thus from (3.3)

$$N(x;q,a) = \sum_{\substack{r \leq y \\ (r,q)=1}} \mu(r)c(r)\left(\frac{x}{qr^2} + O(1)\right) + O(x^2y^{-2}\log x)$$

since by assumption $6 \mid q$ and so $(r,6)=1$. Clearly $c(r) \leq d(r)$, where $d(r)$ is the number of divisors of r, so

$$N(x;q,a) = \frac{x}{q}\sum_{(r,q)=1} \mu(r)c(r)r^{-2} + O(x^2y^{-2}\log x) + O(y\log y) + O(xq^{-1}y^{-1}\log y).$$

Choosing $y = x^{2/3}$ gives the result since by (3.4)

$$c_q = \sum_{(r,q)=1} \mu(r)c(r)r^{-2} = \prod_{p \nmid q}\left(1 - \left(1 + \left(\frac{-3}{p}\right)\right)p^{-2}\right) \geq \prod_{p \geq 7}(1 - 2p^{-2}) > \frac{1}{2}, \tag{3.5}$$

completing the proof of Lemma 2. □

Proof of Theorem 1. We shall apply Lemma 2 with

$$q = \prod_{p \leq y} p,$$

where p runs over primes and $y = \alpha \log x$ for an appropriately chosen α. We want to choose a so that all p with $5 \leq p \leq y$ split completely in K when $t \equiv a \,(\mathrm{mod}\, q)$. To force this, observe that

$$f_t(x) = x^3 - tx^2 - (t+3)x - 1 = 0$$

defines a rational (genus 0) curve over $\mathbb{Q}$ and that for $p \geq 5$ there will always be points (x,t) on this curve over $\mathbf{F}_p$ with $g(t) = t^2+3t+9 \neq 0$. In fact, one computes that the number of $t \,(\mathrm{mod}\, p)$ for which $f_t(x)$ factors into 3 distinct linear factors $(\mathrm{mod}\, q)$ is

$$\begin{cases} \frac{p-4}{3} & \text{if } p \equiv 1 \,(\mathrm{mod}\, 3), \\ \frac{p-2}{3} & \text{if } p \equiv 2 \,(\mathrm{mod}\, 3). \end{cases}$$

Hence, for such t, p will split completely in the Galois extension K. For each p with $5 \leq p \leq y$ fix such a t_p. By the Chinese remainder theorem we can find a with $a \equiv 1 \pmod 6$ and $a \equiv t_p \pmod p$ for each such p. Then $(g(a), q) = 1$ and so by Lemma 2 we conclude that there will exist $\gg x/q$ values of $t \leq x$ for which all primes p with $5 \leq p \leq y$ split completely in K and for which $d = \operatorname{disc}(K) = g(t)^2$ with $g(t)$ square-free, provided that $q \ll x^{1/4}$, say. This can be ensured by taking $y = \frac{1}{5}\log x$. By Lemma 1 we conclude the following.

LEMMA 3. *There exists an absolute constant $c > 0$ so that there are at least $cx^{3/8}$ distinct cubic abelian fields K, each with discriminant $d = m^2$ and conductor m for some m with $1 < m \leq x$, and in which all primes p with $5 \leq p \leq \frac{1}{5}\log x$ split completely and whose regulator R satisfies $R \ll \log^2 d$.*

Returning to (3.2) we next show that the proof of Theorem 1 follows from Lemma 3 and the following result, which is [**16**, Lemma 2].

LEMMA 4 (Montgomery–Weinberger). *Suppose that $0 < \delta < 1$. Then for $(\log m)^\delta \leq y \leq \log m$, and χ a primitive Dirichlet character $\pmod m$ with $m > 1$, we have*

$$\log L(1, \chi) = \sum_{p \leq y} \chi(p) p^{-1} + \mathrm{O}_\delta(1),$$

unless χ lies in an exceptional set $\mathcal{E}(\delta)$. The set $\mathcal{E}(\delta)$ contains $\ll x^\delta$ primitive characters χ with conductor m such that $1 < m \leq x$.

Take $\delta < 3/8$ and $y = \frac{1}{5}\log x$ in Lemma 4 to conclude that there is an absolute constant $c_1 > 0$ so that for at least $c_1 x^{3/8}$ of the fields in Lemma 3 the associated Dirichlet L value satisfies

$$|L(1, \chi)|^2 \gg \log\log d.$$

Theorem 1 now follows from (3.2). □

4. Non-abelian fields and L-functions

For number fields whose normal closure is non-abelian much less is known about our problem. The easiest non-abelian situation in which to produce fields with class numbers as large as predicted in (2.5) is when the Galois group is the full symmetric group, this being the "generic" case in view of well known results of probabilistic Galois theory (see [**11**]). Let $\mathcal{K}_n = \mathcal{K}^+(S_n)$, the set of all totally real number fields K of degree n whose normal closure $\widehat{K}$ has Galois group S_n. As in §2, let $L(s, \chi) = \zeta_K(s)/\zeta(s)$ be the associated Artin L-function. The following result is proven in [**9**] and shows (conditionally) that for $G = S_n$ the upper bound (2.4) is sharp up to the constant.

THEOREM 2. *Fix $n \geq 2$ and assume that each $L(s, \chi)$ is entire and satisfies the GRH. Then there is a constant $c > 0$ depending only on n such that there exist $K \in \mathcal{K}_n$ with arbitrarily large discriminant d for which*

$$h > cd^{1/2}(\log\log d/\log d)^{n-1}.$$

A previously known result in this direction, due to Ankeny–Brauer–Chowla [**1**], states that for any $\varepsilon > 0$ there are infinitely many totally real fields K of degree n with $h > d^{1/2-\varepsilon}$. It imposes no condition on $\operatorname{Gal}(\widehat{K}/\mathbb{Q})$ and is not sharp, but it is unconditional. Later Sprindžuk [**21**][2] strengthened this result by giving an upper

[2] I thank C. Levesque for pointing out this reference.

bound for the density (when measured in terms of the regulator) of totally real number fields of degree n with $h \le d^{1/2-\varepsilon}$. It should be observed, however, that the upper bound (2.4) likely does not hold in general for all totally real fields of degree $n > 3$. In [**8**, Lemma 3] an infinite family of totally real biquadratic fields is constructed whose regulators satisfy $R \ll \log^2 d$ and in this family there are likely infinitely many that violate (2.4) with $n = 4$.

As in the proof of Theorem 1, in order to prove Theorem 2 we construct the needed fields as specializations of a function field. Specifically, we define K by adjoining to $\mathbb{Q}$ a root α of

$$f_t(x) = (x - t)(x - 2^2t)\cdots(x - n^2t) - t.$$

If $t > 1$ is square-free then $f_t(x)$ is an Eisenstein polynomial. This allows us to use the fact that p with $p \mid t$ ramifies completely in K to bound from below the discriminant d of K in terms of $\mathrm{disc}(f_t)$, and hence t. A frequent difficulty in doing so in such problems is due to the index of $\mathbb{Z}[\alpha]$ in the ring of integers in K. This family has another needed feature; if t is sufficiently large, then K is totally real and

$$R \ll_n (\log d)^{n-1}$$

since it is possible to write down a system of $n - 1$ multiplicatively independent units with a determined asymptotic behavior as $t \to \infty$, a property reminiscent of the cyclic cubic example discussed above. A difficulty not present in the cyclic cubic case, since it is automatic, is the need to control the Galois group of $\widehat{K}/\mathbb{Q}$. Here we show that $f(t, x) = f_t(x)$ is irreducible over $\mathbb{C}$ and that the monodromy group of the Riemann surface defined by $f(t, x) = 0$ is S_n, using old techniques of Jordan and Hilbert. Then we apply a result of S. D Cohen [**7**], which is a quantitative form of Hilbert's Irreducibility Theorem, to show that for most integral specializations of t, the Galois group of $\widehat{K}/\mathbb{Q}$ is maximal, that is S_n. The proof of this uses the multi-dimensional large sieve inequality in a way first done by Gallagher [**11**].

The technique used in the abelian cubic case to produce many split primes by restricting the parameter t to certain arithmetic progressions continues to work but the genus of the needed curve is no longer zero and the Riemann Hypothesis for curves must be invoked. The relatively mild condition that t be square-free may be imposed without difficulty.[3] This ingredient in the proof of Theorem 2 is an analogue of a classical result of Bohr and Landau [**5**], as refined by Littlewood [**13**], which is used to find small values of $\zeta(1 + it)$:

CLASSICAL BOHR–LANDAU. *If y is sufficiently large then there is a positive integer t with $\log t \ll y$ so that p^{it} is uniformly close to -1 for each prime $p \le y$, the implied constants depending on how good the approximation is required to be.*

This part of the proof is generalized in [**9**] to families of continuous Galois representations

$$\rho : \mathrm{Gal}(\overline{\mathbb{Q}}/\mathbb{Q}) \to \mathrm{GL}(n, \mathbb{C})$$

whose image is a fixed finite group G with property Gal_T. To say that G has property Gal_T means that G is the Galois group of a finite regular extension E of $\mathbb{Q}(t)$, so $\overline{Q} \cap E = \mathbb{Q}$.[4] Here as usual $\overline{Q}$ is a fixed algebraic closure of $\mathbb{Q}$. The "size"

[3]The construction of Theorem 2 in case $n = 2$ provides a simplification of that of [**16**] since it is unnecessary to force d to be square-free, only t.

[4]It has been conjectured (see [**18**, p. 35]) that every finite group G has property Gal_T.

of such a representation is measured by its conductor. Choose $a(p)$ to be, for each prime p, the trace of an arbitrary element of G. Then we have

BOHR–LANDAU FOR GALOIS. *If y is sufficiently large, then there is a Galois representation of conductor N whose image is G and such that $\log N \ll y$ and $\chi(\sigma_p) = a(p)$ for $1 \ll p \leq y$, where σ_p is any Frobenius element over p.*

The final ingredient needed to prove Theorem 2 is an approximation of $\log L(1,\chi)$ as a short sum over primes. This can be done for any Artin L-function associated to ρ. Denoting the character of ρ by χ, this is given for $\mathrm{Re}(s) > 1$ by the formula

$$\log L(s,\chi) = \sum_p \sum_{m=1}^{\infty} \frac{1}{m} \chi(p^m) p^{-ms}$$

where

$$\chi(p^m) = \frac{1}{|I_p|} \sum_{\iota \in I_p} \chi(\sigma_p^m \iota)$$

where I_p is the inertia group of any of the primes lying over p. In general, $L(s,\chi)$ is known to be meromorphic [**6**] and to satisfy the functional equation [**2**]

$$\Lambda(1-s,\overline{\chi}) = \varepsilon_\chi N^{s-1/2} \Lambda(s,\chi)$$

where $|\varepsilon_\chi| = 1$ and

$$\Lambda(s,\chi) = \pi^{ns/2} \Gamma(s/2)^{(n+\ell)/2} \Gamma\big((s+1)/2\big)^{(n-\ell)/2} L(s,\chi)$$

where ℓ is the value of χ on complex conjugation and N is the conductor of ρ. Artin conjectured that $L(s,\chi)$ is entire unless ρ contains trivial components, in which case it should only have a pole at $s = 1$. Assuming this and GRH for $L(s,\chi)$, it is shown in [**9**] that (1.3) generalizes in a straightforward way to give

$$\log L(1,\chi) = \sum_{p \leq (\log N)^{1/2}} \chi(p) p^{-1} + \mathrm{O}_n(1),$$

which then allows the completion of the proof of Theorem 2.

One is led naturally to consider the extreme values of such a family of Artin L-functions at $s = 1$ and this can be done, at least conditionally. We immediately get the upper bound (under GRH)

$$L(1,\chi) \ll_n (\log\log N)^n \tag{4.1}$$

and prove in [**9**] the following result showing that this is best possible, up to the constant.

THEOREM 3. *Suppose that $G \subset \mathrm{GL}(n,\mathbb{C})$ is non-trivial, irreducible and has property Gal_T. Assume that every Artin L-function $L(s,\chi)$ whose Galois representation has image G is entire and satisfies GRH. Then there is a constant $c > 0$, depending only on G, so that there exist such L-functions with arbitrarily large conductor N that satisfy*

$$L(1,\chi) > c(\log\log N)^n. \tag{4.2}$$

5. Further prospects

It would obviously be desirable to prove Theorems 2 and 3 unconditionally, at least in some new cases.[5] Clearly what is needed is a generalization of Lemma 4 to families of Artin L-functions. The proof of Lemma 4 is based on zero density estimates for Dirichlet L-functions. Proofs of these estimates involve orthogonality properties of distinct Dirichlet characters. This aspect can sometimes be found on the side of automorphic representations. In particular, it would be interesting to establish zero density estimates for families of automorphic L-functions chosen to include, at least conjecturally, the Artin L-functions. In some cases, for example non-abelian cubic extensions, the degree two Artin L-functions that arise in (2.2) are known to come from classical automorphic forms of weight 1 after Hecke, and the type of orthogonality needed would be over varying levels. Similarly, for certain quartic extensions, the associated degree three Artin L-functions that arise are known after Jacquet, Piatetski–Shapiro and Shalika [**12**] to come from GL(3) automorphic forms. Langlands has conjectured that every Artin L-function should come from an automorphic representation on GL(n) for some n.

Another direction of interest is to consider other groups G than the symmetric group for the problem outlined in §2. For such generalizations the construction of fields with small regulators seems to be quite an interesting problem. As we have seen, the existence of Artin L-functions in rather general families with extremal values can be proven conditionally. There might also be some interest in obtaining good constants in (4.1) and (4.2) as was done in case $n = 2$ by Littlewood, and in studying the finer behavior of these L-values along the lines of [**15**]

References

1. N. C. Ankeny, R. Brauer, and S. Chowla, *A note on the class-numbers of algebraic number fields*, Amer. J. Math. **78** (1956), 51–61.
2. E. Artin, *Zur Theorie der L-Reihen mit allgemeinen Gruppencharakteren*, Hamb. Abh. **8** (1930), 292–306; Collected Papers, 165–179.
3. S. Chowla, *On the k-analogue of a result in the theory of the Riemann zeta function*, Math. Z. **38** (1934), 483–487.
4. ———, *On the class-number of the corpus $P(\sqrt{-k})$*, Proc. Nat. Inst. Sci. India **13** (1947), 197–200.
5. H. Bohr and E. Landau, *Über das Verhalten von $1/\zeta(s)$ auf der Geraden $\sigma = 1$*, öttinger Nachrichten, 1924, pp. 168–172.
6. R. Brauer, *On Artin's L-series with general group characters*, Ann. of Math. (2) **48**, (1947), 502–514.
7. S. D. Cohen, *The distribution of Galois groups and Hilbert's irreducibility theorem*, Proc. London Math. Soc. (3) **43** (1981), no. 2, 227–250.
8. T. W. Cusick, *Lower bounds for regulators*, Number Theory (Noordwijkerhout, 1983), Lecture Notes in Math., vol. 1068, Springer, Berlin, 1984, pp. 63–73.
9. W. Duke, *Extreme values of Artin L-functions and class numbers*, Compositio Math. **136** (2003), no. 1, 103–115.
10. T. Estermann, *Einege Sätze über quadratfrei Zahlen*, Math. Ann. **105** (1931), 653–662.
11. P. X. Gallagher, *The large sieve and probabilistic Galois theory*, Analytic Number Theory (St. Louis, 1972), Proc. Sympos. Pure Math., vol. 24, Amer. Math. Soc., Providence, R.I., 1973, pp. 91–101.
12. H. Jacquet, Ilja I. Piatetski–Shapiro, and J. Shalika, *Relévement cubique non normal*, C. R. Acad. Sci. Paris Sér. I Math. **292** (1981), no. 12, 567–571.

[5] Added in proof: In case $n = 3$, Theorem 2 has recently been made unconditional by Ryan Daileda in his UCLA thesis.

13. J. E. Littlewood, *On the function* $1/\zeta(1+ti)$, Collected Papers of J. E. Littlewood, vol. I, The Clarendon Press, Oxford University Press, New York, 1982.
14. ______, *On the class number of the corpus* $P(\sqrt{-k})$, Collected Papers of J. E. Littlewood, vol. II, The Clarendon Press, Oxford University Press, New York, 1982.
15. H. L. Montgomery and R. C. Vaughan, *Extreme values of Dirichlet L-functions at* 1, Number Theory in Progress (Zakopane-Kościelisko, 1997), vol. 2 de Gruyter, Berlin, 1999, pp. 1039–1052.
16. H.-L. Montgomery and P.-J. Weinberger, *Real quadratic fields with large class number*, Math. Ann. **225** (1977), no. 2, 173–176.
17. R. Remak, *Über Grössenbeziehungen zwischen Diskriminante und Regulator eines algebraischen Zahlkörpers*, Compositio Math. **10** (1952), 245–285.
18. J.-P. Serre, *Topics in Galois theory*, Res. Notes Math., vol. 1, Jones and Bartlett Publishers, Boston, MA, 1992.
19. D. Shanks, *The simplest cubic fields*, Math. Comp. **28** (1974), 1137–1152.
20. J.-H. Silverman, *An inequality relating the regulator and the discriminant of a number field*, J. Number Theory **19** (1984), no. 3, 437–442.
21. V. G. Sprindžuk, *"Almost every" algebraic number-field has a large class-number*, Acta Arith. **25** (1973/74), 411–413.
22. E. Thomas, *Fundamental units for orders in certain cubic number fields*, J. Reine Angew. Math. **310** (1979), 33–55.
23. Lawrence C. Washington, *Class numbers of the simplest cubic fields*, Math. Comp. **48** (1987), no. 177, 371–384.
24. Helmut Wielandt, *Finite permutation groups*, Academic Press, New York-London 1964 (German).

Department of Mathematics, University of California, Los Angeles, CA 90095–1555, USA.

E-mail address: duke@math.ucla.edu

Centre de Recherches Mathématiques
CRM Proceedings and Lecture Notes
Volume **36**, 2004

Period Polynomials for $\mathbb{F}_q$ of Fixed Small Degree

Stanley J. Gurak

ABSTRACT. Let $q = p^r$ with p a prime, and e and f be positive integers satisfying $ef + 1 = q$. Let $\mathbb{F}_q$ be the finite field of q elements. The Gauss periods of order e satisfy a period polynomial $F(x)$ of degree e over the rational field $\mathbb{Q}$. In the classical case $q = p$, Gauss showed that $F(x)$ is irreducible over $\mathbb{Q}$ and determined the polynomial for $e = 2$, 3, and 4. Using the methods and results from the theory of cyclotomy, Dickson, Lehmer, Muskat and Whiteman, among others, determined $F(x)$ for $e = 5$, 6, 8, 10, 12, 16, 24 in the case $q = p$. For the general case, Myerson determined $F(x)$ for $e = 2$, 3, and 4, but beyond this very little is explicitly known. Recently I determined $F(x)$ explicitly for the case $q = p^2$ when e divides 8 or 12, by applying work of Berndt and Evans on Gauss sums for $\mathbb{F}_{p^2}$ with characters of order 6, 8, and 12. By extending their results to evaluate Gauss sums for $\mathbb{F}_q$ whenever $e \mid 24$, I am able to determine $F(x)$ explicitly in the general case $q = p^r$ when e divides 24. Here I am content to give this determination for $e = 6$, 8, and 12.

1. Introduction

Let $q = p^r$ with p a prime, and e and f be positive integers such that $ef+1 = q$. Let $\mathbb{F}_q$ denote the field of q elements, $\mathbb{F}_q^*$ its multiplicative group and γ a fixed generator of $\mathbb{F}_q^*$. Then $G = \gamma^{(q-1)/(p-1)}$ generates $\mathbb{F}_p^*$. Let $\mathrm{Tr}\colon \mathbb{F}_q \to \mathbb{F}_p$ be the usual trace map and set $\zeta_m = \exp(2\pi i/m)$ for any positive integer m. Put $\delta = \gcd\big((q-1)/(p-1), e\big)$ and $R = (q-1)/\big(\delta(p-1)\big)$. The (modified) Gauss periods are

$$H_j = \sum_{x\in\mathbb{F}_q} \zeta_p^{\mathrm{Tr}\,\gamma^j x^e} \quad (1 \le j \le e) \tag{1.1}$$

(if $j > e$ then the subscripts in H_j are taken modulo e), and satisfy the period polynomial

$$F(x) = \prod_{j=1}^{e} (x - H_j). \tag{1.2}$$

G. Myerson [**8**] showed that $F(x)$ splits over $\mathbb{Q}$ into δ factors

$$F(x) = \prod_{w=1}^{\delta} F^{(w)}(x), \tag{1.3}$$

2000 *Mathematics Subject Classification.* 11L05, 11T24.
This is the final form of the paper.

each of degree e/δ, where

$$F^{(w)}(x) = \prod_{k=0}^{e/\delta-1} (x - H_{w+k\delta}) \quad (1 \leq w \leq \delta).$$

In the classical case $q = p$ (so $\delta = 1$), Gauss showed that $F(x)$ is irreducible over $\mathbb{Q}$ and determined the polynomial for $e = 2$, 3, and 4. Using the methods and results from the theory of cyclotomy, Dickson, E. Lehmer, Muskat, and Whiteman, among others, determined $F(x)$ for $e = 5$, 6, 8, 10, 12, 16, and 24 in the case $q = p$. For the general case $q = p^r$, Myerson determined the factors $F^{(w)}(x)$ of $F(x)$ for $e = 2$, 3, and 4, but beyond this very little is known. When $q = p^2$, I recently [**6**] applied some results of Berndt and Evans [**1**–**3**] to determine these factors for $e = 6$, 8, and 12. Extending their methods, it is possible to determine each of the factors $F^{(w)}(x)$ of $F(x)$ in (1.3) for $q = p^r$ when $e \mid 24$. Here I wish to give this determination explicitly, at least for the cases $e = 6$, 8, and 12.

2. Preliminaries

The Gauss periods (1.1) are intimately connected with Gauss sums. For χ a character of $\mathbb{F}_q^*$ of order dividing e, define the Gauss sum $G_r(\chi)$ by

$$G_r(\chi) = \sum_{x \in \mathbb{F}_q^*} \chi(x) \zeta_p^{\operatorname{Tr} x}, \tag{2.1}$$

which satisfies

$$G_r(\chi) G_r(\chi^{-1}) = \chi(-1) p^r \quad \text{for } \chi \neq 1. \tag{2.2}$$

(We will write $G(\chi)$ for $G_1(\chi)$.) For $r > 1$, the value of G_r can be expressed in terms of the Eisenstein sum

$$E_r(\chi) = \sum_{\substack{x \in \mathbb{F}_q^* \\ \operatorname{Tr} x = 1}} \chi(x); \tag{2.3}$$

namely (chiefly, [**3**, Theorem 12.1.1]), for χ nontrivial

$$G_r(\chi) = \begin{cases} E_r(\chi) G(\chi^*) & \text{if } \chi^* \text{ is nontrivial,} \\ -pE_r(\chi) & \text{if } \chi^* \text{ is trivial,} \end{cases} \tag{2.4}$$

where χ^* denotes the restriction of χ to $\mathbb{F}_p^*$. Normalizing such characters χ so $\chi_j(\gamma) = \zeta_e^j$ for $1 \leq j \leq e$, one has

$$G_r(\chi_j) = \frac{1}{e} \sum_{i=1}^{e} \zeta_e^{ij} H_i \quad (j \neq e); \qquad G_r(\chi_e) = -1, \tag{2.5}$$

$$H_i = \sum_{j=1}^{e-1} \zeta_e^{-ij} G_r(\chi_j) \quad (1 \leq i \leq e). \tag{2.6}$$

As in [**3**], we set $g_r(e) = \sum_{x \in \mathbb{F}_{p^r}} \zeta_p^{\operatorname{Tr} x^e}$ so $H_e = g_r(e)$ from (1.1). (We write $g(e)$ for $g_1(e)$.) For small values of e, the sums $G(\chi^*)$, $g_r(e)$ and $E_r(\chi)$ are known or are easily derived (in particular, $e = 2$, 3, 4, 6, 8, and 12) and are given next.

Set $i^* = 1$ or i as $p \equiv 1$ or 3 (mod 4). For a prime $p \equiv 1 \pmod 3$, write $4p = r_3^2 + 3s_3^2$ with integers r_3 and s_3 uniquely determined by the conditions $r_3 \equiv 1 \pmod 3$, $s_3 \equiv 0 \pmod 3$ and $3s_3 \equiv (2G^{(p-1)/3} + 1) r_3 \pmod p$. Put

$Z = \operatorname{ind}_G 2$ and $T = \operatorname{ind}_G 3$. Then the quantity $a_3 + ib_3\sqrt{3} = \zeta_3^{2Z}(r_3 + i\sqrt{3}s_3)/2$ satisfies $a_3^2 + 3b_3^2 = p$ for integers a_3 and b_3 with $a_3 \equiv -1 \pmod 3$ and $3b_3 \equiv (2G^{(p-1)/3} + 1)a_3 \pmod p$. Similarly, for a prime $p \equiv 1 \pmod 4$ write $p = a_4^2 + b_4^2$ with integers a_4 and b_4 uniquely determined by the condition $a_4 \equiv -(-1)^Z \pmod 4$ and $b_4 \equiv a_4 G^{(p-1)/4} \pmod p$. For a prime $p \equiv 1 \pmod{12}$ set $a_{12} + ib_{12} = (-1)^{T/2+Z}(a_4 + ib_4)$. (Note that 3 is a quadratic residue modulo p here, so T is even.) If c_{12} is the unique 4th root of unity satisfying

$$c_{12} = \begin{cases} \pm 1 \text{ with } c_{12} \equiv -a_4 \pmod 3 & \text{if } 3 \mid b_4, \\ \pm i \text{ with } c_{12} \equiv -ib_4 \pmod 3 & \text{if } 3 \mid a_4, \end{cases} \tag{2.7}$$

then (chiefly, [**3**, Lemma 3.5.1])

$$a_{12} + ib_{12} = c_{12}^2(a_4 + ib_4). \tag{2.8}$$

Also, if $p \equiv 5 \pmod{12}$, set $\beta = \pm 1$ to satisfy $\beta \equiv a_4 b_4 \pmod 3$. Finally for a prime $p \equiv 1$ or $3 \pmod 8$, write $p = a_8^2 + 2b_8^2$ with integers a_8 and b_8 satisfying $a_8 \equiv -1 \pmod 4$ and $2b_8 \equiv (G^{(p-1)/8} + G^{3(p-1)/8})a_8 \pmod p$ if $p \equiv 1 \pmod 8$, or with $a_8 \equiv (-1)^{(p-3)/8} \pmod 4$ and $2b_8 \equiv (\gamma^{(q-1)/8} - \gamma^{(1-q)/8})a_8 \pmod p$ if $p \equiv 3 \pmod 8$. Assume for convenience that Ψ_e denotes the (normalized) character of $\mathbb{F}_p^*$ satisfying $\Psi_e(G) = \zeta_e$ when $p \equiv 1 \pmod e$. Then $G(\Psi_2) = i^*\sqrt{p}$ and the modified Gauss period

$$H_j(2) = \begin{cases} i^*(-1)^{(p-1)(r-1)/4+j}p^{(r-1)/2}\sqrt{p} & \text{if } r \text{ is even}, \\ -(-1)^{(p-1)r/4+j}p^{r/2} & \text{if } r \text{ is odd}, \end{cases} \tag{2.9}$$

from (2.4), (2.6) and [**3**, Theorem 12.1.5]. For $e = 3$, the modified Gauss sum $g = g(3)$ satisfies $x^3 - 3px - pr_3 = 0$ with r_3 and s_3 as defined above. The correct choice of root for $g(3)$ was determined by Matthews and is described in [**3**]. In terms of $g(3)$, one has [**6**]

$$G_3 = G(\Psi_3) = \big(g + (g'' - g')/(i\sqrt{3})\big)/2 \tag{2.10}$$

$$\widehat{G}_3 = G(\Psi_3^2) = \big(g - (g'' - g')/(i\sqrt{3})\big)/2, \tag{2.11}$$

where the conjugates g' and g'' of g are given by $g' = \big(g^2 - 2p - (s_3 + r_3)g/2\big)/s_3$ and $g'' = \big(2p - g^2 - (s_3 - r_3)g/2\big)/s_3$. In particular, from (2.4), (2.6) and [**3**, Theorem 12.3.1], if $p \equiv 1 \pmod 3$ then

$$H_j(3) = \begin{cases} -(-1)^r p^{r/3}(\zeta_6^{-2j}\lambda^{r/3} + \zeta_6^{2j}\bar{\lambda}^{r/3}) & \text{if } r \equiv 0 \pmod 3, \\ -(-1)^r p^{(r-1)/3}(\zeta_6^{-2j}\lambda^{(r-1)/3}G_3 + \zeta_6^{2j}\bar{\lambda}^{(r-1)/3}\widehat{G}_3) & \\ & \text{if } r \equiv 1 \pmod 3, \\ -(-1)^r p^{(r-2)/3}(\zeta_6^{-2j}\lambda^{(r+1)/3}\widehat{G}_3 + \zeta_6^{2j}\bar{\lambda}^{(r+1)/3}G_3) & \\ & \text{if } r \equiv 2 \pmod 3, \end{cases} \tag{2.12}$$

where $\lambda = (r_3 + i\sqrt{3}s_3)/2$, or if $p \equiv 2 \pmod 3$

$$H_j(3) = -(\zeta_6^{-2j} + \zeta_6^{2j})(-1)^{r/2}p^{r/2} \tag{2.13}$$

with r even. For $e = 6$, one has

$$G_6 = G(\Psi_6) = i^*\zeta_6^{4Z}\lambda\widehat{G}_3/\sqrt{p} \quad \text{and} \quad \widehat{G}_6 = G(\Psi_6^5) = i^*\zeta_6^{2Z}\bar{\lambda}G_3/\sqrt{p}. \tag{2.14}$$

Set

$$W_j = \zeta_6^{-j}G_r(\chi_1) + \zeta_6^j G_r(\chi_5), \tag{2.15}$$

so

$$H_j(6) = W_j + H_j(3) + H_j(2). \tag{2.16}$$

Then again from (2.4), (2.6) and also [**3**, Theorem 12.6.1], if $p \equiv 1 \pmod 6$, then

$$W_j = \begin{cases} -(-1)^{(p-1)r/4} p^{r/6}(\zeta_6^{-j}\lambda^{2r/3} + \zeta_6^j\bar{\lambda}^{2r/3}) & \text{if } r \equiv 0 \pmod 6, \\ (-1)^{(p-1)(r-1)/4} p^{(r-1)/6}(\zeta_6^{-j}\lambda^{(2r-2)/3}G_6 + \zeta_6^j\bar{\lambda}^{(2r-2)/3}\widehat{G}_6) & \\ & \text{if } r \equiv 1 \pmod 6, \\ -(-1)^{(p-1)r/4} p^{(r-2)/6}(\zeta_6^{2Z-j}\lambda^{(2r-1)/3}G_3 + \zeta_6^{4Z+j}\bar{\lambda}^{(2r-1)/3}\widehat{G}_3) & \\ & \text{if } r \equiv 2 \pmod 6, \\ (-1)^{(p-1)(r-1)/4} p^{(r-3)/6}(\zeta_6^{-j}\lambda^{2r/3} + \zeta_6^j\bar{\lambda}^{2r/3})i^*\sqrt{p} & \\ & \text{if } r \equiv 3 \pmod 6, \\ -(-1)^{(p-1)r/4} p^{(r-4)/6}(\zeta_6^{4Z-j}\lambda^{(2r+1)/3}\widehat{G}_3 + \zeta_6^{2Z+j}\bar{\lambda}^{(2r+1)/3}G_3) & \\ & \text{if } r \equiv 4 \pmod 6, \\ (-1)^{(p-1)(r-1)/4} p^{(r-5)/6}(\zeta_6^{-j}\lambda^{(2r+2)/3}\widehat{G}_6 + \zeta_6^j\bar{\lambda}^{(2r+2)/3}G_6) & \\ & \text{if } r \equiv 5 \pmod 6; \end{cases} \tag{2.17}$$

whereas if $p \equiv 5 \pmod 6$

$$W_j = -(-1)^{r(p-1)/4}(\zeta_6^{-j} + \zeta_6^j)p^{r/2} \quad \text{with } r \text{ even.} \tag{2.18}$$

For $e = 4$ with $p \equiv 1 \pmod 4$ one has

$$G_4 = G(\Psi_4) = \epsilon(A + iB) \tag{2.19}$$

$$\widehat{G}_4 = G(\Psi_4^3) = (-1)^Z\epsilon(A - iB), \tag{2.20}$$

where $A = \sqrt{(p + (-1)^Z a_4\sqrt{p})/2}$, $B = (-1)^Z b_4\sqrt{(p - (-1)^Z a_4\sqrt{p})/2}/|b_4|$ and $\epsilon = \pm 1$. The correct choice of sign for ϵ was determined by Matthews and is described in [**3**]. Now set $R_j = \zeta_4^{-j}G_r(\chi_1) + \zeta_4^j G_r(\chi_3)$ so

$$H_j(4) = R_j + H_j(2). \tag{2.21}$$

In particular, from (2.4), (2.6) and [**3**, Theorem 12.4.1] one finds if $p \equiv 1 \bmod 4$ then

$$R_j = \begin{cases} p^{(r-1)/4}(\zeta_4^{-j}\pi^{(r-1)/2}G_4 + \zeta_4^j\bar{\pi}^{(r-1)/2}\widehat{G}_4) & \text{if } r \equiv 1 \pmod 4, \\ (-1)^{(p+3)/4}p^{(r-2)/4}\sqrt{p}(\zeta_4^{-j}\pi^{r/2} + \zeta_4^j\bar{\pi}^{r/2}) & \text{if } r \equiv 2 \pmod 4, \\ (-1)^Z p^{(r-3)/4}(\zeta_4^{-j}\pi^{(r+1)/2}\widehat{G}_4 + \zeta_4^j\bar{\pi}^{(r+1)/2}G_4) & \\ & \text{if } r \equiv 3 \pmod 4, \\ -p^{r/4}(\zeta_4^{-j}\pi^{r/2} + \zeta_4^j\bar{\pi}^{r/2}) & \text{if } r \equiv 0 \pmod 4, \end{cases} \tag{2.22}$$

where $\pi = a_4 + ib_4$, or if $p \equiv 3 \pmod 4$

$$R_j = \begin{cases} -2(-1)^{r(p-3)/8+j/2}p^{r/2} & \text{if } j \text{ is even,} \\ 0 & \text{if } j \text{ is odd,} \end{cases} \tag{2.23}$$

with r even. For $e = 12$, one has from [**7**, Proposition 3]

$$\begin{aligned} G(\Psi_{12}) &= \bar{c}_{12}G_3\widehat{G}_4/\bar{\pi}, & G(\Psi_{12}^5) &= \bar{c}_{12}\widehat{G}_3\widehat{G}_4/\bar{\pi}, \\ G(\Psi_{12}^7) &= c_{12}G_3G_4/\pi, & G(\Psi_{12}^{11}) &= c_{12}\widehat{G}_3G_4/\pi \end{aligned} \tag{2.24}$$

in terms of the Gauss sums of orders 3 and 4. Set

$$Z_j = \zeta_{12}^{-j} G_r(\chi_1) + \zeta_{12}^{-5j} G_r(\chi_5) + \zeta_{12}^{-7j} G_r(\chi_7) + \zeta_{12}^{-11j} G_r(\chi_{11}) \tag{2.25}$$

so

$$H_j(12) = Z_j + R_j + H_j(6). \tag{2.26}$$

Arguing as in the proof of Proposition 4 in [**7**] using (2.4), (2.6), (2.24) and [**7**, Theorem 1] one finds the following generalization of (31)–(34) in [**7**]. For $p = 12k+1$,

$$(-1)^j \frac{Z_j p^{[(r+3)/4]}}{H_j(3)} = \begin{cases} -(\zeta_4^{-j}\pi^{r/2} + \zeta_4^j \bar{\pi}^{r/2}) & \text{if } r \equiv 0 \pmod 4, \\ \bar{c}_{12}\zeta_4^{-j}\pi^{(r+1)/2}\widehat{G}_4 + c_{12}\zeta_4^j\bar{\pi}^{(r+1)/2}G_4 & \\ & \text{if } r \equiv 1 \pmod 4, \\ (-1)^{T/2}(\zeta_4^{-j}\pi^{r/2} + \zeta_4^j\bar{\pi}^{r/2})\sqrt{p} & \text{if } r \equiv 2 \pmod 4, \\ (-1)^k(c_{12}\zeta_4^{-j}\pi^{(r-1)/2}G_4 + \bar{c}_{12}\zeta_4^j\bar{\pi}^{(r-1)/2}\widehat{G}_4) & \\ & \text{if } r \equiv 3 \pmod 4. \end{cases} \tag{2.27}$$

In particular, Proposition 4 in [**7**] generalizes to yield

PROPOSITION 2.1. *For* $p = 12k + 1$

$$Z_j = \begin{cases} c_{12}^r H_j(3) R_j / H_j(2) & \text{if } r \text{ even or } 3 \mid b_4, \\ \omega H_j(3) P_r / R_j & \text{if } r \text{ odd and } 3 \mid a_4, \end{cases}$$

where $\omega = \pm 1$ *satisfies* $\omega \equiv (-1)^{(r+1)/2+k} b_4 \pmod 3$ *and* P_n *is the Lucas sequence given by* $P_n = -i(\pi^n - \bar{\pi}^n)$ *for* $n \geq 0$.

Now set $e_j = \zeta_6^{-j} + \zeta_6^j$ for any integer j. For $p = 12k + 5$ one sees from (2.4), (2.6) and [**7**, Theorem 1] that

$$Z_j = \begin{cases} -(-1)^{r/4} e_j p^{r/4} (\zeta_4^{-j}\pi^{r/2} + \zeta_4^j\bar{\pi}^{r/2}) & \text{if } 4 \mid r, \\ (-1)^{k+(r-2)/4} e_j \beta p^{(r-2)/4} \sqrt{p} (-i(\zeta_4^{-j}\pi^{r/2} - \zeta_4^j\bar{\pi}^{r/2})) & \text{if } 2 \parallel r, \end{cases} \tag{2.28}$$

using the fact that $G_r(\chi_{5i}) = G_r(\chi_i)$ for any integer i and that $\zeta_{12}^j + \zeta_{12}^{5j} = e_j \zeta_4^j$. For $p = 12k + 7$, arguing as in the proof of Theorem 2 in [**7**] one finds the following generalization of (36) in [**7**]. Namely,

$$Z_j = \begin{cases} 2(-1)^{r(k+1)/2+j/2} H_j(3) & \text{if } j \text{ is even}, \\ 0 & \text{if } j \text{ is odd}, \end{cases} \tag{2.29}$$

where r is even. Finally for $p = 12k + 11$, it follows from (2.4), (2.6) and [**7**, Theorem 1] that

$$Z_j = \begin{cases} -2e_{j/2}(-1)^{kr/2} p^{r/2} & \text{if } j \text{ is even}, \\ 0 & \text{if } j \text{ is odd}, \end{cases} \tag{2.30}$$

where r again is even.

For $e = 8$ with r odd one has the following product relations for the Gauss sums $G(\Psi_8^{ir})$ (chiefly, [**3**, Section 4.3]). Namely,

$$G(\Psi_8^r)^2 = (-1)^{Z/2}\sigma_r G(\Psi_8^{2r}), \tag{2.31}$$

$$G(\Psi_8^r)G(\Psi_8^{3r}) = (-1)^k\sqrt{p}\sigma_r, \tag{2.32}$$

$$G(\Psi_8^r)G(\Psi_8^{5r}) = (-1)^{Z/2}G(\Psi_8^{2r})\sqrt{p}, \tag{2.33}$$

$$G(\Psi_8^r)G(\Psi_8^{7r}) = (-1)^k p, \tag{2.34}$$

where σ_r equals $\sigma = a_8 + ib_8\sqrt{2}$ if $r \equiv 1,\ 3 \pmod 8$ or $\bar{\sigma} = a_8 - ib_8\sqrt{2}$ if $r \equiv 5,\ 7 \pmod 8$. Set $Y_j = \zeta_8^{-j}G_r(\chi_1) + \zeta_8^{-3j}G_r(\chi_3) + \zeta_8^{-5j}G_r(\chi_5) + \zeta_8^{-7j}G_r(\chi_7)$ so

$$H_j(8) = Y_j + H_j(4) = Y_j + R_j + H_j(2). \tag{2.35}$$

For $p = 8k + 1$ with $r \equiv 2 \pmod 8$, Y_j is seen to equal

$$-p^{(r-2)/8}\big(\sigma^{r/2} + (-1)^j\bar{\sigma}^{r/2}\big)\big(\zeta_8^{2Z-j}\pi^{(r-2)/4}G_4 + \zeta_8^{6Z-3j}\bar{\pi}^{(r-2)/4}\widehat{G}_4\big)$$

from (2.4), (2.6) and [**3**, Theorem 12.7.1]. Arguing similarly, one finds

$$\begin{aligned}(2.36)\quad & Y_j/(\sigma^{r/2} + (-1)^j\bar{\sigma}^{r/2}) \\ &= \begin{cases} -p^{r/8}(\zeta_8^{-j}\pi^{r/4} + \zeta_8^{-3j}\bar{\pi}^{r/4}) & \text{if } r \equiv 0 \pmod 8, \\ -p^{(r-2)/8}(\zeta_8^{2Z-j}\pi^{(r-2)/4}G_4 + \zeta_8^{6Z-3j}\bar{\pi}^{(r-2)/4}\widehat{G}_4) & \\ & \text{if } r \equiv 2 \pmod 8, \\ -p^{(r-4)/8}\sqrt{p}(\zeta_8^{-j}\pi^{r/4} + \zeta_8^{-3j}\bar{\pi}^{r/4}) & \text{if } r \equiv 4 \pmod 8, \\ -p^{(r-6)/8}(\zeta_8^{2Z-j}\pi^{(r+2)/4}\widehat{G}_4 + \zeta_8^{6Z-3j}\bar{\pi}^{(r+2)/4}G_4) & \\ & \text{if } r \equiv 6 \pmod 8, \end{cases}\end{aligned}$$

when r is even. For r odd, from (2.4) and [**3**, Section 12.7] one finds

$$Y_j = \pm p^{\alpha}\big(\zeta_8^{-j}\pi^{\beta}\sigma^{\delta}G(\Psi_8^r) + \zeta_8^{-3j}\bar{\pi}^{\beta}\sigma^{\delta}G(\Psi_8^{3r}) + \zeta_8^{-5r}\pi^{\beta}\bar{\sigma}^{\delta}G(\Psi_8^{5r}) + \zeta_8^{-7r}\bar{\pi}^{\beta}\bar{\sigma}^{\delta}G(\Psi_8^{7r})\big)$$

with α, β and δ as in Theorem 12.7.1 in [**3**]. Squaring one eventually finds

$$\begin{aligned}Y_j^2 = (-1)^{Z/2}p^{2\alpha}&\big(\sigma^{2\delta}\sigma_r + \bar{\sigma}^{2\delta}\bar{\sigma}_r + 2(-1)^j p^{\delta}\sqrt{p}\big) \\ &\cdot\big(\zeta_4^{-j}\pi^{2\beta}G(\Psi_8^{2r}) + \zeta_4^{j}\bar{\pi}^{2\beta}G(\Psi_8^{6r}) + 2(-1)^{j+k+Z/2}p^{\beta}\sqrt{p}\big)\end{aligned}$$

from (2.31)–(2.34), or

$$Y_j^2 = (-1)^{Z/2}\big(T_r + 2H_j(2)\big)\big(R_j + 2(-1)^{k+Z/2}H_j(2)\big) \tag{2.37}$$

from (2.9) and (2.22), where we set $T_n = \sigma^n + \bar{\sigma}^n$ for any integer $n \geq 0$ and note that $\sigma^{2\delta}\sigma_r + \bar{\sigma}^{2\delta}\bar{\sigma}_r = T_r$ or pT_r as $r \equiv 1, 3 \pmod 8$ or $r \equiv 5, 7 \pmod 8$ accordingly.

For $p = 8k + 3$, one finds $\chi(G) = \zeta_8^{2r}$, so with r even

$$Y_j = \begin{cases} (-1)^{(r+2)/4}p^{(r-2)/4}\big((\zeta_8^{-j} + \zeta_8^{-3j})\sigma^{r/2} + (\zeta_8^{j} + \zeta_8^{3j})\bar{\sigma}^{r/2}\big) & \text{if } 2 \parallel r, \\ -(-1)^{r/4}p^{r/4}\big((\zeta_8^{-j} + \zeta_8^{-3j})\sigma^{r/2} + (\zeta_8^{j} + \zeta_8^{3j})\bar{\sigma}^{r/2}\big) & \text{if } 4 \mid r \end{cases} \tag{2.38}$$

from (2.9) and (2.23).

For $p = 8k+5$, one finds $\chi(G) = \zeta_8^{3r}$, so with r even $Y_j = (\zeta_8^{-j} + \zeta_8^{-5j})G_r(\chi_1) + (\zeta_8^{j} + \zeta_8^{5j})G_r(\chi_3) = 0$ if j is odd. For even j, one finds from [**3**, Theorem 12.7.1] and (2.6) that

$$(2.39)\quad Y_j = \begin{cases} -2p^{3r/8}(\zeta_4^{-j/2}\pi^{r/4} + \zeta_4^{j/2}\bar{\pi}^{r/4}) & \text{if } r \equiv 0 \pmod 8, \\ 2(-1)^{(Z-1)/2}p^{(3r-6)/8}(\zeta_4^{-j/2+1}\pi^{(r+2)/4}\widehat{G}_4 + \zeta_4^{j/2-1}\bar{\pi}^{(r+2)/4}G_4) & \\ & \text{if } r \equiv 2 \pmod 8, \\ -2p^{(3r-4)/8}\sqrt{p}(\zeta_4^{-j/2}\pi^{r/4} + \zeta_4^{j/2}\bar{\pi}^{r/4}) & \text{if } r \equiv 4 \pmod 8, \\ 2(-1)^{(Z-1)/2}p^{(3r-2)/8}(\zeta_4^{-j/2+1}\pi^{(r-2)/4}G_4 + \zeta_4^{j/2-1}\bar{\pi}^{(r-2)/4}\widehat{G}_4) & \\ & \text{if } r \equiv 6 \pmod 8. \end{cases}$$

For $p = 8k+7$ one has r even with $Y_j = -(-1)^{rk/2}p^{r/2}(\zeta_8^{-j}+\zeta_8^{-3j}+\zeta_8^{-5j}+\zeta_8^{-7j})$ or

$$(2.40)\qquad Y_j = \begin{cases} -(-1)^{rk/2+j/4}4p^{r/2} & 4 \mid j, \\ 0 & 4 \nmid j \end{cases}$$

again from [**3**, Theorem 12.7.1] and (2.6).

3. Factors $F^{(w)}(x)$ of $F(x)$ for $e = 6$, 8, 12

The interest here is to explicitly evaluate the factor $F^{(w)}(x)$ of the period polynomial $F(x)$ for the finite field $\mathbb{F}_{p^r}$ for $e \mid 24$ (or at least give a zero of each irreducible factor). To simplify the discussion I shall give complete results for the cases $e = 6$, 8, and 12, but first restate the result of Myerson for $e = 4$ [**8**] in a form more convenient to use here. To this end define the Lucas sequences

$$(3.1)\qquad V_n = \lambda^n + \bar{\lambda}^n, \qquad U_n = (\lambda^n - \bar{\lambda}^n)/(i\sqrt{3}),$$

$$(3.2)\qquad Q_n = \pi^n + \bar{\pi}^n, \qquad P_n = -i(\pi^n - \bar{\pi}^n)$$

$$(3.3)\qquad T_n = \sigma^n + \bar{\sigma}^n, \qquad S_n = (\sigma^n - \bar{\sigma}^n)/(i\sqrt{2})$$

for any integer $n \geq 0$, and the following related sequences

$$V_{j,n} = \zeta_6^{-j}\lambda^n + \zeta_6^j\bar{\lambda}^n, \qquad U_{j,n} = (\zeta_6^{-j}\lambda^n - \zeta_6^j\bar{\lambda}^n)/(i\sqrt{3})$$

$$Q_{j,n} = \zeta_4^{-j}\pi^n + \zeta_4^j\bar{\pi}^n, \qquad P_{j,n} = -i(\zeta_4^{-j}\pi^n - \zeta_4^j\bar{\pi}^n)$$

for any integer j. Note that $V_{j,n} = V_n$ and $U_{j,n} = U_n$ when $6 \mid j$, and also that $Q_{j,n} = Q_n$ and $P_{j,n} = P_n$ when $4 \mid j$.

For $e = 4$, the factors of $F(x)$ are determined by

PROPOSITION 3.1 (Myerson). *(i) If $p = 4k + 3$, then*

$$F(x) = \left(x - (-1)^{r/2}p^{r/2}\right)^3\left(x + 3(-1)^{r/2}p^{r/2}\right).$$

(ii) *If $p = 4k + 1$, then*

(a) *for $r \equiv 0 \pmod 4$, $H_j = -p^{r/4}Q_{j,r/2} - (-1)^j p^{r/2}$ $(1 \leq j \leq 4)$;*

(b) *for $r \equiv 2 \pmod 4$, $H_j = -(-1)^k Q_{j,r/2}p^{(r-2)/4}\sqrt{p} - (-1)^j p^{r/2}$ satisfies*

$$F^{(j)}(x) = x^2 + (-1)^j 2p^{r/2}x - p^{r/2}\left(p^{r/2} + (-1)^j Q_r\right) \quad (j = 1, 2);$$

(c) *for r odd,*

$$F(x) = F^{(1)}(x) = \begin{cases} x^4 - 6p^r x^2 - 4p^r Q_r x + p^r(p^r - Q_r^2) & \text{if } k \text{ is even}, \\ x^4 + 2p^r x^2 + 4p^r Q_r x + p^r(9p^r - Q_r^2) & \text{if } k \text{ is odd}. \end{cases}$$

For $e = 6$, the factors of $F(x)$ are given by

PROPOSITION 3.2. (i) *If* $p = 6k+5$, *then*

$$F(x) = \left(x - (-1)^{r/2}p^{r/2}\right)^5\left(x + 5(-1)^{r/2}p^{r/2}\right).$$

(ii) *If* $p = 6k+1$, *then*

(a) *for* $r \equiv 0 \pmod 6$ *and* $1 \le j \le 6$,

$$H_j = -(-1)^{kr/2}p^{r/6}V_{j,2r/3} - p^{r/3}V_{2j,r/3} - (-1)^{j+kr/2}p^{r/2};$$

(b) *for* $r \equiv 3 \pmod 6$ *and* $1 \le j \le 3$,

$$H_j = p^{r/3}V_{2j,r/3} + (-1)^k p^{(r-3)/6} i^* \sqrt{p}\left(V_{j,2r/3} + (-1)^j p^{r/3}\right)$$

satisfies

$$F^{(j)}(x) = x^2 - 2p^{r/3}V_{2j,r/3}x + p^{2r/3}V_{2j,r/3}^2 - (-1)^k p^{r/3}\left(V_{j,2r/3} + (-1)^j p^{r/3}\right)^2;$$

(c) *for* $r \equiv 2$ *or* $4 \pmod 6$ *and* $j = 1, 2$,

$$F^{(j)}(x) = x^3 + (-1)^{j+kr/2}3p^{r/2}x^2 - 3x\left(p^r + (-1)^{j+kr/2}p^{r/2}V_{2rZ,r}\right)$$
$$- (-1)^{j+kr/2}\left(p^{3r/2}(5 - 3V_{rZ,0}) + (-1)^{j+kr/2}p^r\left(3V_{2rZ,r} - V_r - 3V_{rZ,r}\right) - p^{r/2}V_{2r}\right);$$

(d) *for* $r \equiv 1$ *or* $5 \pmod 6$, $F(x)$ *is the minimal polynomial for*

$$H_6 = (-1)^{k(r-1)/2} i^* \frac{p^{(r-\alpha)/6}}{2\sqrt{p}}\left(gV_{2rZ,(2r+\alpha)/3} - \alpha(g'' - g')U_{2rZ,(2r+\alpha)/3}\right)$$
$$+ p^{[r/3]}\left(gV_{(r-\alpha)/3} + \alpha(g'' - g')U_{(r-\alpha)/3}\right)/2 + (-1)^{k(r-1)/2}p^{(r-1)/2} i^* \sqrt{p},$$

where $\alpha = \pm 1$ *chosen so* $\alpha \equiv r \pmod 6$.

PROOF. When $p = 6k+5$, r is even and $\delta = 6$. One finds that $H_j = (-1)^{r/2}p^{r/2}$ when $j \not\equiv 0 \pmod 3$, $H_3 = \left(3(-1)^{kr/2} - 2(-1)^{r/2}\right)p^{r/2}$ and $H_6 = (-3(-1)^{kr/2} - 2(-1)^{r/2})p^{r/2}$ from (2.9), (2.13) and (2.18).

When $p = 6k+1$ and $3 \mid r$, then $\delta = 3$ and the results (ii)(a) and (b) readily follow from (2.9), (2.12), and (2.19). If $r \equiv 2$ or $4 \pmod 6$ then $\delta = 2$. For $r \equiv 2 \pmod 6$ and $i = 1$ or 2,

$$H_{3i} = (-1)^i\left(G_r(\chi_1) + G_r(\chi_5)\right) + \left(G_r(\chi_2) + G_r(\chi_4)\right) + (-1)^i G_r(\chi_3)$$
$$= -(-1)^{i+kr/2}\left(p^{r/2} + p^{(r-2)/6}(\zeta_6^{2Z}\lambda^{(2r-1)/3}G_3 + \zeta_6^{-2Z}\bar{\lambda}^{(2r-1)/3}\widehat{G}_3)\right)$$
$$- p^{(r-2)/3}(\widehat{G}_3\lambda^{(r+1)/3} + G_3\bar{\lambda}^{(r+1)/3})$$

from (2.13) and (2.18). For $r \equiv 4 \pmod 6$ and $i = 1$ or 2

$$H_{3i} = -(-1)^{i+kr/2}\left(p^{r/2} + p^{(r-4)/6}(\zeta_6^{-2Z}\lambda^{(2r+1)/3}\widehat{G}_3 + \zeta_6^{2Z}\bar{\lambda}^{(2r+1)/3}G_3)\right)$$
$$- p^{(r-1)/3}\left(G_3\lambda^{(r-1)/3} + \widehat{G}_3\bar{\lambda}^{(r-1)/3}\right)$$

from (2.11) and (2.18). The combined expression for the two cases $r \equiv 2$ or $4 \pmod 6$ is

$$(3.4)\quad H_{3i} = -(-1)^{i+kr/2}p^{r/2} - g\left((-1)^{i+kr/2}p^{[r/6]}V_{2rZ,(2r-\alpha)/3} + p^{[r/3]}V_{0,(r+\alpha)/3}\right)/2$$
$$- \alpha(g'' - g')\left((-1)^{i+rk/2}p^{[r/6]}U_{2rZ,(2r-\alpha)/3} - p^{[r/3]}U_{0,(r+\alpha)/3}\right)/2,$$

where $\alpha = \pm 1$ chosen so $2\alpha \equiv r \pmod 6$. It is a routine matter to compute the minimum polynomial for H_{3i} in (3.4) above to obtain the expression for $F^{(1)}(x)$ and $F^{(2)}(x)$ in (ii)(c). Note that the expressions for the factors $F^{(1)}(x)$ and $F^{(2)}(x)$ are merely interchanged when $kr/2$ changes parity.

If $r \equiv 1$ or $5 \pmod 6$ then $\delta = 1$. For $r \equiv 1 \pmod 6$ and $i = 1$ or 2, one finds from (2.12), (2.14) and (2.17) that

$$\begin{aligned}H_{3i} &= (-1)^i\big(G_r(\chi_1) + G_r(\chi_5)\big) + \big(G_r(\chi_2) + G_r(\chi_4)\big) + (-1)^i G_r(\chi_3)\\ &= (-1)^{i+k(r-1)/2} i^* p^{(r-1)/2}\sqrt{p} + p^{(r-1)/3}(\lambda^{(r-1)/3}G_3 + \bar{\lambda}^{(r-1)/3}\widehat{G}_3)\\ &\qquad + (-1)^{i+k(r-1)/2}\frac{i^* p^{(r-1)/6}}{\sqrt{p}}(\zeta_6^{-2Z}\lambda^{(2r+1)/3}\widehat{G}_3 + \zeta_6^{2Z}\bar{\lambda}^{(2r+1)/3}G_3).\end{aligned}$$

For $r \equiv 5 \pmod 6$ and $i = 1$ or 2 one finds similarly that

$$\begin{aligned}H_{3i} &= (-1)^{i+k(r-1)/2} i^* p^{(r-1)/2}\sqrt{p} + p^{(r-2)/3}(\lambda^{(r+1)/3}\widehat{G}_3 + \bar{\lambda}^{(r+1)/3}G_3)\\ &\qquad + (-1)^{i+k(r-1)/2}\frac{i^* p^{(r+1)/6}}{\sqrt{p}}(\zeta_6^{2Z}\lambda^{(2r-1)/3}G_3 + \zeta_6^{-2Z}\bar{\lambda}^{(2r-1)/3}\widehat{G}_3).\end{aligned}$$

Combining these expressions for $r \equiv 1$ or $5 \pmod 6$ one obtains H_{3i} equals

$$\begin{aligned}&(-1)^{i+k(r-1)/2} p^{(r-1)/2} i^* \sqrt{p}\\ &\quad + (-1)^{i+k(r-1)/2}\frac{i^* p^{(r-\alpha)/6}}{2\sqrt{p}}\big(gV_{2rZ,(2r+\alpha)/3} - \alpha(g'' - g')U_{2rZ,(2r+\alpha)/3}\big)\\ &\qquad + p^{[r/3]}\big(gV_{0,(r-\alpha)/3} + \alpha(g'' - g')U_{0,(r-\alpha)/3}\big)/2,\end{aligned}$$

where $\alpha = \pm 1$ chosen so $\alpha \equiv r \pmod 6$. This completes the proof of the proposition. □

For $e = 8$, the factors of $F(x)$ are given by

PROPOSITION 3.3. (i) *If $p = 8k + 7$, then*

$$F(x) = \big(x - (-1)^{r/2}p^{r/2}\big)^7\big(x + 7(-1)^{r/2}p^{r/2}\big).$$

(ii) *If $p = 8k + 5$, then*

(a) *for $r \equiv 0 \pmod 8$, $\delta = 8$ with*

$$H_j = \begin{cases} p^{r/2} - (-1)^{(j-1)/2}p^{r/4}P_{r/2} & \text{if } j \text{ odd},\\ -p^{r/2} - p^{r/4}Q_{r/2} - (-1)^{j/4}2p^{3r/8}Q_{r/4} & \text{if } 4 \mid j,\\ -p^{r/2} + p^{r/4}Q_{r/2} - (-1)^{(j-2)/4}2p^{3r/8}P_{r/4} & \text{if } 2 \parallel j;\end{cases}$$

(b) *for $r \equiv 4 \pmod 8$, $\delta = 4$ with*

$$F^{(j)}(x) = \begin{cases} (x - p^{r/2} + (-1)^{(j-1)/2}p^{r/4}P_{r/2})^2 & \text{if } j = 1, 3,\\ \big(x + p^{r/4}(p^{r/4} - Q_{r/2})\big)^2 - 4p^{3r/4}P_{r/4}^2 & \text{if } j = 2,\\ \big(x + p^{r/4}(p^{r/4} + Q_{r/2})\big)^2 - 4p^{3r/4}Q_{r/4}^2 & \text{if } j = 4;\end{cases}$$

(c) *for $r \equiv 2 \pmod 4$, $\delta = 2$ with*

$$F^{(1)}(x) = (x^2 - 2p^{r/2}x + p^r - p^{r/2}P_{r/2}^2)^2$$

$$\begin{aligned}F^{(2)}(x) &= x^4 + 4p^{r/2}x^3 + 2p^{r/2}x^2(11p^{r/2} - Q_{r/2}^2)\\ &\qquad + 4p^r x(9p^{r/2} - 5Q_{r/2}^2) + p^r(9p^{r/2} - Q_{r/2}^2)^2.\end{aligned}$$

(iii) *If $p = 8k + 3$, then*

(a) *for $r \equiv 0 \pmod 4$, $\delta = 8$ with*

$$F(x) = (x - p^{r/2})^2(x^2 - 2p^{r/2}x + p^r - 4p^{r/2}S_{r/2}^2)^2(x^2 + 6p^{r/2}x + 9p^r - 4p^{r/2}T_{r/2}^2)$$

(b) *for* $r \equiv 2 \pmod 4$,

$$F^{(1)}(x) = F^{(3)}(x) = x^2 + 2p^{r/2}x + p^r + 4p^{r/2}S_{r/2}^2,$$

$$F^{(2)}(x) = (x - 3p^{r/2})^2, \qquad F^{(4)}(x) = x^2 + 2p^{r/2}x + p^r + 4p^{r/2}T_{r/2}^2;$$

(iv) *If* $p = 8k+1$, *then*

(a) *for* $r \equiv 0 \pmod 8$, $\delta = 8$ *with*

$$H_j + (-1)^j p^{r/2} - p^{r/4}Q_{j,r/2} - p^{r/8}AB \quad (1 \le j \le 8);$$

(b) *for* $r \equiv 4 \pmod 8$, $\delta = 4$ *with*

$$F^{(j)}(x) = x^2 + 2p^{r/4}(p^{r/4} + Q_{j,r/2})x + p^{r/2}(p^{r/4} + Q_{j,r/2})^2 - p^{r/4}A^2B^2$$

for $1 \le j \le 4$, *where*

$$\begin{cases} A = Q_{j/2,r/4},\ B = T_{r/2} & \text{if } j \text{ is even,} \\ A = (-1)^{[j/4]}Q_{0,r/4} + (-1)^{[(j-2)/4]}P_{0,r/4},\ B = S_{r/2} & \text{if } j \text{ is odd;} \end{cases}$$

(c) *for* $r \equiv 2 \pmod 4$, $\delta = 2$ *with*

$$\begin{aligned} F^{(1)}(x) = x^4 &- 4p^{r/2}x^3 - 2p^{r/2}x^2(P_{r/2}^2 + 4S_{r/2}^2 - 3p^{r/2}) \\ &- 4p^{r/2}x\big(p^r - p^{r/2}(P_{r/2}^2 + 4S_{r/2}^2) + 2P_{r/2}^2S_{r/2}^2\big) \\ &+ p^r(p^{r/2} + P_{r/2}^2 - 4S_{r/2}^2)^2 - 4p^{r/2}(p^{r/2} - S_{r/2}^2)^2P_{r/2}^2 \\ F^{(2)}(x) = x^4 &+ 4p^{r/2}x^3 - 2p^{r/2}x^2(Q_{r/2}^2 + 2T_{r/2}^2 - 3p^{r/2}) \\ &+ 4p^{r/2}x\big(p^r - p^{r/2}(Q_{r/2}^2 + 2T_{r/2}^2) + Q_{r/2}^2T_{r/2}^2\big) \\ &+ p^r(p^{r/2} + Q_{r/2}^2 - 2T_{r/2}^2)^2 - p^{r/2}(2p^{r/2} - T_{r/2}^2)^2Q_{r/2}^2; \end{aligned}$$

(d) *for* r *odd*, $F(x)$ *is the minimal polynomial satisfied by*

$$H_8 = g_r(8) = R_4 + g_r(2) \pm i^{Z/2}\sqrt{\big(T_r + g_r(2)\big)\big(R_4 + 2(-1)^{k+Z/2}g_r(2)\big)}$$

with R_4 *as in* (2.22).

(Note: One choice of sign in the expression for H_8 in (iv)(d) above gives its conjugate H_4. No way is known to determine the correct sign for H_8.)

PROOF. When $p = 8k+7$, r is even and $\delta = 8$. One finds that

$$H_j = (-1)^{r/2}p^{r/2} \quad \text{if } j \not\equiv 0 \pmod 4,$$

$$H_4 = \big(4(-1)^{rk/2} - 3(-1)^{r/2}\big)p^{r/2}, \qquad H_8 = \big(-4(-1)^{rk/2} - 3(-1)^{r/2}\big)p^{r/2}$$

from (2.9), (2.23) and (2.40). When $p = 8k+5$ and $r \equiv 0 \pmod 4$,

$$H_j = -p^{r/4}(-i^j\pi^{r/2} + i^j\bar{\pi}^{r/2}) + p^{r/2} = p^{r/2} - (-1)^{(j-1)/2}p^{r/4}P_{r/2}$$

if j is odd, from (2.10) and (2.22) since $Y_j = 0$. For even j,

$$\begin{aligned} H_j &= -2p^{3r/8}(\zeta_4^{-j/2}\pi^{r/4} + \zeta_4^{j/2}\bar{\pi}^{r/4}) + (-1)^{j/2}(-p^{r/4}Q_{r/2}) - p^{r/2} \\ &= -(-1)^{j/2}p^{r/4}Q_{r/2} - p^{r/2} - 2p^{3r/8}Q_{j/2,r/4} \end{aligned}$$

if $r \equiv 0 \pmod 8$ or

$$\begin{aligned} H_j &= -2p^{(3r-4)/8}\sqrt{p}(\zeta_4^{-j/2}\pi^{r/4} + \zeta_4^{j/2}\bar{\pi}^{r/4}) - p^{r/2} - (-1)^{j/2}p^{r/4}Q_{r/2} \\ &= -2p^{(3r-4)/8}\sqrt{p}Q_{j/2,r/4} - (-1)^{j/2}p^{r/4}Q_{r/2} - p^{r/2} \end{aligned}$$

if $r \equiv 4 \pmod 8$ from (2.9), (2.22) and (2.38). Thus the results in (ii)(a) et (b) hold for j odd or even. If $r \equiv 2 \pmod 4$ then $\delta = 2$ and one finds for odd j that $H_j = Y_j + R_j + H_j(2) = p^{(r-2)/4}\sqrt{p}(\zeta_4^{-j}\pi^{r/2} + \zeta_4^{j}\bar{\pi}^{r/2}) + p^{r/2}$ or $(-1)^{(j-1)/2}p^{(r-2)/4}\sqrt{p}P_{r/2}+p^{r/2}$ from (2.9) and (2.22) since $Y_j = 0$. Thus $F^{(1)}(x) = (x^2 - 2p^{r/2}x + p^r - p^{r/2}P_{r/2}^2)^2$. For even j in this case, one finds for $r \equiv 2 \pmod 8$ that

$$
\begin{aligned}
H_j - H_j(2) &= H_j + p^{r/2} \\
&= 2i^{Z-1}p^{(3r-6)/8}(\zeta_4^{1-j/2}\pi^{(r+2)/4}\widehat{G}_4 + \zeta_4^{j/2-1}\bar{\pi}^{(r+2)/4}G_4) \\
&\qquad + p^{(r-2)/4}\sqrt{p}Q_{r/2}
\end{aligned}
$$

so

$$H_8 = -2i(-1)^{(Z-1)/2}p^{(3r-6)/8}(AQ_{(r+2)/4} + BP_{(r+2)/4}) + p^{(r-2)/4}\sqrt{p}Q_{r/2} - p^{r/2}.$$

Similarly for $r \equiv 6 \pmod 8$ and j even,

$$H_j + p^{r/2} = 2i^{Z-1}p^{(3r-2)/8}(\zeta_4^{1-j/2}\pi^{(r-2)/4}G_4 + \zeta_4^{j/2-1}\bar{\pi}^{(r-2)/4}\widehat{G}_4) + p^{(r-2)/4}\sqrt{p}Q_{r/2}$$

so

$$H_8 = 2i(-1)^{(Z-1)/2}p^{(3r-2)/8}(AQ_{(r-2)/4} - BP_{(r-2)/4}) + p^{(r-2)/4}\sqrt{p}Q_{r/2} - p^{r/2}.$$

A straightforward calculation gives the expression for $F^{(2)}(x)$ in (ii)(c) whether $r \equiv 2$ or $6 \pmod 8$.

When $p = 8k+3$ and $r \equiv 0 \pmod 8$, $\delta = 8$ and for odd j,

$$
\begin{aligned}
H_j &= -p^{r/4}((\zeta_8^{-j} + \zeta_8^{-3j})\sigma^{r/2} + (\zeta_8^{-5j} + \zeta_8^{-7j})\bar{\sigma}^{r/2}) + p^{r/2} \\
&= -(-1)^{[j/4]}2p^{r/2}S_{r/2} + p^{r/2}
\end{aligned}
$$

from (2.9), (2.23), and (2.38). Whereas for even j, $H_2 = H_6 = 2p^{r/2} - p^{r/2} = p^{r/2}$ from (2.9) and (2.23) since $Y_j = 0$ in (2.35). Also $H_4 = -2p^{r/4}T_{r/2} - 3p^{r/2}$ and $H_8 = 2p^{r/4}T_{r/2} - 3p^{r/2}$ from (2.9), (2.23) and (2.38) when $r \equiv 0 \pmod 8$.

If $r \equiv 4 \pmod 8$ then again $\delta = 8$, but the signs of $G_r(\chi^\nu)$ are opposite those of the case $r \equiv 0 \pmod 8$ for ν odd. Thus $H_j = (-1)^{[j/4]}2p^{r/2}S_{r/2} + p^r$ for j odd, $H_2 = H_6 = p^{r/2}$, $H_4 = 2p^{r/4}T_{r/2} - 3p^{r/2}$ and $H_8 = -2p^{r/4}T_{r/2} - 3p^{r/2}$. Combining the results from the cases $r \equiv 0$ or $4 \pmod 8$ gives the factorization of $F(x)$ as indicated.

If $r \equiv 2 \pmod 4$ then $\delta = 4$ and one finds that for odd j,

$$
\begin{aligned}
H_j &= Y_j + R_j + H_j(2) \\
&= (-1)^{(j-1)/2}p^{(r-2)/4}(-1)^{(r+2)/4}i\sqrt{2}(i\sqrt{p})(\sigma^{r/2} - \bar{\sigma}^{r/2}) - p^{r/2} \\
&= (-1)^{(j+1)/2}p^{(r-2)/4}(-1)^{(r+2)/4}2i\sqrt{p}S_{r/2} - p^{r/2}
\end{aligned}
$$

from (2.9), (2.23) and (2.38), so

$$F^{(1)}(x) = F^{(3)}(x) = x^2 + 2p^{r/2}x + p^r + 4p^{r/2}S_{r/2}^2.$$

Also $H_2 = H_6 = 3p^{r/2}$ from (2.9), (2.23), and (2.38), and

$$H_4 = -2p^{r/2} + p^{r/2} + 2i(-1)^{(r+2)/4}p^{(r-2)/4}\sqrt{p}(\sigma^{r/2} + \bar{\sigma}^{r/2})$$

or

$$2i(-1)^{(r+2)/4}p^{(r-2)/4}\sqrt{p}T_{r/2} - p^{r/2}.$$

Similarly, $H_8 = -2i(-1)^{(r+2)/4}p^{(r-2)/4}\sqrt{p}T_{r/2} - p^{r/2}$. Thus

$$F^{(2)}(x) = (x - 3p^{r/2})^2 \quad \text{and} \quad F^{(4)}(x) = x^2 + 2p^{r/2}x + p^r + 4p^{r/2}T_{r/2}^2.$$

When $p = 8k+1$ and $r \equiv 0 \pmod 8$, $\delta = 8$ and

$$H_j = -p^{r/8}(\zeta_8^{-j}\pi^{r/4} + \zeta_8^{-3j}\bar{\pi}^{r/4})(\sigma^{r/2} + (-1)^j\bar{\sigma}^{r/2}) - p^{r/4}Q_{j,r/2} - p^{r/2}$$

from (2.9), (2.23) and (2.36). If j is even, then

$$H_j = -p^{r/8}T_{r/2}Q_{j/2,r/4} - p^{r/4}Q_{j,r/2} - p^{r/2}$$

else

$$H_j = -p^{r/8}S_{r/2}(i\sqrt{2})(\zeta_8^{-j}\pi^{r/4} + \zeta_8^{-3j}\bar{\pi}^{r/4}) - p^{r/4}Q_{j,r/2} - p^{r/2}$$

or

$$-p^{r/8}S_{r/2}(xQ_{r/4} + yP_{r/4}) - p^{r/4}Q_{j,r/2} - p^{r/2},$$

where $x, y = \pm 1$ are determined by $i\sqrt{2}\zeta_8^{-3j} = x + iy$. In either case the results (iv)(a) and (b) readily follow with $r \equiv 0 \pmod 8$. The same argument gives the result for $r \equiv 4 \pmod 8$ though now $\delta = 4$.

If $r \equiv 2 \pmod 4$ then $\delta = 2$ with H_1 satisfying $F^{(1)}(x)$ and H_8 satisfying $F^{(2)}(x)$. For $r \equiv 2 \pmod 8$, one has from (2.36) that

$$Y_j^2 = \left(\sigma^{r/2} + (-1)^j\bar{\sigma}^{r/2}\right)^2 \cdot p^{(r-2)/4}\left(\zeta_4^{-j}\pi^{(r-2)/2}G_4^2 + \zeta_4^j\bar{\pi}^{(r-2)/2}\widehat{G}_4^2 + 2(-1)^j p^{(r-2)/4}G_4\widehat{G}_4\right)$$

or

$$Y_j^2 = \begin{cases} p^{(r-2)/4}T_{r/2}^2(Q_{j,r/2}\sqrt{p} + 2p^{(r+2)/4}) & \text{if } j \text{ is even,} \\ -2p^{(r-2)/4}S_{r/2}^2(Q_{j,r/2}\sqrt{p} - 2p^{(r+2)/4}) & \text{if } j \text{ is odd.} \end{cases} \tag{3.5}$$

Similarly one finds for $r \equiv 6 \pmod 8$ that

$$Y_j^2 = \left(\sigma^{r/2} + (-1)^j\bar{\sigma}^{r/2}\right)^2 \cdot p^{(r-6)/4}\left(\zeta_4^{-j}\pi^{(r+2)/2}\widehat{G}_4^2 + \zeta_4^j\bar{\pi}^{(r+2)/2}G_4^2 + 2(-1)^j p^{(r+2)/4}G_4\widehat{G}_4\right)$$

which is seen to equal (3.5) above. Thus (3.5) holds whenever $r \equiv 2 \pmod 4$. One can now routinely verify that for $r \equiv 2$ or $6 \pmod 8$, H_1 satisfies the explicit polynomial given for $F^{(1)}(x)$ and H_8 satisfies that for $F^{(2)}(x)$ in (iv)(c). Finally, if r is odd then $\delta = 1$ so $F^{(1)}(x) = F(x)$ is satisfied by H_8 as given in (iv)(d) by (2.9), (2.22) and (2.36). □

The following example illustrates the above result when $p = 8k+1$ with $r = 6$.

EXAMPLE 3.4. Consider the finite field $\mathbb{F}_q$ with $q = 17^6$, where one may take $\gamma = 3 + \zeta_7$ to generate $\mathbb{F}_q^*$ with $G = \gamma^{1508598} = 3$. One finds $a_4 = -1$, $b_4 = 4$ and $a_8 = 3$, $b_8 = 2$ from this choice with $Z = 14$, so $Q_3 = 94$, $P_3 = -104$ in (3.2) and $T_3 = -90$, $S_3 = 76$ in (3.3). One finds from Proposition 3.3(iv)(c) that

$$F^{(1)}(x) = x^4 - 19652x^3 - 188472506x^2 + 345186574268x + 1154552320494017$$

$$F^{(2)}(x) = x^4 + 19652x^3 - 101178322x^2 - 536356120748x + 15678509519801.$$

These agree with the computed factors of the period polynomial in (1.3). Indeed, for the choice of generator γ here one finds (chiefly, [**5**, Proposition 4]) that the modified Gauss sums

$$H_1 = 10177 + 8096\psi_2 + 6992\psi_3 + 5968\psi_4$$
$$H_8 = -615 + 7336\psi_1 + 4456\psi_3 + 5400\psi_4$$

in terms of the Gauss circulants

$$\psi_1 = \zeta_{17}^3 + \zeta_{17}^5 + \zeta_{17}^{12} + \zeta_{17}^{14}, \qquad \psi_2 = \zeta_{17}^2 + \zeta_{17}^8 + \zeta_{17}^9 + \zeta_{17}^{15},$$
$$\psi_3 = \zeta_{17}^6 + \zeta_{17}^7 + \zeta_{17}^{10} + \zeta_{17}^{11}, \qquad \psi_4 = \zeta_{17}^1 + \zeta_{17}^4 + \zeta_{17}^{13} + \zeta_{17}^{16}.$$

Direct calculation shows H_1 and H_8 have minimal polynomials $F^{(1)}(x)$ and $F^{(2)}(x)$ as above.

For $e = 12$, the factors of $F(x)$ are determined by

PROPOSITION 3.5. (i) *If $p = 12k + 11$, then*

$$F(x) = \left(x - (-1)^{r/2}p^{r/2}\right)^{11}\left(x + 11(-1)^{r/2}p^{r/2}\right).$$

(ii) *If $p = 12k + 7$, then*

(a) *for $r \equiv 0 \pmod 6$, $\delta = 12$ with H_j equal to*

$$\begin{cases} -(-1)^{r/2}p^{r/6}V_{j,2r/3} - p^{r/3}V_{2j,r/3} + (-1)^{r/2}p^{r/2} & \text{if } j \text{ is odd,} \\ -(-1)^{r/2}p^{r/6}V_{j,2r/3} + p^{r/3}V_{2j,r/3}(\rho - 1) + p^{r/2}\left(\rho - (-1)^{r/2}\right) & \text{if } 2 \parallel j, \\ -(-1)^{r/2}p^{r/6}V_{j,2r/3} - p^{r/3}(\rho + 1)V_{2j,r/3} - p^{r/2}\left(\rho + (-1)^{r/2}\right) & \text{if } 4 \mid j, \end{cases}$$

where $\rho = 2(-1)^{r(k+1)/2}$.

(b) *For r even, but $r \not\equiv 0 \pmod 6$, $\delta = 4$ with*

$$F^{(1)}(x) = F^{(3)}(x) = x^3 - (-1)^{r/2}3p^{r/2}x^2 - 3x\left(p^r - (-1)^{r/2}p^{r/2}V_{2rZ,r}\right)$$
$$+ (-1)^{r/2}\left(p^{3r/2}(5 - 3V_{rZ,0}) - (-1)^{r/2}p^r(3V_{2rZ,r} - V_r - 3V_{rZ,r}) - p^{r/2}V_{2r}\right).$$

If $r(k+1)/2$ is even, then $F^{(2)}(x) = P\left(x - p^{r/2}(2 - (-1)^{r/2})\right)$ where

$$P(x) = x^3 - 3p^{r/2}x\left(2p^{r/2} - (-1)^{r/2}V_{2rZ,r}\right)$$
$$+ p^{r/2}\left(-p^{r/2}V_r + 3(-1)^{r/2}p^rV_{rZ,0} - 3p^{r/2}V_{rZ,r} + (-1)^{r/2}V_{2r}\right)$$

and $F^{(4)}(x) = Q\left(x + p^{r/2}(2 + (-1)^{r/2})\right)$ where

$$Q(x) = x^3 - 3p^{r/2}x\left(10p^{r/2} + 3(-1)^{r/2}V_{2rZ,r}\right)$$
$$+ p^{r/2}\left(27p^{r/2}V_r + 27(-1)^{r/2}p^rV_{rZ,0} + 9p^{r/2}V_{rZ,r} + (-1)^{r/2}V_{2r}\right).$$

(If $r(k+1)/2$ is odd, then the expressions for $F^{(2)}(x)$ and $F^{(4)}(x)$ above are swapped.)

(iii) *If $p = 12k + 5$, then*

(a) *for $r \equiv 0 \pmod 4$, $\delta = 12$ with*

$$H_j = \begin{cases} (-1)^{j/2}Q_{r/2}p^{r/4}\left((-1)^{r/4} - 1\right) + p^{r/2} & \text{if } j = 2, 4, 8, 10, \\ (-1)^{(j+1)/2}P_{r/2}p^{r/4}\left((-1)^{r/4} + 1\right) + p^{r/2} & \text{if } j = 1, 5, 7, 11, \\ (-1)^{(j-1)/2}P_{r/2}p^{r/4}(2(-1)^{r/4} - 1) + p^{r/2} & \text{if } j = 3, 9, \\ (-1)^{(j+2)/2}Q_{r/2}p^{r/4}(2(-1)^{r/4} + 1) - 5p^{r/2} & \text{if } j = 6, 12. \end{cases}$$

(b) *For* $r \equiv 2$ *or* $10 \pmod{12}$, $\delta = 6$ *with*

$$F^{(1)} = F^{(2)} = F^{(4)} = F^{(5)} = (x + p^{r/2})^2 - p^{r/2}\big(P_{r/2} + (-1)^{(r+2)/4}\beta Q_{r/2}\big)^2$$

$$F^{(3)}(x) = (x - 5p^{r/2})^2 - p^{r/2}\big(2(-1)^{(r-2)/4}\beta Q_{r/2} + P_{r/2}\big)^2$$

$$F^{(6)}(x) = (x + p^{r/2})^2 - p^{r/2}\big(2(-1)^{(r-2)/4}\beta P_{r/2} + Q_{r/2}\big)^2.$$

Here $\beta = \pm 1$ *with* $\beta \equiv a_4 b_4 \pmod{3}$.

(iv) *If* $p = 12k + 1$, *then*

(a) *for* $r \equiv 0 \pmod{12}$, $\delta = 12$ *and for* $1 \le j \le 12$,

$$H_j = -p^{r/12} Q_{j,r/2} V_{-j,r/3} - p^{r/4} Q_{j,r/2} - p^{r/6} V_{j,2r/3} - p^{r/3} V_{2j,r/3} - (-1)^j p^{r/2}.$$

(b) *For* $r \equiv 6 \pmod{12}$, $\delta = 6$ *and for* $1 \le j \le 6$,

$$F^{(j)}(x) = \big(x + p^{r/6}(V_{j,2r/3} + p^{r/6} V_{2j,r/3} + (-1)^j p^{r/3})\big)^2 - p^{r/6}(c_{12}^2 Q_{j,r/2} V_{5j,r/3} - p^{r/6} Q_{j,r/2})^2.$$

(c) *For* $r \equiv 4$ *or* $8 \pmod{12}$, $\delta = 4$ *with* $F^{(j)}(x) = P(x + p^{r/4} Q)$, *where*

$$P(x) = x^3 - 3x(p^{r/2}Q^2 + p^{r/4} Q V_{2rZ,r} + p^r) + (-1)^j p^{r/4}(Q^3 V_r + 3p^{3r/4} Q^2 V_{2rZ,0} + 3p^{r/2} Q V_{4rZ,r} + p^{r/4} V_{2r})$$

with $Q = Q_{j,r/2} + (-1)^j p^{r/4}$ $(1 \le j \le 4)$.

(d) *For* $r \equiv 3 \pmod{6}$, $\delta = 3$ *with*

$$F^{(j)}(x) = x^4 - 4p^{r/3} V_{2j,r/3} x^3 + x^2\big(2C + 4p^{r/3}(p^{2r/3} - (-1)^j V_{j,2r/3} p^{r/3} - V_{j,2r/3}^2)\big) + 4x\big(D((-1)^j p^{(r+1)/2} + p^{(r+3)/6} V_{j,2r/3}) - C V_{2j,r/3} p^{r/3}\big) + C^2 - pD^2 \qquad (j = 1, 2, 3)$$

where

$$C = \begin{cases} (-1)^j p^{2r/3} V_{j,2r/3}\big(3 - 2(-1)^k\big) + 3p^r\big(1 - 2(-1)^k\big) \\ \qquad + p^{r/3}\big(V_{j,2r/3}^2 - (-1)^k 2c_{12} Q_r V_{2j,r/3}\big) & \text{if } 3 \mid b_4, \\ (-1)^j p^{2r/3} V_{j,2r/3}\big(3 - 2(-1)^k\big) + 3p^r\big(1 - 2(-1)^k\big) \\ \qquad + p^{r/3}(V_{j,2r/3}^2 - 2\omega P_r V_{2j,r/3}) & \text{if } 3 \mid a_4, \end{cases}$$

and similarly

$$D = \begin{cases} (-1)^j p^{(r-1)/2}\big(2V_r - (-1)^k 3Q_r\big) - (-1)^k p^{(r-3)/6} Q_r V_{j,2r/3} \\ \qquad + (-1)^j 4p^{(5r-3)/6} V_{2j,r/3}\big(1 - (-1)^k c_{12}\big) & \text{if } 3 \mid b_4, \\ (-1)^j p^{(r-1)/2}\big(2V_r + (-1)^k Q_r\big) + (-1)^k p^{(r-3)/6} Q_r V_{j,2r/3} \\ \qquad + (-1)^j 4p^{(5r-3)/6} V_{2j,r/3} & \text{if } 3 \mid a_4. \end{cases}$$

(e) *For* $r \equiv 2, 10 \pmod{12}$, $\delta = 2$ *with* $F^{(1)}(x)$ *satisfied by*

$$H_3 = p^{r/2} + (-1)^k p^{(r-2)/4}\sqrt{p} P_{r/2} + \frac{g}{2}\big(-(-1)^{T/2} p^{(r-2\alpha)/12} P_{r/2} V_{(r+\alpha)/3}/\sqrt{p} + p^{[r/6]} V_{2rZ,(2r-\alpha)/3} - p^{[r/3]} V_{(r+\alpha)/3}\big) + \alpha h\bigg(\frac{(-1)^{T/2}}{\sqrt{p}} p^{(r-2\alpha)/12} P_{r/2} U_{(r+\alpha)/3} + p^{[r/6]} U_{2rZ,(2r-\alpha)/3} + p^{[r/3]} U_{(r+\alpha)/3}\bigg)$$

and $F^{(2)}(x)$ satisfied by

$$H_{12} = -p^{r/2} - (-1)^k p^{(r-2)/4}\sqrt{p}Q_{r/2} - \frac{g}{2}\big((-1)^{T/2}p^{(r-2\alpha)/12}Q_{r/2}V_{(r+\alpha)/3}/\sqrt{p} + p^{[r/6]}V_{2rZ,(2r-\alpha)/3} + p^{[r/3]}V_{(r+\alpha)/3}\big) + \alpha h\Big(\frac{(-1)^{T/2}}{\sqrt{p}}p^{(r-2\alpha)/12}Q_{r/2}U_{(r+\alpha)/3} - p^{[r/6]}U_{2rZ,(2r-\alpha)/3} + p^{[r/3]}U_{(r+\alpha)/3}\Big)$$

where $h = (g'' - g')/2$ and $\alpha = \pm 1$ chosen so $2\alpha \equiv r \pmod{12}$.

(f) *For $r \equiv 1, 5, 7$ or $11 \pmod{12}$, $\delta = 1$ with $F(x) = F^{(1)}(x)$ satisfied by*

$$H_{12} = g_r(12) = \begin{cases} g_r(6) + R_4\big(1 + c_{12}^r g_r(3)/g_r(2)\big) & \text{if } 3 \mid b_4, \\ g_r(6) + R_4 + \omega g_r(3)P_r/R_4 & \text{if } 3 \mid a_4 \end{cases}$$

with ω as in Proposition 2.1.

PROOF. When $p = 12k + 11$, r is even and $\delta = 12$. One finds that $H_j = (-1)^{r/2}p^{r/2}$ when $j \not\equiv 0 \pmod 6$, $H_6 = \big(6(-1)^{kr/2} - 5(-1)^{r/2}\big)p^{r/2}$ and $H_{12} = \big(-6(-1)^{kr/2} - 5(-1)^{r/2}\big)p^{r/2}$ from (2.9), (2.26), (2.30), and Proposition 3.2.

When $p = 12k + 7$, r is even. Also $H_j = H_j(6)$ for j odd, since $R_j = Z_j = 0$ in (2.26). If $r \equiv 0 \pmod 6)$, then $\delta = 12$ and $H_j = -(-1)^{r/2}p^{r/6}V_{j,2r/3} - p^{r/3}V_{2j,r/3} + (-1)^{r/2}p^{r/2}$ from Proposition 3.2(ii) when j is odd. When j is even, $H_j = -2(-1)^{r(k+1)/2}p^{r/3}V_{j/2,r/3} + H_j(6) - 2(-1)^{r(k+1)/2+j/2}p^{r/2}$ from (2.23) and (2.29), yielding the desired expressions in (ii)(a) again by Proposition 3.2. If $r \equiv 2$, 4, 8, 10 (mod 12), $\delta = 4$ with $H_1 = H_7 = H_1(6)$ so $F^{(1)} = F^{(3)}$ is the same polynomial appearing in (ii)(c) of Proposition 3.2 with $j = 1$ (and "k" there taken odd). It follows from (2.12) and (2.29) that when j is even,

$$Z_j = \begin{cases} -(-1)^{r(k+1)/2}2p^{(r-2)/3}(\zeta_6^{-j/2}\lambda^{(r+1)/3}\widehat{G}_3 + \zeta_6^{j/2}\bar{\lambda}^{(r+1)/3}G_3) & \text{if } r \equiv 2 \pmod 6, \\ -(-1)^{r(k+1)/2}2p^{(r-1)/3}(\zeta_6^{-j/2}\lambda^{(r-1)/3}G_3 + \zeta_6^{j/2}\bar{\lambda}^{(r-1)/3}\widehat{G}_3) & \text{if } r \equiv 4 \pmod 6. \end{cases}$$

Thus one finds from (2.26) that

$$H_{6i} = g_r(6) - (-1)^{r(k+1)/2+i}\big(2p^{r/2} + p^{[r/3]}(gV_{(r+\alpha)/3} - \alpha(g'' - g')U_{(r+\alpha)/3})\big)$$

for $i = 1, 2$ in either case, where $\alpha = \pm 1$ satisfies $2\alpha \equiv r \pmod 6$ as (3.4). A straightforward, albeit tedious, computation using (3.4) gives the polynomial expressions for $F^{(2)}$ and $F^{(4)}$ in (ii)(b).

When $p = 12k + 5$, r is again even. If $r \equiv 0 \pmod 4$, then $\delta = 12$ with

$$Z_j = \begin{cases} -2i^j(-1)^{r/4}p^{r/4}\big(\pi^{r/2} + (-1)^j\bar{\pi}^{r/2}\big) & \text{if } 3 \mid j, \\ i^j(-1)^{r/4}p^{r/4}\big(\pi^{r/2} + (-1)^j\bar{\pi}^{r/2}\big) & \text{otherwise.} \end{cases}$$

from (2.28). Thus from (2.23), (2.26) and Proposition 3.2,

$$H_j = \begin{cases} -2i^j(-1)^{r/4}p^{r/4}\big(\pi^{r/2} + (-1)^j\bar{\pi}^{r/2}\big) + \big(-3(-1)^j - 2\big)p^{r/2} - p^{r/4}Q_{j,r/2} & \text{if } 3 \mid j, \\ i^j(-1)^{r/4}p^{r/4}\big(\pi^{r/2} + (-1)^j\bar{\pi}^{r/2}\big) + p^{r/2} - p^{r/4}Q_{j,r/2} & \text{otherwise.} \end{cases}$$

The result (iii)(a) follows. If $r \equiv 2,\ 10 \pmod{12}$, then $\delta = 6$ with

$$Z_j = \begin{cases} 2(-1)^{k+1}\beta i^{j+r/2}p^{(r-2)/4}\sqrt{p}\big(\pi^{r/2} - (-1)^j\bar{\pi}^{r/2}\big) & \text{if } 3 \mid j, \\ (-1)^k\beta i^{j+r/2}p^{(r-2)/4}\sqrt{p}\big(\pi^{r/2} - (-1)^j\bar{\pi}^{r/2}\big) & \text{otherwise} \end{cases}$$

from (2.28). Thus again from (2.23), (2.26) and Proposition 3.2,

$$H_j = \begin{cases} 2(-1)^{k+1}\beta i^{j+r/2}p^{(r-2)/4}\sqrt{p}\big(\pi^{r/2} - (-1)^j\bar{\pi}^{r/2}\big) \\ \qquad + \big(-3(-1)^j + 2\big)p^{r/2} + (-1)^k Q_{j,r/2}p^{(r-2)/4}\sqrt{p} & \text{if } 3 \mid j, \\ (-1)^k\beta i^{j+r/2}p^{(r-2)/4}\sqrt{p}\big(\pi^{r/2} - (-1)^j\bar{\pi}^{r/2}\big) \\ \qquad - p^{r/2} + (-1)^k p^{(r-2)/4}Q_{j,r/2}\sqrt{p} & \text{otherwise.} \end{cases}$$

A case by case computation yields the polynomial expressions appearing in (iii)(b).

When $p = 12k+1$ and $r \equiv 0 \pmod{12}$, $\delta = 12$ with $Z_j = -p^{r/12}Q_{j,r/2}V_{-j,r/3}$ from (2.9), (2.12), (2.22) and Proposition 2.1 (and noting that $V_{2j,n} = (-1)^j V_{-j,n}$). Thus

$$H_j = -p^{r/12}Q_{j,r/2}V_{-j,r/3} - p^{r/6}V_{j,2r/3} - p^{r/3}V_{2j,r/3} - p^{r/4}Q_{j,r/2} - (-1)^j p^{r/2}$$

from (2.22) and Proposition 3.2.

If $r \equiv 6 \pmod{12}$, $\delta = 6$ with $Z_j = (-1)^k p^{(r-6)/12}\sqrt{p}c_{12}^2 Q_{j,r/2}V_{-j,r/3}$ arguing similarly as above. In particular,

$$\begin{aligned} H_j = (-1)^k p^{(r-6)/12}&\sqrt{p}c_{12}^2 Q_{j,r/2}V_{-j,r/3} - p^{r/6}V_{j,2r/3} \\ &- p^{r/3}V_{2j,r/3} - (-1)^j p^{r/2} - (-1)^k Q_{j,r/2}p^{(r-2)/4}\sqrt{p} \quad (1 \le j \le 6) \end{aligned}$$

from (2.22) and Proposition 3.2, justifying the expressions for $F^{(j)}(x)$ in (iv)(b).

If $r \equiv 4$ or $8 \pmod{12}$ then $\delta = 4$. For $r \equiv 4 \pmod{12}$ and $1 \le i \le 4$,

$$\begin{aligned} Z_{3i} &= -p^{(r-4)/12}Q_{i,r/2}\big(\lambda^{(r-1)/3}G_3 + \bar{\lambda}^{(r-1)/3}\widehat{G}_3\big) \\ &= -p^{(r-4)/12}Q_{i,r/2}\big(gV_{(r-1)/3} + (g'' - g')U_{(r-1)/3}\big)/2 \end{aligned}$$

again from Proposition 3.2. Similarly, for $r \equiv 8 \pmod{12}$ and $1 \le i \le 4$,

$$\begin{aligned} Z_{3i} &= -p^{(r-8)/12}Q_{i,r/2}\big(\lambda^{(r+1)/3}\widehat{G}_3 + \bar{\lambda}^{(r+1)/3}G_3\big) \\ &= -p^{(r-8)/12}Q_{i,r/2}\big(gV_{(r+1)/3} - (g'' - g')U_{(r+1)/3}\big)/2. \end{aligned}$$

Combining these cases yields for $1 \le i \le 4$ and $r \equiv 4$ or $8 \pmod{12}$,

$$\begin{aligned} H_{3i} = (-1)^{i-1}&(p^{r/2} + p^{r/4}Q_{i,r/2}) \\ &- g\big(p^{[r/12]}Q_{i,r/2}V_{(r-\alpha)/3} + (-1)^i p^{[r/6]}V_{2\alpha Z,(2r+\alpha)/3} + p^{[r/3]}V_{(r-\alpha)/3}\big)/2 \\ &- \alpha(g''-g')\big(p^{[r/12]}Q_{i,r/2}U_{(r-\alpha)/3} - (-1)^i p^{[r/6]}U_{2\alpha Z,(2r+\alpha)/3} + p^{[r/3]}U_{(r-\alpha)/3}\big)/2, \end{aligned}$$

where $\alpha = \pm 1$ chosen so $4\alpha \equiv r \pmod{12}$. A strightforward, albeit tedious, calculation gives the expressions for $F^{(j)}(x)$ in (iv)(c) for $1 \le j \le 4$.

If $r \equiv 3 \pmod 6$, then $\delta = 3$ with H_j satisfying

$$x - p^{r/3}V_{2j,r/3} - p^{(r-3)/6}\sqrt{p}V_{j,2r/3} - (-1)^j p^{(r-1)/2}\sqrt{p} = R_j + Z_j \tag{3.6}$$

from (2.26) and Proposition 3.2. Squaring on the left in (3.6) yields

$$\begin{aligned} x^2 - 2x\big(p^{r/3}V_{2j,r/3} + p^{r/6}\big(V_{j,2r/3} + (-1)^j p^{r/3}\big)\big) + 2p^{r/2}V_{j,2r/3}V_{2j,r/3} \\ + p^{2r/3}V_{2j,r/3}^2 + p^{r/3}V_{j,2r/3}^2 + p^r + 2p^{5r/6}(-1)^j V_{2j,r/3} + 2p^{2r/3}(-1)^j V_{j,2r/3} \end{aligned}$$

or

$$(3.7)\quad x^2 - 2x\big(p^{r/3}V_{2j,r/3} + p^{(r-3)/6}\sqrt{p}V_{j,2r/3} + (-1)^j p^{(r-1)/2}\sqrt{p}\big) + 3p^{2r/3}(-1)^j V_{j,2r/3} + p^{r/3}V_{j,2r/3}^2 + 3p^r + 2(-1)^j p^{(r-1)/2}\sqrt{p}V_r + 4p^{(5r-3)/6}\sqrt{p}(-1)^j V_{2j,r/3},$$

as $V_{j,2r/3}V_{2j,r/3} = (-1)^j(V_r + p^{r/3}V_{2j,r/3})$ and $V_{2j,r/3}^2 = 2p^{r/3} + (-1)^j V_{j,2r/3}$. Since $R^2 = (-1)^k\big(2p^r + (-1)^j p^{(r-1)/2}Q_r\sqrt{p}\big)$ from (2.19), (2.20) and (2.22) for r odd, one finds from Proposition 2.1 that

$$(3.8)\quad \begin{aligned}(R_j + Z_j)^2 &= R_j^2(1 + 2Z_j/R_j + Z_j^2/R_j^2)\\ &= 2(-1)^k(c_{12}p^{r/3}Q_r V_{2j,r/3} + p^r + p^{2r/3}V_{2j,r/3}^2)\\ &\quad + (-1)^{j+k}\sqrt{p}(p^{(r-1)/2}Q_r + p^{(r-3)/6}Q_r V_{2j,r/3}^2 + 4c_{12}p^{(5r-3)/6}V_{2j,r/3})\end{aligned}$$

if $3 \mid b_4$. Equating expressions (3.7) and (3.8) leads to the equation

$$(3.9)\quad x^2 - 2p^{r/3}V_{2j,r/3}x + C = \sqrt{p}\big(2x\big((-1)^j p^{(r-1)/2} + p^{(r-3)/6}V_{j,2r/3}\big) - D\big)$$

satisfied by H_j, where

$$\begin{aligned}C &= (-1)^j p^{2r/3}V_{j,2r/3}\big(3 - 2(-1)^k\big) + 3p^r\big(1 - 2(-1)^k\big)\\ &\quad + p^{r/3}\big(V_{j,2r/3}^2 - 2(-1)^k c_{12}Q_r V_{2j,r/3}\big),\\ D &= (-1)^j p^{(r-1)/2}\big(2V_r - 3(-1)^k Q_r\big) - p^{(r-3)/6}(-1)^k Q_r V_{j,2r/3}\\ &\quad + 4(-1)^j p^{(5r-3)/6}V_{2j,r/3}\big(1 - (-1)^k c_{12}\big).\end{aligned}$$

A routine calculation now yields the expression for $F^{(j)}(x)$ in (iv)(d) for $1 \le j \le 3$ in the case $3 \mid b_4$. A similar argument in the case $3 \mid a_4$ leads to the polynomial expressions for $F^{(j)}(x)$ with the choices for C and D as indicated.

If $r \equiv 2$ or $10 \pmod{12}$, then $\delta = 2$. For $r \equiv 2 \pmod{12}$,

$$Z_3 = (-1)^{T/2}p^{(r-14)/12}\sqrt{p}P_{r/2}(\lambda^{(r+1)/3}\widehat{G}_3 + \bar{\lambda}^{(r+1)/3}G_3)$$

by Proposition 2.1 and (2.9), (2.12) and (2.22), noting that $Q_{3,n} = -P_n$ and $c_{12}^2 = (-1)^{k+T/2}$. Similarly $Z_{12} = -(-1)^{T/2}p^{(r-14)/12}\sqrt{p}Q_r(\lambda^{(r+1)/3}\widehat{G}_3 + \bar{\lambda}^{(r+1)/3}G_3)$. So

$$\begin{aligned}H_3 &= R_3 + H_3(6) + Z_3\\ &= (-1)^k p^{(r-2)/4}\sqrt{p}P_{r/2} + p^{(r-2)/6}(\zeta_6^{2Z}\lambda^{(2r-1)/3}G_3 + \zeta_6^{-2Z}\bar{\lambda}^{(2r-1)/3}\widehat{G}_3)\\ &\quad + \big((-1)^{T/2}p^{(r-14)/12}\sqrt{p}P_{r/2} - p^{(r-2)/3}\big)(\lambda^{(r+1)/3}\widehat{G}_3 + \bar{\lambda}^{(r+1)/3}G_3) + p^{r/2},\\ H_{12} &= R_4 + g_r(6) + Z_{12}\\ &= -(-1)^k p^{(r-2)/4}\sqrt{p}Q_{r/2} - p^{(r-2)/6}(\zeta_6^{2Z}\lambda^{(2r-1)/3}G_3 + \zeta_6^{-2Z}\bar{\lambda}^{(2r-1)/3}\widehat{G}_3)\\ &\quad - \big((-1)^{T/2}p^{(r-14)/12}\sqrt{p}Q_{r/2} + p^{(r-2)/3}\big)(\lambda^{(r+1)/3}\widehat{G}_3 + \bar{\lambda}^{(r+1)/3}G_3) - p^{r/2}\end{aligned}$$

from (2.22) and Proposition (3.2). For $r \equiv 10 \pmod{12}$, one finds that

$$\begin{aligned}H_3 &= (-1)^k p^{(r-2)/4}\sqrt{p}P_{r/2} + p^{(r+2)/6}(\zeta_6^{-2Z}\lambda^{(2r+1)/3}\widehat{G}_3 + \zeta_6^{2Z}\bar{\lambda}^{(2r+1)/3}G_3)\\ &\quad + \big((-1)^{T/2}p^{(r-10)/12}\sqrt{p}P_{r/2} - p^{(r-1)/3}\big)(\lambda^{(r-1)/3}G_3 + \bar{\lambda}^{(r-1)/3}\widehat{G}_3) + p^{r/2}\end{aligned}$$

$$H_{12} = -(-1)^k p^{(r-2)/4}\sqrt{p}Q_{r/2} - p^{(r+2)/6}(\zeta_6^{-2Z}\lambda^{(2r+1)/3}\widehat{G}_3 + \zeta_6^{2Z}\bar{\lambda}^{(2r+1)/3}G_3)$$
$$- ((-1)^{T/2}p^{(r-10)/12}\sqrt{p}Q_{r/2} + p^{(r-1)/3})(\lambda^{(r-1)/3}G_3 + \bar{\lambda}^{(r-1)/3}\widehat{G}_3) - p^{r/2}$$

in an analogous manner. Combining the cases $r \equiv 2$ or $10 \pmod{12}$, yields the expressions for H_3 and H_{12} in (iv)(e) in terms of the sequences (3.1) and (3.2).

If $r \equiv 1, 5, 7$ or $11 \pmod{12}$, then $\delta = 1$ with $F(x) = F^{(1)}(x)$ satisfied by $g_r(12)$ as given in (iv)(f) by Proposition 2.1 (or [**7**, Theorem 2]). This completes the proof of the proposition. □

I conclude the discussion here with illustrations of the last result when $p = 12k + 1$ with $r = 3$ and 4.

EXAMPLE 3.6. Consider the finite field $\mathbb{F}_q$ with $q = 13^4$, where one may take $\gamma = 3(2 - \zeta_5)$ to generate $\mathbb{F}_q^*$ with $G = \gamma^{2380} = 2$. One finds $r_3 = -5$, $s_3 = -3$ and $a_4 = -3$, $b_4 = 2$ from this choice with $Z = 1$. Then $V_1 = -5$, $V_2 = -1$, ... in (3.1) satisfy $V_{n+2} = -5V_{n+1} - 13V_n$, with $V_{2,1} = -2$, $V_{2,2} = 23$ and $V_{4,1} = 7$, $V_{4,2} = -22$. Also $Q = Q_{j,2} + (-1)^j 13$ equals -37, 3, 11 or 23 for $j = 1, 2, 3$ or 4 respectively, in Proposition 3.5(iv)(c). One finds here that

$$F^{(1)}(x) = x^3 - 1443x^2 + 124995x + 94874975$$
$$F^{(2)}(x) = x^3 + 117x^2 - 102765x + 8300279$$
$$F^{(3)}(x) = x^3 + 429x^2 - 148317x - 34246225$$
$$F^{(4)}(x) = x^3 + 897x^2 - 216645x - 1783474781.$$

These agree, of course, with the computed factors of the period polynomial $F(x)$ in (1.3). Indeed, for the choice of generator γ above one finds (chiefly, [**5**, Proposition 4]) that the modified Gauss sums $H_3 = 1 + 240\psi_1 + 192\psi_2$, $H_6 = 1 - 24\psi_2 + 144\psi_3$, $H_9 = 721 + 384\psi_2 + 336\psi_3$, and $H_{12} = -119 + 156\psi_2 + 384\psi_3$ in terms of the Gauss circulants

$$\psi_1 = \zeta_{13}^2 + \zeta_{13}^3 + \zeta_{13}^{10} + \zeta_{13}^{11}, \qquad \psi_2 = \zeta_{13}^4 + \zeta_{13}^6 + \zeta_{13}^7 + \zeta_{13}^9,$$
$$\psi_3 = \zeta_{13}^1 + \zeta_{13}^5 + \zeta_{13}^8 + \zeta_{13}^{12}.$$

Directly calculating the minimal polynomials satisfied by H_9, H_6, H_3 and H_{12}, one obtains $F^{(1)}(x)$, $F^{(2)}(x)$, $F^{(3)}(x)$ and $F^{(4)}(x)$, respectively, as above.

EXAMPLE 3.7. Consider the finite field $\mathbb{F}_q$ where $q = 37^3$, where one may take $\gamma = 3^{1/3} - 1$ to generate $\mathbb{F}_q^*$ with $G = \gamma^{1407} = 2$. One finds $r_3 = -11$, $s_3 = 3$ and $a_4 = 1$, $b_4 = -6$ from this choice with $Z = 1$ and $T = 26$. From (2.7), $c_{12} = -1$. In addition, $V_1 = -11$, $V_2 = 47$, ... in (3.1) satisfy $V_{n+2} = -11V_{n+1} - 37V_n$ with $V_{2,1} = 10$, $V_{2,2} = -73$ and $V_{4,1} = 1$ and $V_{4,2} = 26$, and $Q_3 = -214$ in (3.2). Since $3 \mid b_4$, one finds here from Proposition 3.5(iv)(d) that

$$F^{(1)}(x) = x^4 - 1480x^3 + 1594626x^2 - 1558742512x + 615932128693,$$
$$F^{(2)}(x) = x^4 - 148x^3 + 152070x^2 + 61655468x + 18832192993,$$
$$F^{(3)}(x) = x^4 + 1628x^3 + 988566x^2 + 593842748x + 404282845393,$$

where the pairs $(C, D) = (817219, 37458)$, $(169201, -16272)$, and $(685129, -41952)$ respectively. Again these agree with the computed factors of the period polynomial $F(x)$ in (1.3). Indeed, for the choice of generator γ above one finds (from [**5**, Proposition 4]) that the modified Gauss sums $H_4 = 288\psi_1 + 180\psi_2 + 360\psi_4 + 577$,

$H_8 = 36\psi_1 - 144\psi_2 - 36\psi_3 + 1$, and $H_{12} = 216\psi_1 + 324\psi_2 + 228\psi_4 - 215$ in terms of the Gauss circulants

$$\psi_1 = \zeta_{37}^2 + \zeta_{37}^{14} + \zeta_{37}^{15} + \zeta_{37}^{18} + \zeta_{37}^{20} + \zeta_{37}^{24} + \zeta_{37}^{29} + \zeta_{37}^{31} + \zeta_{37}^{32},$$
$$\psi_2 = \zeta_{37}^3 + \zeta_{37}^{4} + \zeta_{37}^{11} + \zeta_{37}^{21} + \zeta_{37}^{25} + \zeta_{37}^{27} + \zeta_{37}^{28} + \zeta_{37}^{30} + \zeta_{37}^{36},$$
$$\psi_3 = \zeta_{37}^5 + \zeta_{37}^{6} + \zeta_{37}^{8} + \zeta_{37}^{13} + \zeta_{37}^{17} + \zeta_{37}^{19} + \zeta_{37}^{23} + \zeta_{37}^{24} + \zeta_{37}^{25},$$
$$\psi_4 = \zeta_{37}^1 + \zeta_{37}^{7} + \zeta_{37}^{9} + \zeta_{37}^{10} + \zeta_{37}^{12} + \zeta_{37}^{16} + \zeta_{37}^{26} + \zeta_{37}^{33} + \zeta_{37}^{34}.$$

Directly calculating the minimal polynomials satisfied by H_4, H_8, and H_{12}, one obtains $F^{(1)}(x)$, $F^{(2)}(x)$ and $F^{(3)}(x)$, respectively, as above.

References

1. B. C. Berndt and R. J. Evans, *Sums of Gauss, Jacobi and Jacobsthal*, J. Number Theory **11** (1979), no. 3, 349–398.
2. ———, *Sums of Gauss, Eisenstein, Jacobi, Jacobsthal and Brewer*, Illinois J. Math. **23** (1979), no. 3, 374–437.
3. B. C. Berndt, R. J. Evans, and K. S. Williams, *Gauss and Jacobi sums*, Canad. Math. Soc. Ser. Monogr. Adv. Texts, Wiley, New York, 1998.
4. S. Gurak, *Factors of period polynomials for finite fields*. I, The Rademacher Legacy to Mathematics (University Park, PA, 1992), Contemp. Math., vol. 166, Amer. Math. Soc., Providence, RI, 1994, pp. 309–333.
5. ———, *Factors of period polynomials for finite fields*. II, Finite Fields: Theory, Applications, and Algorithms (Las Vegas, NV, 1993), Contemp. Math., vol. 168, Amer. Math. Soc., Providence, RI, 1994, pp. 127–138.
6. ———, *Period polynomials for $\mathbb{F}_{p^2}$ of fixed small degree*, Finite fields and applications (Augsburg, 1999), Springer, Berlin, 2000, pp. 196–207.
7. ———, *Gauss and Eisenstein sums for order twelve*, Canad. Math. Bull. **46** (2003), no. 3, 344–355.
8. G. Myerson, *Period polynomials and Gauss sums for finite fields*, Acta Arith. **39** (1981), no. 3, 251–264.

DEPARTMENT OF MATHEMATICS AND COMPUTER SCIENCE, UNIVERSITY OF SAN DIEGO, SAN DIEGO, CA 92110, USA.

E-mail address: `gurak@sandiego.edu`

Centre de Recherches Mathématiques
CRM Proceedings and Lecture Notes
Volume **36**, 2004

Some Recent Results about (Ternary) Quadratic Forms

Jonathan Hanke

Introduction

This paper is meant to be a survey of some recent results in the theory of quadratic forms, specifically the question of representing integers by a positive definite form in 3 variables. The question of representability by a ternary form is particularly interesting because it exhibits both analytic and algebraic features, in contrast to the case of forms in ≥ 4 variables.

While we will briefly discuss the theory for forms in $n \geq 3$ variables, we will not do justice to its long and rich history. For a more complete picture of the history of the problem we recommend the books [**20, 9, 5**], as well as the survey papers [**13, 31, 49**] which focus on the analytic approach used to resolve it. A more algebraic perspective is described in [**55**].

We would especially like to thank William Duke for his excellent survey paper [**13**] which introduced the author to the subject, and hope this article will be a natural continuation of that paper.

1. Main questions

We will be interested in studying an integral quadratic form Q in n variables, by which we mean a homogeneous polynomial

$$Q(\vec{x}) = \sum_{1 \leq i \leq j \leq n} a_{ij} x_i x_j$$

of degree 2 with coefficients a_{ij} in $\mathbb{Z}$. Specifically, we will be interested in which integers $m \in \mathbb{Z}$ are **represented** by Q, meaning that we can find some vector $\vec{x} \in \mathbb{Z}^n$ so that $Q(\vec{x}) = m$. Some familiar answers to this question can be found in the following theorems:

THEOREM (Lagrange, 1772). *An integer m can be written as a sum of four integer squares* $\iff$ $m \geq 0$.

2000 *Mathematics Subject Classification.* 11-02, 11E12, 11E20.

The author would like to thank W. Duke, J. Hsia, P. Sarnak, and H. Stark for their helpful comments, M. Bhargava, H. Iwaniec, and R. Schulze–Pillot for their very careful reading and extensive comments, and finally, E. Goren and the CNTA for organizing the wonderful conference leading up to this volume.

This is the final form of the paper.

THEOREM (Legendre, 1798). *An integer m can be written as a sum of three integer squares* $\iff m \geq 0$ *and* $m \neq 4^a(8k+7)$ *for any* $a, k \in \mathbb{Z}$.

THEOREM (Fermat, 1640). *A prime number* $p > 2$ *can be written as a sum of two integer squares* $\iff p \equiv 1 \pmod 4$.

While we can ask about which integers are represented by a given quadratic form Q, it is also interesting to ask in *how many ways* is m represented by Q? If we let $r_Q(m)$ count the number of ways of representing m by Q (i.e., the number of vectors $\vec{x}$), then we are asking for a description of the function $r_Q(m)$. In these terms, the question of which m are represented asks when $r_Q(m)$ is > 0. In some instances one can actually find exact formulas for $r_Q(m)$, though as we will see later such formulas are quite rare. However, one familiar example of an exact formula is given by:

THEOREM (Jacobi, 1828). *If* $Q = x^2 + y^2 + z^2 + w^2$ *and* $m \in \mathbb{Z} > 0$, *then*

$$r_Q(m) = 8 \sum_{\substack{d \mid m \\ 4 \nmid d > 0}} d.$$

Given the scarcity of exact formulas for $r_Q(m)$, one might instead hope to understand the *approximate size* of $r_Q(m)$. This somewhat more modest goal is still quite useful since a reasonable answer provides enough information to determine which numbers are represented by Q. In the next few sections, we will focus on answering these two main questions:

QUESTION 1. Can we describe which integers m are represented by Q?
(When is $r_Q(m) > 0$?)

QUESTION 2. In about how many ways is m represented by Q?
(How big is $r_Q(m)$?)

2. Some local considerations

One early idea to understand which numbers m could be represented by Q was to look at which numbers are represented modulo M for every $M > 0$ and also to notice restrictions on the sign of m. By the Chinese Remainder Theorem, it is enough to consider congruence conditions modulo p^α for all powers α of every prime number p, and also conditions on the sign of m.

These conditions are called "local" conditions and can be thought of as conditions associated either to a prime number p (for p-power congruences) or to ∞ (for the sign). This language comes from the different ways of completing the integers to fill in the gaps, and can be thought of as conditions at each of these completed places. The choice of a prime p defines the p-adic integers $\mathbb{Z}_p$, and the other choice ∞ gives the real numbers $\mathbb{Z}_\infty = \mathbb{R}$. In general, we refer to statements over $\mathbb{Z}_v$ as **local** and the corresponding statements over $\mathbb{Z}$ as **global**. Much of our effort will be in understanding the passage between these two realms, referred to as a **local-global principle**.

When Q is the sum of 2, 3, and 4 squares, this idea recovers the conditions of the first three theorems. It turns out that in those cases the local conditions are actually enough to guarantee that m is represented by Q, but this is a very special occurrence. In fact, it is a similar kind of local characterization that allows one to write down exact formulas for $r_Q(m)$. We state this as our first strategy toward solving Question 1.

STRATEGY 1. Check if m is represented by Q locally.
(Can we solve $Q(\vec{x}_v) = m$ with $\vec{x}_v$ in $\mathbb{Z}_v$ at all places v?)

When this approach succeeds, it often solves both Questions 1 and 2. However, the problem is that by looking at Q locally we may not in general be able to distinguish it from other quadratic forms. Therefore any local statement would at best be making a statement about all quadratic forms which locally look the same. We call the set $\mathrm{Gen}(Q)$ of such forms the **genus** of Q.

To understand this better, let's look at how many forms are in the genus. One important thing to notice is that two quadratic forms Q_1 and Q_2 look the same (so we will call them **equivalent**) if they differ from each other by some invertible (linear) change of variables with integer coefficients. For convenience, we write this equivalence as $Q_1 \sim_{\mathbb{Z}} Q_2$. Notice that if Q_1 and Q_2 are equivalent then $r_{Q_1}(m) = r_{Q_2}(m)$ for all integers m. (Earlier when we said that Q_1 and Q_2 are locally equivalent, we meant that $Q_1 \sim_{\mathbb{Z}_v} Q_2$ at all places v.) It will also be important to keep track of how many ways a form Q_1 is equivalent to itself. We call these self-equivalences the **automorphisms** of Q_1, and collectively refer to them as $\mathrm{Aut}(Q_1)$.

Since for answering our two main questions we cannot distinguish between equivalent quadratic forms, it makes sense to divide the genus into **classes** of equivalent forms. The following fundamental result in the theory of quadratic forms tells us that our local strategy reduces to asking about only finitely many possibilities.

THEOREM (Hermite). *There are only finitely many classes of forms Q_i in the genus of Q.*

With this, we are tempted to think of our local conditions as making a statement jointly about all forms in the genus of Q. However, we have yet to exhibit a concrete connection between local representability by Q and (global) representability by some form Q_i in the genus. By trying to establish just such a quantitative link, Siegel proved the following remarkable result which allows us to think of our local information as a *weighted average* of (global) information over the classes in the genus. For simplicity, we only state his theorem assuming Q is positive definite.[1]

THEOREM (Siegel). *If the genus of Q is a (disjoint) union of classes represented by forms Q_i, then*

$$r_{\mathrm{Gen}(Q)}(m) := \frac{\sum_{Q_i} r_{Q_i}(m)/\#\,\mathrm{Aut}(Q_i)}{\sum_{Q_i} 1/\#\,\mathrm{Aut}(Q_i)} = \prod_v \beta_v(m) \tag{2.1}$$

where $\beta_v(m)$ are local representation densities at the place v which describe how many local representations m has in $\mathbb{Z}_v$.

Siegel's Theorem allows us to conclude that Strategy 1 succeeds whenever there is only one class in the genus of Q, since if m is locally represented then the local representation densities $\beta_v(m)$ will all be > 0.[2]

---

[1]Since Q is positive definite, then so are all forms Q_i in its genus, hence $\#\,\mathrm{Aut}(Q_i)$ is finite.

[2]Since $\prod_v \beta_v(m)$ is an infinite product, we cannot immediately conclude that the product is nonzero when each $\beta_v(m) > 0$. However by computing the local factors $\beta_v(m)$ almost everywhere

COROLLARY 1. *If the genus of Q has only one class, then*

$$m \text{ is locally represented by } Q \iff m \text{ is (globally) represented by } Q.$$

Since the genus of the sum of 2, 3, and 4 squares forms all have only one class, Corollary 1 instantly proves the first 3 theorems of this section. In fact, by computing the local densities at all places, we can also recover Jacobi's exact formula for $r_Q(m)$ when Q is the sum of four squares. However, this happy coincidence between the genus and the class of Q is quite infrequent. In fact, there are only finitely many positive definite forms Q where this holds.[3] Therefore while Strategy 1 is a good first step, some further tools are needed to answer our two main questions in general.

3. Theta functions and modular forms

One useful approach used to understand a sequence of numbers is to think of them as coefficients of some appropriate series, and to study the transformation properties of the series *as a whole*. For our purposes we are interested in the numbers $r_Q(m)$ for some positive definite quadratic form Q, so it is natural for us to study its **theta function**, defined by the Fourier series

$$\Theta_Q(z) := \sum_{m \geq 0} r_Q(m) e^{2\pi i m z},$$

which is a holomorphic function in the complex upper half-plane $\mathfrak{H} = \{z \in \mathbb{C} \mid \mathrm{Im}(z) > 0\}$. Since $\mathfrak{H}$ has a natural action of the group $\mathrm{SL}_2(\mathbb{Z})$, given by $\gamma = \left[\begin{smallmatrix} a & b \\ c & d \end{smallmatrix}\right] \mapsto (az+b)/(cz+d) =: \gamma z$, we can ask about how $\Theta_Q(\gamma z)$ is related to $\Theta_Q(z)$. It turns out that there is a "congruence" subgroup $\Gamma_0(N)$ of finite index in $\mathrm{SL}_2(\mathbb{Z})$ where the theta function has the particularly simple multiplicative transformation property

$$\Theta_Q(\gamma z) = \chi_Q(d)(cz+d)^{n/2}\Theta_Q(z), \tag{3.1}$$

for some character χ_Q on $(\mathbb{Z}/N\mathbb{Z})^\times$.[4]

Rather than study $\Theta_Q(z)$ by itself, it is more convenient to study $\Theta_Q(z)$ as one of many functions with similar transformation properties. In particular, $\Theta_Q(z)$ is an example of a "modular form" of weight $n/2$ for $\Gamma_0(N)$, and we now turn our attention to describing these in some detail.[5] This digression may appear unwarranted, but the general features and subtleties of modular forms, especially those of half-integral weight, will play a central role in answering our main questions about quadratic forms in small numbers of variables.

---

one can check that it does converge when $n \geq 2$ (though conditionally when $n = 2$ or 3), so our conclusion is justified. (See [**35**, Prop. 5.6.2, p. 124] for details.)

[3] The program of determining all genera with one class has been extensively pursued by Watson in a series of papers starting with [**64**].

[4] Since $n/2$ is not an integer when n is odd, this relationship only makes sense with some appropriate choice of square root of $cz + d$ for each $\gamma \in \Gamma_0(N)$, so that $\Theta(\gamma_1(\gamma_2 z)) = \Theta((\gamma_1\gamma_2)z)$. If such a consistent choice of square roots exists, we call it a **multiplier system** for $\Gamma_0(N)$. It is a nonobvious fact that multiplier systems exist for all $\Gamma_0(N)$ mentioned above.

[5] For a more detailed introduction to modular forms of integral weight on $\Gamma_0(N)$ one can refer to [**44, 57, 29, 39**]. Half-integral weight forms are treated from an analytic point of view in [**29**, Chaps. 1–5] and from an arithmetic point of view in [**39**, Chap. IV].

Modular forms for $\Gamma_0(N)$. Let $\Gamma_0(N)$ denote the subgroup of $SL_2(\mathbb{Z})$ whose reduction mod N looks like $\left[\begin{smallmatrix} * & * \\ 0 & * \end{smallmatrix}\right]$. We say that a holomorphic function $f(z)$ on $\mathfrak{H}$ is a **modular form** of **weight** k and **level** N for some character χ on $(\mathbb{Z}/N\mathbb{Z})^\times$ if:

(1) $f(\gamma z) = \chi(d)(cz+d)^k f(z)$ for all $\gamma \in \Gamma_0(N)$,
(2) $f(z)$ is holomorphic at all "cusps" and has "mild growth" near them.

The first condition is just the transformation formula (3.1), and the second condition has to do with the growth properties of $f(z)$ as $\mathrm{Im}(z) \to \infty$ and as $z \to \alpha \in \mathbb{Q}$. In particular, condition 2 implies that we can assign a value to $f(z)$ at the boundary points $\mathbb{Q} \cup \infty$ of $\mathfrak{H}$. By condition 1, we see that these values are determined by specifying them at one point in each $\Gamma_0(N)$-orbit of the boundary. We refer to this finite collection of representative boundary points as the **cusps** of $\Gamma_0(N)$.

We now quickly sketch some important features of the space of modular forms of fixed weight k and level N:

Finite-dimensionality: The space of modular forms of fixed weight and level form a finite dimensional vector space over $\mathbb{C}$.

Fourier expansions: Since $\left[\begin{smallmatrix} 1 & 1 \\ 0 & 1 \end{smallmatrix}\right] \in \Gamma_0(N)$, we see that $f(z) = f(z+1)$ from condition 1, so $f(z)$ has a Fourier expansion $f(z) = \sum_{m \geq 0} a(m) e^{2\pi i m z}$.

Cusp forms and Eisenstein series: There is a subspace of functions vanishing at all cusps called **cusp forms** and a naturally perpendicular space of functions called **Eisenstein series**. The Fourier coefficients of an Eisenstein series tend to be larger than those of a cusp form. Specifically, the Fourier coefficients $a_E(m)$ of an Eisenstein series are either zero or about as big as m^{k-1}, while the Fourier coefficients $a(m)$ of a cusp form are bounded by some constant multiple of $m^{k/2}$. (The bound $|a(m)| \ll m^{k/2}$ is often referred to as the **trivial bound** for cusp forms.) Also, if $f(z)$ is a cusp form then $a(0) = 0$.

Hecke operators: For every $m \in \mathbb{N}$ with $\gcd(m, N) = 1$, we have a linear operator T_m acting on the space of cusp forms. These **Hecke operators** have the following two very useful properties:

(1) T_m is a normal operator.
(2) If $\gcd(m_1, m_2) = 1$, then $T_{m_1 m_2} = T_{m_1} T_{m_2}$.

As a result, we can find a basis of (Hecke) eigenforms $\{f_i(z)\}$ for all T_m and can write any cusp form $f(z)$ as a linear combination $f(z) = \sum_i \gamma_i f_i(z)$.

If $f_i(z)$ has integral weight (i.e., $k \in \mathbb{Z}$), then its Hecke eigenvalues are connected to the Fourier coefficients by $a_i(1) T_m f_i(z) = a_i(m) f_i(z)$. If $f_i(z)$ has half-integral weight, then the Hecke operators T_m are identically zero unless m is a square. In this case, for the nonzero Hecke operators T_{m^2} we still have $a_i(t) T_{m^2} f_i(z) = a_i(tm^2) f_i(z)$ for any square-free number t.

Since the eigenforms can be freely multiplied by any constant, we will always assume that they are **normalized** in some appropriate sense. For integral weight forms we require $a_i(1) = 1$, while for half-integral weight we require the forms to have norm 1 (under the natural inner product).[6]

The analytic approach when $n \geq 5$. Now that we have some basic facts about modular forms in general, we can apply this knowledge to better understand

[6]Strictly speaking, for integral weight the "theory of newforms" only guarantees that $a_i(1) \neq 0$ if $f_i(z)$ does not arise as $f'(rz)$ for some form $f'(z)$ of lower level. However, for such (old) forms we can just normalize the associated (newform) $f'(z)$ of smallest level using $a'(1) = 1$.

our specific modular form $\Theta_Q(z)$. Firstly, from the general decomposition into cusp forms and Eisenstein series, we can write

$$\Theta_Q(z) = E(z) + f(z)$$

as the sum of an Eisenstein series $E(z)$ and a cusp form $f(z)$.[7] By looking at the mth Fourier coefficients of each, we have

$$r_Q(m) = a_E(m) + a(m).$$

Since we expect $a_E(m) = 0$ or $\sim m^{n/2-1}$ and we know $|a(m)| \ll m^{n/4}$, when $n > 4$ these combine to show that $r_Q(m) > 0$ when $a_E(m) \neq 0$ and m is sufficiently large. This is the heart of the analytic approach to this problem, which has its origins in the circle method of Ramanujan, Hardy, and Littlewood. Therefore, if we could prove that the Eisenstein series had large enough Fourier coefficients, we could answer our main questions when $n \geq 5$. Thankfully, Siegel has given us a beautiful interpretation of the coefficients $a_E(m)$ in terms of the local behavior of Q when $n \geq 5$. He showed that $a_E(m)$ is exactly given by the weighted average/infinite product expression in (2.1). By using this interpretation of $a_E(m)$, we can prove the following theorem:

THEOREM (Tartakowsky). *Suppose Q is a positive definite integral quadratic form in $n \geq 5$ variables. Then an integer m is represented by Q if m is locally represented by Q and sufficiently large.*

3.1. The analytic approach when $n = 4$. One can try to modify the analytic approach above to prove a similar type of theorem when $n = 4$, however there are several new complications and features which must be dealt with. We again only discuss positive definite forms, though we will revisit indefinite forms in Section 4.

The existence of anisotropic primes. The first notable difference between forms in 4 variables and their higher dimensional counterparts is that under certain circumstances the Eisenstein coefficients $a_E(m)$ don't necessarily grow as m grows. Using Siegel's interpretation of $a_E(m)$ in (2.1) (which still holds when $n = 4$ as we will see below), we can analyze the growth of $a_E(m)$ locally at each of the local factors $\beta_v(m)$. What we find is that there are at most finitely many "anisotropic" primes $p \mid N$ where $\beta_p(m) = \beta_p(p^2 m)$ when m is sufficiently p-divisible. The presence of these primes presents a problem for our analytic approach, since allowing m to grow by introducing powers of anisotropic primes p will not cause our main term to grow.

The term **anisotropic prime** comes from the fact that Q is anisotropic over the p-adic numbers $\mathbb{Q}_p$, meaning that the only representation $Q(\vec{x}) = 0$ with $\vec{x} \in \mathbb{Q}_p^n$ is when $\vec{x} = \vec{0}$. This geometric property means that when m is sufficiently divisible by p (i.e., p-adically close enough to zero) there are no new representations of m, causing the local factor not to grow as it become more p-divisible. Notice that the condition for a prime to be anisotropic is harder to satisfy as n gets larger; in fact when $n \geq 5$ there are no anisotropic primes.

The presence of anisotropic primes when $n = 4$ requires us to only consider m which are large but not too highly divisible by these primes. This will add the

[7]The Eisenstein series $E(z)$ is actually a linear combination of the "standard" Eisenstein series naturally associated to the cusps of $\Gamma_0(N)$.

additional condition of "a priori bounded divisibility at the anisotropic primes" to any analytic results we may obtain.

Convergence of the Eisenstein series. Siegel's identification of the coefficients $a_E(m)$ with the expression (2.1) when $n \geq 5$ depends on having a convergent series expression for the weight k Eisenstein series $E_k(z) := E(z)$. For small weights the defining series does not converge, but one can recover $E_k(z)$ as the value at $s = k = n/2$ of the generalized Eisenstein series $E(z, s)$ whose extra parameter $s \in \mathbb{C}$ is related to the weight. The series defining $E(z, s)$ is only absolutely convergent when $\mathrm{Re}(s) > 2$, however it is known to have meromorphic continuation to all $s \in \mathbb{C}$.

When $n = 4$ the Eisenstein series $E_2(z)$ no longer lives in the region of absolute convergence, so more care is needed. However, since it is on the edge of the region of absolute convergence, one can use limiting arguments to show that the result holds in this case as well.[8]

Need for an improved error estimate. Even after understanding the growth of $a_E(m)$ via Siegel's formula, there is a further challenge when $n = 4$. Our estimates show that, avoiding growth at the anisotropic primes, we have $a_E(m) = 0$ or $\sim m^{n/2-1}$ and $|a(m)| \ll m^{n/4}$. However when $n = 4$ this shows that $a_E(m) \sim m$ and $|a(m)| \ll m$, which is not very useful since this bound is as large as the main term as m grows.

To salvage the situation, we are led to ask if one can improve on the trivial bound for $a(m)$. The best possible bound for the Fourier coefficients of a normalized cusp form of integral weight $k = n/2$ was conjectured by Ramanujan and Petersson to be $|a(m)| \leq \tau(m)m^{(k-1)/2}$, and this was established by Deligne [**10**] using deep methods in algebraic geometry.[9] However historically, an improvement over the trivial bound was first achieved by Kloosterman using his stronger version of the circle method. This leads to the following theorem:

THEOREM (Kloosterman). *Suppose Q is a positive definite integral quadratic form in 4 variables. Then an integer m is represented by Q if it is locally represented by Q and sufficiently large with a priori bounded divisibility at the anisotropic primes.*

4. Refined local considerations

At the outset, we focused our attention on two main questions about the integers m represented by a quadratic form Q. Our first strategy for answering them was to check the representability of m locally at all places, then by using modular forms we were able to bound the error in our local strategy. However in this process we made the simplifying assumption that Q was positive definite, which ignores the question of representability by an indefinite form. In this section we will refine our initial strategy, and in the process we recover information about indefinite forms as well.

From the point of view of quadratic forms as homogeneous polynomials of degree 2, it may seem that Strategy 1 is the best one can do when speaking about

[8] In fact (2.1) is still meaningful when $n = 3$. This was justified explicitly in [**52**, Korollar 1, p. 289], and follows more generally from the work of Kudla and Rallis [**40, 41**] on the Siegel–Weil formula.

[9] This was first established in weight $k = 2$ by Eichler [**16**], Shimura [**56**], and Igusa [**27**].

local equivalence. We defined a quadratic form as a particular polynomial, and then spoke of *equivalent* forms. In this language, the class and genus of a quadratic form are its equivalence classes under the actions of the groups $\mathrm{GL}_n(\mathbb{Z})$ and $\prod_v \mathrm{GL}_n(\mathbb{Z}_v)$ respectively.

Another equally good (and perhaps better) point of view is to think of a quadratic form as a lattice $L \cong \mathbb{Z}^n$ embedded in a fixed vector space V equipped with a fixed quadratic form φ (in the first sense). This point of view has two advantages. The first is that equivalences of lattices inside (V, φ) are defined by an action of the orthogonal group O^φ carrying one to another. The second advantage is that over a number field K we can speak of a lattice even when it does not have a basis over the ring of integers. Such a basis is required to write down a homogeneous polynomial, but may not exist when the class number of K is > 1.

Since we lose nothing by doing so, we will instead take the determinant 1 groups above, so we are looking at equivalences under the groups SL_n and SO^φ respectively. From either point of view we can define the notion of a class and genus of forms by using the relevant group over $\mathbb{Z}$ or over $\mathbb{Z}_v$ for all places v. An interesting feature of the second point of view is that SO^φ is not simply connected when $n \geq 3$, and admits a "double cover" called the **spin group** and denoted by Spin^φ. This is particularly interesting because this gives us a natural geometric refinement of our local equivalence by using Spin^φ instead of SO^φ. This refined equivalence is called **spinor equivalence** and its equivalence classes are called **spinor genera** which are intermediate to our notions of class and genus.

Since spinor equivalence is also geometrically defined and local, it is still relatively simple to answer representability questions for the spinor genus (though it is more involved than using the genus). This gives the following refined local strategy:

STRATEGY 2. Check if m is represented by the spinor genus of Q.

For this to be as useful as Strategy 1, we need a version of Siegel's Theorem for the (weighted) average number of representations $r_{\mathrm{Spn}(Q)}(m)$ of m by the spinor genus, defined analogously to $r_{\mathrm{Gen}(Q)}(m)$ of (2.1). One approach to understanding the $r_{\mathrm{Spn}(Q)}(m)$ is to see how they behave as we vary over all forms Q_i in a genus. To do this, one considers not only spinor equivalence of lattices, but also of the representations themselves. These can be geometrically realized as vectors in the associated lattice, which are also naturally acted upon by the spin group.

When $n \geq 4$ there are enough orthogonal transformations fixing a given representation vector to show that any representation by the genus can be transformed into a representation by any of the spinor genera it contains. Concretely, this means that $r_{\mathrm{Spn}(Q)}(m)$ is the same for all spinor genera in a genus, which together with (2.1) implies $r_{\mathrm{Spn}(Q)}(m) = r_{\mathrm{Gen}(Q)}(m)$. When $n = 3$ there are fewer spin transformations and the genus breaks into two half-genera (each containing half of the spinor genera) where $r_{\mathrm{Spn}(Q)}(m)$ is constant. In fact there is a quadratic character $\psi = \psi_m$ which distinguishes between these two half-genera (for any given m).

By incorporating this character ψ into Siegel's weighted average, Schulze–Pillot [**51**] has shown:

THEOREM (Schulze–Pillot). *Suppose Q is a positive definite integral quadratic form in 3 variables. Then*

$$r_{\mathrm{Spn}(Q)}(m) - r_{\mathrm{Gen}(Q)}(m) = \prod_v \widetilde{\beta}_v(m) \tag{4.1}$$

where $\widetilde{\beta}_v(m)$ are local spinor representation densities at the place v which describe how many local representations m has by the spinor genus in $\mathbb{Z}_v$.

This, together with Siegel's Theorem, allows us to compute $r_{\mathrm{Spn}(Q)}(m)$ based purely on local information, and to ask about spinor genera, which may be smaller than the genus given by Strategy 1. If the spinor genus happens to have only one class then we can completely answer our main questions. While this is not true in general, it is true for indefinite forms in $n \geq 3$ variables:[10] [11]

THEOREM (Eichler). *If Q is an indefinite quadratic form in 3 or more variables, then its spinor genus consists of only one class.*

5. The connection with the Shimura lift

In dealing with our main questions when n is odd, we are led naturally to study the Fourier coefficients of modular forms of half-integral weight. By using Siegel's Theorem and the general decomposition of modular forms into Eisenstein series and cusp forms, we can further restrict our attention to understanding the coefficients of cusp forms.

Because of the close connection between the Fourier coefficients of eigenforms and their Hecke eigenvalues, it seems reasonable to study the (nonzero) Hecke operators T_{m^2} when $\gcd(m, N) = 1$. While doing this, Shimura noticed that the relationships between the Hecke operators T_{m^2} acting on half-integral weight forms closely resembled the relationships between the Hecke operators T_m for integral weight forms. He further showed that (for every square-free number t) there was in fact a way of **lifting** an eigenform of half-integral weight k to a form of integral weight $2k-1$, and that this lifting commutes with the action of the Hecke operators (in the sense just described). Also, when $n > 3$ the lifting actually gives a cusp form. (See [**58**] for details.) This **Shimura lift** allows one to apply Deligne's theorem to obtain a sharp upper bound for the growth of the Fourier coefficients of an eigenform within a given square class $t\mathbb{Z}^2$.

For our purposes, the issue of deciding whether the Shimura lift is a cusp form when $n = 3$ is of great importance, since Eisenstein series have much larger Fourier coefficients than cusp forms. Using a variety of techniques, Flicker, Niwa, Cipra, and others have characterized the subspace whose Shimura lift is an Eisenstein series as linear combinations of particularly simple cusp forms

$$h_\psi(tz) = \sum_{m \in \mathbb{Z}} \psi(m) m e^{2\pi i t m^2 z}$$

for some odd Dirichlet character ψ, whose Fourier coefficients vanish outside of the square class $t\mathbb{Z}^2$. However, this does not answer the question of whether these simple cusp forms appear in the theta function $\Theta_Q(z)$. In fact they do appear, and this feature is closely related to the appearance of half-genera in the genus of Q.

More precisely, when $n = 3$ we can write $\Theta_Q(z)$ as

$$\Theta_Q(z) = E(z) + H(z) + f(z),$$

where $E(z)$ is an Eisenstein series, both $H(z)$ and $f(z)$ are cusp forms, and the Shimura lifts $\mathrm{Shi}(H;t)$ and $\mathrm{Shi}(f;t)$ (for any square-free t) are respectively an

[10] This is a consequence of the strong approximation property for the associated spin groups.

[11] A nice overview of the arithmetic of indefinite forms and spinor genera is given in [**23**].

Eisenstein series and a cusp form. At the level of Fourier coefficients, this says

$$r_Q(m) = a_E(m) + a_H(m) + a(m).$$

The connection between these Fourier coefficients and the original quadratic form Q is given by the following (local) theorem, due to Schulze–Pillot [**52**].

THEOREM (Schulze–Pillot). *Suppose Q is a positive definite quadratic form in 3 variables. Then*

$$a_E(m) = r_{\mathrm{Gen}(Q)}(m) \qquad \text{and} \qquad a_H(m) = r_{\mathrm{Spn}(Q)}(m) - r_{\mathrm{Gen}(Q)}(m).$$

The local nature of this result can be seen from (2.1) and (4.1), and its proof involves explicitly computing the action of the Hecke operators.

6. Analytic estimates for Fourier coefficients

While the Hecke eigenvalues will play an important role in answering our main questions, they are substantially weaker tools for understanding coefficients of half-integral weight forms than for their integral weight counterparts. In particular, we only have the (nonzero) Hecke operators T_{m^2} which leaves completely open the question of understanding the square-free Fourier coefficients. To establish upper bounds for these, the only available techniques are analytic in nature, and involve estimating the size of exponential sums or special values of L-functions. In this section we will discuss some features of this approach, and some results related to our main questions.

As we mentioned when discussing the case $n = 4$, Kloosterman needed to improve the trivial bound for weight 2 forms to say something about the numbers represented by Q. Since in these estimates we are more interested in the exponents than the constants, it suffices to only make a statement about a *basis* for the space of cusp forms. For arithmetic purposes one is most interested in the basis of Hecke eigenforms, but for analytic pursuits it is often more convenient to work with a basis of **Poincaré series**. The advantage of doing this is that their Fourier coefficients can be expressed as sums involving the **Kloosterman sums**[12]

$$K(u, v, c) = \sum_{d \in (\mathbb{Z}/c\mathbb{Z})^\times} \vartheta(\gamma) e^{2\pi i (ud + v\bar{d})/c} \quad \text{where } \gamma = \left[\begin{smallmatrix} a & b \\ c & d \end{smallmatrix}\right] \in \Gamma_0(N).$$

The trivial estimate $|K(u, v, c)| \leq c$ for individual Kloosterman sums recovers the trivial bound $|a(m)| \ll m^{k/2+\varepsilon}$ for Fourier coefficients of a weight k cusp form. Kloosterman's improvement was equivalent to showing that $|K(u, v, c)| \ll c^{3/4+\varepsilon}$, which was further refined by Weil and Salié to $|K(u, v, c)| \ll c^{1/2+\varepsilon}$. This refinement is best possible, and it produces the bound $|a(m)| \ll m^{k/2-1/4+\varepsilon}$. When $k = \frac{3}{2}$, we see that this bound is achieved by the simple cusp forms $h_\psi(tz)$, so no further progress can be made without somehow excluding these forms. We therefore restrict our attention to the coefficients $a(t)$ where t is square-free. To obtain an improved estimate for $|a(t)|$, we must exhibit some further cancellation between the individual Kloosterman sums. This is particularly important for our main questions when $n = 3$ since the Eisenstein series $E(z)$ has coefficients $a_E(m) \sim m^{1/2}$, so only a slight improvement is needed to ensure this is the main term.

[12] Here the factor $\vartheta(\gamma)$ is related to the multiplier system at γ, and $\bar{d} \in \mathbb{Z}$ is a representative for the class $d^{-1} \pmod{c}$. (Multiplier systems are defined in footnote 4.)

This breakthrough was accomplished by Iwaniec for weights $k \geq \frac{5}{2}$ in [**28**], proving

$$|a(t)| \ll t^{k/2-2/7+\varepsilon},$$

and was extended to weight $\frac{3}{2}$ by Duke [**12**]. Together with the Hecke eigenvalue bound from the Shimura lift, this provides an estimate for the coefficients $a(m)$ of the cusp form $f(z)$ for all m. This bound is sufficient to settle the case $n = 3$ provided m is not in one of the finitely many **exceptional-type**[13] square classes where the coefficients of $H(z)$ are nonzero and as large as $E(z)$. This gives the following result:

THEOREM (Duke–Schulze–Pillot). *Suppose Q is a positive definite integral ternary quadratic form. Then every sufficiently large integer m locally represented by Q with a priori bounded divisibility at the anisotropic primes not lying within one of finitely many exceptional-type square classes is represented by Q.*

7. The exceptional-type square classes

While the analytic methods of the previous section provide an adequate answer to our main questions for most numbers when $n = 3$, they fail to make any statement about the representability of numbers within an exceptional-type square class $t\mathbb{Z}^2$. There are two methods of investigating this question, each with slightly different merits, which give the following theorem:

THEOREM (Schulze–Pillot, Hanke). *Suppose Q is a positive definite integral quadratic form in 3 variables. Then any sufficiently large integer m locally represented by the spinor genus of Q with bounded divisibility at the anisotropic primes is represented by Q except for at most finitely many infinite sequences $m = Mp^2$, where M is anti-exceptional*[14] *with $r_Q(M) = 0$, and p runs over all primes $p \nmid NM$ with $\psi_t(p) = -1$.*

The main feature of all of our theorems so far has been that all but finitely many numbers which are locally represented by Q are actually represented by Q. In slightly different language, one would say that a local-global principle[15] holds for all but finitely many numbers. Interestingly, our latest theorem casts some doubt as to whether this is still true when $n = 3$, and under certain circumstances (if they ever happen) guarantees that even a refined local-global principle based on the spinor genus fails infinitely often.

7.1. Primitive spinor representations. In this approach, one keeps track not only of the number $r_{\mathrm{Spn}(Q)}(m)$ of representations by the spinor genus, but also the number $r^*_{\mathrm{Spn}(Q)}(m)$ of **primitive representations** by $\mathrm{Spn}(Q)$, meaning that they are representations of m which do not come from stretching a representation of some smaller integer m/r^2. This is also a local quantity, and one can establish local product formulas analogous to (2.1) and (4.1) whose factors count local primitive representations of m by $\mathrm{Spn}(Q)$.

[13]The term "exceptional-type" is synonymous with the phrase "spinor exceptional" often found in the literature.

[14]The term "anti-exceptional" is defined in Section 7.2.

[15]In this context we mean that a local-global principle for representation over the integers as described in Section 2. If we instead speak about rational representations, then Hasse (in 1923–24) has showed that a *rational* local-global principle holds for *all* (rational) numbers.

Since the growth of $r_{\mathrm{Spn}(Q)}(m)$ comes primarily from these primitive representations, it is useful to know which m have no primitive representations by some spinor genus (hence half-genus) in $\mathrm{Gen}(Q)$. These are called **primitive spinor exceptions**, and they can be characterized by local conditions on the image of the spinor norm map. Using these, one can show that if m is a primitive spinor exception, then so is mp^2 for all primes $p \nmid Nm$ with $\psi_t(p) = -1$. By keeping track of which spinor genera (half-genus) has no primitive representations, one can show that if m is not represented by Q then neither is mp^2 (since such a representation would be primitive). This approach was carried out in [**53**].[16]

The advantage of this approach is that it is purely local, so it provides a conceptually simple way of answering our first question. However, in principle it could be extended to provide asymptotics for $r_Q(m)$.

7.2. The structure of the Shimura lift. In the second approach, one tries to show that the term $a_E(m) + a_H(m)$ grows more quickly than $a(m)$ as m gets large. This is done by computing local densities $\beta_v(m)$ at *all* places to produce a lower bound for $a_E(m)$, then using the explicit Fourier expansion for $H(z)$ to establish a lower bound for the main term. Notice that since $a_E(m)$ and $|a_H(m)|$ are both $\sim m^{1/2}$, significant cancellation may occur.

To describe the situation more fully, we say that m is **exceptional** if $a_E(m) + a_H(m) = 0$, and **anti-exceptional** if $a_E(m) = a_H(m)$.[17] From our local computations, we see that the main term either grows more quickly than $|a(m)|$ or is constant, and it is constant when either m stays exceptional (where it is zero, hence not represented by the spinor genus) or for numbers mp^2 where m is anti-exceptional and p runs over primes $p \nmid Nm$ with $\psi_t(p) = -1$.

When the main term fails to grow, the behavior of $r_Q(m)$ depends only on the Fourier coefficients of the cusp form $f(z)$, which a priori seems difficult to understand. (In general $f(z)$ is not even an eigenform!) However, by taking advantage of the fact that $r_Q(m) \geq 0$, one can show that the coefficients of $f(z)$ behave reasonably.

Since half-integral weight forms are difficult to work with, it is much more convenient to consider the weight 2 Shimura lift $g(z) := \mathrm{Shi}(f; t)$. This can be written as a linear combination of eigenforms $g_i(z)$, and by an argument with the associated Galois representations ρ_i one shows that the eigenforms $g_i(z)$ appear together as a difference of a form and its twist by ψ_t. This shows that many of the Fourier coefficients of $g(z)$ vanish, and when translated back to $f(z)$, shows that $a(m) = a(mp^2)$ when m is anti-exceptional and $\psi_t(p) = -1$. This shows that when m is anti-exceptional and not represented by Q, then neither is mp^2 for all primes $p \nmid Nm$ with $\psi_t(p) = -1$. This approach was carried out in [**22**].

Because of its analytic nature, this approach answers both the question of representability and provides an estimate of the size of $r_Q(m)$. It also describes some of the structure of the cusp form $f(z)$ and its Shimura lift $g(z)$. However it is technically more complicated and involves global tools such as the Shimura

[16]Partial representation results based on primitive representability by the spinor genus within a square class have been established by many authors. Most recently, analytically in [**14**, Theorem 3] and arithmetically in [**25**, Section 4, Theorem].

[17]In general we have the inequality $|a_H(m)| \leq a_E(m)$, so the terms "exceptional" and "anti-exceptional" characterize the extremal cases.

lift and Galois representations (or alternatively some weak form of the Sato–Tate Conjecture).

7.3. An example. We now consider an example first described by Jones and Pall in 1939. Consider the genus of forms

$$S_1 \begin{cases} Q_1 = x^2 + 48y^2 + 144z^2 \\ Q_2 = 4x^2 + 48y^2 + 49z^2 + 4xz + 48yz \end{cases}$$

$$S_2 \begin{cases} Q_3 = 9x^2 + 16y^2 + 48z^2 \\ Q_4 = 16x^2 + 25y^2 + 25z^2 + 16xz + 16xy + 14yz \end{cases}$$

of level $N = 576 = 2^6\, 3^2$ and determinant $D = 6912 = 2^8\, 3^3$. This genus breaks into two spinor genera S_i which are their own half-genera. By checking local spinor norms one can show that the only exceptional-type square class is $\mathbb{Z}^2$, whose associated quadratic character is $\psi_1(\cdot) = \left(\frac{-tD}{\cdot}\right) = \left(\frac{-3}{\cdot}\right)$. Also, this genus is anisotropic only at 3 and ∞. We now apply the above theorem to understand the representation behavior of the forms in this genus.

In the spinor genus S_1, one can check that only the numbers 1 and 4 are anti-exceptional. Since Q_1 represents both of them, a local-global principle holds for all but finitely many integers m. However Q_2 misses the number 1, so $r_{Q_2}(p^2) = 0$ for all primes $p \neq 2$ with $p \equiv 2 \pmod 3$. *This gives an example of a failure of the local-global principle for infinitely many numbers* m!

In the spinor genus S_2 there are no anti-exceptional numbers, and 1 and 4 are exceptional. Therefore, for each of these forms a local-global principle holds for all but finitely many numbers.

It is interesting to notice that by global arithmetic methods Jones and Pall showed that for the forms Q_1 and Q_3, the local global principle

$$r_{\mathrm{Spn}(Q)}(m) > 0 \implies r_Q(m) > 0$$

holds *for all integers*, though they gave no results for Q_2 and Q_4. For more precise results about them one would need an effective version of the theorem above. Unfortunately, making it effective depends on some rather deep and unproven conjectures about L-functions, which follow from the Riemann hypothesis for these L-functions.

8. Applications and generalizations

In closing, we describe some further questions and results related to our main themes.

Effective lower bounds. In our goal to characterize which numbers are represented by a quadratic form Q, we have been satisfied with answers applying to "sufficiently large" integers. However, this vague kind of answer becomes less satisfying when one seeks to explicitly characterize or list those numbers represented by a given Q. For this purpose, the statement "sufficiently large" begs the question "how large?" To establish this more precise (effective) answer, one must care not just about the exponents in our bounds, but also about the constants.

When $n \geq 4$ one can establish effective bounds by making precise local computations and noting the size of the coefficients of the eigenforms in the cusp form, however there is an additional subtlety when $n = 3$. The problem here becomes that the main term $a_E(t)$ (for square-free numbers t) is essentially as large as $L(1,\chi_t)$, for which we know only Siegel's ineffective lower bound $L(1,\chi_t) \gg t^{\varepsilon}$. This issue is equivalent to the possibility of a "Siegel zero" for the associated Dirichlet L-functions. This is one of the major outstanding problems in analytic number theory, and would follow from the Riemann hypothesis for such L-functions. Since this is only an issue for square-free numbers, by allowing m to grow within only finitely many square classes $t\mathbb{Z}^2$ we may again establish an effective bound.

When $n \geq 5$ effective bounds have been established by Watson [**63**] and Hsia–Icaza [**24**] by algebraic methods. When $n \geq 4$ one can apply the analytic ideas above to make Kloosterman's original argument effective [**38**], or use the modular forms directly as in [**54**] and [**21**]. This was carried out in [**21**] to show that $x^2 + 3x^2 + 5z^2 + 7w^2$, a form which has resisted algebraic methods, represents all positive integers except 2 and 22. When $n = 3$, by assuming various Riemann hypotheses, Ono and Soundararajan [**46**] have precisely characterized all numbers represented by the form $Q = x^2 + y^2 + 10z^2$.

Generalizations of the 15-theorem. One interesting application of sharp effective results for quadratic forms is that they are useful in proving finiteness theorems for quadratic forms. The original theorem along these lines is the following theorem of Conway and Schneeberger:[18]

15-THEOREM. *Suppose Q is a positive definite classically integral quadratic form.*[19] *Then*

$$\begin{matrix} Q \text{ represents all} \\ \text{positive integers} \end{matrix} \iff \begin{matrix} Q \text{ represents the numbers} \\ 1,2,3,5,6,7,10,14,15. \end{matrix}$$

Recently, Bhargava [**2**] has shown that these finiteness theorems are quite common, and that in fact for any set $\mathbb{S} \subset \mathbb{N}$ there is a unique minimal finite subset $\mathbb{S}_0$ such that

$$Q \text{ represents } \mathbb{S} \iff Q \text{ represents } \mathbb{S}_0.$$

Finding the set $\mathbb{S}_0$ requires one to understand which numbers are represented by a certain finite list of quadratic forms (depending on $\mathbb{S}$). When $\mathbb{S} = \mathbb{N}$ only 1,850 forms arise and one can use various arithmetic methods to understand them, however in general these are not sufficient to settle the question. In some cases, it is reasonable to use analytic methods to supplement this approach.

As an example, by reformulating the result when $\mathbb{S} = 2\mathbb{N}$, one can hope to find the finite set $\mathbb{T}_0 \subset \mathbb{N}$ for the following theorem:

290-THEOREM. *Suppose Q is a positive definite integral quadratic form. Then*

$$Q \text{ represents all positive integers} \iff Q \text{ represents } \mathbb{T}_0.$$

It is likely that the largest number in $\mathbb{T}_0$ is 290, and the calculations required to prove this are currently underway [**3**]. In general, the complexity of the problem grows quickly with the number of small integers not in $\mathbb{S}$, thus for very thin sets the computations are not very feasible. The special case of these theorems for forms in 4 variables is known as "Ramanujan's universal quaternary forms problem,"

[18] A sketch of the original proof appears in [**50**], however a simpler proof is described in [**1**].

[19] Q is **classically integral** if Q is integral and the numbers a_{ij} are required to be even integers when $i \neq j$.

and its resolution is a crucial step in proving precise versions of the previous two theorems.[20]

Binary forms. On the surface, it may seem as though we have been unfair in postponing discussion of our main questions for quadratic forms in 2 variables until this point. However the nature of the question in this case is purely algebraic, and requires completely different techniques. There is currently no completely satisfactory answer to this question, though for **principal forms** (i.e., those corresponding to the identity of the ideal class group of a quadratic extension) one can answer this in terms of the Hilbert class field of the associated quadratic number field.[21]

From an analytic point of view, the theta function $\Theta_Q(z)$ is a modular form of weight 1. We can still decompose this into an Eisenstein $E(z)$ and cuspidal part $f(z)$, however we expect their Fourier coefficients to be bounded by some constant (since $a_E(m) \sim m^{k-1}$ and $|a(m)| \ll m^{(k-1)/2}$). Since these coefficients do not grow as m grows, we cannot make any statements about the representation behavior of "sufficiently large" numbers.

Quadratic forms over number fields. Given our success with integral quadratic forms over $\mathbb{Q}$, one might be tempted to ask about quadratic forms over a number field F. In this case, the definition of integral requires the coefficients a_{ij} lie in its ring of integers $\mathfrak{o}_F$. As with $\mathbb{Q}$, the main distinction to be made is between definite and indefinite forms, however here there are more places v "at infinity." If Q is definite at all of these infinite places, we say Q is **totally definite**, otherwise Q is said to be **indefinite**.

The theories for totally definite and indefinite forms are the same as when $F = \mathbb{Q}$, however we note that Q can only by totally definite if F is **totally real** (i.e., at all infinite places $F_v = \mathbb{R}$). Local computations in this setting are very similar to those over $\mathbb{Q}$, with the additional technical complication that the primes $\mathfrak{p}$ in $\mathfrak{o}_F$ over 2 may be ramified.[22] When Q is totally *positive* definite[23] the associated $\Theta_Q(z)$ is a **Hilbert modular form** which is now a function on a finite product of upper half-planes. One can pursue all of our previous analytic strategies for Hilbert modular forms, but it becomes technically more complicated. As a consequence, results in that setting are less complete. For instance, the best possible bound for Hecke eigenvalues has not been established in full generality (see [**4**]), and only very recently was a subconvexity estimate established for square-free Fourier coefficients of half-integral weight forms [**7, 8**]. However, Shimura has generalized his lifting to this setting [**59, 60**].

Representing forms by forms. One very interesting feature of our main questions is that they can be reformulated in the much broader context of embeddings of quadratic forms into each other. From this perspective one fixes a quadratic form Q in n variables and some positive integer $r < n$, and is interested

---

[20]A positive definite integral quadratic form is said to be **universal** if it represents all positive integers. Briefly, this problem starts with Ramanujan's 1916 paper [**47**] enumerating all (positive definite) universal diagonal quaternary forms (with one mistake). Later Willerding [**65**], in her 1948 thesis, listed 168 classically integral universal quaternary forms. This list was corrected and completed in 2000 by Conway, Schneeberger, and Bhargava [**1**] to a final list of 204 such forms.

[21]For a complete and very readable description of the situation for principal forms, see [**9**].

[22]For some related explicit formulas, see Donkar's thesis [**11**].

[23]Q is **totally positive definite** when Q is positive definite at all infinite places. If Q is totally definite, one may arrange for this by multiplying Q by some appropriate integer.

in the number of ways it represents all smaller forms q in r variables. This question was first investigated by Siegel, and led to the theory of **Siegel modular forms** for congruence subgroups of the symplectic group $\mathrm{Sp}_{2r}(\mathbb{Z})$.

In this context there are analogues of (2.1) and (4.1), and the (local) features found when $n = 3$ now appear in the "codimension 2" case $n = r + 2$. As expected, the local calculations are still feasible though there are some difficulties when $p = 2$.

The asymptotic approach is also tractable in this setting, though current results there are analogous to the case $n \geq 5$ and here the statement "m is sufficiently large" is replaced by requiring "the minimal length vector in q is sufficiently large." (See [**26, 34, 36**] for details. A nice survey is given in [**37**].)

The main new analytic feature in this case is that Fourier coefficients here are symmetric matrices and the Eisenstein series are stratified into $E_i(z)$ defined by the vanishing of Fourier coefficients of rank $\leq i$. The local information is contained completely in $E_0(z)$, while the theta function decomposes as

$$\Theta_Q(z) = E_0(z) + E_1(z) + \cdots + E_{r-1}(z) + f(z),$$

where $f(z)$ is a cusp form. Since we have no additional information about them, the presence of the forms $E_i(z)$ with $1 \leq i \leq r-1$ give rise to a fairly crude bound on the size of the error term. To make any substantial progress, one must somehow understand these more complicated Eisenstein series.

References

1. M. Bhargava, *On the Conway–Schneeberger fifteen theorem*, Quadratic Forms and Their Applications (Dublin, 1999), Contemp. Math., vol. 272, Amer. Math. Soc., Providence, RI, 2000, pp. 27–37.
2. ______, *Finiteness theorems for quadratic forms* (preprint).
3. M. Bhargava and J. Hanke, *On universal quaternary quadratic forms* (in progress).
4. J.-L. Brylinski and J.-P. Labesse, *Cohomologie d'intersection et fonctions L de certaines variétés de Shimura*, Ann. Sci. École Norm. Sup. (4) **17** (1984), no. 3, 361–412.
5. J. W. S. Cassels, *Rational quadratic forms*, London Math. Soc. Monogr., vol. 13, Academic Press, London–New York, 1978.
6. B. A. Cipra, *On the Niwa–Shintani theta-kernel lifting of modular forms*, Nagoya Math. J. 91 (1983), 49–117.
7. J. W. Cogdell, *On sums of three squares*, Proceedings of the XXII Journées Arithmétiques (Lille, 2001), J. Théor. Nombres Bordeaux (to appear).
8. J. W. Cogdell, I. I. Piatetski-Shapiro, and P. Sarnak, *Estimates on the critical line for Hilbert modular L-functions and applications* (in preparation).
9. D. Cox, *Primes of the form $x^2 + ny^2$. Fermat, class field theory, and complex multiplication*, Wiley-Intersci. Publ., John Wiley & Sons, Inc., New York, 1989.
10. P. Deligne, *La conjecture de Weil.* I, Inst. Hautes Études Sci. Publ. Math. **43** (1974), 273–307.
11. E. Donkar, *On sums of three squares in algebraic number fields*, Ph.D. thesis, Princeton University, Princeton, NJ, 1973.
12. W. Duke, *Hyperbolic distribution problems and half-integral weight Maass forms*, Invent. Math. **92** (1988), no. 1, 73–90.
13. ______, *Some old problems and new results about quadratic forms*, Notices Amer. Math. Soc. **44** (1997), no. 2, 190–196.
14. W. Duke and R. Schulze–Pillot, *Representation of integers by positive ternary quadratic forms and equidistribution of lattice points on ellipsoids*, Invent. Math. **99** (1990), no. 1, 49–57.
15. M. Eichler, *Quadratische Formen und orthogonale Gruppen*, Grundlehren Math. Wiss., vol. 53, Springer-Verlag, Berlin–Göttingen–Heidelberg, 1952.
16. ______, *Quaternäre quadratische Formen und die Riemannsche Vermutung für die Kongruenzzetafunktion*, Arch. Math. **5** (1954), 355–366.
17. ______, *Quadratische Formen und orthogonale Gruppen*, Grundlehren Math. Wiss., vol. 63, Springer-Verlag, Berlin–New York, 1974.

18. P. Fermat, *Œuvres de Pierre Fermat.* II, Gauthier-Villars, Paris, 1891–1896, pp. 212, 432.
19. Y. Flicker, *Automorphic forms on covering groups of* GL(2), Invent. Math. **57** (1980), no. 2, 119–182.
20. E. Grosswald, *Representations of integers as sums of squares*, Springer-Verlag, New York, 1985.
21. J. Hanke, *Local densities and explicit bounds for representability by a quadratic form*, Duke Math. J. (accepted).
22. ———, *On a local-global principle for integral quadratic forms* (preprint).
23. J. S. Hsia, *Arithmetic of indefinite quadratic forms*, Integral Quadratic Forms and Lattices (Seoul, 1998), Contemp. Math., vol. 249, Amer. Math. Soc., Providence, RI, 1999, pp. 1–15.
24. J. S. Hsia and M. I. Icaza, *Effective version of Tartakowsky's theorem*, Acta Arith. **89** (1999), no. 3, 235–253.
25. J. S. Hsia and M. Jöchner, *Almost strong approximations for definite quadratic spaces*, Invent. Math. **129** (1997), no. 3, 471–487.
26. J. S. Hsia, Y. Kitaoka, and M. Kneser, *Representations of positive definite quadratic forms*, J. Reine Angew. Math. **301** (1978), 132–141.
27. J. Igusa, *Kroneckerian model of fields of elliptic modular functions*, Amer. J. Math. **81** (1959), 561–577.
28. H. Iwaniec, *Fourier coefficients of modular forms of half-integral weight*, Invent. Math. **87** (1987), no. 2, 385–401.
29. ———, *Topics in classical automorphic forms*, Grad. Stud. Math., vol. 17, Amer. Math. Soc., Providence, RI, 1997.
30. ———, *Introduction to the spectral theory of automorphic forms*, Bibl. Rev. Mat. Iberoamericana, Revista Matemática Iberoamericana, Madrid, 1995.
31. ———, *Spectral theory of automorphic functions and recent developments in analytic number theory*, Proceedings of the International Congress of Mathematicians, Vol. 1, 2 (Berkeley, 1986), Amer. Math. Soc., Providence, RI, 1987, pp. 444–456.
32. C. G. J. Jacobi, *Note sur la décomposition d'un nombre donné en quatre carrés*, J. Reine Angew. Math. **3** (1828), 191.
33. B. W. Jones and G. Pall, *Regular and semi-regular positive ternary quadratic forms*, Acta. Math. **70** (1939), no. 3-4, 165–191.
34. Y. Kitaoka, *Lectures on Siegel modular forms and representation by quadratic forms*, Tata Inst. Fund. Res. Lectures on Math. and Phys., vol. 77, Springer-Verlag, Berlin, 1986.
35. ———, *Arithmetic of quadratic forms*, Cambridge Tracts in Math., vol. 106, Cambridge Univ. Press, Cambridge, 1993.
36. ———, *The minimum and the primitive representation of positive definite quadratic forms*, Nagoya Math. J. **133** (1994), 127–153
37. ———, *Representations of positive definite quadratic forms and analytic number theory*, Sūgaku **43** (1991), no. 2, 115–127; English transl., Sugaku Expositions **7** (1994), no. 1, 1–17.
38. H. D. Kloosterman, *On the representation of numbers in the form* $ax^2 + by^2 + cz^2 + dt^2$, Acta. Math. **49** (1926), 407–464.
39. N. Koblitz, *Introduction to elliptic curves and modular forms*, 2nd ed., Grad. Texts in Math., vol. 97, Springer-Verlag, New York, 1993.
40. S. S. Kudla and S. Rallis, *On the Siegel–Weil formula*, J. Reine Angew. Math. **387** (1988), 1–68.
41. ———, *On the Siegel–Weil formula.* II. *The isotropic convergent case*, J. Reine Angew. Math. **391** (1988), 65–84.
42. J.-L. Lagrange, *Sur une nouvelle espèce de calcul relatif à la différentiation et à l'intégration des quantités variables*, Nouveau Mémoire de l'Académie Royale des Sciences et Belles-Lettres, Berlin, 1772, pp. 123–133; Œuvres de Lagrange, vol. 3, Gauthier-Villars, Paris, 1869, pp. 189–201.
43. A.-M. Legendre, *Théorie des nombres*, 1798; reprinted, Albert Blanchard, Paris, 1955.
44. T. Miyake, *Modular forms*, Springer-Verlag, Berlin, 1989.
45. S. Niwa, *Modular forms of half integral weight and the integral of certain theta-functions*, Nagoya Math. J. **56** (1975), 147–161.
46. K. Ono and K. Soundararajan, *Ramanujan's ternary quadratic form*, Invent. Math. **130** (1997), no. 3, 415–454.

47. S. Ramanujan, *On the expression of a number in the form $ax^2 + by^2 + cz^2 + du^2$*, Proc. Cambridge Philos. Soc. **19** (1916), no. 1, 11–21.
48. P. Sarnak, *Some applications of modular forms*, Cambridge Tracts in Math., vol. 99, Cambridge Univ. Press., Cambridge, 1990.
49. ———, *Kloosterman, quadratic forms and modular forms*, Nieuw Arch. Wiskd. (5) **1** (2000), no. 4, 385–389.
50. W. A. Schneeberger, *Arithmetic and geometry of integral lattices*, Ph.D. thesis, Princeton University, Princeton, NJ, 1995.
51. R. Schulze–Pillot, *Darstellungsmaße von Spinorgeschlechtern ternärer quadratischer Formen*, J. Reine Angew. Math. **352** (1984), 114–132.
52. ———, *Thetareihen positiv definiter quadratischer Formen*, Invent. Math. **75** (1984), no. 2, 283–299.
53. ———, *Exceptional integers for genera of ternary positive definite quadratic forms*, Duke Math. J. **102** (2000), no. 2, 351–357.
54. ———, *On explicit versions of Tartakovski's theorem*, Arch. Math. (Basel) **77** (2001), no. 2, 129–137.
55. ———, *Representation by integral quadratic forms—a survey*, Proceedings of the International Conference on the Algebraic and Arithmetic Theory of Quadratic Forms (Talca, Chile) (to appear).
56. G. Shimura, *Correspondances modulaires et les fonctions ζ de courbes algébriques*, J. Math. Soc. Japan **10** (1958), 1–28.
57. ———, *Introduction to the arithmetic theory of automorphic functions*, Publ. Math. Soc. Japan, vol. 11, Iwanami Shoten, Publishers, Tokyo; Princeton Univ. Press, Princeton, NJ, 1971; reprinted, Princeton Univ. Press, Princeton, NJ, 1994.
58. ———, *On modular forms of half-integral weight*, Ann. of Math. (2) **97** (1973), 440–481.
59. ———, *On Hilbert modular forms of half-integral weight*, Duke Math. J. **55** (1987), no. 4, 765–838.
60. ———, *On the Fourier coefficients of Hilbert modular forms of half-integral weight*, Duke Math. J. **71** (1993), no. 2, 501–557.
61. C. L. Siegel, *Über die analytische Theorie der quadratischen Formen*, Ann. of Math. (2) **36** (1935), no. 3, 527–606.
62. W. Tartakowsky, *Die Gesamtheit der Zahlen, die durch eine positive quadratische Form $F(x_1, x_2, \ldots, x_s)$ $(s \geq 4)$ darstellbar sind.* I, Bull. Acac. Sci. Leningrad **2** (1929), no. 7, 111–122; II, 165–196.
63. G. L. Watson, *Quadratic Diophantine equations*, Philos. Trans. Roy. Soc. London Ser. A **253** (1960/1961), 227–254.
64. ———, *The class-number of a positive quadratic form*, Proc. London Math. Soc. (3) **13** (1963), 549–576.
65. M. E. Willerding, *Determination of all classes of positive quaternary quadratic forms which represent all (positive) integers*, Ph.D. thesis, St. Louis University, St. Louis, MO, 1948.

Department of Mathematics, Princeton University, Fine Hall, Washington Road, Princeton, NJ 08544–1000, USA.

E-mail address: jonhanke@math.princeton.edu

Centre de Recherches Mathématiques
CRM Proceedings and Lecture Notes
Volume **36**, 2004

An Extension of a Theorem of D. H. Lehmer

Lenny Jones

ABSTRACT. Given a set of primes $q_1 < q_2 < \cdots < q_t$, let Q be the set of all numbers of the form $q_1^{\alpha_1} q_2^{\alpha_2} \cdots q_t^{\alpha_t}$, where $\alpha_i \geq 0$. Lehmer has given necessary and sufficient conditions to determine when both of the integers S and $S + k$, with $k = 1$, 2 or 4, belong to Q. In addition, he has provided upper bounds for the number of pairs of such integers contained in Q, and for the largest value of S. In this paper we investigate extensions of Lehmer's results to the case when k is an arbitrary positive integer.

1. Introduction

Throughout this paper we let $q_1 < q_2 < \cdots < q_t$ be a given set of primes and let $Q = \{q_1^{\alpha_1} q_2^{\alpha_2} \cdots q_t^{\alpha_t} \mid \alpha_i \geq 0\}$. In 1897, Størmer [**39**] proved that there are only finitely many pairs $(S, S+1)$ of numbers such that both S and $S+1$ belong to Q. This result alone can be deduced from Thue's Theorem as Thue [**40**] himself noted in 1908. Thue's method, however, is nonconstructive while Størmer gave an explicit method, which entails solving $3^t - 2^t$ Pell equations, for finding all such pairs. In addition, Størmer showed that the number of such pairs, η, cannot exceed $3^t - 2^t$. Because of his work on consecutive power residues, D. H. Lehmer [**5, 8, 21, 22**] was interested in an efficient way to determine these pairs of numbers. Lehmer [**18**] improved upon Størmer's solution: first by reducing the number of Pell equations to $2^t - 1$ that one needs to solve to recover all pairs; and secondly, by showing that η cannot exceed $(q_t + 1)(2^t - 1)/2$ when $q_t > 3$. Lehmer's approach allowed him to generalize to pairs $(S, S+k)$, where $k = 2$ or 4. He also produced an upper bound, in terms of the primes in Q, for the largest such pair, which we denote $(S_1, S_1 + k)$. The bounds achieved by Lehmer arose from his exploitation of the beautiful divisibility properties present in a Lucas sequence associated with a Pell equation. In this article we investigate extending Lehmer's results to pairs $(S, S+k)$, where k is an arbitrary positive integer. In our investigation, non-Lucas binary linear recurrence sequences come into play, and unfortunately, little is known about the divisibility properties of these sequences. Nonetheless, various methods can be employed to achieve effective bounds, although these bounds are typically not as easy to compute or as sharp. For example, applying a theorem of Evertse [**13**],

2000 *Mathematics Subject Classification.* Primary: 11B37, 11B39; Secondary: 11D59.
We thank the referees for the valuable suggestions.
This is the final form of the paper.

in which he has modified the hypergeometric methods of Thue [**41**] and Siegel [**32**], to approximate cube roots of unity by numbers from a fixed algebraic number field, we have

$$\eta \leq 3 \cdot 7^{2t+3}. \tag{1.1}$$

While the bound in (1.1) has the advantage of being independent of q_t, one might suspect that a sharper bound could be attained, at least in certain cases, by considering q_t in the calculations. Following Tijdeman [**42**] and directly applying linear forms in logarithms to the pair $(S_1, S_1 + k)$ gives

$$\log S_1 < \log k + c_1^{t^4} (\log q_t)^{14t^2} \log\log S_1, \tag{1.2}$$

where c_1 is an effectively computable constant. Utilizing another theorem in [**42**], which incorporates a method of Baker [**2**] on the solutions of the Diophantine equation $y^2 = x^3 + k$, we get

$$\log S_1 \leq (10ke^{q_t})^{10^5}. \tag{1.3}$$

It should be pointed out that Marshall Hall, Jr. [**7**] conjectured that if x and y are integers with $x \neq y$, satisfying $y^2 = x^3 + k$, then $|x^3 - y^2| > |x|^{1/2}$ for $x > x_0$. If this conjecture were true, then e^{q_t} in (1.3) could be replaced by a constant power of q_t, greatly improving the bound. For recent progress on this conjecture, the interested reader is directed to [**12**].

Due in part to the fact that the bounds in (1.1), (1.2), and (1.3) are growing exponentially, they are, in general, quite weak. For this reason we choose a slightly different avenue of attack. To calculate the bounds on η and S_1 in this paper, we invoke a theorem of Yu and Hung [**52**], which applies techniques of linear forms in logarithms directly to linear recurrence sequences. This approach reduces somewhat the exponential growth of the bounds for η and S_1, in the sense that we achieve sharper bounds when $q_t \ll t^{c_2}$, where c_2 is an appropriate constant. In addition, our computation of the bounds indicates explicitly the procedure for calculating the constants involved. The main results of this paper are given below in Theorems A, B, and C.

THEOREM A. *If $S(S+k) \in Q$, then $S = (u-k)/2$, where (u,v) is a solution to*

$$u^2 - Dv^2 = k^2 \tag{1.4}$$

in which

$$\begin{aligned} &D \in Q \text{ is square-free;} \\ &(u,v) = (u_n, v_n) \text{ with } v_n \text{ even, } v_n/2 \in Q, \text{ and } 1 \leq n \leq M, \\ &\qquad \text{where } M \text{ is an effectively computable number.} \end{aligned} \tag{1.5}$$

Conversely, if (u,v) is a solution to (1.4) satisfying the conditions in (1.5), then $S(S+k) \in Q$ where $S = (u-k)/2$.

THEOREM B. *Let $\mu = \max(3, (q_t+1)/2)$ and let η denote the number of pairs $(S, S+k)$ of positive integers such that $S(S+k) \in Q$. Then*

$$\eta \leq (2^t - 1)\left(\prod_{i=1}^{m}(2a_i + 1) - 1\right) N + (2^t - 1)\mu + M_k, \tag{1.6}$$

where M_k and N are effectively computable numbers, and $\prod_{i=1}^{m} p_i^{a_i}$ is the factorization of k into distinct prime powers.

THEOREM C. *Let $P = q_1 q_2 \cdots q_t$ and let M be as in Theorem* A. *If S_1 is the largest value of S such that $S(S+k) \in Q$, then*

$$\log S_1 < \log k + (M+1)\bigl(2 + \log(8P)\bigr)\sqrt{2P} - \log 2. \tag{1.7}$$

2. Preliminaries

We first make the trivial observation that the pair $(S, S+k)$ is contained in Q if and only if the product $S(S+k)$ is contained in Q. As mentioned in the introduction, using Lehmer's result to find all pairs $(S, S+1)$ such that $S(S+1) \in Q$ requires the solutions of certain Pell equations, while our extended theorem involves the solutions of the generalized Pell equations:

$$u^2 - Dv^2 = k^2 \tag{2.1}$$

where D is a positive square-free integer. Solutions of (2.1) with $D \neq 1$ are, of course, intimately related to solutions of

$$x^2 - Dy^2 = 1. \tag{2.2}$$

Finding the solutions to (2.1) is not difficult, but for our purposes, we need to categorize the solutions for further analysis. We assume for now that $D \neq 1$. The concepts that follow, which are essentially due to Nagell [**27**], are crucial for our study.

DEFINITION 2.1. Two solutions (u_1, v_1) and (u_2, v_2) of (2.1) are said to be *associates* if $u_1 + v_1\sqrt{D} = (u_2 + v_2\sqrt{D})(x + y\sqrt{D})$ for some solution (x, y) of (2.2).

Definition 2.1 defines an equivalence relation on the set of all solutions of (2.1) and therefore, given a solution (u, v) of (2.1), we define the *class K of (u, v)* to be the set of all associates of (u, v). We will occasionally abuse this notation and say that $u + v\sqrt{D} \in K$ when $(u, v) \in K$. We also denote $\overline{K}$, called the *conjugate class of* K, to be the class of $(u, -v)$. If $K \neq \overline{K}$, we call K *unambiguous* and we call the element (u, v) in K the *fundamental solution of* K if v is the smallest nonnegative value among all elements in K. Note that the value of u is uniquely determined since $(-u, v) \in \overline{K}$, and in this case $(-u, v)$ is the fundamental solution of $\overline{K}$. When $K = \overline{K}$, we call K *ambiguous* and the fundamental solution of K is chosen uniquely by imposing the further restriction that $u \geq 0$. The following three facts, due to Stolt [**36**], are slightly enhanced versions of Nagell's results [**27**].

PROPOSITION 2.2. *Two solutions (u_1, v_1) and (u_2, v_2) of* (2.1) *are associates if and only if k^2 divides $u_1 v_2 - u_2 v_1$.*

PROPOSITION 2.3. *Let (u, v) be the fundamental solution of the class K. Then $K = \overline{K}$ if and only if k^2 divides $2uv/k^2$.*

THEOREM 2.4. *If (u_0, v_0) is the fundamental solution of a class of* (2.1) *and (x_1, y_1) is the fundamental solution of* (2.2), *then*

$$0 \leq v_0 \leq \frac{k y_1}{\sqrt{2(x_1+1)}} \tag{2.3}$$

$$0 < |u_0| \leq \frac{k\sqrt{2(x_1+1)}}{2}. \tag{2.4}$$

Conversely, each solution (u, v) *of* (2.1) *with* $u > 0$ *satisfying* (2.3) *and* (2.4) *determines a distinct class.*

Note that conditions (2.3) and (2.4) of Theorem 2.4 are equivalent. The next theorem follows easily from the work of Stolt [**37**].

THEOREM 2.5. *Let* k *be a positive integer and let* $\prod_{i=1}^{m} p_i^{a_i}$ *be the factorization of* k *into distinct prime powers. Then equation* (2.1) *has at most*

$$\prod_{i=1}^{m}(2a_i + 1)$$

distinct classes of solutions and has exactly one ambiguous class.

The bound in Theorem 2.5 is actually the best possible for certain values of k and cannot, in general, be greatly improved without a case by case consideration. We examine briefly now the special case of $D = 1$ in (2.1).

PROPOSITION 2.6. *Let* k *be a positive integer and let* $\prod_{i=1}^{m} p_i^{a_i}$ *be the factorization of* k *into distinct prime powers. Then the exact number of distinct solutions to equation* (2.1) *with* $D = 1$ *is*

$$\begin{cases} \dfrac{\prod_{i=1}^{m}(2a_i+1) - 1}{2} & \text{if } k \text{ is odd,} \\ \dfrac{(2a_1 - 1)\prod_{i=2}^{m}(2a_i+1) - 1}{2} & \text{if } k \text{ is even.} \end{cases}$$

PROOF. When k is odd, each solution of (2.1) with $D = 1$ arises from a pair (c, d), of integers, such that $c < d$ and $cd = k^2$. The total number of divisors of k^2 is $\prod_{i=1}^{m}(2a_i + 1)$. Subtracting off the situation when $c = d = k$ and dividing by 2 to guarantee that $c < d$ yields the result. Now suppose that k is even, and let $p_1 = 2$. Each solution of (2.1) with $D = 1$ arises from a pair (c, d), of even integers, such that $c < d$ and $cd = k^2$. The total number of even divisors of k^2 equals the total number of divisors of $k^2/4$, which is $(2a_1 - 1)\prod_{i=2}^{m}(2a_i + 1)$. As before, we subtract off the situation when $c = d = k$ and divide by 2 to guarantee that $c < d$ which completes the proof. □

NOTE. When $D = 1$, we use the convention of ordering the solutions (u, v) of (2.1) in terms of the size of v. That is, for any two solutions (u_i, v_i) and (u_j, v_j), we have $v_i < v_j$ if and only if $i < j$, with (u_1, v_1) being the smallest.

To understand more clearly equation (2.1) when $D > 1$ and k is an arbitrary positive integer, we give now a detailed analysis of the description of the classes of solutions to (2.1) in the special case when k is prime. This analysis seems to appear nowhere in the current literature. For the remainder of this section we let $D > 1$ be square-free, p be prime and we focus on the equation

$$u^2 - Dv^2 = p^2. \tag{2.5}$$

LEMMA 2.7. *If equation* (2.5) *has a solution* (u, v) *with* $v > 0$ *satisfying* (2.3), *then* p *does not divide* v.

PROOF. Suppose that p divides v and write $v = p\hat{v}$. Then, from (2.3), we have

$$0 < p\hat{v} = v \leq \frac{y_1 p}{\sqrt{2(x_1+1)}},$$

and so

$$0 < \hat{v} \le \frac{y_1}{\sqrt{(2x_1+1)}} < y_1,$$

contradicting the fact that (x_1, y_1) is the fundamental solution of (2.2). □

THEOREM 2.8. *Equation* (2.5) *has at most one solution* (u, v) *with* $u, v > 0$ *in which* u *and* v *satisfy the inequalities* (2.3) *and* (2.4).

PROOF. Suppose that (u_1, v_1) and (u_2, v_2), with $u_1, v_1, u_2, v_2 > 0$, are solutions to (2.5) satisfying (2.3) and (2.4). Then

$$u_1^2 - Dv_1^2 = p^2 \quad \text{and} \quad u_2^2 - Dv_2^2 = p^2. \tag{2.6}$$

Eliminating D in (2.6) gives

$$(u_1v_2 - u_2v_1)(u_1v_2 + u_2v_1) \equiv 0 \pmod{p^2}.$$

If p divides both $(u_1v_2 - u_2v_1)$ and $(u_1v_2 + u_2v_1)$, then p divides either u_1 or v_2. But p cannot divide v_2 by Lemma 2.7. Thus, p divides u_1, so that p^2 divides u_1^2. By Lemma 2.7, p cannot divide v_1. Consequently, p^2 divides D, contradicting the fact that D is square-free. Hence, p^2 divides exactly one of $(u_1v_2 - u_2v_1)$ and $(u_1v_2 + u_2v_1)$. Suppose that p^2 divides $u_1v_2 + u_2v_1$. Observe that the product of the equations in (2.6) can be written as

$$\left(\frac{u_1u_2 + Dv_1v_2}{p^2}\right)^2 - D\left(\frac{u_1v_2 + u_2v_1}{p^2}\right)^2 = 1.$$

Therefore

$$\left(\frac{u_1u_2 + Dv_1v_2}{p^2}, \frac{u_1v_2 + u_2v_1}{p^2}\right)$$

is a solution to (2.2). Hence,

$$u_1v_2 + u_2v_1 \ge p^2y_1 \tag{2.7}$$

since (x_1, y_1) is fundamental. If the solutions (u_1, v_1) and (u_2, v_2) to (2.5) are distinct, then we can assume, without loss of generality, that $v_1 < v_2$. But then, from Theorem 2.4, we have that

$$u_1v_2 + u_2v_1 < u_1v_2 + u_2v_2 \le 2\left(\frac{p\sqrt{2(x_1+1)}}{2}\right)\left(\frac{py_1}{\sqrt{2(x_1+1)}}\right) = p^2y_1$$

contradicting (2.7). Thus, $(u_1, v_1) = (u_2, v_2)$. A similar argument can be used if p^2 divides $u_1v_2 - u_2v_1$. □

Although the following corollary is a special case of Theorem 2.5, we present a proof based on the proof of Theorem 2.8.

COROLLARY 2.9. *Equation* (2.5) *has at most three distinct classes of solutions.*

PROOF. Suppose that (u, v) is a solution of (2.5) satisfying (2.3) and (2.4). Let (r, s) be any solution to (2.5). From the proof of Theorem 2.8, exactly one of $(us - vr)/p^2$, $(us + vr)/p^2$ and s/p is an integer. The corollary then follows from Proposition 2.2. □

COROLLARY 2.10. *Equation* (2.5) *has either exactly one or exactly three classes of solutions according to the following:*

- *Equation* (2.5) *has exactly three classes of solutions if and only if equation* (2.5) *has a solution* (u, v) *such that* p *does not divide* v *if and only if there exists a solution* (u, v) *with* $v > 0$ *satisfying* (2.3) *and* (2.4).
- *Equation* (2.5) *has exactly one class of solutions if and only if every solution* (u, v) *is such that* p *divides* v.

PROOF. If (u, v) is a solution to equation (2.5) such that p does not divide v, then (u, v) is in a separate class by Corollary 2.9. From Proposition 2.3, we conclude that $(u, -v)$ represents a third class. □

We have immediately:

PROPOSITION 2.11. *If equation* (2.5) *has exactly three classes of solutions, then they can be described as follows*:

- $K_1 = pH$;
- $K_2 = (u_0 + v_0\sqrt{D})H$;
- $K_3 = \overline{K_2} = (u_0 - v_0\sqrt{D})H$,

where H *is the set of all solutions to equation* (2.2) *and* $u_0 + v_0\sqrt{D}$ *is the fundamental solution of the class* K_2 *of equation* (2.5).

PROPOSITION 2.12. *If equation* (2.5) *has exactly three classes of solutions, then* $\left(\frac{D}{p}\right) = 1$.

PROOF. If equation (2.5) has exactly three classes of solutions, then, by Corollary 2.10, there exists a solution (u, v) such that p does not divide v. If u were divisible by p, then D would be divisible by p^2, contradicting the fact that D is square-free. Hence, $u^2 \equiv Dv^2 \pmod{p}$ and $D \equiv \left(\frac{u}{v}\right)^2 \pmod{p}$. □

It is interesting to note that the converse of Proposition 2.12 is false. For example, $x^2 - 34y^2 = 9$ has only one class of solutions despite the fact that $\left(\frac{34}{3}\right) = 1$.

Recall that for any solution (x, y) to (2.2) with $x, y > 0$, there exists a positive integer n such that $x + y\sqrt{D} = (x_1 + y_1\sqrt{D})^n$; and conversely, for all positive integers n, (x, y) is a solution to (2.2) with $x, y > 0$, where $x + y\sqrt{D} = (x_1 + y_1\sqrt{D})^n$. Then, utilizing Proposition 2.11, we can describe exactly the solutions to (2.5) that we seek.

PROPOSITION 2.13. *The solutions of* (2.5) *that we seek are the solutions* $(u, v) = (u_n, v_n)$ *with* $v_n \in Q$ *from the various classes as described below.*

- *From* K_1 *(provided* $p \in Q$*)*:

$$u_n + v_n\sqrt{D} = p(x_1 + y_1\sqrt{D})^n, \quad n = 1, 2, \ldots, M_1.$$

- *From* K_2:

$$u_n + v_n\sqrt{D} = (u_0 + v_0\sqrt{D})(x_1 + y_1\sqrt{D})^n, \quad n = 0, 1, 2, \ldots, M_2.$$

- *From* $K_3 = \overline{K_2}$:

$$u_n + v_n\sqrt{D} = (u_0 - v_0\sqrt{D})(x_1 + y_1\sqrt{D})^n, \quad n = 1, 2, \ldots, M_3.$$

Where (x_1, y_1) *is the fundamental solution of* (2.2), (u_0, v_0) *is the fundamental solution of the class* K_2 *of* (2.5) *and* M_1, M_2, M_3 *are effectively computable numbers.*

The constants M_1, M_2 and M_3 tell us how far "out" we need to go to get all desired pairs. The computation of these constants is linked to prime divisors of related binary linear recurrence sequences. In Lehmer's original study of the pairs $(S, S+1)$, there is only one class of solutions, corresponding in Proposition 2.13 to K_1 with $p = 1$. Therefore, the computation of only the single constant M_1 is required. In addition, the related linear recurrence sequence in this case is a Lucas sequence with many divisibility properties which can be used to compute the bound quite readily. We present some standard notation and well-known facts about binary linear recurrence sequences.

DEFINITION 2.14. Let r and s be integers with $r^2 + 4s^2 \neq 0$. Let w_0 and w_1 be integers and define

$$w_n = rw_{n-1} + sw_{n-2}$$

for $n = 2, 3, \ldots$. The sequence $\{w_n\}_{n=0}^{\infty}$ is called a *binary linear recurrence sequence.* If r and s are relatively prime with $w_0 = 0$ and $w_1 = 1$, then the sequence $\{w_n\}_{n=0}^{\infty}$ is called a *Lucas sequence of the first kind.*

PROPOSITION 2.15. *Let $\{w_n\}_{n=0}^{\infty}$ be a binary linear recurrence sequence. If α and β are the roots of the equation $z^2 - rz - s = 0$, with $\alpha \neq \beta$, then*

$$w_n = a\alpha^n + b\beta^n$$

where

$$a = \frac{w_0\beta - w_1}{\beta - \alpha} \quad \text{and} \quad b = \frac{w_1 - w_0\alpha}{\beta - \alpha}.$$

DEFINITION 2.16. The binary linear recurrence sequence $\{w_n\}_{n=0}^{\infty}$ is said to be *nondegenerate* if $ab\alpha\beta \neq 0$ and α/β is not a root of unity.

DEFINITION 2.17. Let $\{w_n\}_{n=0}^{\infty}$ be a binary linear recurrence sequence. Define $P(w_n)$ to be the largest prime divisor of w_n.

DEFINITION 2.18. Let $\{w_n\}_{n=0}^{\infty}$ be a binary linear recurrence sequence. We call a prime divisor q of w_n a *primitive divisor of w_n* if q does not divide w_m for any $m < n$.

In [**18**], Lehmer uses divisibility properties of the Lucas sequence $\{L_n\}_{n=0}^{\infty}$ defined by

$$L_n = \frac{y_n}{y_1} = \frac{\alpha^n - \beta^n}{\alpha - \beta},$$

where $\alpha = x_1 + y_1\sqrt{D}$, $\beta = x_1 - y_1\sqrt{D}$, (x_1, y_1) is the fundamental solution to (2.2) and $x_n + y_n\sqrt{D} = (x_1 + y_1\sqrt{D})^n$, to prove:

PROPOSITION 2.19. *If $n > 3$, then $P(y_n) \geq 2n - 1$.*

Proposition 2.19 allows us to easily compute M_1 in Proposition 2.13. If we let

$$M_1 = \max\bigl(3, (q_t + 1)/2\bigr),$$

then for $n > M_1$, $(u, v) \in K_1$, we have that

$$P(v) = P(py_n) = \max\bigl(p, P(y_n)\bigr) \geq \max(p, 2n-1) > \max(p, q_t) \geq q_t$$

implying that $v \notin Q$. The computation of M_2 and M_3 is a bit more treacherous. Let (u_0, v_0) be the fundamental solution of class K_2 and for $n \geq 0$, let $u_n + v_n\sqrt{D} =$

$(u_0 + v_0\sqrt{D})(x_1 + y_1\sqrt{D})^n$. Then the sequence $\{v_n\}_{n=0}^{\infty}$ is a nondegenerate binary linear recurrence sequence which satisfies, for $n \geq 2$,

$$v_n = 2x_1 v_{n-1} - v_{n-2}.$$

The characteristic polynomial $z^2 - 2x_1 z + 1$ has roots $\alpha = x_1 + y_1\sqrt{D}$ and $\beta = x_1 - y_1\sqrt{D}$. For $n \geq 0$, we also have the representations

- $v_n = a\alpha^n + b\beta^n$,
- $v_n = v_1 L_n - v_0 L_{n-1}$,
- $v_n = u_0 y_1 L_n + v_0 x_n$,

where $a = (v_0\beta - v_1)/(\beta - \alpha)$, $b = (v_1 - v_0\alpha)/(\beta - \alpha)$ and $\{L_n\}_{n=-1}^{\infty}$ is the Lucas sequence defined by $L_{-1} = -1$ and $L_n = (\alpha^n - \beta^n)/(\alpha - \beta)$ for $n \geq 0$. The situation is similar for $\overline{K_2}$.

Unfortunately, the sequence $\{v_n\}_{n=0}^{\infty}$, which arises here, is not as well-behaved or well-understood as a Lucas sequence. Nevertheless, there is an abundance of literature concerning $P(v_n)$. See, for example, [**28, 30, 31, 33–35, 52**]. Three of the strongest known results are given below in Theorems 2.20, 2.21, and 2.22. Theorem 2.20, proved independently by Šparlinskiĭ [**33**] and Stewart [**34**], is applicable in our situation since, for the sequence $\{v_n\}_{n=0}^{\infty}$, we have $|\alpha| > |\beta|$. In fact, Theorem 2.20 is true for more general linear recurrence sequences as long as the characteristic polynomial has a single root of largest modulus. Theorem 2.21, due to Shorey [**30**], is an improvement of a result of Stewart [**34**]. Theorem 2.22, due to Yu and Hung [**52**], is our theorem of choice to compute the bounds in Theorems A, B, and C, since it improves upon the bound given in Theorem 2.21. We should remark that Theorem 2.22 can be applied to more general linear recurrences. In particular, Theorem 2.22 holds for binary recurrence sequences whose terms are algebraic integers in an algebraic number field. The effectively computable constant that arises in this theorem, however, is dependent, not only on the sequence, but also on the exponent of a certain ideal in the class group of the algebraic number field. Fortunately, in our situation, this turns out to be 1. The proofs of Theorems 2.20, 2.21 and 2.22 depend upon estimates for linear forms in the logarithms of algebraic numbers, due to Baker [**1, 3**] in the complex case, and to van der Poorten [**23, 43**] and Yu [**49–51**] in the p-adic case.

THEOREM 2.20. *There exists an effectively computable constant C_3 such that $P(v_n) \geq C_3 \log(n)$.*

THEOREM 2.21. *For $n > m$, $n \geq C_4$, $v_m \neq 0$,*

$$P\left(\frac{v_n}{(v_m, v_n)}\right) > C_5\left(\frac{n}{\log n}\right)^{1/3}$$

for some effectively computable numbers $C_4, C_5 > 0$ depending only on the sequence $\{v_n\}_{n=0}^{\infty}$.

THEOREM 2.22. *For $n > m$, $n \geq C_6$, $v_m \neq 0$,*

$$P\left(\frac{v_n}{(v_m, v_n)}\right) > C_7 n^{1/3}$$

for some effectively computable numbers $C_6, C_7 > 0$ depending only on the sequence $\{v_n\}_{n=0}^{\infty}$.

3. Proofs of the main results

Throughout this section we let k be a positive integer and let $\prod_{i=1}^{m} p_i^{a_i}$ be its factorization into distinct prime powers.

PROOF OF THEOREM A. Suppose first that (u, v) satisfies (1.4) subject to the conditions in (1.5). Since v is even, it follows that u and k have the same parity. Let $S = (u-k)/2$. Thus

$$S(S+k) = \frac{u^2-k^2}{4} = D\left(\frac{v}{2}\right)^2 \in Q.$$

Conversely, suppose that $S(S+k) \in Q$, and write

$$4S(S+k) = Dv^2 \tag{3.1}$$

where D is square-free (possibly $D = 1$). In any case, $D \in Q$. Then rewrite equation (3.1) as

$$(2S+k)^2 - Dv^2 = k^2$$

which produces the solution $(u, v) = (2S+k, v)$ of (1.4). Hence, each such S leads to some solution (u, v) of (1.4) in which the conditions in (1.5) hold. The bound M can be arrived at in the following way. For each of the $2^t - 1$ equations (1.4) with $D > 1$, we have, from Theorem 2.5, at most

$$\prod_{i=1}^{m}(2a_i+1)$$

distinct classes of solutions with exactly one ambiguous class. We have then a total of at most

$$(2^t-1)\left(\prod_{i=1}^{m}(2a_i+1) - 1\right)$$

unambiguous classes of solutions. For each of these classes we have a pair of computable numbers C_6 and C_7 from Theorem 2.22. Let C_8 be the minimum of all values of C_6 and let C_9 be the maximum of all values of C_7 so obtained. Let $N \geq C_9$ be such that $C_8 n^{1/3} > q_t$ for all $n \geq N$. When $D = 1$, we know from Proposition 2.6 the exact number of solutions to equation (1.4). Denote this number as M_k. Finally, define $M = \max(3, (q_t+1)/2, N, M_k)$ and the theorem is proved. □

PROOF OF THEOREM B. There are $2^t - 1$ equations where $D > 1$, with each of these equations having at most $\prod_{i=1}^{m}(2a_i+1) - 1$ unambiguous classes from Theorem 2.5. For each of these classes we have at most N solutions of interest. Hence, we get the first term on the right hand side of (1.6). The second term arises from the count of the number of solutions in the ambiguous class with $D > 1$. Finally, the third term on the right hand side of (1.6) is the number of solutions when $D = 1$, from Proposition 2.6. □

To compute a bound on the largest pair, we use the following result of Lehmer [**18**] which follows from a theorem of Hua [**14**].

LEMMA 3.1. *Let* $x_n + y_n\sqrt{D} = (x_1 + y_1\sqrt{D})^n$. *Then*

$$\log(x_n + y_n\sqrt{D}) < n\big(2 + \log(4D)\big)\sqrt{D}.$$

PROOF OF THEOREM C. Let (u_0, v_0) be a solution to (1.4) which is the fundamental solution of a class K. Using Theorem 2.4 we have that

$$u_0 + v_0\sqrt{D} \le k\left(\frac{x_1 + y_1\sqrt{D} + 1}{\sqrt{2(x_1+1)}}\right) < k(x_1 + y_1\sqrt{D}).$$

From Proposition 2.13, every solution (u_n, v_n) in K that we seek can be written as

$$u_n + v_n\sqrt{D} = (u_0 + v_0\sqrt{D})(x_1 + y_1\sqrt{D})^n.$$

Since $S_1 = (u_n - k)/2$ for some n by Theorem A, it follows that

$$\begin{aligned} 2S_1 = u_n - k &< (u_0 + v_0\sqrt{D})(x_1 + y_1\sqrt{D})^n \\ &\le (u_0 + v_0\sqrt{D})(x_1 + y_1\sqrt{D})^M \\ &< k(x_1 + y_1\sqrt{D})^{M+1} \\ &= k(x_{M+1} + y_{M+1}\sqrt{D}). \end{aligned}$$

The result follows from Lemma 3.1. □

Because of the methods used to obtain the bounds in Theorems A, B and C, the values of M and N are quite large. The weakness of the bounds is inherent in the technique of linear forms in logarithms. Some strategies, each having many drawbacks, that could improve the bounds are:

(1) Assume the *abc*-conjecture is true.
(2) Prove the *abc*-conjecture.
(3) Prove the conjecture of Marshall Hall, Jr. [**7**].
(4) Develop alternate methods using Galois representations and modular forms. For an example of such methods, see [**4**] (`http://www.math.ubc.ca/~bennett/publ.html`).
(5) Prove that v_n has a primitive divisor for all $n > N$, where N is some reasonably small number. The corresponding monumental theorem for Lucas sequences has recently been proved in [**6**].

References

1. A. Baker, *A sharpening of the bounds for linear forms in logarithms.* II, Acta Arith. **24** (1973), 33–36.
2. ———, *Contributions to the theory of Diophantine equations.* II. *The Diophantine equation* $y^2 = x^3 + k$, Philos. Trans. Roy. Soc. London Ser. A **263** (1967/1968), 193–208.
3. ———, *The theory of linear forms in logarithms*, Transcendence Theory: Advances and Applications (Cambridge 1976), Academic Press, London, 1977.
4. M. A. Bennett and C. Skinner, *Ternary Diophantine equations via Galois representations and modular forms*, Canad. J. Math. (to appear).
5. R. G. Bierstedt and W. H. Mills, *On the bound for a pair of consecutive quartic residues of a prime*, Proc. Amer. Math. Soc. **14** (1963), 628–632.
6. Y. Bilu, G. Hanrot, and P. M. Voutier, *Existence of primitive divisors of Lucas and Lehmer numbers*, J. Reine Angew. Math. **539** (2001), 75–122 (with an appendix by M. Mignotte).
7. B. J. Birch, S. Chowla, M. Hall, Jr., and A. Schinzel, *On the difference* $x^3 - y^2$, Norske Vid. Selsk. Forh.(Trondheim) **38** (1965), 65–69.
8. J. Brillhart, D. H. Lehmer, and E. Lehmer, *Bounds for consecutive seventh and higher power residues*, Math. Comp. **18** (1964), 397–407.
9. R. D. Carmichael, *On the numerical factors of the arithmetic forms* $\alpha^n \pm \beta^n$, Ann. of Math. (2) **15** (1913/1914), no. 1-4, 30–70.
10. J. H. H. Chalk, *An estimate for the fundamental solutions of a generalized Pell equation*, Math. Ann. **132** (1956), 263–276.

11. L. E. Dickson, *History of the theory of numbers. II. Diophantine analysis*, Carnegie Inst. of Washington, Washington, DC, 1920, Chapter 12.
12. N. D. Elkies, *Rational points near curves and small nonzero $|x^3 - y^2|$ via lattice reduction*, Algorithmic Number Theory (Leiden, 2000), Lecture Notes in Comput. Sci., vol. 1838, Springer, Berlin, 2000, pp. 33–63.
13. J.-H. Evertse, *On equations in S-units and the Thue–Mahler equation*, Invent. Math. **75** (1984), no. 3, 561–584.
14. L. K. Hua, *On the least solution of Pell's equation*, Bull. Amer. Math. Soc. **48** (1942), 731–735.
15. M. Hall, Jr., *Divisors of second-order sequences*, Bull. Amer. Math. Soc. **43** (1937), 78–80.
16. C. G. Lekkerkerker, *Prime factors of the elements of certain sequences of integers*. I, Nederl. Akad. Wetensch. Proc. Ser. A **56** (1953), 265-277; II, 277–280.
17. D. H. Lehmer, *An extended theory of Lucas' functions*, Ann. of Math. (2) **31** (1930), no. 3, 419–448
18. ———, *On a problem of Størmer*, Illinois J. Math., **8** (1964), 57–79.
19. ———, *The prime factors of consecutive integers*, Amer. Math. Monthly **72** (1965) no. 2, part II, 19–20.
20. ———, *The mathematical work of Morgan Ward*, Math. Comp. **61** (1993), no. 203, 307–311.
21. D. H. Lehmer and E. Lehmer, *On runs of residues*, Proc. Amer. Math. Soc. **13** (1962), 102–106.
22. D. H. Lehmer, E. Lehmer, and W. H. Mills, *Pairs of consecutive power residues*, Canad. J. Math. **15** (1963), 172–177.
23. J. H. Loxton and A.J. van der Poorten, *On the growth of recurrence sequences*, Math. Proc. Cambridge Philos. Soc. **81** (1977), no. 3, 369–376.
24. K. Mahler, *A remark on recursive sequences*, J. Math. Sci. **1** (1966), 12–17.
25. M. Mignotte, *A note on linear recursive sequences*, J. Austral. Math. Soc. **20** (1975), no. 2, 242–244.
26. T. Nagell, *Sur une classes d'équations exponentielles*, Ark. Mat. **3** (1958), 569-582.
27. ———, *Introduction to number theory*, 2nd ed., Chelsea Publishing Co., New York, 1964.
28. A. Pethö, *On the greatest prime factor and divisibility properties of linear recursive sequences*, Indag. Math. (N.S.) **1** (1990), no. 1, 85–93.
29. A. Schinzel, *On two theorems of Gelfond and some of their applications*, Acta Arith. **13** (1967/1968), 177–236.
30. T. N. Shorey, *Linear forms in members of a binary recursive sequence*, Acta Arith. **43** (1984), no. 4, 317–331.
31. T. N. Shorey and R. Tijdeman, *Exponential Diophantine equations*, Cambridge Tracts in Math., vol. 87, Cambridge Univ. Press, Cambidge, 1986.
32. C. L. Siegel, *Die Gleichung $ax^n - by^n = c$*, Math. Ann. **114** (1937), 57–68; Gesammelte Abhandlungen, vol. 2, Springer-Verlag, Berlin-New York, 1966, pp. 8–19.
33. I. E. Šparlinskiĭ, *Prime divisors of recursive sequences*, Izv. Vyssh. Uchebn. Zaved. Mat. (1980), no. 4, 100–103 (Russian).
34. C. L. Stewart, *On divisors of terms of linear recurrence sequences*, J. Reine Angew. Math. **333** (1982), 12–31.
35. ———, *On the greatest prime factor of terms of a linear recurrence sequence*, Rocky Mountain J. Math. **15** (1985), no. 2, 599–608.
36. B. Stolt, *On the Diophantine equation $u^2 - Dv^2 = \pm 4N$*, Ark. Mat. **2** (1952), 1–23.
37. ———, *On the Diophantine equation $u^2 - Dv^2 = \pm 4N$. II*, Ark. Mat. **2** (1952), 251–268.
38. ———, *On the Diophantine equation $u^2 - Dv^2 = \pm 4N$. III*, Ark. Mat. **3** (1955), 117–132.
39. C. Størmer, *Quelques théorèmes sur l'équation de Pell $x^2 - Dy^2 = \pm 1$ et leurs applications*, Christiania Vidensk. Selsk. Skr. 1897, no. 2.
40. A. Thue, *Bermerkungen über gewisse Näherungsbruche algebraischer Zahlen*, Christiania Vidensk. Selsk. Skr. 1908, no. 3; Selected Mathematical Papers, Universitetsforlaget, Oslo, 1977, pp. 159–190.
41. ———, *Berechnungaller Lösungen gewisser Gleichungen von der Form $ax^r - by^r = f$*, Christiania Vidensk. Selsk. Skr. 1918, no. 4; Selected Mathematical Papers, Universitetsforlaget, Oslo, 1977, pp. 565-572.
42. R. Tijdeman, *On integers with many small prime factors*, Compositio Math. **26** (1973), 319-330

43. A. J. van der Poorten, *Linear forms in logarithms in the p-adic case*, Transcendence Theory: Advances and Applications (Cambridge, 1976), Academic Press, London, 1977, pp. 29–57.
44. A. J. van der Poorten and J. H. Loxton, *Computing the effectively computable bounds in Baker's inequality for linear forms in logarithms*, Bull. Austral. Math. Soc. **15** (1976), no. 1, 33–57, corrigenda and addenda, **17** (1977), no. 1, 151–155.
45. ———, *Multiplicative relations in number fields*, Bull. Austral. Math. Soc. **16** (1977), n. 1, 83–98
46. M. Ward, *An arithmetical property of recurring series of the second order*, Bull. Amer. Math. Soc., **40** (1934), 825.
47. ———, *Prime divisors of second order recurring sequences*, Duke Math. J. **21** (1954), 607–614.
48. ———, *The intrinsic divisors of Lehmer numbers*, Ann. of Math. (2) **62** (1955), 230–236.
49. K. R. Yu, *Linear forms in p-adic logarithms*, Acta. Arith, **53** (1989), no. 2, 107-186
50. ———, *Linear forms in p-adic logarithms.* II, Compositio Math. **74** (1990), no. 1, 15–113
51. ———, *Linear forms in p-adic logarithms.* III, Compositio Math. **91** (1994), no. 3, 241–276
52. K. R. Yu and L.-k. Hung, *On binary recurrence sequences*, Indag. Math (N.S.) **6** (1995), no. 3, 341–354

Department of Mathematics, Shippensburg University, 1871 Old Main Drive, Shippensburg, PA 17257, USA

E-mail address: lkjone@ship.edu

Centre de Recherches Mathématiques
CRM Proceedings and Lecture Notes
Volume **36**, 2004

On a Family of Real Bicyclic Biquadratic Fields

Shin-ichi Katayama, Claude Levesque, and Toru Nakahara

Abstract. This is the sequel of a previous paper whose purpose was to exhibit a new family of real bicyclic biquadratic fields K for which we can write the Hasse unit index of the group generated by the units of the three quadratic subfields in the unit group E_K of K. In that previous paper, the family was given as a function of odd integers M. In this note, we consider a certain family involving even integers M_1.

1. Introduction

Denote by $e \in \{1, -1\}$ the norm of a fixed positive power $\eta_1 = \varepsilon_1^s$ of the fundamental unit $\varepsilon_1(> 1)$ of a given real quadratic field k_1, with the restriction that *the integer* s *is odd when* $N_{k_1/\boldsymbol{Q}}(\varepsilon_1) = 1$. Then there exists an integer $M \geq 1$ such that

$$\eta_1 = \tfrac{1}{2}M + \tfrac{1}{2}\sqrt{M^2 - 4e} \quad \text{where} \quad e = N_{k_1/\boldsymbol{Q}}(\eta_1).$$

Defining $g_n = g_n(M, e)$ and $h_n = h_n(M, e)$ for $n \in \boldsymbol{N} = \{0, 1, 2, \dots\}$ by

$$\eta_1^n = \tfrac{1}{2}g_n + \tfrac{1}{2}h_n\sqrt{M^2 - 4e}$$

leads to the two sequences $\{g_n\}_{n\in\boldsymbol{N}}$ and $\{h_n\}_{n\in\boldsymbol{N}}$ which can be written as

$$g_n = \eta_1^n + \bar{\eta}_1^n, \quad h_n = \frac{\eta_1^n - \bar{\eta}_1^n}{\sqrt{M^2 - 4e}} \quad \text{with} \quad \bar{\eta}_1 = \tfrac{1}{2}M - \tfrac{1}{2}\sqrt{M^2 - 4e},$$

and which satisfy the order 2 linear recurrences

$$g_{n+2} = Mg_{n+1} - eg_n \quad \textit{and} \quad h_{n+2} = Mh_{n+1} - eh_n$$

with initial terms $g_0 = 2$, $g_1 = M$ and $h_0 = 0$, $h_1 = 1$. Since $N_{k_1/\boldsymbol{Q}}(\eta_1^n) = e^n$, we deduce

$$g_n^2 - (M^2 - 4e)h_n^2 = 4e^n, \tag{1.1}$$

whereupon

$$(M^2 - 4e)(h_{2n+1}^2 - 1) = g_{2n+1}^2 - M^2.$$

2000 *Mathematics Subject Classification.* 11R27, 11R29, 11R16, 11B37.

The second author was supported by NSERC and FCAR grants, while the third author was supported by a grant (#11640036) from the Japan Society for the Promotion of Science. The three authors wish to express their gratitude to the referee for his (her) remarks.

This is the final form of the paper.

In a previous paper, we concentrated on the compositum $K = k_1k_2 = k_2k_3 = k_1k_3$ of the real quadratic fields

$$k_1 = \boldsymbol{Q}\left(\sqrt{M^2-4e}\right), \quad k_2 = \boldsymbol{Q}\left(\sqrt{h_{2n+1}^2-1}\right) \quad \text{and} \tag{1.2}$$
$$k_3 = \boldsymbol{Q}\left(\sqrt{g_{2n+1}^2-M^2}\right)$$

and exhibited a fundamental system of units of K out of the fundamental units of the three quadratic subfields of K, under the assumption that M was an odd positive integer. So the biquadratic field K was defined with the help of some integers $g_{2n+1}(M,e)$ and $h_{2n+1}(M,e)$ involving some fixed indices $2n+1 \in \boldsymbol{N}$ and some fixed odd integers M defined by means of a fixed power $\eta_1 = \varepsilon_1^s$ of the fundamental unit ε_1 of k_1. In this paper, we plan to do the same with *certain even integers* M_1. This will be achieved by taking the even integer $M_1 \in \boldsymbol{N}$ and the integer $e_1 \in \{1,-1\}$ defined by

$$\begin{cases} M_1 = g_{3m}(M,e) \quad \text{with} \quad M \,\text{odd} \geq 1, \\ e_1 = N_{k_1\boldsymbol{Q}}(\eta_1^{3m}) = e^{3m} = e^m, \end{cases} \tag{1.3}$$

with the constraint that that m is odd when $N_{k_1/\boldsymbol{Q}}(\varepsilon_1) = 1$, and by considering the biquadratic field

$$K = \boldsymbol{Q}\left(\sqrt{D_1}, \sqrt{g_{2n+1}^2(M_1,e_1)-M_1^2}\right) = \boldsymbol{Q}\left(\sqrt{D_1}, \sqrt{h_{2n+1}^2(M_1,e_1)-1}\right)$$

with $D_1 = M_1^2 - 4e_1 > 1$. So the field K involves this time some fixed judiciously chosen even integers M_1 whose definition is related this time to η_1^{3m}.

It is worth mentioning at this point that for all s,t,M, we have the formulas

$$g_{st}(M,e) = g_s\big(g_t(M,e),e^t\big) \quad \text{and} \quad h_{st}(M,e) = h_s\big(g_t(M,e),e^t\big)h_t(M,e).$$

Therefore

$$\begin{aligned} M_1^2 - 4e_1 &= g_{3m}^2(M,e) - 4e^{3m} = (M^2-4e)h_{3m}^2(M,e), \\ g_{2n+1}^2(M_1,e_1) - M_1^2 &= (M^2-4e)\big(h_{3m}^2(M,e)\big)\left[h_{2n+1}^2(g_{3m}(M,e),e_1)-1\right] \\ &= (M^2-4e)[h_{(2n+1)(3m)}^2(M,e) - h_{3m}^2(M,e)], \end{aligned}$$

and we conclude that the field K can also be described as

$$K = \boldsymbol{Q}\left(\sqrt{D}, \sqrt{h_{(2n+1)(3m)}^2(M,e) - h_{3m}^2(M,e)}\right) \text{with} \quad D = M^2-4e.$$

Since we proceed as in [**3**], we refer the reader to that paper for the details. Even if $D_1 = M_1^2 - 4e_1 = (M^2-4e)h_{3m}^2(M,e)$ has a square factor, let us emphasize that $D = M^2-4e$ is not required to be square-free. The argument given at the end of [**3**, Section 2] shows that $[K:\boldsymbol{Q}] = 4$ for $n \geq 1$.We suppose that $\varepsilon_i(>1)$ is the fundamental unit of the subfield k_i $(i=1,2,3)$ of class number h_{k_i} and we use again the following three units of the subfields of K:

$$\eta_1 = \tfrac{1}{2}M + \tfrac{1}{2}\sqrt{M^2-4e} \in k_1, \quad \eta_2 = h_{2n+1}(M_1,e_1) + \sqrt{h_{2n+1}^2(M_1,e_1)-1} \in k_2,$$

$$\eta_3 = \frac{g_{2n+1}(M_1,e_1)}{M_1} + \sqrt{\left(\frac{g_{2n+1}(M_1,e_1)}{M_1}\right)^2 - 1} \in k_3.$$

The determination of the index of the group generated by the units $\{\varepsilon_1, \varepsilon_2, \varepsilon_3\}$ in the unit group E_K of K comes down (see for instance [**4**]) to the determination of the units of

$$S = \{\varepsilon_1, \varepsilon_2, \varepsilon_3, \varepsilon_1\varepsilon_2, \varepsilon_1\varepsilon_3, \varepsilon_2\varepsilon_3, \varepsilon_1\varepsilon_2\varepsilon_3\}$$

which are of norm 1 and which have a square root in K. In practice, those units (of norm $+1$) can be written in the form

$$x + \sqrt{x^2 - r^2} = \frac{(x + r + \sqrt{x^2 - r^2})^2}{2(x+r)},$$

and the question boils down to deciding whether $2(x+r)$ has a square root in K or not. The magic is that the element $x + r$ (in its additive form) can be written in the (multiplicative) form $g_s g_t$ or $g_s h_t$ or $g_s h_t D$.

We consider that it is a nice achievement to be able to determine those square roots without explicitly knowing these elements ε_1, ε_2, ε_3, since we impose no squarefree hypothesis on the discriminats of the subfields. The determination of the perfect squares in the linear binary recurrences when M is even is a very difficult problem, and the only significant results obtained in 1972 by J. H. E. Cohn [**1**] dictated the choice of our infinite family of biquadratic fields. As was stressed in the paragraph following [**3**, Proposition 4.8], the key ingredient in dealing with eventual square roots in K is to investigate whether ot not $g_r = g_r(M_1, e_1) \sim 1, 2, M_1$ or $2M_1$.

We know from [**3**] the following result which, as a matter of fact, holds true whatever the parity of M_1 is.

THEOREM 1. *There exists a computable constant N_0 such that for all $n > N_0$, $\{\varepsilon_1, \varepsilon_2, \varepsilon_3\}$ is a fundamental system of units of K and $h_K = \frac{1}{4} h_{k_1} h_{k_2} h_{k_3}$.*

The purpose of the paper is to establish the following.

THEOREM 2. *Let M_1 be the even positive integer defined by $M_1 = g_{3m}(M, e)$ with M odd ≥ 1, let $e_1 = e^m$ where $e = N_{k_1/\mathbf{Q}}(\eta_1)$ with $\eta_1 = \varepsilon_1^s$, and suppose m is odd when $N_{k_1/\mathbf{Q}}(\varepsilon_1) = 1$. Then the unit group E_K of K is given by*

$$E_K = \begin{cases} \langle -1, \varepsilon_1, \varepsilon_2, \sqrt{\varepsilon_1\varepsilon_2\varepsilon_3} \rangle & \text{if } (n, m, e_1, M_1) = (1, 1, -1, 4) \quad \text{or} \quad (1, 1, -1, 140), \\ \langle -1, \varepsilon_1, \varepsilon_2, \sqrt{\varepsilon_2\varepsilon_3} \rangle & \text{if } n = 1 \quad \text{and} \quad (m, e_1, M_1) \neq (1, -1, 4), (1, -1, 140), \\ \langle -1, \varepsilon_1, \varepsilon_2, \varepsilon_3 \rangle & \text{otherwise}, \end{cases}$$

and the class number h_K of K verifies $h_K = \begin{cases} \frac{1}{2} h_{k_1} h_{k_2} h_{k_3} & \text{if } n = 1, \\ \frac{1}{4} h_{k_1} h_{k_2} h_{k_3} & \text{if } n \neq 1. \end{cases}$

2. Preliminaries

In this paper, we will use again the notation $b \sim c$ for two integers b and c if there exist positive integers y and z such that $by^2 = cz^2$. We will also use the following useful properties of $g_r = g_r(M_1, e_1)$ and $h_r = h_r(M_1, e_1)$.

PROPOSITION 1. *For $r \geq 0$, the following equalities hold true:*

(i) $g_r h_r = h_{2r}$;
(ii) $(g_r, g_{r+1}) = 2$;
(iii) $(h_r, h_{r+1}) = 1$;
(iv) $(g_r, h_r) = (2, h_r) = (2, r)$;

(v) $(g_r, h_{r+1}) = \begin{cases} 1 & \textit{if } r \textit{ is even,} \\ M_1 & \textit{if } r \textit{ is odd;} \end{cases}$

(vi) $(h_r, g_{r+1}) = \begin{cases} M_1 & \textit{if } r \textit{ is even,} \\ 1 & \textit{if } r \textit{ is odd;} \end{cases}$

(vii) $(g_r, M_1) = \begin{cases} 2 & \textit{if } r \textit{ is even,} \\ M_1 & \textit{if } r \textit{ is odd;} \end{cases}$

(viii) $(g_r, D_1) = \begin{cases} 2 & \textit{if } r \textit{ is even,} \\ (4, M_1) & \textit{if } r \textit{ is odd;} \end{cases}$

(ix) $(g_r, D) = 1$ *and* $D \sim D_1$.

PROOF. See [**3**, Proposition 2.2]. As far as (viii) is concerned, note that by (1.1), we have $D_1 = M_1^2 - 4e_1 = g_{3m}^2 - 4e^{3m} = (M^2 - 4e)h_{3m}^2(M, e) = Dh_{3m}^2$. Finally, (ix) follows from (viii) because D is odd. □

The basic hypothesis of [**3**] was that $\eta_1 = \varepsilon_1^s$ is the s-th power of the fundamental unit ε_1 with s odd when $N_{k_1/\boldsymbol{Q}}(\varepsilon_1) = 1$. Here are the five cases covered in this paper by our hypotheses on m and s:

	m	s	$N_{k_1/\boldsymbol{Q}}(\varepsilon_1)$	$e = N_{k_1/\boldsymbol{Q}}(\eta_1)$	$e_1 = e^m$
(i) $g_n\big(g_{3m}(M, 1), 1^{3m}\big)$	odd	odd	$+1$	$+1$	$+1$
(ii) $g_n\big(g_{3m}(M, 1), 1^{3m}\big)$	odd	even	-1	$+1$	$+1$
(iii) $g_n\big(g_{3m}(M, -1), (-1)^{3m}\big)$	odd	odd	-1	-1	-1
(iiv) $g_n\big(g_{3m}(M, -1), (-1)^{3m}\big)$	even	odd	-1	-1	$+1$
(v) $g_n\big(g_{3m}(M, 1), 1^{3m}\big)$	even	even	-1	$+1$	$+1$

In the case (iii) we will use [**1**, Theorems 1, 11 and 12], where $e = e_1 = -1$. In the cases (i), (ii) and (v), we will use [**1**, Theorems 3, 14 and 15], where $e = e_1 = 1$. In the case (iv), we write $m = 2t$, so $g_{3m}(M, -1) = g_{6t}(M, -1) = g_{3t}(M^2 + 1, 1)$, whereupon we can use [**1**, Theorem 3] since $M^2 + 2$ is odd.

The following proposition will prove useful in the sequel and shows that the constant N_0 of Theorem 1 can *a posteriori* be taken to be 2.

PROPOSITION 2. *Let M_1 be the even positive integer defined by $M_1 = g_{3m}(M, e)$ with M odd ≥ 1, let $e_1 = e^m$ where $e = N_{k_1/\boldsymbol{Q}}(\eta_1)$ with $\eta_1 = \varepsilon_1^s$, and suppose m odd when $N_{k_1/\boldsymbol{Q}}(\varepsilon_1) = 1$. Then M_1 is even and the following properties of $g_n = g_n(M_1, e_1)$ hold true:*

$$\begin{cases} g_n \sim 1 & \Longleftrightarrow [n = 1, m = 1, e_1 = e = -1, M_1 \in \{4, 36\}, M \in \{1, 3\}]; \\ g_n \sim 2 & \Longleftrightarrow [n = 1, m = 1, e_1 = e = 1, M_1 \in \{18, 19602\}, M \in \{3, 27\}] \\ & \qquad \textit{or } [n = 1, m = 2, e_1 = 1, e = -1, M_1 \in \{18, 19602\}, M \in \{1, 5\}] \\ & \qquad \textit{or } [n = 2, m = 1, e_1 = e = -1, M_1 \in \{4, 140\}, M \in \{1, 5\}]; \\ g_n \sim M_1 & \Longleftrightarrow n = 1; \\ g_n \sim 2M_1 & \Longleftrightarrow [n = 2, m = 1, e_1 = e = -1, M_1 = 4, M = 1]. \end{cases}$$

PROOF. It is easy to see that M_1 is even. One uses [**1**, Theorems 1 and 3] to deal with $g_n(M_1, e_1) \sim 1$ or 2; note that $g_{6t}(M, -1) = g_{3t}(M^2 + 2, 1)$ with $M^2 + 2$ still odd. For the last two equivalences, use [**1**, Theorems 11, 12, 14 and 15] and the fact that

$$g_{3mn}(M, e) = g_n\big(g_{3m}(M, e), e^{3m}\big) = g_n(M_1, e_1)$$

to obtain:

$$\begin{aligned} g_n(M_1,e_1)\sim M_1 &\iff g_{3mn}(M,e)g_{3m}(M,e)\sim 1 \iff n=1,\\ g_n(M_1,e_1)\sim 2M_1 &\iff g_{3mn}(M,e)g_{3m}(M,e)\sim 2\\ &\iff \begin{cases} n=2, m=1, e_1=e=-1\\ \text{and } (M_1,M)=(4,1). \end{cases} \end{aligned}$$

□

We plan to build on the following results, which appeared as [**3**, Propositions 3.1, 3.3 and 3.4].

Proposition 3. *Let M_1 be the even positive integer defined by $M_1 = g_{3m}(M,e)$ with M odd ≥ 1, let $e_1 = e^m$ where $e = N_{k_1/\mathbf{Q}}(\eta_1)$ with $\eta_1 = \varepsilon_1^s$, and suppose m* odd *when $N_{k_1/\mathbf{Q}}(\varepsilon_1) = 1$.*

(i) *Then we have:*

$$\begin{aligned} \eta_2 \in \langle \varepsilon_2^2\rangle &\iff [g_nh_{n+1}\sim 2] \quad or \quad [g_{n+1}h_n\sim 2];\\ \eta_3 \in \langle \varepsilon_3^2\rangle &\iff [g_ng_{n+1}\sim 2M_1] \quad or \quad [h_nh_{n+1}\sim 2M_1D]. \end{aligned}$$

(ii) *In the cases where $\eta_2 \notin \langle\varepsilon_2^2\rangle$ and $\eta_3 \notin \langle\varepsilon_3^2\rangle$, we have:*

$$\begin{aligned} \sqrt{\varepsilon_2}\in E_K &\iff [g_nh_{n+1}\sim 2D] \ or\ [g_{n+1}h_n\sim 2D];\\ \sqrt{\varepsilon_3}\in E_K &\iff [g_ng_{n+1}\sim 2M_1D] \quad or \quad [h_nh_{n+1}\sim 2M_1];\\ \sqrt{\varepsilon_2\varepsilon_3}\in E_K &\iff [g_nh_n\sim M_1 \quad or \quad M_1D] \quad or \quad [g_{n+1}h_{n+1}\sim M_1 \ or\ M_1D]. \end{aligned}$$

(iii) *Suppose that $N_{k_1/\mathbf{Q}}(\varepsilon_1) = 1$ and that $\eta_2 \notin \langle\varepsilon_2^2\rangle$ and $\eta_3 \notin \langle\varepsilon_3^2\rangle$; we also have:*

$$\begin{aligned} \sqrt{\varepsilon_1}\in E_K &\iff g_ng_{n+1}h_nh_{n+1}\sim M_1+2 \ or\ M_1-2;\\ \sqrt{\varepsilon_1\varepsilon_2}\in E_K &\iff \begin{cases} [g_nh_{n+1}\sim 2(M_1+2)\ or\ 2(M_1-2)]\\ or\\ [g_{n+1}h_n\sim 2(M_1+2)\ or\ 2(M_1-2)]; \end{cases}\\ \sqrt{\varepsilon_1\varepsilon_3}\in E_K &\iff \begin{cases} [g_ng_{n+1}\sim 2M_1(M_1+2)\ or\ 2M_1(M_1-2)]\\ or\\ [h_nh_{n+1}\sim 2M_1(M_1+2)\ or\ \ 2M_1(M_1-2)]; \end{cases}\\ \sqrt{\varepsilon_1\varepsilon_2\varepsilon_3}\in E_K &\iff \begin{cases} [g_nh_n\sim M_1(M_1+2\ or\ M_1(M_1-2)]\\ or\\ [g_{n+1}h_{n+1}\sim M_1(M_1+2)\ or\ M_1(M_1-2)]. \end{cases} \end{aligned}$$

3. Proof of the main result

The first step is a crucial one and is taken care of by the following.

Proposition 4. *Let $n = 1$, $e_1 = -1$ and $M_1 \in \{4, 140\}$; then $\eta_2 \in \langle\varepsilon_2^2\rangle$ and $\eta_3 \in \langle\varepsilon_3^2\rangle$. Otherwise, we have $\eta_2 \notin \langle\varepsilon_2^2\rangle$ and $\eta_3 \notin \langle\varepsilon_3^2\rangle$.*

Proof. (a) Suppose $\eta_2 \in \langle\varepsilon_2^2\rangle$. By Proposition 3(i), this is equivalent to either $[g_nh_{n+1}\sim 2]$ or $[g_{n+1}h_n\sim 2]$.

(i) Consider first n even. Suppose $g_n h_{n+1} \sim 2$. Then by Proposition 1(v), $g_n \sim 1$ or 2; by Proposition 2 and because n is even, we have $n = 2$, $m = 1$, $e_1 = -1$, $M_1 \in \{4, 140\}$. If $g_{n+1} h_n \sim 2$, then by Proposition 1(vi), $g_{n+1} \sim M_1$ or $2M_1$ and by Proposition 2, we have $n + 1 = 1$ (i.e., $n = 0$, which is ruled out) or we have $n + 1 = 2$, i.e., $n = 1$ but n should be even.

(ii) Next consider n odd. If $g_n h_{n+1} \sim 2$, then this time $g_n \sim M_1$ or $2M_1$, and either we have $n = 1$ (which is ruled out since $2 \sim g_n h_{n+1} = g_1 h_2 = M_1^2$) or we have $n = 2$ (but n should be odd). Suppose now $g_{n+1} h_n \sim 2$, whereupon $g_{n+1} \sim 1$ or 2. A first possibility could be $n + 1 = 1$, but the case $n = 0$ is ruled out. A second possibility is $n + 1 = 2$, $m = 1$, $e_1 = -1$, $M_1 \in \{4, 140\}$.

(b) Suppose $\eta_3 \in \langle \varepsilon_3^2 \rangle$. By Proposition 3(i), this is equivalent to either $[g_n g_{n+1} \sim 2M_1]$ or $[h_n h_{n+1} \sim 2M_1 D]$. By Proposition 1, we have $(g_n, g_{n+1}) = 2$, $(h_n, h_{n+1}) = 1$, $(g_n, M_1) \in \{2, M_1\}$.

(i) Consider first n even. So $(g_{n+1}, M_1) = M_1$. Suppose $g_n g_{n+1} \sim 2M_1$. Then $g_{n+1} \sim 2$ or $2M_1$, whereupon $n + 1 \in \{1, 2\}$, i.e., $n \in \{0, 1\}$, a contradiction. Suppose now $h_n h_{n+1} = (g_{n/2} h_{n/2}) h_{n+1} \sim 2M_1 D$. Here $(g_{n/2}, h_{n/2}) = (2, n/2)$. By Proposition 1(ix), $(g_{n/2}, D) = 1$. Therefore we have $g_{n/2} \sim 1$ or 2 or M_1 or $2M_1$ and because of Proposition 2, there are six possibilities to investigate.

- $n/2 = 1$, $n = 2$, $m = 1$, $e_1 = -1$, $M_1 \in \{4, 36\}$, and $h_n h_{n+1} = h_2 h_3 \sim 2M_1 D$, i.e., $M_1(M_1^2 + 1) \sim 2M_1(M_1^2 + 4)$; this leads to a contradiction when $M_1 \in \{4, 36\}$;
- $n/2 = 1$, $n = 2$, $m = 1$, $e_1 = 1$, $M_1 \in \{18, 19602\}$ and $h_n h_{n+1} = h_2 h_3 \sim 2M_1 D$, i.e., $M_1(M_1^2 - 1) \sim 2M_1(M_1^2 - 4)$; this leads to a contradiction when $M_1 \in \{18, 19602\}$;
- $n/2 = 1$, $n = 2$, $m = 2$, $e_1 = 1$, $M_1 \in \{18, 19602\}$ and $h_n h_{n+1} = h_2 h_3 \sim 2M_1 D$, which leads again to another contradiction;
- $n/2 = 2$, $n = 4$, $m = 1$, $e_1 = -1$, $M_1 \in \{4, 140\}$ and $h_n h_{n+1} = h_4 h_5 \sim 2M_1 D$, i.e., $(M_1^3 + 2M_1)(M_1^4 + 3M_1^2 + 1) \sim 2M_1(M_1^2 + 4)$; this is again impossible when $M_1 \in \{4, 140\}$.
- $n/2 = 1, n = 2$, whereupon $h_n h_{n+1} = h_2 h_3 \sim 2M_1 D$, i.e., $M_1(M_1^2 - e_1) \sim 2M_1(M_1^2 - 4e_1)$, i.e., $(M_1^2 - e_1)(M_1^2 - 4e_1) \sim 2$. Now

$$(M_1^2 - 1, M_1^2 - 4) = \begin{cases} 3 & \text{if } M_1 \equiv 2 \text{ or } 4 \pmod 6, \\ 1 & \text{if } M_1 \equiv 0 \pmod 6; \end{cases}$$

$$(M_1^2 + 1, M_1^2 + 4) = \begin{cases} 1 & \text{if } M_1 \equiv 2 \text{ or } 4 \pmod 6, \\ 3 & \text{if } M_1 \equiv 0 \pmod 6. \end{cases}$$

We are then led to four possibilities:

$$\text{either} \begin{cases} M_1^2 - e_1 = x^2, \\ M_1^2 - 4e_1 = 2y^2, \end{cases} \quad \text{or} \quad \begin{cases} M_1^2 - e_1 = 2x^2, \\ M_1^2 - 4e_1 = y^2, \end{cases}$$

$$\text{or} \begin{cases} M_1^2 - e_1 = 3x^2, \\ M_1^2 - 4e_1 = 6y^2, \end{cases} \quad \text{or} \quad \begin{cases} M_1^2 - e_1 = 6x^2, \\ M_1^2 - 4e_1 = 3y^2. \end{cases}$$

The first equation in the first case forces $M_1 \in \{0, 1\}$, a contradiction. The first equation in the second case and in the fourth case implies $2 \mid e_1$, a contradiction. Since it is impossible to have $M_1^2 + 1 = 3x^2$ in the third case, we assume $e_1 = 1$, whereupon

$$\begin{cases} M_1^2 - 3x^2 = 1, \\ x^2 - 2y^2 = 1, \end{cases} \quad \text{i.e.,} \quad \begin{cases} M_1^2 - 3x^2 = 1, \\ (2y)^2 - 2x^2 = -2. \end{cases}$$

Using Rickert's procedure as in [**3**, Section 6], we find that this system has no positive integral solution.

- $n/2 = 2$, $n = 4$, $m = 1$, $e_1 = -1$, $M_1 = 4$, whereupon $h_n h_{n+1} = h_4 h_5 = (M_1^3 + 2M_1)(M_1^4 + 3M_1^2 + 1) \sim 2M_1 D = 2M_1(M_1^2 + 4)$, a contradiction for $M_1 = 4$.

(ii) Next consider n odd. Then $(g_{n+1}, M_1) = 2$. Suppose first $g_n g_{n+1} \sim 2M_1$. Then $g_{n+1} \sim 1$ or 2, whereupon $n + 1 \in \{1, 2\}$, i.e., $n = 1$ (since n is odd). So $2M_1 \sim g_n g_{n+1} = g_1 g_2 = M_1 g_2$, i.e., $g_2 \sim 2$. By Proposition 2, $m = 1, e_1 = -1$ and $M_1 \in \{4, 140\}$.
Suppose $h_n h_{n+1} = h_n g_{(n+1)/2} h_{(n+1)/2} \sim 2M_1 D$. Then $(g_{(n+1)/2}, h_{(n+1)/2}) \in \{1, 2\}$ and $(g_{(n+1)/2}, D) = 1$. Hence $g_{(n+1)/2} \sim 1$ or 2 or M_1 or $2M_1$. By Proposition 2, $(n + 1)/2 \in \{1, 2\}$. If $(n + 1)/2 = 1$, then $n = 1$ and $2M_1 D \sim h_n h_{n+1} = h_1 h_2 = M_1$, i.e., $D \sim 2$, a contradiction to the hypothesis that D is odd. If $(n + 1)/2 = 2$, i.e., $n = 3$, then (see Proposition 2 we have $m = 1$, $e_1 = -1 = e$, $M_1 \in \{4, 140\}$, whereupon $h_n h_{n+1} = h_3 h_4 = (M_1^2 + 1)(M_1^3 + 2M_1) \sim 2M_1 D$, which is impossible for $M_1 \in \{4, 140\}$.

□

Let us secure Theorem 2 in the case where $\eta_2 \in \langle \varepsilon_2^2 \rangle$ and $\eta_3 \in \langle \varepsilon_3^2 \rangle$. We are then dealing with the two cases $(n, m, e_1, M_1) = (1, 1, -1, 4), (1, 1, -1, 140)$, and it is easy to show that

$$\{\varepsilon_1, \varepsilon_2, \sqrt{\varepsilon_1 \varepsilon_2 \varepsilon_3}\} = \begin{cases} \frac{1}{2}\left(3 + \sqrt{2} + \sqrt{5} + \sqrt{10}\right) & \text{if } K = \boldsymbol{Q}\left(\sqrt{2}, \sqrt{5}\right), \\ \frac{1}{2}\left(2319\sqrt{2} + 5\sqrt{29} + 3\sqrt{58}\right) & \text{if } K = \boldsymbol{Q}\left(\sqrt{2}, \sqrt{29}\right), \end{cases}$$

is a fundamental system of units of K.

From now on, let us assume that $\eta_2 \notin \langle \varepsilon_2^2 \rangle$ and $\eta_3 \notin \langle \varepsilon_3^2 \rangle$, i.e.,

$$\boxed{(n, m, e_1, M_1) \neq (1, 1, -1, 4), (1, 1, -1, 140)}.$$

This has the nice consequence that there exist non-negative integers a and b such that

$$\eta_2 = \varepsilon_2^{2a+1} \quad \text{and} \quad \eta_3 = \varepsilon_3^{2b+1},$$

whereupon the existence of $\sqrt{\varepsilon_2}$ (resp. $\sqrt{\varepsilon_3}$) in K is equivalent to the existence of $\sqrt{\eta_2}$ (resp. $\sqrt{\eta_3}$) in K.

PROPOSITION 5. *Let $\eta_2 \notin \langle \varepsilon_2^2 \rangle, \eta_3 \notin \langle \varepsilon_3^2 \rangle$. Then the following holds true:*

$$\sqrt{\varepsilon_2} \notin E_K; \quad \sqrt{\varepsilon_3} \notin E_K; \quad \sqrt{\varepsilon_2 \varepsilon_3} \in E_K \iff n = 1.$$

PROOF. (i) Suppose $\sqrt{\varepsilon_2} \in E_K$. Because of Proposition 3(ii), either $[g_n h_{n+1} \sim 2D]$ or $[g_{n+1} h_n \sim 2D]$. Using Proposition 1, we get either $g_n \sim 1$, 2, M_1 or $2M_1$ (i.e., $n \in \{1, 2\}$) or $g_{n+1} \sim 1$, 2, M_1 or $2M_1$ (i.e., $n = 1$). Looking at the values of $g_1 h_2$, $g_2 h_3$ and $g_2 h_1$ leads to contradictions.

(ii) Suppose $\sqrt{\varepsilon_3} \in E_K$. Then either, $[g_n g_{n+1} \sim 2M_1 D]$ or $[h_n h_{n+1} \sim 2M_1]$. However, we cannot have $g_n g_{n+1}/(2M_1) \sim D$, because $(g_n g_{n+1}, D) = 1$. Moreover, if $(g_{n/2} h_{n/2}) h_{n+1} \sim 2M_1$ or if $h_n g_{(n+1)/2} h_{(n+1)/2} \sim 2M_1$, then

$n/2 \in \{1,2\}$ or $(n+1)/2 \in \{1,2\}$, but the values of h_2h_3, h_4h_5, h_1h_2, h_3h_4 will lead to contradictions.

(iii) Suppose $\sqrt{\varepsilon_2\varepsilon_3} \in E_K$. Then $[g_nh_n \sim M_1$ or $M_1D]$ or $[g_{n+1}h_{n+1} \sim M_1$ or $M_1D]$. Because $(g_n, D) = (g_{n+1}, D) = 1$, we have $[g_n \sim 1$ or $M_1]$ or $[g_{n+1} \sim 1$ or $M_1]$. It is then easy to conclude $\sqrt{\varepsilon_2\varepsilon_3} \in E_K \iff n = 1$. □

If $N_{k_1/\boldsymbol{Q}}(\varepsilon_1) = -1$, then ε_1, $\varepsilon_1\varepsilon_2$, $\varepsilon_1\varepsilon_3$ and $\varepsilon_1\varepsilon_2\varepsilon_3$ are not totally positive and hence $\sqrt{\varepsilon_1}$, $\sqrt{\varepsilon_1\varepsilon_2}$, $\sqrt{\varepsilon_1\varepsilon_3}$, $\sqrt{\varepsilon_1\varepsilon_2\varepsilon_3} \notin E_K$. Let us treat the case $N_{k_1/\boldsymbol{Q}}(\varepsilon_1) = 1$ (so $e_1 = 1$).

PROPOSITION 6. *$N_{k_1/\boldsymbol{Q}}(\varepsilon_1) = 1$ and suppose $\eta_2 \notin \langle\varepsilon_2^2\rangle, \eta_3 \notin \langle\varepsilon_3^2\rangle$. Then $\sqrt{\varepsilon_1} \notin E_K$, $\sqrt{\varepsilon_1\varepsilon_2} \notin E_K$, $\sqrt{\varepsilon_1\varepsilon_2\varepsilon_3} \notin E_K$, $\sqrt{\varepsilon_1\varepsilon_3} \notin E_K$.*

PROOF. We proceed by contradiction and use Propositions 3(iii) and 2.

(i) Let $\sqrt{\varepsilon_1} \in E_K$, so $g_ng_{n+1}h_nh_{n+1} \sim M_1 + 2$ or $M_1 - 2$.

Let n be odd. By Proposition 1, $(g_{n+1}, g_n) = 2$, $(g_{n+1}, h_n) = 1$, $(g_{n+1}, h_{n+1}) = 2$, $(g_{n+1}, D_1) = \big(g_{n+1}, (M_1-2)(M_1+2)\big) = 2$. Therefore, $g_{n+1} \sim 1$ or 2, i.e., $n+1 = 2$, $e_1 = -1$, a contradiction to $e_1 = 1$.

Let n be even. Then $(g_{n+1}, g_n) = 2$, $(g_{n+1}, h_n) = M_1$, $(g_{n+1}, h_{n+1}) = 1$, $(g_{n+1}, D_1) \in \{2,4\}$. Therefore, $g_{n+1} \sim M_1$ or $2M_1$, i.e., $n+1 = 2$, $e_1 = -1$, a contradiction to $e_1 = 1$.

(ii) Let $\sqrt{\varepsilon_1\varepsilon_2} \in E_K$. Then $[g_{n+1}h_n \sim 2(M_1+2)$ or $2(M_1-2)]$ or $[g_nh_{n+1} \sim 2(M_1+2)$ or $2(M_1-2)]$.

Let n be odd. Now $(g_{n+1}, D_1) = 2$, $(g_{n+1}, h_n) = 1$, and $(g_{n+1}, D_1) \in \{2,4\}$, $(g_n, h_{n+1}) = M_1$. On the one hand, we may have $g_{n+1} \sim 1$ or 2, i.e., $n+1 = 2$, $e_1 = -1$, a contradiction. On the other hand, we may have $g_n \sim M_1$ or $2M_1$, i.e., $n = 1$; however, $g_1h_2 = M_1^2 \sim 2(M_1+2)$ or $2(M_1-2)$; this means $M_1 + 2 \sim 2$ or $M_1 - 2 \sim 2$. Since $M_1 = g_{3m}(M,1) = g_3(x,1) = x^3 - 3x$ for $x = g_m(M,1)$, the equality $M_1 \pm 2 = 2y^2$ implies $(x \mp 1)^2(x \pm 2) = 2y^2$, so $x + 2 \sim 2$ or $x - 2 \sim 2$ for some $x < M_1$. Continuing with the same argument, we get $g_{3t+r}(M,1) + 2 \sim 2$ or $g_{3t+r}(M,1) - 2 \sim 2$ with $r = 1$ or 2. However, since M is odd, it is easy to see that that $g_{3t+r}(M,1)$ is odd for $r = 1, 2$. Hence we have a contradiction.

Let n be even. Then $(g_{n+1}, D_1) \in \{2,4\}$, $(g_{n+1}, h_n) = M_1$, and $(g_n, D_1) = 2$, $(g_n, h_{n+1}) = 1$. On the one hand, we may have $g_{n+1} \sim M_1$ or $2M_1$, i.e., $n+1 = 1$ (ruled out), or $n+1 = 2$ with $e_1 = -1$, a contradiction. On the other hand, we may have $g_n \sim 1$ or 2, i.e., $n = 1$, $M_1 \in \{18, 19602\}$, but this leads to contradictions.

(iii) Suppose $\sqrt{\varepsilon_1\varepsilon_2\varepsilon_3} \in E_K$. Then $[g_nh_n \sim M_1(M_1+2)$ or $M_1(M_1-2)]$ or $[g_{n+1}h_{n+1} \sim M_1(M_1+2)$ or $M_1(M_1-2)]$.

Let n be odd. Now $(g_n, h_n) = 1$, $(g_n, M_1) = M_1$, $(g_n, D_1) \in \{2,4\}$ and $(g_{n+1}, h_{n+1}) = 2$, $(g_{n+1}, M_1) = 2$, $(g_{n+1}, D_1) = 2$. On the one hand, we may have $g_n \sim M_1$, i.e., $n = 1$. So $g_1h_1 = M_1 \sim M_1(M_1+2)$ or $M_1(M_1-2)$. Therefore, $M_1 + 2 = x^2$ or $M_1 - 2 = x^2$, i.e., $2M_0 = M_1 = x^2 \mp 2 = 4x_0^2 \mp 2$, i.e., $M_0 = 2x_0^2 \mp 1$. Hence

$$\eta_1 = M_0 + \sqrt{M_0^2 - 1} = 2x_0^2 \mp 1 + \sqrt{4x_0^4 \mp 4x_0^2} = \left(x_0 + \sqrt{x_0^2 \mp 1}\right)^2,$$

i.e., $\eta_1 \in \langle\varepsilon_1^2\rangle$, a contradiction. On the other hand, we may have $g_{n+1} \sim 1$ or 2, i.e., $n+1 = 2$, $n = 1$, $M_1 \in \{18, 19602\}$, but $g_2h_2 = (M_1^2+2)M_1 \sim M_1(M_1+2)$, $g_2h_2 \sim M_1(M_1-2)$ give contradictions.

Let n be even. Now $(g_n, h_n) = 2$, $(g_n, M_1) = 2$, $(g_n, D_1) = 2$ and $(g_{n+1}, h_{n+1}) = 1$, $(g_{n+1}, M_1) = 2$, $(g_{n+1}, D_1) = 2$. On the one hand, we may have $g_n \sim 1$ or 2, i.e., $n = 1$, a contradiction. On the other hand, we may have $g_{n+1} \sim M_1$, i.e., $n+1 = 1$, a contradiction.

(iv) Finally, suppose $\sqrt{\varepsilon_1\varepsilon_3} \in E_K$. It is easy to see that it is impossible to have $g_n g_{n+1} \sim 2M_1(M_1+2)$ or $2M_1(M_1-2)$. Now let us investigate whether or not we can have $h_n h_{n+1} \sim 2M_1(M_1+2)$ or $2M_1(M_1-2)$. After having considered the four possible cases corresponding to n even with $n/2$ odd or even, and n odd with $(n+1)/2$ odd or even and used the equality $h_{2r} = g_r h_r$, we look at the four possibilities

$$\frac{h_n h_{n+1}}{M_1} \in \left\{ \frac{g_{n/2}}{M_1} h_{n/2} h_{n+1}, g_{n/2}\frac{h_{n/2}}{M_1} h_{n+1}, h_n \frac{g_{(n+1)/2}}{M_1} h_{(n+1)/2}, h_n g_{(n+1)/2}\frac{h_{(n+1)/2}}{M_1} \right\}$$

and can see that $n = 1$ or 2. This leads to $[h_1h_2 = M_1 \sim 2M_1(M_1+2)$ or $2M_1(M_1-2)]$ or to $[h_2h_3 = M_1(M_1^2-1) \sim 2M_1(M_1+2)$ or $2M_1(M_1-2)]$. We have shown above that $M_1 \pm 2 \not\sim 2$ for $M_1 = g_{3m}(M,1)$. So $M_1^2 - 1 \sim 2(M_1+2)$ or $2(M_1-2)$, and there are four possibilities to investigate.

Suppose first $(M_1^2-1, M_1+2) = 1$ or $(M_1^2-1, M_1-2) = 1$. So $M_1^2 - 1 \sim 1$, i.e., $M_1^2 - 1 = y^2$, a contradiction. Suppose next $(M_1^2-1, M_1+2) = 3$. Then $M_1 - 1 = 3z^2$, $M_1 + 1 = x^2$, whereupon $x^2 - 3z^2 = 2$, which is impossible modulo 3. Finally, suppose $(M_1^2-1, M_1-2) = 3$; then

$$\begin{cases} M_1 - 1 = x^2, \\ M_1 - 2 = 6y^2, \\ M_1 + 1 = 3z^2, \end{cases} \quad \text{so} \quad \begin{cases} x^2 - 3z^2 = -2, \\ z^2 - 2y^2 = 1, \end{cases} \quad \text{i.e.,} \quad \begin{cases} x^2 - 3z^2 = -2, \\ (2y)^2 - 2z^2 = -2, \end{cases}$$

whose non-negative solutions (with Rickert's technique in [**5**]) are $(x, y, z, M_1) = (1, 0, 1, 2)$, $(5, 2, 3, 24)$. But $24^2 - 1 \not\sim 2(24+2), 2(24-2)$. Now $M_1 = 2$ cannot occur, since we want $M^2 - 4 > 0$. □

This concludes the proof of the last two parts of Theorem 2.

4. Assuming the ABC conjecture

Let $\boldsymbol{N}^*$ stand for the set $\{1, 2, \dots\}$. Define $\mathcal{S}(M)$ and $\mathcal{S}(M_1)$ by

$$\mathcal{S}(M) = \left\{ \boldsymbol{Q}\left(\sqrt{M^2-4e}, \sqrt{h_{2n+1}^2(M,e)-1}\right) \,\middle|\, n \in \boldsymbol{N}^* \right\},$$

$$\mathcal{S}(M_1) = \left\{ \boldsymbol{Q}\left(\sqrt{M_1^2-4e^m}, \sqrt{h_{2n+1}^2(M_1,e^m)-1}\right) \,\middle|\, n \in \boldsymbol{N}^* \right\},$$

where $M_1 = g_{3m}(M,e)$ with M odd, and where $e \in \{1,-1\}$ for m odd and $e = -1$ for m even. It is easy to show that the families $\mathcal{S}(M)$ and $\mathcal{S}(M_1)$ are both infinite families of biquadratic fields, even if $\boldsymbol{Q}\left(\sqrt{M^2-4e}\right) = \boldsymbol{Q}\left(\sqrt{M_1^2-4e^m}\right)$.

Let us prove that $\mathcal{S}(M) \cap \mathcal{S}(M_1)$ is finite. For this purpose, we will assume the ABC conjecture. In [**3**], we defined in each field K of $\mathcal{S}(M)$ the units

$$\begin{cases} \eta_2 = h_{2n+1}(M,e) + \sqrt{h_{2n+1}^2(M,e)-1}, \\ \eta_3 = \dfrac{g_{2n+1}(M,e)}{M} + \sqrt{\left(\dfrac{g_{2n+1}(M,e)}{M}\right)^2 - 1}, \end{cases}$$

and at the beginning of this paper, we defined the units

$$\begin{cases} \eta_2 = h_{2n+1}(M_1, e^m) + \sqrt{h_{2n+1}^2(M_1, e^m) - 1}, \\ \eta_3 = \dfrac{g_{2n+1}(M_1, e^m)}{M_1} + \sqrt{\left(\dfrac{g_{2n+1}(M_1, e^m)}{M_1}\right)^2 - 1}. \end{cases}$$

We want to use S.-I. Katayama's next result [**2**].

PROPOSITION 7. *Let ε_1 (> 1) be the fundamental unit of a fixed real quadratic field k_1 and let $\eta_1 = \varepsilon_1^s$ be a fixed unit where $\eta_1 = (M + \sqrt{M^2 - 4e})/2$ with M odd. Assume the* **ABC** *conjecture and define η_2 and η_3 as above. Then $\{\varepsilon_1, \eta_2, \eta_3\}$ is a fundamental system of units of $K = \boldsymbol{Q}\left(\sqrt{M_1^2 - 4e^m}, \sqrt{h_{2n+1}^2(M_1, e^m) - 1}\right)$ except for finitely many indices m.*

Using this proposition, we shall show that the set $S(M_1) \cap S(M)$ is finite, i.e., we will have constructed (under the **ABC** conjecture) a new family of bicyclic biquadratic fields with an explicit fundamental sytem of units. If it happens that the field $K = \boldsymbol{Q}\left(\sqrt{M_1^2 - 4e^m}, \sqrt{h_{2n+1}^2(M_1, e^m) - 1}\right)$ of $\mathcal{S}(M_1)$ also belongs to $\mathcal{S}(M)$, then (under the **ABC** conjecture) we have (by comparing units)

$$h_{2n+1}(M_1, e^m) = h_{2k+1}(M, e) \quad \text{or} \quad \frac{g_{2k+1}(M, e)}{M} \tag{4.1}$$

for some k, except for finitely many indices n. Writing $w^2 = (M^2 - 4e)(M_1^2 - 4e^m)$, we get the following two cases:

$x = g_{2n+1}(M_1, e^m)$ $y = h_{2n+1}(M_1, e^m) = h_{2k+1}(M, e)$ $z = g_{2k+1}(M, e)$	$x = g_{2n+1}(M_1, e^m)$ $y = h_{2n+1}(M_1, e^m) = \frac{g_{2k+1}(M,e)}{M}$ $z = h_{2k+1}(M, e)$
$\begin{cases} x^2 - (M_1^2 - 4e^m)y^2 = 4e^m \\ z^2 - (M^2 - 4e)y^2 = 4e \end{cases}$	$\begin{cases} x^2 - (M_1^2 - 4e^m)y^2 = 4e^m \\ M^2y^2 - (M^2 - 4e)z^2 = 4e \end{cases}$
$Y^2 = (X^2 + A)(X^2 + B)$ with $\begin{cases} Y = wxz \\ X = wy \\ A = 4(M^2 - 4e)e^m \neq 0 \\ B = 4(M_1^2 - 4e^m)e \neq 0 \end{cases}$	$Y^2 = (X^2 + A)(X^2 + B)$ with $\begin{cases} Y = M(M_1^2 - 4e^m)wxz \\ X = M(M_1^2 - 4e^m)y \\ A = 4M^2(M_1^2 - 4e^m)e^m \neq 0 \\ B = 4(M_1^2 - 4e^m)^2e \neq 0 \end{cases}$

Since $Y^2 = (X^2 + A)(X^2 + B)$ has only finitely many integer solutions ($A \neq B$ in both cases), we conclude that (4.1) can hold only finitely often and that indeed the set $\mathcal{S}(M) \cap \mathcal{S}(M_1)$ is finite.

REMARK 1. It is well known that there are infinitely many real quadratic fields of Richaud–Degert type, i.e., fields of the form $\boldsymbol{Q}(\sqrt{M^2 - 4e})$. Thus, assuming the **ABC** conjecture, we conclude that there are infinitely many bicyclic biquadratic fields whose fundamental system of units is explicitly given by $\{\varepsilon_1, \eta_2, \eta_3\}$.

References

1. J. H. E. Cohn, *Squares in some recurrent sequences*, Pacific J. Math. **41** (1972), 631–646.
2. S.-I. Katayama, *The* **ABC** *conjecture and the fundamental system of units of certain real bicyclic biquadratic fields*, Proc. Japan Acad. Ser. A **75** (1999), 198–199.

3. S.-I. Katayama, C. Levesque, and T. Nakahara, *On the unit group and the class number of certain composita of two real quadratic number fields*, Manuscripta Math. **105**, (2001), 85–101.
4. T. Kubota, *Über den bizyklischen biquadratischen Zahlkörper*, Nagoya. Math. J. **10** (1956), 65–85.
5. J. H. Rickert, *Simultaneous rational approximations and related diophantine equations*, Math. Proc. Cambridge Philos. Soc. **113** (1993), 461–472.

Department of Mathematical Science, University of Tokushima, Tokushima 770-8502, Japan.

CICMA, Département de Mathématiques et de Statistique, Univ. Laval, Québec, QC G1K 7P4, Canada.

Department of Mathematics, Faculty of Science and Engineering, Saga University, Saga 840-8502, Japan.

E-mail address, S.-I. Katayama: katayama@ias.tokushima-u.ac.jp
E-mail address, C. Levesque: Claude.Levesque@mat.ulaval.ca
E-mail address, T. Nakahara: nakahara@ms.saga-u.ac.jp

Centre de Recherches Mathématiques
CRM Proceedings and Lecture Notes
Volume **36**, 2004

Belyi Parametrisations of Elliptic Curves and Congruence Defects

Chandrashekhar Khare

ABSTRACT. We consider refined integral questions about modular parametrisations of elliptic curves and raise questions about refinements of Belyi's result in the context of parametrisations of elliptic curves defined over number fields.

Given an elliptic curve E defined over a number field K it follows from Belyi's theorem (cf. [**11**, p. 71]) that there is a finite index subgroup Γ of $\mathrm{SL}_2(\mathbf{Z})$ and a surjective mapping $\mathcal{H}/\Gamma \cup \{\text{cusps}\} := X_\Gamma \to E$ of Riemann surfaces with $\mathcal{H}$ the upper half-plane. Further given any finite set of points S of E we can choose Γ and the mapping so that S is contained in the image of the cusps. This is called a *hyperbolic parametrisation* of E in Mazur's article [**8**].

We will say that such a Γ *covers* E. One may choose a Γ to have an isomorphism $\mathcal{H}/\Gamma \cup \{\text{cusps}\} := X_\Gamma \simeq E$ of Riemann surfaces. For us the more flexible notion of Γ covering E will be more useful.

In a sense the result of Belyi takes in a sense no heed of the properties of the number field K over which E is defined. We can ask for a deepening of Belyi's result by asking if we can choose Γ covering E so that it reflects some of the arithmetical properties of E, like its field of definition. It is the goal of this brief note to raise some questions about such a deepening of Belyi's theorem.

The first clue that such a deepening might exist is the Shimura–Taniyama–Weil conjecture (now a theorem, cf. [**1**]) that says in the case that E is defined over $\mathbf{Q}$ one can choose a Γ that covers E to be a *congruence subgroup* of $\mathrm{SL}_2(\mathbf{Z})$. This is what is called a *hyperbolic parametrisation of arithmetic type* in [**8**]. In the appendix to that article it is shown that if E defined over $\mathbf{Q}$ has a hyperbolic parametrisation of arithmetic type then there is an N and a non-constant map $X_0(N) \to E$ *defined over* $\mathbf{Q}$.

We begin by studying in Section 1 the better-chartered territory of modular parametrisations of elliptic curves where certain issues arise that are easy to see "up

2000 *Mathematics Subject Classification.* Primary: 11F32; Secondary: 11F23.

The author was led to ask himself the "Belyi questions" considered here when thinking about Henri Darmon's conjectures in [**2**] on "$\mathrm{SL}_2(\mathbf{Z}[1/p])$-parametrisations" of elliptic curves. The references [**8**] and [**15**] have been very helpful in formulating the questions of this note. He is grateful to Bas Edixhoven, Barry Mazur, Dipendra Prasad, Jean-Pierre Serre and Glenn Stevens for helpful correspondence.

This is the final form of the paper.

to isogeny", and that we can respond to in certain situations (see Propositions 1 and 2 below) but that we are unable to resolve satisfactorily on the nose. These issues are related to results in [**9**] controlling kernels of degeneracy maps between Jacobians of modular curves.

Throughout this note we assume the elliptic curves considered do not have CM, as in the CM case the issues we deal with about "Belyi parametrisations" in Section 2 are known (see [**13**]), and for the considerations of Section 1 this is a simplifying assumption.

1. Modular parametrisations of elliptic curves

Given a positive integer M we consider the modular curve $X_1(M)$ with its canonical model defined over $\mathbf{Q}$ (see [**12**]) in which the 0-cusp is a rational point.

DEFINITION 1. (1) Let E be an elliptic curve defined over a number field K. A K-modular parametrisation of E is a non-constant algebraic map $\phi\colon X_1(M) \to E$, for some positive integer M, defined over K that we normalise by requiring that ϕ sends the 0-cusp to the origin.

(2) Let E be an elliptic curve defined over $\overline{\mathbf{Q}}$. A $\overline{\mathbf{Q}}$-modular parametrisation of E is a K-modular parametrisation of E for some number field K.

REMARK. Our definitions are inspired by those in [**15**]. We have relaxed the condition in [**15**] that the pull-back of a differential of E is a non-zero multiple of the differential on $X_1(N)$ associated to a newform (see [**15**] for fine conjectures about these constants). This is because later we want to consider "Belyi parametrisations" in which case there is no notion of newforms. The condition that the 0-cusp be sent to the origin is included for ease of comparison with [**15**].

1.1. Q-modular parametrisations. Let E be a non-CM elliptic curve defined over $\mathbf{Q}$. It is now known that there is a $\mathbf{Q}$-modular parametrisation $X_1(N) \to E$, where N is the conductor of the elliptic curve. This induces by Picard functoriality a map $E \to J_1(N)$: this map that is induced by a $\mathbf{Q}$-modular parametrisation will be referred to again as a $\mathbf{Q}$-modular parametrisation. (We employ a similar abuse of notation for $\overline{\mathbf{Q}}$-modular parametrisations.) This is justified by noting that starting from a non-trivial homomorphism of abelian varieties $E \to J_1(N)$ defined over $\mathbf{Q}$, we obtain a map $J_1(N) \to E$ by duality and then a map $X_1(N) \to E$ defined over $\mathbf{Q}$ that is the composition $X_1(N) \to J_1(N) \to E$ where $X_1(N) \to J_1(N)$ is with respect to the 0-cusp. This map is non-trivial as the first map induces an isomorphism of cotangent spaces.

If there is a $\mathbf{Q}$-modular parametrisation $\phi\colon X_1(M) \to E$ for some positive integer M, then N divides M. For each divisor d of M/N recall that there is a map $\alpha_d^*\colon J_1(N) \to J_1(M)$ induced by Picard functoriality from the standard degeneracy map $\alpha_d\colon X_1(M) \to X_1(N)$.

In [**15**, Theorem 1.9] it is proved that any modular parametrisation $X_1(M) \to E$ (in the sense of [**15**]) factors through a modular parametrisation $X_1(N) \to E$. In loc. cit. this is proved using the requirement asked of a modular parametrisation there that the pull-back of a differential of E is the differential of a newform up to scalars (which gives the result "up to isogeny"), and the fact that the natural map $J_1(N) \to J_1(M)$ is injective (which gives the result "on the nose"). We would expect a similar property for the $\mathbf{Q}$-modular parametrisations considered here. Namely we would expect that any $\mathbf{Q}$-modular parametrisation $E \to J_1(M)$, factors through a

Q-modular parametrisation $E \to J_1(N)$ via a map $\sum_{d|M/N} a_d\alpha_d^*\colon J_1(N) \to J_1(M)$ with d divisors of M/N, and $a_d \in \mathbf{Z}$. But our relaxing one of the conditions of modular parametrisations in [**15**] introduces certain complications, as we have to contend with more endomorphisms of the Jacobian of the pro-modular curve $\widehat{X}$ of [**15**] than in loc. cit.: more explicitly what we have to control is kernels of sums of certain degeneracy maps $J_1(N) \times \cdots \times J_1(N) \to J_1(M)$. We can prove a universal property at the moment only under the additional assumption that M is square-free.

PROPOSITION 1. *Let M be a square-free integer. Any **Q**-modular parametrisation $E \to J_1(M)$ factors through a **Q** modular parametrisation $E \to J_1(N)$ via a map $\sum_{d|M/N} a_d\alpha_d^*\colon J_1(N) \to J_1(M)$ with the a_d's integers and d divisors of M/N.*

PROOF. The proposition follows "up to isogeny" easily from Eichler–Shimura theory in [**12**, Chapter 7], Carayol's theorem of the equality of the geometric and analytic conductor and the Atkin–Lehner theory of newforms: namely the decomposition up to isogeny of the Jacobians $J_1(N), J_1(M)$ into blocks formed of simple **Q**-abelian varieties A_f's with f's running through $G_{\mathbf{Q}}$ conjugacy classes of newforms, and the determination of the Galois action on the Tate modules of A_f in terms of the Fourier expansion of the newform f. The blocks arise from packets of old forms attached to f, which in turn arises from the degeneracy maps α_d recalled above. For example from this general theory it is clear that the image of E is contained in the abelian subvariety of $J_1(M)$ which is the image of $E' \times \cdots \times E' \subset J_1(N)^{\sigma_0(M/N)}$ (with E' the optimal curve in the **Q**-isogeny class of E and $\sigma_0(M/N)$ the number of divisors of M/N) in $J_1(M)$ under the sum of the degeneracy maps $\sum_{d|M/N} \alpha_d^* J_1(N) \to J_1(M)$. The more refined statement as claimed in the proposition will follow (see [**15**, Theorem 1.9]) if we show that the degeneracy map $\sum_{d|M/N} \alpha_d^* J_1(N) \to J_1(M)$ is injective. This follows from [**9**] which shows that the degeneracy map $J_1(N')^2 \to J_1(N'p)$ is injective if p is a prime that does not divide N', for any positive integer N'. □

REMARKS. (1) By using the method of [**15**, proof of Theorem 1.6] one may show that for arbitrary M a **Q**-modular parametrisation $\phi\colon X_1(M) \to E$ is such that $\phi^*(\omega_E) = (\sum_{d|M/N} a_d\alpha_d^*)(f)$ where f is the newform associated to E and ω_E a Neron differential of E and $a_d \in \mathbf{Z}$.

(2) The reason that we have to assume that M is square-free (or what is really needed to apply [**9**], the assumption that $(N, M/N) = 1$ and M/N is square-free) is that in [**5**] (where the analog of the result of [**9**] is proven in the case when $p \mid N'$ employing the notation of the proof above) the groups of connected components of kernels of the degeneracy maps considered there are controlled only up to Eisenstein "errors". If this result could be refined as suggested there ([**5**, Remark 2.4, p. 641]) to show that these groups of connected components are images of Shimura subgroups (see [**7**]) we could drop the assumption that M is square-free.

1.2. $\overline{\mathbf{Q}}$-modular parametrisations. Let E be a (non-CM) elliptic curve now defined over $\overline{\mathbf{Q}}$. $\overline{\mathbf{Q}}$-modular parametrisations are considered in Ribet's article [**10**]. There it is shown that if E has a $\overline{\mathbf{Q}}$-modular parametrisation then E is a **Q**-curve (see loc. cit. for the definition).

As a partial converse Ribet showed that any $\mathbf{Q}$-curve arises as a factor of a GL_2-type abelian variety defined over $\mathbf{Q}$ (see loc. cit.): for this the main ingredient in [**10**] is the result of Tate that for the absolute Galois group G_K of a number field K, $H^2(G_K, \overline{\mathbf{Q}}^*)$ vanishes where the coefficients have trivial G_K-action. Conjecturally (for instance Serre's conjectures imply this) GL_2-type abelian varieties over $\mathbf{Q}$ are modular, and thus it would follow that all $\mathbf{Q}$-curves E have $\overline{\mathbf{Q}}$-modular parametrisations. Recent work of Hida [**4**, Chapter 5.2] and Ellenberg–Skinner, cf. [**3**], proves much of this conjecture.

By the result of Tate and the fact that E is non-CM, for a sufficiently small open subgroup G_K of $G_{\mathbf{Q}}$ the p-adic representation $G_K \to \mathrm{GL}_2(\mathbf{Q}_p)$ attached to a $\mathbf{Q}$-curve E extends to a representation $G_{\mathbf{Q}} \to \overline{\mathbf{Q}_p}^* \mathrm{GL}_2(\mathbf{Q}_p)$ that is unique up to twists. Let E have a $\overline{\mathbf{Q}}$-modular parametrisation. Let $\rho_E \colon G_{\mathbf{Q}} \to \overline{\mathbf{Q}_p}^* \mathrm{GL}_2(\mathbf{Q}_p)$ be a representation that extends the G_K-representation afforded by the p-adic Tate module of E and that has minimal conductor (say N) amongst its twists. Recall that given a newform f and a Dirichlet character χ we can form its twist $f \otimes \chi$ (which may no longer be a newform). The Galois representations "associated" to these modular forms (that are eigenforms for almost all Hecke operators) are related by twisting by χ: this map (up to a Gauss sum factor) is induced by a certain map of Jacobians of modular curves R_χ (see [**14**]). The following proposition is known (for the second part see the appendix of [**8**]).

Proposition 2. *Let E be a non-*CM *elliptic curve defined over a number field K.*

(1) *If $\phi \colon X_1(M) \to E$ is a $\overline{\mathbf{Q}}$-modular parametrisation then M is divisible by N and the representation ρ_E is attached to a newform f. Further there is such a parametrisation with $M = N$. The pull-back of a differential on E under ϕ is in the $\overline{\mathbf{Q}}$-linear span of images of f under compositions of the degeneracy maps α_d^* and twisting maps R_χ.*
(2) *If E has a $\overline{\mathbf{Q}}$-modular parametrisation $X_1(M') \to E$ for some integer M', then it also has a K-modular parametrisation $X_1(M) \to E$ for some (possibly different) integer M.*

Proof. The first part follows from what was said earlier, E being non-CM and the Eichler–Shimura theory of the decomposition of the Jacobians of modular curves as under the hypothesis ρ_E is the Eichler–Shimura representation attached to a newform. We now take up the second part. Let $X_1(M') \to E$ be a $\overline{\mathbf{Q}}$-modular parametrisation for some integer M'. Then as $J_1(M')$ breaks (up to isogeny) over K into the product of abelian varieties A_f attached to conjugacy classes of Hecke eigenforms (which may not be simple) we deduce from Eicheler–Shimura theory that there is a weight 2 Hecke eigenform f such that the $\wp$-adic G_K-representation $\rho_{f,\wp}$ attached to f is isomorphic to the p-adic representation $\rho_{E,p}$ associated to E on restriction to an open subgroup of $G_{\mathbf{Q}}$ with $\wp$ a place above p. By the hypothesis on E it follows that this restriction is irreducible and thus the G_K-representations $\rho_{f,\wp}$ and $\rho_{E,p}$ are isomorphic up to twisting. Then considering a certain twist g of f we get that $\mathrm{Hom}_{G_K}(\mathrm{Ta}_p(E), \mathrm{Ta}_p(J_1(M)))$ is non-zero for some M and using Falting's isogeny we get that there is a morphism defined over K from $E \to J_1(M)$. Dualising this map and (pre)composing with the morphism $X_1(M) \to J_1(M)$ with respect to the 0-cusp the proposition follows. □

One would want to prove that any $\overline{\mathbf{Q}}$-modular parametrisation $\phi\colon X_1(M) \to E$ of $\mathbf{Q}$-curves E induces a map $E \to J_1(M)$ that factors through a $\overline{\mathbf{Q}}$-modular parametrisation $E \to J_1(N)$ via a morphism $J_1(N) \to J_1(M)$ (recall that N is the least conductor of twists of $G_{\mathbf{Q}}$-representations which extend the p-adic representation attached to E of G_K where K is any field over which E gets defined): this is again easy to prove "up to isogeny". By considering the pull-back ϕ^* on differentials we see that this morphism will be a "$\overline{\mathbf{Q}}$-linear combination" of compositions of morphisms α_d^*'s and R_χ's for some integers d and Dirichlet characters χ.

In the context of Proposition 1 we can prove a weaker universal property without the assumption that M is square-free. We state this result while omitting its proof which is along the lines of Proposition 1 and uses the results of both [**5, 9**] (see also the remarks after Proposition 1), and some of the considerations in this section. We would hope that it will be possible to prove a cleaner result in the future.

PROPOSITION 3. *Let K be a number field whose normal closure does not contain any non-trivial abelian extension of $\mathbf{Q}$. Let E be a non-*CM *elliptic curve over K and p be a "good" prime, i.e., the semi-simplification of $\rho_{E,p}$, does not come by restriction from an abelian $G_{\mathbf{Q}}$-representation. Consider a K-modular parametrisation $E \to J_1(M)$ and the induced G_K-equivariant map $\mathrm{Ta}_p(E) \to \mathrm{Ta}_p(J_1(M))$ on p-adic Tate modules. Then the latter map factors through a G_K-equivariant map $\mathrm{Ta}_p(J_1(N)) \to \mathrm{Ta}_p(J_1(M))$.*

REMARKS. (1) This is an easy result for $\mathbf{Q}$-parametrisations as the result is non-trivial only for primes p such that the mod p representation is not irreducible as a G_K-module. The result is again easy up to isogeny, i.e., up to tensoring with $\mathbf{Q}_p$, but integrally does not follow from generalities as in non-trivial cases the G_K representation afforded by the p-adic Tate module of E is residually reducible.

(2) The condition that K does not contain non-trivial abelian extensions of $\mathbf{Q}$ avoids difficulties arising from the twisting operators R_χ. We cannot do better than this as we do not have exact control of kernels of twisting maps R_χ (see [**6**] for a result in this direction), and even less of kernels of sums of twisting maps and degeneracy maps. We would expect a universal property as in the proposition above to hold without any assumptions on K (i.e., for $\overline{\mathbf{Q}}$-modular parametrisations). This time the "good" p would be those such that the semi-simplification of the $G_{K\mathbf{Q}^{\mathrm{ab}}}$-module $E[p]$ does not come via restriction from an abelian $G_{\mathbf{Q}^{\mathrm{ab}}}$-representation: note that the maps R_χ are defined over $\mathbf{Q}^{\mathrm{ab}}$.

2. Belyi parametrisations of elliptic curves

Now leaving the highly structured world of modular curves we consider curves that arise from arbitrary finite index subgroups of $\mathrm{SL}_2(\mathbf{Z})$. As seen above this is forced on us when considering hyperbolic parametrisations of non-$\mathbf{Q}$ elliptic curves E defined over $\overline{\mathbf{Q}}$.

DEFINITION 2. Let E be an elliptic curve defined over $\overline{\mathbf{Q}}$.

(1) A Belyi parametrisation of E is a non-constant algebraic morphism $X_\Gamma \to E$ (defined over $\overline{\mathbf{Q}}$), with X_Γ the projective curve (defined over $\overline{\mathbf{Q}}$) associated

to a subgroup of finite index Γ of $\mathrm{SL}_2(\mathbf{Z})$, that we normalise by requiring that the 0-cusp is sent to the origin. We say that such a Γ covers E.
(2) Given a finite index subgroup Γ of $\mathrm{SL}_2(\mathbf{Z})$ its *congruence hull* Γ^c is the smallest congruence subgroup which contains Γ: thus Γ^c is the intersection of all congruence subgroups of $\mathrm{SL}_2(\mathbf{Z})$ which contain Γ. The congruence defect cd_Γ of Γ is the index $[\Gamma^c : \Gamma]$.
(3) The congruence defect cd_E of E is the smallest congruence defect of a finite index subgroup Γ of $\mathrm{SL}_2(\mathbf{Z})$ that covers E.

By Belyi's theorem, a Belyi parametrisation of E always exists and thus cd_E is a well-defined invariant associated to the isomorphism class of E (or equivalently to its j-invariant): in fact cd_E depends only on the $\overline{\mathbf{Q}}$-isogeny class of E.

QUESTION 1. Given a number field K is there a constant c_K, such that for all elliptic curves E defined over K, $cd_E \leq c_K$?

The STW conjecture answers this affirmatively when $K = \mathbf{Q}$: all elliptic curves that are defined over $\mathbf{Q}$ have congruence defect 1. This is the only piece of evidence for an affirmative answer to the question. We have seen above that one has a (partly conjectural) characterisation of the class of E's with congruence defect 1 as precisely the class of $\mathbf{Q}$-curves. From what we have seen above the appropriate field when considering these questions is:

DEFINITION 3. The $\mathbf{Q}$-field of an elliptic curve E defined over $\overline{\mathbf{Q}}$ is the fixed field of the open subgroup of $G_{\mathbf{Q}}$ which fixes the $\overline{\mathbf{Q}}$-isogeny class of E.

Thus the $\mathbf{Q}$-field of E is $\mathbf{Q}$ precisely when E is a $\mathbf{Q}$-curve. A refinement of the question above would be:

QUESTION 2. Given a number field K is there a constant c'_K, such that any elliptic curve with $\mathbf{Q}$-field K, $cd_E \leq c'_K$?

REMARKS. (1) Our questions are naive from the point of view of modular parametrisations (especially from the modern automorphic viewpoint) as say when wanting to parametrise E defined over a totally real number field K of odd degree by quotients of the upper half plane one switches to a quaternion algebra defined over K ramified at all but one infinite place and considers Shimura curves arising from congruence subgroups of this quaternion algebra. Note on the other hand that when the $\mathbf{Q}$-field of E is not totally real one does not have in a direct fashion an "automorphic parametrisation".
(2) It would be of interest to gather computational evidence towards an answer to the questions above. J.-P. Serre in an e-mail to the author in response to a message posing the first question above suggested testing out the question in the following situation: Consider $K = \mathbf{Q}(i)$, and a specific elliptic curve over K, say one with its j-invariant $1/(n+i)$, $n \geq 2$. Such a curve cannot be uniformized by a congruence subgroup, as since its j-invariant is $1/(n+i)$ it has multiplicative reduction at primes dividing $n+i$ and hence cannot be geometrically isogenous, i.e., isogenous over $\overline{\mathbf{Q}}$, to its conjugate. What kind of function of n is the congruence defect of such a curve?

(3) It will be interesting to see if answers to the questions above have interesting diophantine consequences similar to the rich diophantine consequences of existence of modular parametrisations of elliptic curves.

(4) In the enriched situation of modular parametrisations that we considered earlier, by the associated Eichler–Shimura theory, all such paramaterisations were "related up to isogeny". There is no question of expecting such a property of the ensemble of Belyi parametrisations of an elliptic curve E as there is just too much freedom (in for instance the choice of the map ϕ above, or in choosing the set S of E which will be in the image of the cusps under such a parametrisation). In other words we cannot expect that there is one Belyi parametrisation $X_\Gamma \to E$ which "generates" all others.

References

1. C. Breuil, B. Conrad, F. Diamond, and R. Taylor, *On the modularity of elliptic curves over* **Q**: *wild 3-adic exercises*, J. Amer. Math. Soc. **14** (2001), 843–939.
2. H. Darmon, *Integration on $\mathcal{H}_p \times \mathcal{H}$ and arithmetic applications*, Ann. of Math. (2) **154** (2001), no. 3, 589–639.
3. J. Ellenberg and C. Skinner, *On the modularity of* **Q**-*curves*, Duke Math. J. **109** (2001), 97–122.
4. H. Hida, *Geometric modular forms and elliptic curves*, World Scientific Publishing Co., Inc., River Edge, NJ, 2000.
5. C. Khare, *Congruences between cusp forms: the (p, p) case*, Duke Math. J. **80** (1995), 631–667.
6. ———, *Maps between Jacobians of modular curves*, J. Number Theory **62** (1997), 107–114.
7. S. Ling and J. Oesterlé, *The Shimura subgroup of $J_0(N)$*, Astérisque **196–197** (1991), 171–203.
8. B. Mazur, *Number theory as gadfly*, Amer. Math. Monthly **98** (1991), 593–610.
9. K. Ribet, *Congruence relations between modular forms*, Proceedings of the International Congress of Mathematicians, vol. 1, 2 (Warsaw, 1983), PWN, Warsaw, 1984, pp. 503–514.
10. ———, *Abelian varieties over* **Q** *and modular forms*, Algebra and Topology 1992 (Taejŏn), Korea Adv. Inst. Sci. Tech., Taejŏn, 1992, pp. 53–79,
11. J.-P. Serre, *Lectures on the Mordell–Weil theorem*, Aspects of Mathematics, E15. Friedr. Vieweg & Sohn, Braunschweig, 1989.
12. G. Shimura, *Introduction to the arithmetic theory of automorphic functions*, Kanô Memorial Lectures, no. 1. Publ. Math. Soc. of Japan, no. 11, Iwanami Shoten, Publishers, Tokyo, Princeton University Press, Princeton, N.J., 1971.
13. ———, *On elliptic curves with complex multiplication as factors of the Jacobians of modular function fields*, Nagoya Math. J. **43** (1973), 199–208.
14. ———, *On the factors of the Jacobian variety of a modular function field*, J. Math. Soc. Japan **25** (1973), 523–544.
15. G. Stevens, *Stickelberger elements and modular parametrisations of elliptic curves*, Invent. Math. **98** (1989), 75–106.

School of Mathematics, TIFR, Homi Bhabha Road, Mumbai 400 005, India.

Current address: Department of Mathematics, University of Utah, 155 South 1400 East, Salt Lake City, UT 84112–0090, USA.

E-mail address: shekhar@math.utah.edu; shekhar@math.tifr.res.in

Centre de Recherches Mathématiques
CRM Proceedings and Lecture Notes
Volume **36**, 2004

An Application of Exterior Square Functoriality of GL_4; Asai Lift

Henry H. Kim

1. Introduction

Let K/F be a quadratic extension of number fields with the rings of adeles $\mathbb{A}_K, \mathbb{A}_F$, resp. The functoriality of the Asai lift from cuspidal representations of GL_2/K to automorphic representations of GL_4/F, has been proved by Ramakrishnan [**11**] by integral representation technique, and by Krishnamurthy [**8**] using Langlands–Shahidi method in his thesis. In this paper we prove it as an application of the exterior square lift of GL_4, which was established in [**6**].[1] We thank the referee for several useful comments.

2. Asai lift

Let $\mathrm{Gal}(K/F) = \{1, \theta\}$ be the Galois group. Let $K_v = K \otimes_F F_v$. If v remains a prime, K_v/F_v is a quadratic extension. If v splits in K, $K_v \simeq F_v \oplus F_v$. In this case, we define the "Galois" automorphism $x = (x_1, x_2) \mapsto \overline{x} = (x_2, x_1)$. Let $\omega = \omega_{K/F}$ be the quadratic grössencharacter of F attached to K/F by class field theory. Then $\omega_v = 1$ if v splits. If v remains a prime, it is defined in the following way: let ϖ_{K_v}, ϖ_{F_v} be uniformizers of K_v, F_v, resp. Then $\varpi_{K_v} = \varpi_{F_v}$, and $|\varpi_{K_v}|_{K_v} = q_{K_v}^{-1}$, and $|\varpi_{F_v}|_k = q_{F_v}^{-1}$. We have $q_{K_v} = q_{F_v}^2$. Hence for any $x \in K_v$, $|x|_{K_v} = |N_{K_v/F_v}(x)|_{F_v}$, and ${F_v}^\times/N_{K_v/F_v}({K_v}^\times)$ has order 2. Let $\omega_v = \omega_{K_v/F_v}$ be the character of ${F_v}^\times$ such that $\omega_{K_v/F_v} = \begin{cases} 1, & \text{on } N_{K_v/F_v}(K_v^\times) \\ -1, & \text{otherwise} \end{cases}$.

Let $\mathbf{G} = \mathrm{Res}_{K/F}\,\mathrm{GL}_n$ be the quasi-split group obtained by the restriction of scalars. Then the L-group is

$$^L\mathbf{G} = \big(\mathrm{GL}_n(\mathbb{C}) \times \mathrm{GL}_n(\mathbb{C})\big) \rtimes \mathrm{Gal}(K/F),$$

where θ acts by interchanging the two factors. One defines a representation

$$r\colon {}^L\mathbf{G} \to \mathrm{GL}(\mathbb{C}^n \otimes \mathbb{C}^n) \simeq \mathrm{GL}_{n^2}(\mathbb{C}) = {}^L\,\mathrm{GL}_{n^2},$$

2000 *Mathematics Subject Classification.* Primary: 11F70; Secondary: 22E55.

This is the final form of the paper.

Partially supported by an NSERC grant.

[1] D. Ramakrishnan informed me that he also obtained the same result along the line of this paper besides his own earlier result.

by setting, for all $x, y \in \mathbb{C}^n$,

$$r(g, g'; 1)(x \otimes y) = g(x) \otimes g'(y); \quad r(1, 1; \theta)(x \otimes y) = y \otimes x.$$

Let $\pi = \otimes_v \pi_v$ be a cuspidal automorphic representation of $\mathbf{G}(\mathbb{A}_F) = \mathrm{GL}_n(\mathbb{A}_K)$. Let $\phi_v \colon W_{F_v} \times \mathrm{SL}_2(\mathbb{C}) \to {}^L\mathbf{G}$ be the parametrization of π_v for each v, given by the local Langlands correspondence [**4, 5, 9**]. Then we obtain a map $r \circ \phi_v \colon W_{F_v} \times \mathrm{SL}_2(\mathbb{C}) \to \mathrm{GL}_{n^2}(\mathbb{C})$. Let $\mathrm{As}(\pi_v)$ be the irreducible admissible representation of $\mathrm{GL}_{n^2}(F_v)$, attached to $r \circ \phi_v$ by the local Langlands correspondence. Then $\mathrm{As}(\pi) = \otimes_v \mathrm{As}(\pi_v)$ is an irreducible admissible representation of $\mathrm{GL}_{n^2}(\mathbb{A}_F)$. Langlands' functoriality predicts that $\mathrm{As}(\pi)$ is an automorphic representation.

In this paper, we will be concerned with $n = 2$. In this case, we can make it more explicit: let v be any unramified finite place of F such that π_w is unramified at any place w of K above v. Suppose v splits into (w_1, w_2) in K, and let the semi-simple conjugacy classes for π_{w_1} and π_{w_2} be represented by $\mathrm{diag}(\alpha_{w_1}, \beta_{w_1})$ and $\mathrm{diag}(\alpha_{w_2}, \beta_{w_2})$, resp. Then the semi-simple conjugacy class for $\mathrm{As}(\pi_v)$ is represented by $\mathrm{diag}(\alpha_{w_1}\alpha_{w_2}, \alpha_{w_1}\beta_{w_2}, \beta_{w_1}\alpha_{w_2}, \beta_{w_1}\beta_{w_2})$. Namely, $\mathrm{As}(\pi_v) = \pi_{w_1} \boxtimes \pi_{w_2}$.

Suppose v is inert, and let the semi-simple conjugacy class for π_v be represented by $\mathrm{diag}(\alpha_v, \beta_v)$. Then the semi-simple conjugacy class for $\mathrm{As}(\pi_v)$ is represented by

$$\begin{pmatrix} \alpha_v & 0 & 0 & 0 \\ 0 & 0 & \alpha_v & 0 \\ 0 & \beta_v & 0 & 0 \\ 0 & 0 & 0 & \beta_v \end{pmatrix}.$$

The Asai lift is different from the automorphic induction (cf. [**1**]). Automorphic induction corresponds to the L-group homomorphism

$$r \colon {}^L\mathrm{Res}_{K/F}\,\mathrm{GL}_n = (\mathrm{GL}_n(\mathbb{C}) \times \mathrm{GL}_n(\mathbb{C})) \rtimes \mathrm{Gal}(K/F) \to \mathrm{GL}(\mathbb{C}^n \oplus \mathbb{C}^n) \simeq \mathrm{GL}_{2n}(\mathbb{C}),$$
$$r(g, g'; 1)(x \oplus y) = g(x) \oplus g'(y), \quad r(1, 1; \theta)(x \oplus y) = y \oplus x.$$

Given $\pi = \otimes_v \pi_v$ be a cuspidal representation of $\mathrm{GL}_n(\mathbb{A}_K)$, we can define $I_K^F(\pi)$ as in the case of $\mathrm{As}(\pi)$; $I_K^F(\pi) = \otimes_v I_K^F(\pi_v)$, where $I_K^F(\pi_v)$ is the irreducible admissible representation of $\mathrm{GL}_{2n}(F_v)$, attached to $r \circ \phi_v$ by the local Langlands correspondence. It is known [**1**] that $I_K^F(\pi) = \otimes I_K^F(\pi_v)$ is an automorphic representation of $\mathrm{GL}_{2n}(\mathbb{A}_F)$; $I_K^F(\pi)$ is not cuspidal if and only if $\pi \simeq \pi \circ \theta$. Here $\pi \circ \theta$ denotes the cuspidal representation of $\mathrm{GL}_n(\mathbb{A}_K)$ such that if $\pi_w = \pi(\mu_1, \ldots, \mu_n)$ is an unramified representation, then $(\pi \circ \theta)_w = \pi(\mu_1 \circ \theta, \ldots, \mu_n \circ \theta)$. If $\pi \simeq \pi \circ \theta$, then $\pi = \sigma_K$, the base change of a cuspidal representation σ of $\mathrm{GL}_n(\mathbb{A}_F)$ to K. And $I_K^F(\sigma_K) = \sigma \boxplus (\sigma \otimes \omega_{K/F})$.

Let us make it more explicit when $n = 2$. Let v be any unramified finite place of F such that π_w is unramified at any place w of K above v. Suppose v splits into (w_1, w_2) in K, and let the semi-simple conjugacy classes for π_{w_1} and π_{w_2} be represented by $\mathrm{diag}(\alpha_{w_1}, \beta_{w_1})$ and $\mathrm{diag}(\alpha_{w_2}, \beta_{w_2})$, resp. Then the semi-simple conjugacy class for $I_K^F(\pi_v)$ is represented by $\mathrm{diag}(\alpha_{w_1}, \beta_{w_1}, \alpha_{w_2}, \beta_{w_2})$. Namely, $I_K^F(\pi_v) = \pi_{w_1} \boxplus \pi_{w_2}$.

Suppose v is inert, and let the semi-simple conjugacy class for π_v be represented by $\mathrm{diag}(\alpha_v, \beta_v)$. Then the semi-simple conjugacy class for $I_K^F(\pi_v)$ is represented by

$$\begin{pmatrix} 0 & 0 & \alpha_v & 0 \\ 0 & 0 & 0 & \beta_v \\ 1 & 0 & 0 & 0 \\ 0 & 1 & 0 & 0 \end{pmatrix}.$$

In other words, if $\pi_v = \pi(\mu, \nu)$, then

$$I_K^F(\pi_v) = \pi\big(\mu|_{F_v^\times}^{1/2}\omega_{K_v/F_v}(\mu|_{F_v^\times})^{1/2}, \nu|_{F_v^\times}^{1/2}\omega_{K_v/F_v}(\nu|_{F_v^\times})^{1/2}\big).$$

Hence it satisfies the adjoint relation: for any representation π' of $\mathrm{GL}_2(F_v)$,

$$L\big(s, \pi' \times I_K^F(\pi_v)\big) = L(s, \pi'_{K_v} \times \pi_v),$$

where π'_{K_v} is the base change of π' to K_v.

3. Main result

We are motivated by the following identity from [**11**]: if $\sigma\colon \mathrm{Gal}(\overline{F}/K) \to \mathrm{GL}_2(\mathbb{C})$ is an irreducible representation,

$$\wedge^2\big(\mathrm{Ind}_K^F(\sigma)\big) = \big(\mathrm{As}(\sigma) \otimes \omega_{K/F}\big) \oplus \mathrm{Ind}_K^F(\det \sigma),$$

where $\mathrm{Ind}_K^F(\sigma)$, $\mathrm{Ind}_K^F(\det \sigma)$ are the induced representations.

Let π be a cuspidal representation of $\mathrm{GL}_2(\mathbb{A}_K)$. Let $I_K^F(\pi)$ be the automorphic induction. It is an automorphic representation of $\mathrm{GL}_4(\mathbb{A}_F)$. Let $I_K^F(\omega_\pi)$ be the automorphic induction which is an automorphic representation of $\mathrm{GL}_2(\mathbb{A}_F)$. In light of the cuspidality criterion in [**11**], we can assume that $I_K^F(\pi), I_K^F(\omega_\pi)$ are cuspidal representations. Recall the exterior square functoriality from [**6**]: by the local Langlands' correspondence, we can obtain an irreducible admissible representation $\wedge^2\big(I_K^F(\pi)\big) = \otimes_v \wedge^2\big(I_K^F(\pi_v)\big)$ of $\mathrm{GL}_6(\mathbb{A}_F)$. Langlands' functoriality predicts that $\wedge^2\big(I_K^F(\pi)\big)$ is an automorphic representation of $\mathrm{GL}_6(\mathbb{A}_F)$. In [**6**], we proved a weaker result. Namely,

THEOREM 1 ([**6**]). *Let T be a set of places such that $v \mid 2, 3$ and $I_K^F(\pi_v)$ is a supercuspidal representation of $\mathrm{GL}_4(F_v)$. Then there exists an automorphic representation $\Pi = \otimes_v \Pi_v$ of $\mathrm{GL}_6(\mathbb{A}_F)$ such that $\Pi_v \simeq \wedge^2\big(I_K^F(\pi_v)\big)$ for all $v \notin T$.*

Let S be a finite set of places, including archimedean places, such that everything is unramified outside S. We prove

PROPOSITION 1. *The L-function $L_S\big(s, I_K^F(\omega_\pi^{-1}) \otimes I_K^F(\pi), \rho_2 \otimes \wedge^2\rho_4\big)$ has a pole at $s = 1$. Here ρ_m is the standard representation of $\mathrm{GL}_m(\mathbb{C})$.*

PROOF. We prove the adjoint formula:

$$L_S\big(s, I_K^F(\omega_\pi^{-1}) \otimes I_K^F(\pi), \rho_2 \otimes \wedge^2\rho_4\big) = L_S\big(s, \widetilde{\pi} \times (\pi \circ \theta)\big) L_S(s, 1) L_S\big(s, \omega_\pi^{-1}(\omega_\pi \circ \theta)\big),$$

where $L_S(s, 1)$ and $L_S\big(s, \omega_\pi^{-1}(\omega_\pi \circ \theta)\big)$ are Hecke L-functions of K.

Since $L_S(s, 1)$ has a pole at $s = 1$ and the other L-functions do not have zeros at $s = 1$, our assertion follows. For this, we only need to prove the local identity for each unramified v:

$$L_v\big(s, I_K^F(\omega_\pi^{-1}) \otimes I_K^F(\pi), \rho_2 \otimes \wedge^2\rho_4\big) = L_v\big(s, \widetilde{\pi} \times (\pi \circ \theta)\big) L_v(s, 1) L_v\big(s, \omega_\pi^{-1}(\omega_\pi \circ \theta)\big).$$

First, suppose v splits into (w_1, w_2) in K. Let the semi-simple conjugacy classes for π_{w_1} and π_{w_2} be represented by $\mathrm{diag}(\alpha_{w_1}, \beta_{w_1})$ and $\mathrm{diag}(\alpha_{w_2}, \beta_{w_2})$, resp. Then the semi-simple conjugacy class for $I_K^F(\pi_v)$ is represented by

$$\mathrm{diag}(\alpha_{w_1}, \beta_{w_1}, \alpha_{w_2}, \beta_{w_2}).$$

The semi-simple conjugacy class for $I_K^F(\omega_{\pi_v}^{-1})$ is represented by

$$\mathrm{diag}\big((\alpha_{w_1}\beta_{w_1})^{-1}, (\alpha_{w_2}\beta_{w_2})^{-1}\big).$$

Our assertion follows from direct computation, by noting that the right hand side is

$$L(s, \widetilde{\pi}_{w_1} \times \pi_{w_2})L(s, \pi_{w_1} \times \widetilde{\pi}_{w_2})L_{w_1}(s, 1)L_{w_2}(s, 1)L(s, \omega_{\pi_{w_1}}^{-1}\omega_{\pi_{w_2}})L(s, \omega_{\pi_{w_2}}^{-1}\omega_{\pi_{w_1}}).$$

Second, v is inert. Let the semi-simple conjugacy class for π_v be represented by $\mathrm{diag}(\alpha_v, \beta_v)$. Then the semi-simple conjugacy class for $I_K^F(\pi_v)$ is represented by $\mathrm{diag}(\alpha_v^{1/2}, -\alpha_v^{1/2}, \beta_v^{1/2}, -\beta_v^{1/2})$ after diagonalization. Likewise, the semi-simple conjugacy class for $I_K^F(\omega_{\pi_v}^{-1})$ is represented by $\mathrm{diag}\big((\alpha_v\beta_v)^{-1/2}, -(\alpha_v\beta_v)^{-1/2}\big)$. Our assertion follows from direct computation, by noting that the right hand side is

$$L(s, \widetilde{\pi}_v \times \pi_v)L_v(s, 1)^2. \qquad \square$$

Proposition 1 implies that $\Pi = \Pi(\pi) \boxplus I_K^F(\omega_\pi)$, where $\Pi(\pi)$ is an automorphic representation of $\mathrm{GL}_4(\mathbb{A}_F)$. Also we see easily that $\Pi(\pi)_v \simeq \mathrm{As}(\pi_v) \otimes \omega_{K_v/F_v}$ if π_v is unramified. We now prove

THEOREM 2. $\Pi(\pi) \simeq \mathrm{As}(\pi) \otimes \omega_{K/F}$. *Hence* $\mathrm{As}(\pi)$ *is an automorphic representation of* $\mathrm{GL}_4(\mathbb{A}_F)$.

PROOF. In light of the result in [**6**], we only need to verify that $\Pi(\pi)_v \simeq \mathrm{As}(\pi_v) \otimes \omega_{K_v/F_v}$, when $v \mid 2, 3$ and $I_K^F(\pi_v)$ is a supercuspidal representation of $\mathrm{GL}_4(F_v)$. In this case π_v is a supercuspidal representation of $\mathrm{GL}_2(K_v)$. We need to show by the local converse theorem [**2**] that, for every supercuspidal representation σ_v of $\mathrm{GL}_m(F_v)$, $m = 1, 2$,

$$\gamma\big(s, \sigma_v \times \Pi(\pi)_v, \psi_v\big) = \gamma\big(s, \sigma_v \times \mathrm{As}(\pi_v) \otimes \omega_{K_v/F_v}, \psi_v\big).$$

We follow the proof of [**6**, Proposition 5.1.4]. We use the following setup: let k be a non-archimedean local field of characteristic zero, and E/k be a quadratic extension. Since $m = 1$ is easy, we deal with $m = 2$.

Let η_1, η_2 be supercuspidal representations of $\mathrm{GL}_2(k), \mathrm{GL}_2(E)$, resp. with corresponding parametrizations $\tau_1 \colon W_k \to \mathrm{GL}_2(\mathbb{C}), \tau_2 \colon W_E \to \mathrm{GL}_2(\mathbb{C})$, resp. We can think of τ_1, τ_2 as a representation of $\mathrm{Gal}(\bar{k}/k), \mathrm{Gal}(\bar{k}/E)$, resp. We need to show that

$$\gamma(s, \eta_1 \otimes (I_E^k \eta_2), \rho_2 \otimes \wedge^2 \rho_4, \psi) = \gamma(Fs, \tau_1 \otimes (\mathrm{Ind}_E^k \tau_2), \rho_2 \otimes \wedge^2 \rho_4, \psi),$$

As in the proof of [**6**, Proposition 5.1.4], by using [**10**, Lemma 3, p. 320], we can find a quadratic extension of number fields K/F with $E = K_v, k = F_v$ and irreducible 2-dimensional representations σ_1, σ_2 of $\mathrm{Gal}(\overline{F}/F)$, $\mathrm{Gal}(\overline{F}/K)$ with solvable image such that $\sigma_{iv} = \tau_i$. (Let us be more precise. Let L_1/k, L_2/E denote the Galois extensions cut out by the kernel of τ_1, τ_2, resp. Let L be the compositum of L_1, L_2, and let $\tilde{L}/k$ be the Galois closure of L/k. Applying [**10**, Lemma 3, p. 320], we obtain a Galois extension $\widetilde{M}/F$ of number fields with local extension $\tilde{L}/k$ such that $\mathrm{Gal}(\widetilde{M}/F) = \mathrm{Gal}(\tilde{L}/k)$. Denote the fixed fields in $\widetilde{M}$ of $\mathrm{Gal}(\tilde{L}/L)$, $\mathrm{Gal}(\tilde{L}/L_1)$, $\mathrm{Gal}(\tilde{L}/L_2)$, $\mathrm{Gal}(\tilde{L}/E)$ by M, M_1, M_2, K, resp. Then we can see that L_1, L_2 are local completions of M_1, M_2 with $\mathrm{Gal}(L_1/k) = \mathrm{Gal}(M_1/F)$, $\mathrm{Gal}(L_2/E) = \mathrm{Gal}(M_2/K)$. Thus we obtain continuous, irreducible representations σ_1, σ_2 of $\mathrm{Gal}(\overline{F}/F)$, $\mathrm{Gal}(\overline{F}/K)$, acting via the respective quotients $\mathrm{Gal}(M_1/F)$, $\mathrm{Gal}(M_2/K)$ such that $\sigma_{iv} = \tau_i$.) The global Langlands correspondence is available for those representations with solvable image, and hence we can find corresponding cuspidal representations π_1, π_2 of $GL_2(\mathbb{A}_F)$, $GL_2(\mathbb{A}_K)$ such that $\pi_{iv} = \eta_i$. We compare the functional equations of $L\big(s, \pi_1 \otimes (I_K^F \pi_2), \rho_2 \otimes \wedge^2 \rho_4\big)$ and $L\big(s, \sigma_1 \otimes (\mathrm{Ind}_K^F \sigma_2), \rho_2 \otimes \wedge^2 \rho_4\big)$. Even though we do not know the holomorphy of $L\big(s, \sigma_1 \otimes (\mathrm{Ind}_K^F \sigma_2), \rho_2 \otimes \wedge^2 \rho_4\big)$, the functional equation is known and it suffices for our purpose. Since $L\big(s, \pi_{1w} \otimes (I_K^F \pi_2)_w, \rho_2 \otimes \wedge^2 \rho_4\big) = L\big(s, \sigma_{1w} \otimes (\mathrm{Ind}_K^F \sigma_2)_w, \rho_2 \otimes \wedge^2 \rho_4\big)$ for unramified places, we have an equality

$$\prod_{u \in S} \gamma\big(s, \pi_{1u} \otimes (I_K^F \pi_2)_u, \rho_2 \otimes \wedge^2 \rho_4, \psi_u\big) = \prod_{u \in S} \gamma\big(s, \sigma_{1u} \otimes (\mathrm{Ind}_K^F \sigma_2)_u, \rho_2 \otimes \wedge^2 \rho_4, \psi_u\big).$$

By choosing a grössencharacter which is highly ramified at all the ramified places except v and which is 1 at v, we obtain

$$\gamma\big(s, \pi_{1v} \otimes (I_K^F \pi_2)_v, \rho_2 \otimes \wedge^2 \rho_4, \psi_v\big) = \gamma\big(s, \sigma_{1v} \otimes (\mathrm{Ind}_K^F \sigma_2)_v, \rho_2 \otimes \wedge^2 \rho_4, \psi_v\big).$$

Our assertion follows. □

We refer [**11**] for the cuspidality criterion of $\mathrm{As}(\pi)$ and the application of the Asai lift to the strong Artin conjecture for solvable Galois representations of $GO(4)$ type.

REMARK 1. We emphasize here that for the functoriality of Π in Theorem 1, we do not need the full functoriality of the exterior square of GL_4 as we remarked in the proof of the functoriality of symmetric fourth [**6**, Remark 7.2]. Namely, we do not need the descent argument in [**6**, Appendix 1] due to D. Ramakrishnan, and also the comparison of Hecke conjugacy classes in [**6**, Section 4.1]. The reason is that first, $I_K^F(\pi)$ satisfies the weak Ramanujan property, and hence we can use Proposition 4.1.2 in [**6**]. Secondly, the reason we needed the descent method, was that we could not verify Proposition 4.2 in [**6**] in the case of supercuspidal representations. But we can prove the equality of γ-functions as in [**6**, Remark 7.2]. Namely, we can prove the case of $m = 1$ in the proof of Theorem 2: the setup is the same. Let η be supercuspidal representations of $GL_2(E)$ with the corresponding parametrization $\tau \colon W_E \to GL_2(\mathbb{C})$. We can think of τ as a representation of $\mathrm{Gal}(\bar{k}/E)$. We need to show that

$$\gamma\big(s, I_E^k(\eta), \wedge^2 \rho_4 \otimes \chi, \psi\big) = \gamma\big(s, \mathrm{Ind}_E^k(\tau), \wedge^2 \rho_4 \otimes \chi, \psi\big),$$

for any character χ of $k^\times$, which we identify it as a character of $\mathrm{Gal}(\bar{k}/k)$. By appealing to [**10**, Lemma 3, Section 4], we can find a quadratic extension of number fields K/F with $E = K_v, k = F_v$ and irreducible 2-dimensional representations σ of $\mathrm{Gal}(\overline{F}/K)$ with solvable image such that $\sigma_v = \tau$ and σ_u is unramified for $u \mid 2$, $u \neq v$. Let π be the cuspidal representation of $GL_2(\mathbb{A}_E)$ such that $\pi_v = \eta$, given by the global Langlands correspondence. Take a grössencharacter μ such that $\mu_v = \chi$. By comparing the functional equations of $L\big(s, I_E^k(\pi), \wedge^2 \rho_4 \otimes \mu\big)$ and $L\big(s, \mathrm{Ind}_E^k(\sigma), \wedge^2 \rho_4 \otimes \mu\big)$, we obtain the equality, by noting that if $u \mid 2$, $u \neq v$, π_u is unramified. Hence we can apply the converse theorem [**6**, Theorem 2.1] to $I_K^F(\pi)$

as in [**6**, Section 4.1] and obtain a weak lift, and follow Section 5.2 in [**6**] to obtain the strong lift.

REMARK 2. We would like to make a correction on [**6**]: in the proof of [**6**, Theorem 4.2.3], we need to take the grössenchacter $\chi(j)$ such that $\chi(j)_v$ is trivial at archimedean places (this is certainly possible in the application of Grunwald–Wang theorem). This is necessary so that each archimedean places may split in F_j. Then the local components of the automorphic descent Π at archimedean places are the local exterior square lift of π_v. This has been pointed out by M. Krishnamurthy.

REMARK 3. We note that if π is a cuspidal representation of $\mathrm{GL}_2(\mathbb{A}_F)$ and π_K is the base change to K, then $\mathrm{As}(\pi_K) = \mathrm{Sym}^2(\pi) \oplus (\omega_\pi \omega_{K/F})$. We also note that if π is a cuspidal representation of $\mathrm{GL}_2(\mathbb{A}_K)$ and $\mathrm{As}(\pi)$ is the Asai lift, then $\wedge^2(\mathrm{As}(\pi)) = \omega_\pi|_{\mathbb{A}_F^\times} \otimes I_K^F(\mathrm{Ad}(\pi))$, where $\mathrm{Ad}(\pi)$ is the adjoint square lift of π.

REMARK 4. For $n = 3$, in order to obtain the Asai lift from GL_3/K to GL_9/F, we need the L-function $L(s, \sigma \otimes \pi, \rho_m \otimes r)$, where π is a cuspidal representation of $\mathrm{GL}_3(\mathbb{A}_K)$, $r\colon \big(\mathrm{GL}_3(\mathbb{C}) \times \mathrm{GL}_3(\mathbb{C})\big) \rtimes \mathrm{Gal}(K/F) \to \mathrm{GL}_9(\mathbb{C})$ is given as in Section 2, and σ is a cuspidal representation of $\mathrm{GL}_m(\mathbb{A}_F)$ for $m = 1, 2, \ldots, 7$. Unfortunately, it is available only for $m = 1, 2$. The case $m = 2$ uses the quasi-split group of type E_6.

References

1. J. Arthur and L. Clozel, *Simple algebras, base change, and the advanced theory of the trace formula*, Ann. of Math. Stud., vol. 120, Princeton University Press, Princeton, NJ, 1989.
2. J. Cogdell and I. I. Piatetski–Shapiro, *Converse theorems for* GL_n. II, J. Reine Angew. Math. **507** (1999), 165–188.
3. M. Harris, *The local Langlands conjecture for* $\mathrm{GL}(n)$ *over a p-adic field,* $n < p$, Invent. Math. **134** (1998), 177–210.
4. M. Harris and R. Taylor, *The geometry and cohomology of some simple Shimura varieties*, Ann. of Math. Stud., vol. 151, Princeton University Press, Princeton, NJ, 2001.
5. G. Henniart, *Une preuve simple des conjectures de Langlands pour* $\mathrm{GL}(n)$ *sur un corps p-adique*, Invent. Math. **139** (2000), 439–455.
6. H. Kim, *Functoriality for the exterior square of* GL_4 *and symmetric fourth of* GL_2, J. Amer. Math. Soc. **16** (2003), no. 1, 139–183.
7. H. Kim and F. Shahidi, *Functorial products for* $\mathrm{GL}_2 \times \mathrm{GL}_3$ *and symmetric cube for* GL_2, Ann. of Math. (2) **155**, 837–893.
8. M. Krishnamurthy, *The* Asai *transfer to* GL_4: *via the Langlands–Shahidi method*, Int. Math Res. Not. **2003**, no. 41, 2221–2254.
9. R. P. Langlands, *On the classification of irreducible representations of real algebraic groups*, Representation Theory and Harmonic Analysis on Semisimple Lie groups (P. J. Sally, Jr. and D. A. Vogan, eds.) Math. Surveys Monogr., vol. 31, 1989, Amer. Math. Soc., Providence, RI, pp. 101–170.
10. D. Prasad and D. Ramakrishnan, *On the global root numbers of* $\mathrm{GL}(n) \times \mathrm{GL}(m)$, Automorphic Forms, Automorphic Representations, and Arithmetic (Fort Worth, TX, 1996), Proc. Sympos. Pure Math., vol. 66, Part 2, Amer. Math. Soc., Providence, RI, 1999, pp. 311–330.
11. D. Ramakrishnan, *Modularity of solvable Artin representations of* $\mathrm{GO}(4)$*-type*, Int. Math. Res. Not. **2002**, no. 1, 1–54.
12. ______, *Modularity of the Rankin–Selberg L-series, and multiplicity one for* $\mathrm{SL}(2)$, Ann. of Math. (2) **152** (2000), no. 1, 45–111.

DEPARTMENT OF MATHEMATICS, UNIV. OF TORONTO, TORONTO, ON M5S 3G3, CANADA.
E-mail address: henrykim@math.toronto.edu

Centre de Recherches Mathématiques
CRM Proceedings and Lecture Notes
Volume **36**, 2004

Totally Real Integral Points on a Plane Algebraic Curve

Michel Laurent

ABSTRACT. Let $F(X,Y) = \sum_{i=0}^{m} \sum_{j=0}^{n} a_{i,j} X^i Y^j$ be an absolutely irreducible polynomial in $\mathbf{Z}[X,Y]$. Suppose that $m \geq 1, n \geq 2$ and that the polynomial $\sum_{j=0}^{n} a_{m,j} Y^j$ is reducible in $\mathbf{Q}[Y]$, has n simple roots and an unique real root. Let L be a totally real number field and let $(\xi, \zeta) \in \mathcal{O}_L \times L$ be such that $F(\xi, \zeta) = 0$. We give an upper bound for the absolute height $H(\xi)$ which depends only upon the polynomial F. Our result may be viewed as a natural extension of Runge's method to arbitrary totally real number fields.

1. Introduction and results

Let $\mathcal{C}$ be a plane algebraic curve rational over $\mathbf{Q}$ defined by an equation

$$F(X,Y) = \sum_{i=0}^{m} \sum_{j=0}^{n} a_{i,j} X^i Y^j = 0$$

for some absolutely irreducible polynomial F in $\mathbf{Z}[X,Y]$. For any number field L, denote by $\mathcal{O}_L$ its ring of integers. Our goal is to obtain an uniform upper bound of the absolute height of points

$$(\xi, \zeta) \in \mathcal{O}_L \times L \quad \text{with} \quad F(\xi, \zeta) = 0,$$

rational over some totally real number field L, in relation with Runge's Theorem. Runge's method is most frequently worked out for rational points $(\xi, \zeta) \in \mathbf{Z} \times \mathbf{Q}$. See [**9**] for sharp quantitative upper bounds of $|\xi|$ in this frame. Here we show that under simple conditions relevant to Runge's Theorem, similar bounds hold likewise for the absolute height of any totally real integral point (ξ, ζ) as above, if we assume moreover that the curve $\mathcal{C}$ has exactly one real point at infinity. In collaboration with Dimitrios Poulakis, we had already obtained in [**5**] such a result under the stronger condition that this real point at infinity is rational over $\mathbf{Q}$. The proof of our theorem relies on the same techniques.

Recall briefly some notations from [**5**]. We denote by

$$F_\infty(Y) = \sum_{j=0}^{n} a_{m,j} Y^j$$

2000 *Mathematics Subject Classification.* Primary: 11D75; Secondary: 11J25.
This is the final form of the paper.

the leading coefficient of F viewed as a polynomial in X. We shall use the naïve heights

$$H(F) = \max_{\substack{0\leq i\leq m\\ 0\leq j\leq n}} (|a_{i,j}|) \quad \text{and} \quad h(F) = \log H(F),$$

for the polynomial F, and we shall indicate by

$$H(\xi) = M(\xi)^{1/[\mathbf{Q}(\xi):\mathbf{Q}]} \quad \text{and} \quad h(\xi) = \log H(\xi)$$

the multiplicative and logarithmic absolute heights of the algebraic number ξ, where $M(\xi)$ stands for its Mahler's measure.

THEOREM. *Suppose that $m \geq 1$, $n \geq 2$, that the polynomial F_∞ has n simple roots and has an unique real root η. Assume moreover that*

$$\delta = [\mathbf{Q}(\eta) : \mathbf{Q}] < n.$$

Let L be a totally real number field and let $(\xi, \zeta) \in \mathcal{O}_L \times L$ with $F(\xi,\zeta) = 0$. Then we have the upper bound

$$H(\xi) \leq H^{4mn^3\delta^2/(n-\delta)^2} \quad \textit{with} \quad H = 8(m+1)^{2n+2}(n+1)^{2n+3}H(F)^{2n-1}.$$

The same upper bound is given in [**5**, Theorem 6] under the stronger assumption that η should be a rational number. See also [**1**, Chapter 5] in the special case of superelliptic equations. Notice however that in the earlier Theorem 6, we considered the more general data of polynomials F whose coefficients belong to an arbitrary totally real number field.

It may be enlightening to translate the assumptions of the theorem in simple geometrical terms, as in [**5**, Section 2.2]. Homogeneize F with respect to X and Y into a bihomogeneous polynomial $\overline{F}$ of bidegree (m, n) and denote by $\overline{\mathcal{C}}$ the curve in $\mathbf{P}^1 \times \mathbf{P}^1$ with equation $\overline{F} = 0$. Then $\overline{\mathcal{C}}$ is the Zariski closure of the curve $\mathcal{C}$ naturally embedded into $\mathbf{P}^1 \times \mathbf{P}^1$. In this setting, the roots of the polynomial F_∞ are in one-to-one correspondence with the points of the intersection $\overline{\mathcal{C}} \cap (\infty \times \mathbf{P}^1)$; so that the main assumption of our theorem asserts that $\overline{\mathcal{C}} \cap (\infty \times \mathbf{P}^1)$ contains an unique real point and that the degree over $\mathbf{Q}$ of this point is $< n$. Note that the hypothesis $\delta < n$ is inherent to Runge's method. It means that the algebraic set $\overline{\mathcal{C}} \cap (\infty \times \mathbf{P}^1)$ is reducible over $\mathbf{Q}$. The unicity of the real point at infinity ensures us that any point $(x, y) \in \mathcal{C}(\mathbf{R})$ which is sufficiently close to this point belongs to its *branch*, according to the terminology of [**5**]. This will be proved in the lemma below. The remaining assumptions on the simplicity of the roots of F_∞ are less essential. They mean that the points of $\overline{\mathcal{C}} \cap (\infty \times \mathbf{P}^1)$ are distinct from $(\infty \times \infty)$ and are unramified in the covering $(x, y) \mapsto x$. Some monomial transformation may often be performed in order to satisfy these last conditions. See for instance Ref. [**5**, Corollary of Section 2.2].

The theorem will be deduced from the same diophantine inequality that was employed in our previous result. We have called in [**5**] this estimate a "Liouville's type inequality on $\mathcal{C}$", since it links the distance of two algebraic distinct points located on $\mathcal{C}$ to their heights. We refer to [**2–4**] and [**8**] for earlier versions of our inequality. It is known that such estimates have interesting corollaries in the domain of diophantine equations. Notably, Bombieri observed in [**2**] that Runge's Theorem can be easily deduced from his inequalities and that this approach provides fine effective results. Explicit upper bounds for the height of the more general

s-integral points were subsequently obtained by P. Debes in [**4**]. It may be worthwhile to mention here that effective versions of Hilbert Irreducibility Theorem can equally be obtained using the same principles. See [**3, 4, 8**] for a large variety of results in this direction.

Since our Liouville's type inequalities involve all possible metrics of a given number field, one can readily notice that a similar theorem holds in a non Archimedean setting, replacing real points by points rational over the p-adic field $\mathbf{Q}_p$. For simplicity, we have restrained our statement to the real case. We reproduce in the next Section the special case of the inequality which is needed here.

2. A Liouville's type inequality on $\mathcal{C}$

In the sequel, we shall view $\overline{\mathbf{Q}}$ as a subfield of $\mathbf{C}$. Let

$$P_1 = (\xi_1, \zeta_1) \in \mathcal{C}(\overline{\mathbf{Q}}) \quad \text{and} \quad P_2 = (\xi_2, \zeta_2) \in \mathcal{C}(\overline{\mathbf{Q}})$$

be two algebraic points $\in \mathcal{C}$ which will be assumed unramified in the covering $(x, y) \mapsto x$. For each $\sigma \in \mathrm{Gal}(\overline{\mathbf{Q}}/\mathbf{Q})$, denote by $y_\sigma(x)$ the implicit function defined in some neighborhood of $\sigma\xi_1$ by the conditions

$$F\big(x, y_\sigma(x)\big) = 0 \quad \text{and} \quad y_\sigma(\sigma\xi_1) = \sigma\zeta_1.$$

Let r_σ be the radius of convergence of the Taylor's series of y_σ at the point $\sigma\xi_1$. Notice that the functions y_σ and the radii r_σ depend only upon the restriction of σ to the field $\mathbf{Q}(\xi_1, \zeta_1)$.

Fix now a number field K such that the points P_1 and P_2 are both rational over K. Let S be the set of embeddings $\sigma \colon K \hookrightarrow \mathbf{C}$ satisfying

$$|\sigma\xi_2 - \sigma\xi_1| < r_\sigma \quad \text{and} \quad y_\sigma(\sigma\xi_2) = \sigma\zeta_2. \tag{2.1}$$

Assume that $\xi_1 \neq \xi_2$. Then we have the upper bound

$$\frac{1}{[K:\mathbf{Q}]} \sum_{\sigma \in S} \log^+ \frac{1}{|\sigma\xi_2 - \sigma\xi_1|} \leq \frac{h(\xi_1) + h(\xi_2)}{n} + \rho + \sqrt{h_1 h_2} + \log 4 \tag{2.2}$$

where

$$\rho = \frac{1}{[K:\mathbf{Q}]} \sum_{\sigma \in S} \log^+ \frac{1}{r_\sigma},$$

and

$$h_i = m(2n-1)h(\xi_i) + (2n-1)h(F) + \log\big(8(m+1)^{2n+2}(n+1)^{2n+3}\big), \quad (i = 1, 2).$$

The estimate (2.2) is a direct consequence of Ref. [**5**, Theorem 1]. Note that (2.1) means that the point P_2 *belongs to the branch* $y_{P_1,\sigma}$ according to the terminology used in [**5**]. Thus our set S is contained in the set $S(K, P_1, P_2)$ introduced in [**5**]. Following Remark 1 after Theorem 1, one can replace in the statement of this theorem the set $S(K, P_1, P_2)$ by any of its subset. Therefore (2.2) holds, since our term ρ is larger than the corresponding one in Theorem 1, the other terms being identical. The same remark shows that (2.2) remains valid with S replaced by any of its subset in both sides of the inequality.

3. Proof of the theorem

We shall employ the estimate (2.2) for points belonging to the plane curve $\widetilde{\mathcal{C}}$ with equation $G(x,y)=0$, where $G(X,Y)=X^mF(X^{-1},Y)$.

Remind that η denotes the real root of the polynomial F_∞. Then $(0,\eta)$ is an algebraic point of $\widetilde{\mathcal{C}}$ smooth in the covering $(x,y)\mapsto x$. Let $y(x)$ be the implicit function defined in a neighborhood of 0 by the conditions

$$G\big(x,y(x)\big)=0 \quad \text{and} \quad y(0)=\eta.$$

LEMMA. *The convergence radius r of the Taylor series of the function y at 0 is $> H^{-n}$. Let x be a real number with $|x|\geq H^n$. Then the polynomial $F(x,Y)$ has n simple roots and $y(x^{-1})$ is its unique real root.*

PROOF. For any root γ of the polynomial F_∞, let Y_γ be the function defined implicitely in a neighborhood of 0 by the conditions

$$G\big(x,Y_\gamma(x)\big)=0 \quad \text{and} \quad Y_\gamma(0)=\gamma.$$

The radius of convergence R_γ of the Taylor's series of Y_γ at the origin can be bounded from below in the following way:

$$\begin{aligned} R_\gamma &\geq \left(\frac{4(m+1)^{2n+2}(n+1)^{2n+3}H(G)^{2n-1}}{|F'_\infty(\gamma)|^2|a_{m,n}|^{2n-3}}\right)^{-1} \\ &\geq \big(4(m+1)^{2n+2}(n+1)^{2n+3}H(F)^{2n-1}\big)^{-[\mathbf{Q}(\gamma):\mathbf{Q}]} > H^{-n}. \end{aligned}$$

The first lower bound is in Ref. [**5**, Corollary of Lemma 4], while the second one follows from the usual Liouville's inequality combined with the upper bounds

$$|F'_\infty(\sigma\gamma)|^2|a_{m,n}|^{2n-3}\leq 4(m+1)^{2n+2}(n+1)^{2n+3}H(F)^{2n-1}, \quad \big(\sigma\in \mathrm{Gal}(\overline{\mathbf{Q}}/\mathbf{Q})\big),$$

which are special cases of the estimation (5) contained in the proof of [**5**, Lemma 6]. Notice that $F'_\infty(\gamma)^2a_{m,n}^{2n-3}$ is an algebraic integer and see Ref. [**5**, Section 6.2] for more details on the computation. Since $y(x)=Y_\eta(x)$ and $r=R_\eta$, we have just proven the first assertion of the lemma.

Observe that for any complex number x with $|x|\geq H^n$, the numbers $Y_\gamma(x^{-1})$, when γ ranges over the n roots of F_∞, are the n distinct roots of the polynomial $F(x,Y)$. Note that the n values $Y_\gamma(x^{-1})$ are necessarily distinct. Otherwise, x should be a root of the resultant R of the polynomials $F(x,Y)$ and $F'_Y(x,Y)$ with respect to the variable Y, and we would obtain the upper bound

$$|x|\leq 1+H(R)\leq 1+\big((m+1)(n+1)\sqrt{n}H(F)\big)^{2n-1}<H^n,$$

using [**6**] for the first inequality and Ref. [**7**, Lemma 4] for the estimation of $H(R)$.

Suppose now that x is real and that $|x|\geq H^n$. Since

$$\overline{Y_\gamma(z)}=Y_{\overline{\gamma}}(\overline{z})$$

for any complex number z with $|z|\leq H^{-n}$, it follows that $Y_\gamma(x^{-1})$ is real if and only if γ is real. Therefore $y(x^{-1})=Y_\eta(x^{-1})$ is the unique real root of the polynomial $F(x,Y)$. □

We are now able to prove the theorem. Fix temporarily $\tau\in\mathrm{Gal}(\overline{\mathbf{Q}}/\mathbf{Q})$ and consider the two algebraic points

$$P_1=(0,\eta) \quad \text{and} \quad P_2=(\tau\xi^{-1},\tau\zeta)$$

which belong to the curve $\widetilde{\mathcal{C}}$. Both points are rational over the field $K = (\tau L)(\eta)$. Denote by S_τ the set of $\mathbf{Q}(\eta)$-embeddings $\sigma\colon K \hookrightarrow \mathbf{C}$. Then

$$\operatorname{Card} S_\tau = [K : \mathbf{Q}(\eta)] = \frac{[K : \mathbf{Q}]}{\delta}.$$

Let us show that the condition (2.1) holds for any $\sigma \in S_\tau$ satisfying

$$|\sigma\tau\xi| \geq H^n.$$

Observe first that $y_\sigma = y$, since $\sigma P_1 = P_1$ for any $\sigma \in S_\tau$. On the other hand, $\sigma\tau\zeta$ is a real root of the polynomial $F(\sigma\tau\xi, Y)$ since the number field L is supposed to be totally real. The preceding lemma tells us that the polynomial $F(\sigma\tau\xi, Y)$ has only one real root provided $|\sigma\tau\xi| \geq H^n$ and that this root is then $y(\sigma\tau\xi^{-1})$. Therefore $y(\sigma\tau\xi^{-1}) = \sigma\tau\zeta$, which gives (2.1).

Now the estimate (2.2) provides us with the upper bound

$$\frac{1}{\delta \operatorname{Card} S_\tau} \sum_{\substack{\sigma \in S_\tau \\ |\sigma\tau\xi| \geq H^n}} \log^+ |\sigma\tau\xi| \leq \frac{h(\xi)}{n} + \frac{1}{\delta \operatorname{Card} S_\tau} \left(\sum_{\substack{\sigma \in S_\tau \\ |\sigma\tau\xi| \geq H^n}} \log^+ \frac{1}{r} \right)$$
$$+ \sqrt{\log H \bigl(\log H + m(2n-1)h(\xi)\bigr)} + \log 4,$$

from which we deduce

$$\text{(3.1)} \qquad \frac{1}{\delta \operatorname{Card} S_\tau} \sum_{\sigma \in S_\tau} \log^+ |\sigma\tau\xi| \leq \frac{h(\xi)}{n} + \frac{n \log H}{\delta}$$
$$+ \sqrt{\log H \bigl(\log H + m(2n-1)h(\xi)\bigr)} + \log 4$$

using the lower bound $r > H^{\;n}$ furnished by the lemma.

Observe that the set of embeddings

$$\sigma\tau\colon L \hookrightarrow \mathbf{C} \quad , \quad (\sigma \in S_\tau),$$

which appear in the left hand side of (3.1), can be naturally identified with the orbit of the coset $\tau \operatorname{Gal}(\overline{\mathbf{Q}}/L)$ under the left action of the group $\operatorname{Gal}\bigl(\overline{\mathbf{Q}}/\mathbf{Q}(\eta)\bigr)$ on the set of complex embeddings

$$E = \{\iota\colon L \hookrightarrow \mathbf{C}\} \simeq \operatorname{Gal}(\overline{\mathbf{Q}}/\mathbf{Q}) / \operatorname{Gal}(\overline{\mathbf{Q}}/L).$$

Denote by ℓ the total number of orbits in E under the action of $\operatorname{Gal}\bigl(\overline{\mathbf{Q}}/\mathbf{Q}(\eta)\bigr)$. Then ℓ is equal to the cardinality of the set of double cosets

$$\operatorname{Gal}\bigl(\overline{\mathbf{Q}}/\mathbf{Q}(\eta)\bigr) \backslash \operatorname{Gal}(\overline{\mathbf{Q}}/\mathbf{Q}) / \operatorname{Gal}(\overline{\mathbf{Q}}/L).$$

Select now a full set of representatives

$$\tau_1 \in \operatorname{Gal}(\overline{\mathbf{Q}}/\mathbf{Q}), \ldots, \tau_\ell \in \operatorname{Gal}(\overline{\mathbf{Q}}/\mathbf{Q})$$

of these double cosets. Each element of E can be uniquely written in the form

$$\sigma\tau_j\colon L \hookrightarrow \mathbf{C} \quad \text{with} \quad j \in \{1, \ldots, \ell\} \quad \text{and} \quad \sigma \in S_{\tau_j}.$$

It follows that

$$\text{(3.2)} \qquad [L : \mathbf{Q}] = \operatorname{Card} S_{\tau_1} + \cdots + \operatorname{Card} S_{\tau_\ell}$$

Multiply now (3.1) by $\mathrm{Card}\, S_\tau/[L : \mathbf{Q}]$ and sum the inequalities obtained for $\tau = \tau_1, \dots, \tau = \tau_\ell$. Using (3.2), we find the estimate

$$\frac{h(\xi)}{\delta} \le \frac{h(\xi)}{n} + \frac{n \log H}{\delta} + \sqrt{\log H\big(\log H + m(2n-1)h(\xi)\big)} + \log 4. \tag{3.3}$$

Write finally

$$h(\xi) = \lambda m n^3 \frac{\delta^2}{(n-\delta)^2} \log H$$

with $\lambda \ge 1$. We easily deduce from (3.3) that

$$\lambda - \sqrt{2\lambda} \le \frac{n/\delta + \log 4/\log H}{mn^2\delta/n - \delta} \le 1.$$

It follows that $\lambda \le 4$, so that

$$H(\xi) \le H^{4mn^3\delta^2/(n-\delta)^2},$$

which was to be proved.

References

1. Y. André, *G-functions and geometry*, Vieweg, Braunschwieg, 1989.
2. E. Bombieri, *On Weil's "Théorème de décomposition"*, Amer. J. Math. **105** (1983), 295–308.
3. P. Debes, *G-fonctions et théorème d'irréductibilité de Hilbert*, Acta Arith. **47** (1986), 371–401.
4. P. Debes, *Hilbert subsets and s-integral points*, Manuscripta Math. **89** (1996), 107–137.
5. M. Laurent and D. Poulakis, *On the global distance between two algebraic points on a curve*, J. Number Theory **104** (2004), 210–254.
6. M. Mignotte, *An inequality on the greatest roots of a polynomial*, Elem. Math. **46** (1991), 85–86.
7. W. Schmidt, *Eisenstein's theorem on power series expansion of algebraic functions*, Acta Arith. **56** (1990), 161–179.
8. V. G. Sprindzuck, *Arithmetic specializations of polynomials*, J. Reine. Angew. Math. **340** (1983), 26–52.
9. P. G. Walsh, *A quantitative version of Runge's theorem on diophantine equations*, Acta Arith. **62** (1992), 157–172.

Institut de Mathématiques de Luminy, CNRS, 163 Avenue de Luminy, Case 907, 13288 Marseille Cédex 9, France.

E-mail address: laurentiml.univ-mrs.fr

Centre de Recherches Mathématiques
CRM Proceedings and Lecture Notes
Volume **36**, 2004

On the Largest Prime Factor of a Mersenne Number

Leo Murata and Carl Pomerance

1. Introduction

Over two millennia ago Euclid demonstrated that a prime p of the form $2^n - 1$ gives rise to the perfect number $2^{n-1}p$, and he found four such primes. Presumably Euclid also knew that if $2^n - 1$ is prime, then so is n prime, and that the converse does not always hold. In the 18th century, Euler showed that Euclid's formula for perfect numbers gives rise to all even examples. (It is conjectured that there are no odd perfect numbers.) Probably because of the connection to perfect numbers, many mathematicians over the ages have been interested in finding primes of the form $2^n - 1$. In the early 17th century, the French monk and mathematician Marin Mersenne made an uncanny guess of which n in the range $29 \leq n \leq 257$ give such primes. He guessed that it was $n = 31$, 67, 127, and 257, while the truth is $n = 31$, 61, 89, 107, and 127. What is uncanny is that he got two right, and that there are indeed so few of them. In fact the sparsity of the Mersenne primes is hardly suggested by the evidence below 31: Seven of ten candidates are prime. In honor of Mersenne and his guess, primes of the form $2^n - 1$ are now called Mersenne primes. Currently (June, 2003) we know 39 Mersenne primes, the largest having exponent $n = 13466917$, and rumors of a 40th are circulating. Nevertheless, it has never been proved that there are infinitely many Mersenne primes, nor has it been proved that there are infinitely many prime numbers n with $2^n - 1$ composite.

We shall call any number of the form $2^n - 1$ a *Mersenne number*, regardless if n is prime. For an integer $m > 1$, let $P(m)$ denote the largest prime factor of m, and let $P(1) = 1$. We are interested in this paper in studying $P(2^n - 1)$.

First, we shall review what is known about $P(2^n - 1)$. Schinzel [**11**] has shown that $P(2^n - 1) \geq 2n + 1$ for all $n > 12$. Remarkably, this result still stands as the best lower bound proved for $P(2^n - 1)$ for all sufficiently large n. It is perhaps reasonable to conjecture that $P(2^n - 1) > n^K$ for every fixed K and all sufficiently

---

2000 *Mathematics Subject Classification.* Primary: 11B83; Secondary: 11J86, 11M26, 11N36.

We thank Ram Murty, Tarlok Shorey, and Cam Stewart for their comments and for their interest in this note. We also acknowledge some helpful comments from an anonymous referee. The first author would like to thank the Number Theory Foundation and the hospitality of Bell Laboratories, and the second author would like to thank the hospitality of RIMS in Kyoto, Japan.

This is the final form of the paper.

large n depending on the choice of K, or maybe even $P(2^n - 1) > 2^{n/\log n}$ for sufficiently large n, but clearly we are very far from proving such assertions.

It is natural, in light of Euclid, to consider the special case of $P(2^p - 1)$ with p prime. Here we have somewhat better results: Stewart [**12**] showed that $P(2^p - 1) > \frac{1}{2}p(\log p)^{1/4}$ for all primes p beyond an effectively computable constant. This was improved by Stewart [**13**], and independently by Erdős and Shorey [**3**], to $P(2^p - 1) > c_1 p \log p$ for all primes $p > c_2$, where c_1, c_2 are effectively computable.

In addition, people have studied $P(2^n - 1)$ ignoring a set of numbers n of asymptotic density 0. For example, consider that $2^n - 1$ is always divisible by a prime that is congruent to 1 mod n, when $n > 1$. Thus, those values of n where the least prime $p \equiv 1 \pmod{n}$ is large give rise to values of n where $P(2^n - 1)$ is large. In particular, it follows easily from the Brun–Titchmarsh inequality that if $\varepsilon(n) > 0$ tends to zero monitonically and arbitrarily slowly, then the least prime $p \equiv 1 \pmod{n}$ satisfies $p > \varepsilon(n) n \log n$ on a set of asymptotic density 1.

PROOF. For a positive integer $j \le \varepsilon(x) \log x$, the number of positive integers n with $n \le x$ and $jn + 1$ prime is $O\big(jx/\varphi(j)\log x\big)$, by the Brun–Titchmarsh inequality. Summing on j, we involve only $O\big(\varepsilon(x)x\big)$ numbers n, and ignoring these, we have for the other numbers $n \le x$ and any prime $p \equiv 1 \pmod{n}$ that $(p-1)/n > \varepsilon(x) \log x \ge \varepsilon(n) \log n$. □

Thus, ignoring a set of integers n of asymptotic density 0, we have $P(2^n - 1) > \varepsilon(n) n \log n$. Improving on this, Stewart [**13**], showed that

$$P(2^n - 1) > \varepsilon(n) n (\log n)^2 / \log\log n \tag{1.1}$$

ignoring a set of integers n of asymptotic density 0. Earlier, Erdős and Shorey [**3**] had shown that

$$P(2^p - 1) > cp(\log p)^2 / (\log\log p \log\log\log p) \tag{1.2}$$

but for $o\big(\pi(x)\big)$ primes $p \le x$.

Most of these results are proved using theorems about linear forms in logarithms of algebraic numbers, in particular estimates of Baker, and subsequent improvements. In a recent paper, Murty and Wong [**8**] replace this deep tool with an even deeper unproved hypothesis, namely the ABC conjecture. This conjecture asserts that for each $\varepsilon > 0$, there are at most finitely many coprime triples a, b, c of positive integers with $a + b = c$ and $c > m^{1+\varepsilon}$, where m is the product of the distinct primes that divide abc. Assuming this hypothesis, Murty and Wong prove that

$$P(2^n - 1) > n^{2-\varepsilon} \tag{1.3}$$

for each fixed $\varepsilon > 0$ and all sufficiently large values of n, depending on the choice of ε. In the case when $n = p$, a prime, this result is easy to see: if m is the largest squarefree divisor of $2^p - 1$, then the ABC conjecture applied with $a = 2^p - 1$, $b = 1$, $c = 2^p$ implies that $2^p < (2m)^{1+\varepsilon}$ for all sufficiently large primes p. But all the prime factors of m are 1 mod p, so if they are all smaller than $p^{2-\varepsilon}$, then there are at most $p^{1-\varepsilon}$ such primes, and so $m < p^{(2-\varepsilon)p^{1-\varepsilon}}$. Hence, assuming $\varepsilon < 1$,

$$2^p < (2m)^{1+\varepsilon} < 4p^{3p^{1-\varepsilon}} = o(2^p),$$

a contradiction. In fact, a slight elaboration of this argument using the Brun–Titchmarsh inequality shows that $P(2^p - 1) > cp^2$ for some effectively computable positive constant c and all sufficiently large primes p.

It is our goal in this paper to also give some conditional results about $P(2^n - 1)$, but the condition we shall assume is not the ABC conjecture, but rather the Generalized Riemann Hypothesis (GRH) for various Kummerian fields of the form $\mathbf{Q}(2^{1/j}, e^{2\pi i/jk})$. In addition, we strongly use some results in Stewart [**13**] and Erdős and Shorey [**3**] (for our first 3 theorems) which depend on estimates for linear forms in logarithms of algebraic numbers. We shall prove the following four theorems:

THEOREM 1.1. (GRH) *The set of natural numbers n with*

$$P(2^n - 1) > n^{4/3}/\log\log n$$

has asymptotic density 1.

THEOREM 1.2. (GRH) *But for $o(\pi(x))$ primes $p \le x$ we have*

$$P(2^p - 1) > p^{4/3}/(\log p)^{2/3}\log\log p.$$

Let $\Phi_n(x)$ denote the nth cyclotomic polynomial.

THEOREM 1.3. (GRH) *Uniformly for each $\varepsilon > 0$ and $x \ge 2$, but for $O(\varepsilon x + x/(\log x)^2)$ integers $n \le x$, every prime factor of $\Phi_n(2)$ exceeds $n^{1+\varepsilon}$. Further, but for $O(\varepsilon\pi(x))$ primes $p \le x$, every prime factor of $2^p - 1$ exceeds $p^{1+\varepsilon}$.*

THEOREM 1.4. (GRH) *Let $M(x)$ denote the number of primes $p \le x$ which divide some $2^q - 1$ with q prime. That is, $M(x)$ is the number of odd primes $p \le x$ such that the order of $2 \bmod p$ is a prime number. Then $M(x) = O(x/(\log x)^2)$.*

It should be noted that the conclusions of Theorems 1.1 and 1.2 are considerably weaker than the conclusion (1.3) of the Murty–Wong theorem. In particular, not only is their inequality stronger, we allow density-0 exceptions, while they allow only finitely many exceptions. However, they assume the ABC conjecture, while we assume the GRH. Of course, since neither conjecture is presently a theorem, it is unclear which is the stronger hypothesis.

Theorem 1.3 appears to be new with no known ABC analog. We mention that Theorem 1.4 is a strengthening of the GRH-conditional estimate $M(x) = O(x\log\log x/(\log x)^2)$ achieved by the second author in [**10**]. In the same paper, an unconditional estimate of $M(x) = O(\pi(x)\log\log\log x/\log\log x)$ is proved. We give below a heuristic argument that Theorem 1.4 is best possible, and we even suggest a candidate number c for which it may be true that $M(x) \sim cx/(\log x)^2$.

We remark that it is likely that our results can be generalized to the largest prime factor of expressions such as $a^n - b^n$, $(\alpha^n - \beta^n)/(\alpha - \beta)$ where α, β are conjugate algebraic numbers of degree 2, and to similar sequences, as was done in most of the referenced papers. We leave these details to another time, and perhaps to another person.

2. Preliminaries

In the following we often use the Vinogradov order notation $\ll$, which means the same as the big-O notation. That is, for positive functions $f(x)$, $g(x)$ we write

$f(x) \ll g(x)$ if and only if $f(x) = O(g(x))$. In addition, we write $f(x) \gg g(x)$ if and only if $g(x) = O(f(x))$. In what follows, all implied constants are absolute.

For an odd integer $m > 0$, let $l(m)$ denote the least positive integer l with $2^l \equiv 1 \pmod{m}$. We say a prime factor p of $\Phi_n(2)$ is *primitive* if $l(p) = n$, and otherwise we say that p is *intrinsic.*

LEMMA 2.1. *The prime p is an intrinsic prime factor of $\Phi_n(2)$ if and only if $n = p^j l(p)$ for some integer $j > 0$. For each integer $n \neq 1$, 6 there is at least one primitive prime factor of $\Phi_n(2)$.*

These results are well-known. A reference for the first is Nagell [**9**], and the second, known as Bang's theorem, has been independently proved multiple times, an early English language reference being [**1**]. Note that as a corollary of the first assertion, an intrinsic prime factor of $\Phi_n(2)$ must be the largest prime factor of n.

LEMMA 2.2. *Let j and k be natural numbers, and let $n_{j,k} = [\mathbf{Q}(2^{1/j}, e^{2\pi i/k}) : \mathbf{Q}]$. Then we have*

$$n_{j,k} = \begin{cases} \frac{1}{2} j\varphi(k), & \text{if } 8 \mid k \\ j\varphi(k), & \text{otherwise.} \end{cases}$$

PROOF. Since $[\mathbf{Q}(2^{1/j}) : \mathbf{Q}] = j$, $[\mathbf{Q}(e^{2\pi i/k}) : \mathbf{Q}] = \varphi(k)$, and

$$n_{j,k} = \frac{[\mathbf{Q}(2^{1/j}) : \mathbf{Q}][\mathbf{Q}(e^{2\pi i/k}) : \mathbf{Q}]}{[\mathbf{Q}(2^{1/j}) \cap \mathbf{Q}(e^{2\pi i/k}) : \mathbf{Q}]},$$

we need only to compute $[\mathbf{Q}(2^{1/j}) \cap \mathbf{Q}(e^{2\pi i/k}) : \mathbf{Q}]$. We remark here that the maximal normal subfield which is contained in $\mathbf{Q}(2^{1/j})$ is $\mathbf{Q}$ if j is odd, and is $\mathbf{Q}(\sqrt{2})$ if j is even. Further, any cyclotomic field which contains $\mathbf{Q}(\sqrt{2})$ also contains $\mathbf{Q}(e^{2\pi i/8})$. Since the field $\mathbf{Q}(2^{1/j}) \cap \mathbf{Q}(e^{2\pi i/k})$ is a normal extension of $\mathbf{Q}$, it is equal to $\mathbf{Q}$ or $\mathbf{Q}(\sqrt{2})$, and the latter happens if and only if $\mathbf{Q}(\sqrt{2}) \subset \mathbf{Q}(e^{2\pi i/k})$ if and only if $\mathbf{Q}(e^{2\pi i/8}) \subset \mathbf{Q}(e^{2\pi i/k})$ if and only if $8 \mid k$. □

This completes the proof of the lemma. Let $\tau(n)$ denote the number of positive divisors of n, and let $\nu(n)$ denote the number of these divisors which are prime. In [**3**, Lemma 2.2], Erdős and Shorey showed that there is an effectively computable positive constant c_0 such that if p is an odd prime, then $P(2^p - 1) < p^2$ implies that $\nu(2^p - 1) > c_0 \log p / \log\log p$. We now consider an analogous result for arbitrary integers n, but here we must allow for some exceptions, and the conclusion is slightly weaker.

LEMMA 2.3. *Let $\varepsilon(n)$ be a positive function that tends to 0 arbitrarily slowly. There is a set of natural numbers of asymptotic density 1 such that if n is in this set and $P(\Phi_n(2)) < n^2$ then $\nu(\Phi_n(2)) > \varepsilon(n) \log n / \log\log n$.*

PROOF. We sketch the proof following the argument given in Stewart [**13**]. We may assume that $\varepsilon(n)$ tends to 0 monotonically. Let $\mathcal{S}$ denote the set of natural numbers n with $\tau(n) < \log n$ and n has two consecutive divisors $d_0 < d_1$ with n/d_0, n/d_1 both squarefree and $d_1/d_0 > n^{\sqrt{\varepsilon(n)}}$. Since the normal value of $\tau(n)$ is $(\log n)^{\log 2 + o(1)}$, the condition that $\tau(n) < \log n$ removes only a density-0 set of numbers n that otherwise would be in $\mathcal{S}$. That $\mathcal{S}$ has asymptotic density 1 now follows from [**13**, Lemma 11] plus a short additional argument based on the easy fact that most numbers n are nearly squarefree, that is, they are not divisible by a large square.

Suppose $n \in \mathcal{S}$ and write

$$R_0 = \prod_{d|n, d \le d_0} (1-2^{-d})^{\mu(n/d)}, \quad R_1 = \prod_{d|n, d \ge d_1} (1-2^{-d})^{\mu(n/d)}.$$

Note that

$$\Phi_n(2) = 2^{\varphi(n)} R_0 R_1$$

and that

$$|\log R_1| \le -\sum_{d|n, d \ge d_1} \log(1-2^{-d}) < 2 \sum_{d \ge d_1} 2^{-d} = 2^{2-d_1}.$$

Also note that from the second assertion of Lemma 2.1, it is easy to see that the rational number R_1 is not 1. It is clear that the rational number R_0 has height at most $2^{d_0 \tau(n)} < 2^{d_0 \log n}$.

Let $\nu(\Phi_n(2)) = k$. We apply to [**13**, Lemma 9] (linear forms in logarithms of algebraic numbers) to the identity $R_1 = 2^{-\varphi(n)} \Phi_n(2) R_0^{-1}$. With this result and the assumption that $P(\Phi_n(2)) < n^2$, there is an absolute constant $c > 0$ such that

$$\log |\log R_1| \ge -(2k^c \log n)^{k+2} d_0.$$

But we above argued that

$$\log |\log R_1| < -d_1 \log 2 + 2 \log 2.$$

Putting these last two assertions together with $d_1/d_0 > n^{\sqrt{\varepsilon(n)}}$, we have

$$k \gg \sqrt{\varepsilon(n)} \log n / \log\log n,$$

from which the lemma immediately follows. □

REMARK. The condition $P(\Phi_n(2)) < n^2$ may be replaced with $P(\Phi_n(2)) \le n^{O(1)}$.

LEMMA 2.4. (GRH) *For all natural numbers j, d and all $x \ge 2$, let $N_{j,d}(x)$ denote the number of primes $p \le jx$ which split completely in the field $\mathbf{Q}(2^{1/j}, e^{2\pi i/jd})$. Then, using the notation of Lemma 2.2, we have uniformly*

$$N_{j,d}(x) = \frac{1}{n_{j,jd}} \mathrm{li}(jx) + O\big((jx)^{1/2} \log(jx) + j^2 d \log(jd)\big). \tag{2.1}$$

This result follows from [**7**, Theorem 1.1] and a calculation for the discriminant of the field $\mathbf{Q}(2^{1/j}, e^{2\pi i/jd})$. One might also compare this theorem with (28) in Hooley [**6**].

LEMMA 2.5. (GRH) *Fix a positive integer j. Let $M_j(x)$ denote the number of primes $p \le jx$ with $p \equiv 1 \pmod{j}$, $2^{(p-1)/j} \equiv 1 \pmod{p}$, and $(p-1)/j$ is prime. Then uniformly for $j \le x^{1/3}/2$,*

$$M_j(x) \ll \frac{j}{\varphi(j)^2} \frac{x}{\log x \log(x^{1/3}/j)} + \frac{x^{5/6}}{j^{1/2}} \big(\log(x^{1/3}/j)\big)^2 \log x.$$

PROOF. We use the form of Selberg's sieve as presented in [**4**, Theorem 5.2]. Let $z = (x^{1/3}/j)^{1/2}$ and let $\mathcal{A}$ denote the set of numbers $(p-1)/j$ where $p \le jx$ is prime and $2^{(p-1)/j} \equiv 1 \pmod{p}$. Further, let $\mathcal{P}$ denote the set of primes that do not divide j. Then clearly,

$$M_j(x) \le \mathcal{S}(\mathcal{A}, \mathcal{P}, z) + \pi(z),$$

where $\mathcal{S}(\mathcal{A}, \mathcal{P}, z)$ denotes the number of members of $\mathcal{A}$ which are not divisible by any member of $\mathcal{P} \cap [1, z]$. Note that from (2.1) we have $X := |\mathcal{A}| = N_{j,1}(x)$ well estimated. Suppose d is squarefree, relatively prime to j, and $d \le z^2$. To apply the sieve we need to compute $\omega(d)$, and to estimate $R_d := |\mathcal{A}_d| - \big(\omega(d)/d\big)|\mathcal{A}|$, where $\mathcal{A}_d$ is the set of members of $\mathcal{A}$ which are divisible by d.

Note that we may assume that j is even, since the only prime of the form $(p-1)/j$ with j odd is 2. For a prime $p \le jx$, $(p-1)/j$ is in $\mathcal{A}_d$ if and only if p splits completely in the field $\mathbf{Q}(2^{1/j}, e^{2\pi i/jd})$. Hence $|\mathcal{A}_d| = N_{j,d}(x)$. Since d is coprime to the even number j, Lemma 2.2 implies that $n_{j,jd} = \varphi(d)n_{j,j}$. Further, by Lemma 2.4, and taking $\omega(d) = d/\varphi(d)$, we have

$$\begin{aligned}|\mathcal{A}_d| &= \frac{1}{\varphi(d)n_{j,j}} \operatorname{li}(jx) + O\big((jx)^{1/2}\log(jdx) + j^2 d\log(jd)\big) \\ &= \frac{\omega(d)}{d}X + O\big((jx)^{1/2}\log x + j^2 d\log x\big).\end{aligned}$$

Note that $j^2 d \le j^2 z^2 = jx^{1/3} \le (jx)^{1/2}$ in our range for j. Thus we may take $R_d \ll (jx)^{1/2}\log x$ for each d, so that by the sieve theorem mentioned, we have

$$\begin{aligned}\mathcal{S}(\mathcal{A}, \mathcal{P}, z) &\ll \frac{j}{\varphi(j)}\frac{X}{\log z} + \sum_{d \le z^2} \mu^2(d) 3^{\nu(d)} (jx)^{1/2}\log x \\ &\ll \frac{j}{\varphi(j)}\frac{X}{\log z} + z^2(\log z)^2(jx)^{1/2}\log x.\end{aligned}$$

The lemma now follows from an estimate for X afforded by Lemma 2.4, and the choice of z. □

REMARK. The proof supports the same inequality for $M_j'(x)$, which is defined exactly as $M_j(x)$, but with the condition that $(p-1)/j$ is prime being replaced with the condition that $(p-1)/j$ has each of its prime factors exceeding $(x^{1/3}/j)^{1/2}$.

3. Proofs of the theorems

PROOF OF THEOREM 1.1. By Lemma 2.4, if $J \le x$, we have

$$\begin{aligned}\sum_{j \le J} N_{j,1}(x) &\ll \sum_{j \le J}\left(\frac{1}{\varphi(j)}\frac{x}{\log x} + (Jx)^{1/2}\log x + J^2\log J\right) \\ &\ll \frac{x\log J}{\log x} + J^{3/2}x^{1/2}\log x + J^3\log x.\end{aligned}$$

Applying this result with $J = x^{1/3}/\log\log x$, we get that

$$\sum_{j \le x^{1/3}/\log\log x} N_{j,1}(x) \ll x\log x/(\log\log x)^{3/2}. \tag{3.1}$$

Let $\varepsilon(x)$ decrease to 0 arbitrarily slowly. It follows from Lemma 2.3 that but for $o(x)$ choices of $n \in (x/2, x]$, if $P(2^n - 1) \le n^{4/3}/\log\log n$, then

$$\nu\big(\Phi_n(2)\big) > 1 + \varepsilon(x)\log x/\log\log x. \tag{3.2}$$

Say there are E numbers $n \in (x/2, x]$ with $P\big(\Phi_n(2)\big) \le n^{4/3}/\log\log n$ and (3.2) holding. Consider such a number n, and let the distinct *primitive* prime factors of $\Phi_n(2)$ be denoted $p_1, \dots, p_k$ where, by Lemma 2.1 and (3.2), we have $k > \varepsilon(x)\log x/\log\log x$. Then each $p_i = j_i n + 1$ for some $j_i \le J$ and $2^{(p_i-1)/j_i} \equiv 1$

(mod p_i). That is, p_i is counted by $N_{j_i,1}(x)$. Since these primitive prime factors of $\Phi_n(2)$ are not primitive prime factors of any other $\Phi_{n'}(2)$ it follows that

$$E\varepsilon(x)\log x/\log\log x \le \sum_{j\le J} N_{j,1}(x),$$

so that from (3.1),

$$E \ll \frac{x}{\varepsilon(x)(\log\log x)^{1/2}}.$$

Since we may take $\varepsilon(x) = 1/\log\log\log x$, say, the theorem follows. □

REMARK. The proof allows a slightly stronger result: for any positive function $\varepsilon(n)$ which decreases to 0, the set of natural numbers n such that $P(\Phi_n(2)) > \varepsilon(n)n^{4/3}/(\log\log n)^{2/3}$ has asymptotic density 1.

PROOF OF THEOREM 1.2. Let E denote the number of primes $p \in (x/2, x]$ with

$$P(2^p - 1) \le p^{4/3}/(\log p)^{2/3}\log\log p.$$

As remarked before Lemma 2.3, it follows from [**3**] that for such a prime p, $\nu(2^p - 1) \gg \log x/\log\log x$. Let $p_1, \dots, p_k$ be the distinct prime factors of $2^p - 1$ for one of these exceptional primes p, so that $k \gg \log x/\log\log x$. Then each $p_i = j_i p + 1$ for some integer $j_i \le J := x^{1/3}/(\log x)^{2/3}\log\log x$. Hence $p = (p_i - 1)/j_i$ is counted by the expression $M_{j_i}(x)$ defined in Lemma 2.5, so that

$$E\log x/\log\log x \ll \sum_{j\le J} M_j(x). \tag{3.3}$$

We have, by Lemma 2.5, that

$$\sum_{j\le x^{1/3}/(\log x)^3} M_j(x) \ll \frac{x\log\log x}{\log x} + \frac{x(\log\log x)^2}{(\log x)^{1/2}},$$

and by (2.1), that

$$\sum_{x^{1/3}/(\log x)^3<j\le J} M_j(x) \le \sum_{x^{1/3}/(\log x)^3<j\le J} N_{j,1}(x) \ll \frac{x\log\log x}{\log x} + \frac{x}{(\log\log x)^{3/2}}.$$

Hence,

$$\sum_{j\le J} M_j(x) \ll \frac{x}{(\log\log x)^{3/2}},$$

so that (3.3) implies that $E = o(\pi(x))$. This completes the proof of the theorem. □

REMARK. The proof gives a slightly stronger result: for almost all primes p (that is, but for a set of primes with relative density 0 in the set of all primes) we have $P(2^p - 1) > \varepsilon(p)p^{4/3}/(\log p\log\log p)^{2/3}$.

PROOF OF THEOREM 1.3. By Lemma 2.1, $\Phi_n(2)$ has an intrinsic prime factor p if and only if $n = p^a l(p)$ for some positive integer a. Since $l(p) \gg \log p$ we have that the number of integers $n \le x$ of the form $p^a l(p)$ for some prime p is $O(x/(\log x)^2)$. Hence it suffices to consider only those numbers n where every prime factor of $\Phi_n(2)$ is primitive. In particular, it is sufficient to show that but for $O(\varepsilon x)$ integers $n \le x$, every *primitive* prime factor of $\Phi_n(2)$ exceeds $n^{1+\varepsilon}$.

We may assume that $\varepsilon < \frac{1}{4}$. If $n \le x$ is such that $\Phi_n(2)$ has a primitive prime factor p with $p \le n^{1+\varepsilon}$, then $p = jn+1$ for some integer $j < p/n \le n^{\varepsilon} \le x^{\varepsilon}$, and $2^{(p-1)/j} \equiv 1 \pmod{p}$. Thus, the number of such numbers $n \le x$ is at most, using (2.1),

$$\sum_{j \le x^{\varepsilon}} N_{j,1}(x) \ll \varepsilon x.$$

The proof for $2^p - 1$, with p prime, is similar. In this case every prime factor of $2^p - 1 = \Phi_p(2)$ is primitive. If $p \le x$ and $q \mid 2^p - 1$ is prime with $q \le p^{1+\varepsilon}$, then p is counted by some $M_j(x)$ (the notation coming from Lemma 2.5), with $j \le \varepsilon$. The theorem now follows since by Lemma 2.5, the number of such primes p is at most

$$\sum_{j \le x^{\varepsilon}} M_j(x) \ll \varepsilon x / \log x. \qquad \square$$

PROOF OF THEOREM 1.4. Suppose q is prime and $\log x < q \le x^{1/5}$. If $l(p)$ is prime and $q|p-1$, then either $l(p) = q$ or $2^{(p-1)/q} \equiv 1 \ (\mathrm{mod} p)$. The number of primes p with $l(p) = q$ is trivially bounded by q (since their product divides $2^q - 1$), and the number of primes $p \le x$ in the second category is $\ll x/q^2 \log x$, using Lemma 2.4, with "x, j, d" of that lemma replaced with x/q, q, 1, respectively. Thus, the number of primes $p \le x$ with $l(p)$ prime and $q \mid p-1$ for some prime q in $(\log x, x^{1/5}]$, is $o\big(x/(\log x)^2\big)$. Hence we may consider only primes $p \le x$ with $p-1$ not divisible by any prime in $(\log x, x^{1/5}]$. Write $p - 1 = mr$, where every prime factor of m is at most $\log x$ and every prime factor of r is $> x^{1/5}$. For $m \le x^{1/5}$, the number of primes $p \le x$, with $p \equiv 1 \ (\mathrm{mod} m)$, $2^{(p-1)/m} \equiv 1 \ (\mathrm{mod} p)$, and $(p-1)/m$ divisible only by primes bigger than $x^{1/5}$, is $\ll x/\varphi(m)^2(\log x)^2$ by the remark following Lemma 2.5. Since $\sum 1/\varphi(m)^2 \ll 1$, we thus may assume that $m > x^{1/5}$. But, using [**2**, Theorem 1], we have that for $t > x^{1/5}$ the number of integers $m \le t$ with $P(m) \le \log x$ is at most $t^{c/\log\log x}$ for some absolute constant c. Hence, by partial summation, we have

$$\sum_{m > x^{1/5}, P(m) \le \log x} \frac{1}{m} \ll x^{-1/5+o(1)},$$

so that the number of integers $mr \le x$ with $m > x^{1/5}$, $P(m) \le \log x$ is negligible. This completes the proof of the theorem. $\square$

Note that Theorem 1.4 improves the estimate $M(x) = O\big(x \log\log x/(\log x)^2\big)$ attained in [**10**]. The new element in Theorem 1.4 that allows the elimination of the factor $\log\log x$ is the use of Lemma 2.5. We also remark that using [**5**, Lemma 1], one has $\gg x/(\log x)^2$ primes p up to x with $l(p)$ either a prime or the product of two primes both larger than $x^{1/4}$. (Indeed, from this lemma we obtain $\gg x/(\log x)^2$ primes $p \le x$ with $p \equiv 7 \ (\mathrm{mod} 8)$ and either $p-1 = 2q$ with q prime or $p-1 = 2rs$ where r, s are primes $> x^{1/4}$. For such a prime p we have $l(p) = q, r, s$ or rs.) Note that it is trivial that there are at least infinitely many primes p with $l(p)$ prime, in fact there are $\gg \log x/\log\log x$ such primes p below x.

It is probably the case that the estimate of Theorem 1.4 is best possible. In particular, using the Hardy–Littlewood version of the prime k-tuples conjecture, the number of primes $p \le x$ with $p \equiv 7 \ (\mathrm{mod} 8)$ and $(p-1)/2$ prime is of order of magnitude $x/(\log x)^2$. But each of these primes p has $l(p) = (p-1)/2$, a prime number. One might conjecture that

$$M(x) \sim cx/(\log x)^2,$$

where the value of c may be evaluated as follows. First, by Hardy–Littlewood, for a fixed even number k, the number of primes $q \le x/k$ with $kq+1$ prime is $\sim c_k x/(\log x)^2$, where

$$c_k = \frac{2\alpha}{k} \prod_{p>2, p|k} \frac{p-1}{p-2},$$

and where α is the twin prime constant $\prod_{p>2}(1 - 1/(p-1)^2)$. If $2\|k$, then heuristically, the further chance that 2 is a kth power modulo the prime $kq+1$ is $1/k$. If $2^2\|k$, then the further chance that 2 is a k-th power modulo the prime $kq+1$ is 0, since 2 is a quadratic nonresidue modulo $kq+1$. And if $2^3 \mid k$, then the further chance that 2 is a kth power modulo the prime $kq+1$ is $2/k$. Thus, a perhaps reasonable guess for the number c is

$$c = \frac{7}{12} \prod_{p>2} \left(1 - \frac{1}{(p+1)(p-1)^2}\right),$$

as per the following calculation:

$$\begin{aligned}
c &= \sum_{2\|k} \frac{c_k}{k} + \sum_{2^3|k} \frac{2c_k}{k} \\
&= \frac{\alpha}{2} \sum_{k\,\text{odd}} \frac{1}{k^2} \prod_{p|k} \frac{p-1}{p-2} + \frac{\alpha}{16} \sum_{k} \frac{1}{k^2} \prod_{p|k, p>2} \frac{p-1}{p-2} \\
&= \frac{\alpha}{2} \prod_{p>2} \frac{p^2-p-1}{p^2-p-2} + \frac{\alpha}{16}\frac{4}{3} \prod_{p>2} \frac{p^2-p-1}{p^2-p-2} \\
&= \frac{7\alpha}{12} \prod_{p>2} \frac{p^2-p-1}{p^2-p-2} \\
&= \frac{7}{12} \prod_{p>2} \left(1 - \frac{1}{(p+1)(p-1)^2}\right).
\end{aligned}$$

Thus, $c \approx 0.53824$.

References

1. G. D. Birkhoff and H. S. Vandiver, *On the integral divisors of $a^n - b^n$*, Ann. of Math. (2) **5** (1904), no. 4, 173–180.
2. N. G. de Bruijn, *On the number of positive integers $\le x$ and free of prime factors $> y$*, II, Nederl. Akad. Wetensch. Proc. Ser. A 69 Indag. Math. (N.S.) **28** (1966), 239–247.
3. P. Erdős and T. N. Shorey, *On the greatest prime factor of $2^p - 1$ for a prime p and other expressions*, Acta Arith. **30** (1976), 257–265.
4. H. Halberstam and H.-E. Richert, *Sieve methods*, London Math. Soc. Monogr. (N.S.), no. 4, Academic Press, London–New York, 1974.
5. D. R. Heath-Brown, *Artin's conjecture for primitive roots*, Quart. J. Math. Oxford Ser. (2) **37** (1986), 27–38.
6. C. Hooley, *On Artin's conjecture*, J. Reine Angew. Math. **225** (1967), 209–220.
7. J. C. Lagarias and A. M. Odlyzko, *Effective versions of the Chebotarev density theorem*, Algebraic Number Fields: L-Functions and Galois Properties (Durham, 1975), Proc. Sympos., Academic Press, London, 1977, pp. 409–464.

8. M. R. Murty and S. Wong, *The* ABC *conjecture and prime divisors of the Lucas and Lehmer sequences*, Number Theory for the Millennium, III (Urbana, IL, 2000), A K Peters, Natick, MA, 2002, pp. 43–54.
9. T. Nagell, *Introduction to number theory*, 2nd ed., Chelsea, New York, 1964.
10. C. Pomerance, *On primitive divisors of Mersenne numbers*, Acta Arith. **46** (1986), 355–367.
11. A. Schinzel, *On primitive prime factors of* $a^n - b^n$, Proc. Cambridge Philos. Soc. **58** (1962), 555–562.
12. C. L. Stewart, *The greatest prime factor of* $a^n - b^n$, Acta Arith. **26** (1974/75), 427–433.
13. ———, *On divisors of Fermat, Fibonacci, Lucas, and Lehmer numbers*, Proc. London Math. Soc. (3) **35** (1977), 425–447.

Department of Mathematics, Faculty of Economics, Meiji Gakuin University, 1-2-37 Shirokanedai, Minato-ku, Tokyo 108-8636, Japan.
E-mail address: leo@eco.meijigakuin.ac.jp

Department of Mathematics, Dartmouth College, Hanover, NH 03755, USA.
E-mail address: carlp@math.dartmouth.edu

Centre de Recherches Mathématiques
CRM Proceedings and Lecture Notes
Volume **36**, 2004

On the p-Adic Series $\sum_{n=1}^{\infty} n^k \cdot n!$

M. Ram Murty and Sarah Sumner

1. Introduction

Let $\mathbb{Q}_p$ be the field of p-adic numbers. It is well known that the sum

$$\sum_{n=1}^{\infty} a_n,$$

with $a_n \in \mathbb{Q}_p$ converges if and only if $|a_n|_p \to 0$ as $n \to \infty$ (see for example, [**6**, p. 87]). Since $|n!|_p \to 0$, we see that $\sum_{n=1}^{\infty} n!$ converges in $\mathbb{Q}_p$, as well as the related sums $\sum_{n=1}^{\infty} n^k \cdot n!$ for k a non-negative integer. Our goal in this paper is to investigate these p-adic sums.

Our first question is whether

$$\alpha = \sum_{n=1}^{\infty} n!$$

is a p-adic irrational or not, a question first asked by Schikhof [**9**, p. 17]. We conjecture that it is. Observe however that

$$\sum_{n=1}^{\infty} n \cdot n! = -1.$$

A simple proof by induction shows that $1\cdot 1! + 2\cdot 2! + 3\cdot 3! + \cdots + m\cdot m! = \sum_{n=0}^{m} n\cdot n! = (m+1)! - 1$. Since $\lim_{m\to\infty} |(m+1)!|_p = 0$, the desired result follows. On the other hand, we show below that

$$\alpha_k = \sum_{n=1}^{\infty} n^k \cdot n! = v_k - u_k \alpha,$$

where $v_k, u_k \in \mathbb{Z}$. We will prove using ideas from combinatorics and number theory that $u_k \neq 0$ for $k \equiv 0$ or $2 \pmod 3$. Thus, the irrationality of α_k hinges on the irrationality of α in these cases. This strengthens the observation made by Dragovich [**5**] that if α is irrational, then so is α_k. We suspect $u_k \neq 0$ for every $k \geq 2$. In this paper, we will assume that α is irrational and study the sequences u_k and v_k.

2000 *Mathematics Subject Classification.* Primary: 11B73; Secondary: 11J72, 11K41.
This is the final form of the paper.

The sequence of integers u_k defined below is of independent combinatorial interest. We hope to address the question of the non-vanishing of u_k for $k \geq 2$ in a future paper. Below, we establish some relationships with Stirling numbers of the second kind that allows us to prove non-vanishing results in some cases. It should be clear from the context that in the summations being considered, which ones are p-adic sums and which are summations in the field of real numbers.

2. Preliminaries

We must first recall some notation and elementary lemmas from combinatorics. The reader who wishes a more detailed review of this material is advised to consult Comtet [**3**].

The rising factorial is

$$\langle x \rangle_n = x(x+1)(x+2)\cdots(x+n-1).$$

The Stirling number of the first kind, denoted by $s(k,j)$, is defined by the rule that $(-1)^{k+j}s(k,j)$ is the number of permutations of $\{1,\ldots,k\}$ with j cycles. The Stirling number of the second kind, denoted by $S(k,j)$, is the number of partitions of $\{1,\ldots,k\}$ into j non-empty parts. Stirling numbers of the first kind are related to the rising factorial by the following relation [**3**, p. 213].

Lemma 1.

$$\langle x \rangle_j = \sum_{j=1}^{k} |s(k,j)| x^j.$$

We will also have need for a very useful inversion lemma [**3**, p. 144].

Lemma 2. *Let $\{f_k\}$ and $\{g_k\}$ be sequences of real numbers. Then,*

$$f_k = \sum_{j=1}^{k} S(k,j) g_j \Leftrightarrow g_k = \sum_{j=1}^{k} s(k,j) f_j.$$

We recall the following basic fact which will be used below:

$$\left(\sum_{n=0}^{\infty} \frac{a_n T^n}{n!}\right)\left(\sum_{n=0}^{\infty} \frac{b_n T^n}{n!}\right) = \sum_{n=0}^{\infty} \frac{c_n T^n}{n!}$$

where

$$c_n = \sum_{k=0}^{n} \binom{n}{k} a_k b_{n-k}.$$

Bell numbers will appear below, and will be denoted B_k. Since B_{k+1} is just the total number of all partitions of the set $\{1,2,\ldots,k+1\}$ [**3**, p. 210], we deduce:

Lemma 3.

$$B_{k+1} = \sum_{j=1}^{k+1} S(k+1,j).$$

3. A generating function for u_k

We must first demonstrate that $\sum_{n=1}^{\infty} n^k \cdot n! = v_k - u_k\alpha$, where $\alpha = \sum_{n=1}^{\infty} n!$ and v_k and u_k are integers. Notice that $\sum_{n=1}^{\infty}\big((n+k)! - n!\big)$ is a telescoping sum that equals $-\sum_{n=1}^{k} n!$. For $k = 2$ this becomes,

$$\sum_{n=1}^{\infty}\big((n+2)! - n!\big) = \sum_{n=1}^{\infty} n!\big((n+2)(n+1) - 1\big),$$

$$\sum_{n=1}^{\infty} n^2 \cdot n! = -\sum_{n=1}^{\infty} n!.$$

An inductive argument now shows that $\sum_{n=1}^{\infty} n^k \cdot n!$ can be written as $v_k - u_k\alpha$ where $v_k, u_k \in \mathbb{Z}$.

Observe now that for the same sum,

$$\begin{aligned}\sum_{n=1}^{\infty}\big((n+k)! - n!\big) &= \sum_{n=1}^{\infty}(n-1)!\{(n+k)(n+k-1)\cdots n - n\}\\ &= \sum_{n=1}^{\infty}(n-1)!\langle n\rangle_{k+1} - \sum_{n=1}^{\infty} n!.\end{aligned}$$

Hence applying Lemma 1,

$$\begin{aligned}-\sum_{n=1}^{k} n! + \alpha &= \sum_{n=1}^{\infty}(n-1)!\langle n\rangle_{k+1}\\ &= \sum_{n=1}^{\infty}(n-1)!\sum_{j=1}^{k+1}|s(k+1,j)|n^j\\ &= \sum_{j=1}^{k+1}|s(k+1,j)|\sum_{n=1}^{\infty} n^{j-1}n!.\end{aligned}$$

Letting $\sum_{n=1}^{\infty} n^{j-1}n! = v_{j-1} - u_{j-1}\alpha$ gives

$$-\sum_{n=1}^{k} n! + \alpha = \sum_{j=1}^{k+1}|s(k+1,j)|\{v_{j-1} - u_{j-1}\alpha\}.$$

If we now suppose that α is irrational, then this yields

$$\sum_{j=1}^{k+1}|s(k+1,j)|v_{j-1} = -\sum_{n=1}^{k} n!, \tag{3.1}$$

$$\sum_{j=1}^{k+1}|s(k+1,j)|u_{j-1} = -1. \tag{3.2}$$

Removing absolute values from equation (3.2) we get

$$\sum_{j=1}^{k+1}(-1)^j s(k+1,j)u_{j-1} = (-1)^k. \tag{3.3}$$

Applying Lemma 2 to equation (3.3) yields one generating function. Let $f_{k+1} = (-1)^{k+1}u_k$ and $g_j = (-1)^{j-1}$ to get

$$\begin{aligned} f_{k+1} = (-1)^{k+1}u_k &= \sum_{j=1}^{k+1} S(k+1,j)g_j \\ &= \sum_{j=1}^{k+1} (-1)^{j-1} S(k+1,j). \end{aligned}$$

Thus,

LEMMA 4.

$$(-1)^k u_k = \sum_{j=1}^{k+1} (-1)^j S(k+1,j).$$

This is a simplification of the method of solving $k+1$ linear equations to determine each u_k outlined by Dragovich [**5**, p. 101]. Using this relation, the first few terms in the sequence $\{u_k\}$ can be easily calculated:

$$\{0, 1, -1, -2, 9, -9, -50, 267, -413, -2180, 17731, -50533, -110176, \\ 1966797, -9938669, 8638718, 278475061, -2540956509, 9816860358 \ldots\}$$

This is the negative of sequence A014182 of Sloane [**10**]. Lemma 4 implies the crucial result

LEMMA 5.

$$u_k \equiv \sum_{j=1}^{k+1} S(k+1,j) \pmod 2.$$

Other generating functions for the u_k also exist. Now we caution the reader that below, the series being considered are not p-adic series but usual series involving real numbers.

LEMMA 6.

$$(-1)^k u_k = e \sum_{n=0}^{\infty} \frac{n^{k+1}}{n!}(-1)^n.$$

PROOF. Recall the known identity [**3**, p. 204]

$$S(k+1,j) = \frac{1}{j!} \sum_{r=0}^{j} (-1)^r \binom{j}{r} (j-r)^{k+1}.$$

Now,

$$\begin{aligned} (-1)^k u_k &= \sum_{j \geq 1} (-1)^j S(k+1,j) \\ &= \sum_{j \geq 1} \frac{(-1)^j}{j!} \sum_{0 \leq r \leq j} (-1)^r \binom{j}{r} (j-r)^{k+1} \\ &= \sum_{r \geq 0} \frac{(-1)^r}{r!} \sum_{j \geq r} \frac{(-1)^j}{j!} \frac{j!}{(j-r)!} (j-r)^{k+1} \end{aligned}$$

$$= \sum_{r\geq 0} \frac{1}{r!} \sum_{j\geq r} \frac{(-1)^{j-r}}{(j-r)!}(j-r)^{k+1}$$

$$= \sum_{r\geq 0} \frac{1}{r!} \sum_{n=0}^{\infty} \frac{n^{k+1}}{n!}(-1)^n$$

$$(-1)^k u_k = e\sum_{n=0}^{\infty} \frac{n^{k+1}}{n!}(-1)^n. \qquad \square$$

LEMMA 7.

$$\sum_{k=0}^{\infty} \frac{(-1)^k u_k T^k}{k!} = -e^{-e^T+T+1}$$

PROOF.

$$\sum_{k=0}^{\infty} \frac{(-1)^k u_k T^k}{k!} = e\sum_{k=0}^{\infty} \frac{T^k}{k!} \sum_{n=0}^{\infty} \frac{(-1)^n n^{k+1}}{n!}$$

$$= e\sum_{n=0}^{\infty}\sum_{k=0}^{\infty} \frac{T^k n^k}{k!} \frac{(-1)^n n}{n!}$$

$$= e\sum_{n=1}^{\infty} \frac{(-1)^n}{n!} n e^{Tn}$$

$$= -e\sum_{n=1} \frac{(-1)^{n-1}}{(n-1)!} e^{T(n-1)} e^T$$

$$= -e \cdot e^{-e^T} \cdot e^T$$

$$\sum_{k=0}^{\infty} \frac{(-1)^k u_k T^k}{k!} = -e^{-e^T+T+1}. \qquad \square$$

Since $e^{e^T-1} = \sum B_k T^k/k!$ [**3**, p. 211], applying the remark made earlier about the multiplication of exponential generating functions and Lemma 7, we see that

$$\left(\sum_{k=0}^{\infty} \frac{B_k T^k}{k!}\right)\left(\sum_{k=0}^{\infty} \frac{(-1)^k u_k T^k}{k!}\right) = -\sum_{k=0}^{\infty} \frac{T^k}{k!}$$

which proves

LEMMA 8.

$$\sum_{k=0}^{n} \binom{n}{k} B_k (-1)^{n-k} u_{n-k} = -1.$$

4. Congruences for $\{u_k\}$

There are several interesting congruence relations for $\{u_k\}$.

LEMMA 9. *For every prime p, $u_{p-1} \equiv -2 \pmod{p}$.*

PROOF. Recall that $S(p,1) = S(p,p) = 1$ and that for $2 \le k \le p-1$, $S(p,k) \equiv 0 \pmod p$, [**3**, p. 219]. Observe,

$$\begin{aligned}(-1)^{p-1}u_{p-1} &= \sum_{j=1}^{p}(-1)^j S(p,j),\\ (-1)^{p-1}u_{p-1} &\equiv (-1)S(p,1) + (-1)S(p,p) \pmod p,\\ u_{p-1} &\equiv -2 \pmod p.\end{aligned}$$

□

LEMMA 10. *For every prime p, $u_p \equiv -1 \pmod p$.*

PROOF. Substitute the recurrence $S(k+1,j) = S(k,j-1)+jS(k,j)$, [**3**, p. 208] into

$$\begin{aligned}(-1)^p u_p &= \sum_{j=1}^{p+1}(-1)^j S(p+1,j),\\ -u_p &= \sum_{j=1}^{p+1}(-1)^j S(p,j-1) + \sum_{j=1}^{p+1}(-1)^j jS(p,j),\\ -u_p &\equiv -u_{p-1} + (-1)S(p,1) \pmod p,\\ u_p &\equiv u_{p-1}+1 \pmod p,\\ u_p &\equiv -2+1 \pmod p,\\ u_p &\equiv -1 \pmod p.\end{aligned}$$

□

Using the fact that for every prime p,

$$\binom{p^r}{k} \equiv 0 \pmod p$$

whenever $1 \le k \le p^r - 1$ along with Lemma 8, we conclude

LEMMA 11. *Let p be a prime number. For every $r \ge 1$,*

$$B_{p^r} - 1 \equiv (-1)^{p^r} u_{p^r} \pmod p.$$

5. A general theorem

THEOREM 1. *If $k \equiv 0$ or $2 \pmod 3$ then $u_k \ne 0$, and hence in these cases α_k will be irrational provided α is.*

Since we have already shown in Lemma 5 that

$$u_k \equiv \sum_{j=1}^{k+1} S(k+1,j) \equiv B_{k+1} \pmod 2,$$

Theorem 1 will follow from the following lemma:

LEMMA 12. *If $k \equiv 2 \pmod 3$ then B_k is even, otherwise B_k is odd.*

PROOF. We will proceed by induction. $B_0 = 1$, $B_1 = 1$ and $B_2 = 2$, so the base cases are clear. Suppose the lemma is true for all $j \le k$. Recall the recursion for B_{k+1},

$$B_{k+1} = \sum_{j=0}^{k}\binom{k}{j}B_j.$$

By the induction hypothesis, if $j \equiv 2 \pmod 3$ then B_j is even, and otherwise B_j is odd. Hence the recursive formula becomes

$$\sum_{j \not\equiv 2 \pmod 3} \binom{k}{j} \equiv \sum_{j \equiv 0 \pmod 3} \binom{k}{j} + \sum_{j \equiv 1 \pmod 3} \binom{k}{j} \pmod 2.$$

Let $\zeta = e^{2\pi i/3}$ be a cube root of unity. From the binomial theorem, we see

$$\sum_{j=0}^{k} \binom{k}{j} x^j = (1+x)^k,$$

$$\sum_{j=0}^{k} \binom{k}{j} \zeta^j x^j = (1+\zeta x)^k,$$

$$\sum_{j=0}^{k} \binom{k}{j} \zeta^{2j} x^j = (1+\zeta^2 x)^k.$$

Adding these together we get

$$\sum_{j=0}^{k} \binom{k}{j} x^j (1+\zeta^j+\zeta^{2j}) = (1+x)^k + (1+\zeta x)^k + (1+\zeta^2 x)^k.$$

Let $x = 1$.

$$\sum_{j=0}^{k} \binom{k}{j} (1+\zeta^j+\zeta^{2j}) = 2^k + (1+\zeta)^k + (1+\zeta^2)^k.$$

Recall

$$1+\zeta^j+\zeta^{2j} = \begin{cases} 3 & \text{if } j \equiv 0 \pmod 3 \\ 0 & \text{otherwise.} \end{cases}$$

So

$$\begin{aligned} 3 \sum_{j \equiv 0 \pmod 3} \binom{k}{j} &= 2^k + (1+\zeta)^k + (1+\zeta^2)^k \\ &= 2^k + (1+\zeta)^k + (1+\zeta^{-1})^k \\ &= 2^k + 2\Re(1+\zeta)^k \\ &= 2^k - 2\Re(\zeta^{2k}). \end{aligned}$$

Let us note that

$$\zeta^{2k} = \begin{cases} 1 & \text{if } k \equiv 0 \pmod 3 \\ \zeta & \text{if } k \equiv 2 \pmod 3 \\ \zeta^2 & \text{if } k \equiv 1 \pmod 3). \end{cases} \tag{5.1}$$

Now for the other sum. Consider

$$\sum_{j=0}^{k}\binom{k}{j}x^j = (1+x)^k,$$

$$\sum_{j=0}^{k}\binom{k}{j}\zeta^{j-1}x^j = \zeta^{-1}(1+\zeta x)^k,$$

$$\sum_{j=0}^{k}\binom{k}{j}\zeta^{2j-2}x^j = \zeta^{-2}(1+\zeta^2 x)^k.$$

Adding these together gives

$$\sum_{j=0}^{k}\binom{k}{j}x^j(1+\zeta^{j-1}+\zeta^{2j-2}) = (1+x)^k+\zeta^{-1}(1+\zeta x)^k+\zeta^{-2}(1+\zeta^2 x)^k.$$

Let $x=1$

$$\sum_{j=0}^{k}\binom{k}{j}(1+\zeta^{j-1}+\zeta^{2j-2}) = 2^k+\zeta^{-1}(1+\zeta)^k+\zeta^{-2}(1+\zeta^2)^k.$$

Now

$$1+\zeta^{j-1}+\zeta^{2j-2} = \begin{cases} 3 & \text{if } j-1\equiv 0 \pmod 3 \\ 0 & \text{otherwise.} \end{cases}.$$

Hence,

$$\begin{aligned} 3\sum_{j\equiv 1 \pmod 3}\binom{k}{j} &= 2^k+\zeta^{-1}(1+\zeta)^k+\zeta(1+\zeta^{-1})^k \\ &= 2^k+2\Re\zeta(1+\zeta^{-1})^k \\ &= 2^k+2\Re\zeta(1+\zeta^2)^k \\ &= 2^k+2\Re\zeta(-\zeta)^k \\ &= 2^k+(-1)^k 2\Re(\zeta^{k+1}). \end{aligned}$$

Now

$$\zeta^{k+1} = \begin{cases} 1 & \text{if } k+1\equiv 0 \pmod 3 \\ \zeta & \text{if } k+1\equiv 1 \pmod 3 \\ \zeta^2 & \text{if } k+1\equiv 2 \pmod 3. \end{cases} \tag{5.2}$$

Combining the information in equations (5.1) and (5.2), we see that if $k\equiv 1 \pmod 3$ then $B_{k+1}\equiv 0 \pmod 2$, and if $k\not\equiv 1 \pmod 3$ then $B_{k+1}\equiv 1 \pmod 2$ proving Theorem 1. □

6. Concluding remarks

The non-vanishing of u_k is a conjecture of Wilf (see [**7**]). In [**11**], it is proved that the number of $k\le x$ with $u_k=0$ is $O(x^{2/3})$.

Other questions deserving attention linked to the study of $\{u_k\}$ concern the existence and properties of p-adic interpolation of functions of the form

$$f(s)=\sum n^s\cdot n!.$$

If the sum is taken over all odd numbers n, then it can be shown that $f(s)$ has a 2-adic interpolation, which raises the question of whether $\{u_k\}$ and $\{v_k\}$ have 2-adic limits or not. Similar questions can be raised about p-adic interpolation and limits for odd primes, which seem to be more involved cases.

References

1. Z. I. Borevich and I. R. Shafarevich, *Number theory*, Academic Press, Toronto, 1966.
2. E. B. Burger and T. Struppeck, *Does $\sum_{n=0}^{\infty} 1/n!$ really converge? Infinite series and p-adic analysis*, Amer. Math. Monthly **103** (1996), no. 9, 565–577.
3. L. Comtet, *Advanced combinatorics*, D. Reidel Publishing Company, 1974.
4. B. G. Dragovich, *Power series everywhere convergent on R and all Q_p*, J. Math. Phys. **34** (1993), no. 3, 95 105.
5. ———, *On some p-adic series with factorials*, p-adic Functional Analysis, (W. H. Schikhof, C. Perez–Garcia and J. Kąkol, eds.), Lecture Notes in Pure and App. Math., vol. 192, Marcel Dekker, Inc., New York, 1997, pp. 95-105.
6. F. Q. Gouvêa, *p-adic numbers: An introduction*, Springer-Verlag, New York, 1993.
7. M. Klazar, *Counting even and odd partitions*, Amer. Math. Monthly **110** (2003), no. 6, 527–532.
8. W. H. Schikhof, C. Perez–Garcia and J. Kąkol (eds.), *p-adic functional analysis*, Lecture Notes in Pure and Appl. Math., vol. 192, Marcel Dekker, Inc., New York, 1997.
9. W. Schikhof, *Ultrametric calculus: an introduction to p-adic analysis*, Cambridge University Press, Cambridge, 1984.
10. N. J. A. Sloane, *The on-line encyclopedia of integer sequences*, Notices Amer. Math. Soc. **50** (2003), no. 8, 912–915; `http://www.research.att.com/~njas/sequences`.
11. Y. Yang, *On a multiplicative partition function*, Electron. J. Combin. **8** (2001), no. 1.
12. L. C. Washington, *Introduction to cyclotomic fields*, Grad. Texts in Math., vol. 83, Springer-Verlag, New York, 1983.

Department of Mathematics and Statistics, Queen's University, Kingston, ON K7L 3N6, Canada.
E-mail address: `murty@mast.queensu.ca`

General Dynamics Canada Ltd., Ottawa, ON K2H 5B7, Canada.
E-mail address: `sarah.sumner@gdcanada.com`

Centre de Recherches Mathématiques
CRM Proceedings and Lecture Notes
Volume **36**, 2004

Hecke Operators and the q-Expansion of Modular Forms

Ken Ono

1. Introduction and statement of results

In a recent paper [**4**], the author, Bruinier, and Kohnen investigated the values of a certain sequence of modular functions in connection with the arithmetic properties of meromorphic modular forms on $\mathrm{SL}_2(\mathbb{Z})$. In this partially expository note we describe some of the combinatorial consequences of these results for the Fourier coefficients of modular forms. Suppose that

$$f(z) = \sum_{n=h}^{\infty} a_f(n)q^n \quad (q := e^{2\pi iz} \text{ throughout})$$

is a weight $k \in 2\mathbb{Z}$ meromorphic modular form on $\mathrm{SL}_2(\mathbb{Z})$. It is well known that $f(z)$ is distinguished by its weight k and its "first few coefficients". The Riemann–Roch Theorem provides the number of coefficients which are sufficient for distinguishing such a form $f(z)$. Here we obtain a combinatorial extension of this classical fact; specifically, we give universal recursion relations which produce the entire Fourier expansion of any meromorphic modular form on $\mathrm{SL}_2(\mathbb{Z})$ in terms of the weight k and its first few coefficients.

To state this result, we require a few preliminaries. If $k \geq 2$ is even, then let $E_k(z)$ denote the usual Eisenstein series

$$E_k(z) := 1 - \frac{2k}{B_k}\sum_{n=1}^{\infty}\sigma_{k-1}(n)q^n. \tag{1.1}$$

Here B_k denotes the usual kth Bernoulli number and $\sigma_{k-1}(n) := \sum_{d|n} d^{k-1}$ (note: throughout this note all divisors of integers are taken to be positive). If $k > 2$, then $E_k(z)$ is a weight k modular form on $\mathrm{SL}_2(\mathbb{Z})$. Although the Eisenstein series

$$E_2(z) = 1 - 24\sum_{n=1}^{\infty}\sigma_1(n)q^n \tag{1.2}$$

2000 *Mathematics Subject Classification.* Primary: 11F03, 11F11, 11F33.

The author thanks the support of a John S. Guggenheim Fellowship, an Alfred P. Sloan Fellowship, a David and Lucile Packard Fellowship, an H. I. Romnes Fellowship and a grant from the National Science Foundation.

This is the final form of the paper.

is not a modular form, it will play an important role. As usual, let $j(z)$ denote the modular function on $\mathrm{SL}_2(\mathbb{Z})$ which is holomorphic on $\mathcal{H}$, the upper half of the complex plane, with Fourier expansion

$$j(z) := q^{-1} + 744 + 196884q + 21493760q^2 + \cdots . \tag{1.3}$$

We require a specific sequence of modular functions $j_m(z)$. To define this sequence, let

$$j_0(z) := 1 \quad \text{and} \quad j_1(z) := j(z) - 744. \tag{1.4}$$

If $m \geq 2$, then define $j_m(z)$ by

$$j_m(z) := j_1(z) \mid T_0(m), \tag{1.5}$$

where $T_0(m)$ is the usual normalized mth weight zero Hecke operator defined by

$$g(z) \mid T_0(m) = \sum_{\substack{d|m \\ ad=m}} \sum_{b=0}^{d-1} g\left(\frac{az+b}{d}\right). \tag{1.6}$$

Each $j_m(z)$ is a monic polynomial in $j(z)$ of degree m. Here we list the first few:

$$\begin{aligned}
&j_0(z) = 1,\\
&j_1(z) = j(z) - 744 = q^{-1} + 196884q + \cdots ,\\
&j_2(z) = j(z)^2 - 1488j(z) + 159768 = q^{-2} + 42987520q + \cdots ,\\
&j_3(z) = j(z)^3 - 2232j(z)^2 + 1069956j(z) - 36866976 = q^{-3} + 2592899910q + \cdots .
\end{aligned}$$

Let $\mathfrak{F}$ denote the usual fundamental domain of the action of $\mathrm{SL}_2(\mathbb{Z})$ on $\mathcal{H}$. By assumption, $\mathfrak{F}$ does not include the cusp at ∞. Throughout, let $i := \sqrt{-1}$ and let $\omega := (1 + \sqrt{-3})/2$. If $\tau \in \mathfrak{F}$, then define e_τ by

$$e_\tau := \begin{cases} \frac{1}{2} & \text{if } \tau = i, \\ \frac{1}{3} & \text{if } \tau = \omega, \\ 1 & \text{otherwise.} \end{cases} \tag{1.7}$$

The next result contains the universal polynomial recursion formulae.

THEOREM 1. *For every positive integer n define $F_n(x_1, \ldots, x_n) \in \mathbb{Q}[x_1, \ldots, x_n]$ by*

$$\begin{aligned}
F_n(x_1, \ldots, x_n) \\
:= -\frac{2x_1\sigma_1(n)}{n} + \sum_{\substack{m_1,\ldots,m_{n-1}\geq 0,\\ m_1+2m_2+\cdots+(n-1)m_{n-1}=n}} (-1)^{m_1+\cdots+m_{n-1}} \frac{(m_1 + \cdots + m_{n-1} - 1)!}{m_1! \cdots m_{n-1}!} \\
\times x_2^{m_1} \cdots x_n^{m_{n-1}}.
\end{aligned}$$

If $f(z) = q^h + \sum_{n=1}^{\infty} a_f(h+n)q^{h+n}$ is a weight k meromorphic modular form on $\mathrm{SL}_2(\mathbb{Z})$, then for every positive integer n we have

$$a_f(h+n) = F_n\big(k, a_f(h+1), \ldots, a_f(h+n-1)\big) - \frac{1}{n} \sum_{\tau \in \mathfrak{F}} e_\tau \operatorname{ord}_\tau(f) \cdot j_n(\tau).$$

The first few polynomials F_n are

$$F_1(x_1) := -2x_1, \tag{1.8}$$

$$F_2(x_1, x_2) := -3x_1 + \frac{x_2^2}{2}, \tag{1.9}$$

$$F_3(x_1, x_2, x_3) := -\frac{8x_1}{3} - \frac{x_2^3}{3} + x_2x_3, \tag{1.10}$$

$$F_4(x_1, x_2, x_3, x_4) := -\frac{7x_1}{2} - x_2^2x_3 + x_2x_4 + \frac{x_2^4}{4} + \frac{x_3^2}{2}. \tag{1.11}$$

The $n = 1$ case of Theorem 1 implies that

$$a_f(h+1) = 60k - 744h - \sum_{\tau \in \mathfrak{F}} e_\tau \operatorname{ord}_\tau(f) \cdot j(\tau).$$

EXAMPLE 1. Since

$$\Delta(z) = \sum_{n=1}^{\infty} \tau(n)q^n = q - 24q^2 + 252q^3 - \cdots,$$

the unique normalized weight 12 cusp form on $\mathrm{SL}_2(\mathbb{Z})$, is nonvanishing in $\mathcal{H}$, Theorem 1 implies that

$$\tau(n+1) = F_n\big(12, \tau(2), \dots, \tau(n)\big).$$

Here we also examine the infinite product expansion of modular forms. In [**2,3**], Borcherds investigated the infinite product expansion of those modular forms possessing a Heegner divisor (note: Zagier [**7**] has recently obtained a more elementary proof). For such forms $f(z)$, he provided a striking description for the infinite product expansion. For example, if the integers $c(n)$ are the exponents in the infinite product expansion of the classical Eisenstein series $E_4(z)$

$$E_4(z) = 1 + 240\sum_{n=1}^{\infty} \sigma_3(n)q^n = (1-q)^{-240}(1-q^2)^{26760}\cdots = \prod_{n=1}^{\infty}(1-q^n)^{c(n)},$$

his theorem implies that there is a weight $\frac{1}{2}$ meromorphic modular form

$$G(z) = \sum_{n \geq -3} b(n)q^n = q^{-3} + 4 - 240q + 26760q^4 + \cdots - 4096240q^9 + \cdots$$

on $\Gamma_0(4)$ with the property that $c(n) = b(n^2)$ for every positive integer n.

In more recent work [**5**], Bruinier and the author have obtained a different type of result for a wider class of forms using p-adic modular forms. However, these results do not provide formulae for the exponents in the infinite product expansion of a generic modular form. The next result provides such formulae.

THEOREM 2. *Suppose that* $f(z) = \sum_{n=h}^{\infty} a_f(n)q^n$ *is a weight* k *meromorphic modular form on* $\mathrm{SL}_2(\mathbb{Z})$ *for which* $a_f(h) = 1$*, and let* $c(n)$ *denote the complex numbers for which*

$$f(z) = q^h \prod_{n=1}^{\infty}(1-q^n)^{c(n)}.$$

If n *is a positive integer, then*

$$c(n) = 2k + \frac{1}{n}\sum_{\tau \in \mathfrak{F}} e_\tau \operatorname{ord}_\tau(f) \sum_{d \mid n} \mu(n/d) j_d(\tau).$$

EXAMPLE 2. Let $f(z) = \Delta(z) = q - 24q^2 + 252q^3 - \cdots$ be the unique normalized cusp form of weight 12 on $\mathrm{SL}_2(\mathbb{Z})$. Since it is nonvanishing on $\mathcal{H}$, Theorem 2 easily gives the classical infinite product formula

$$\Delta(z) = q \prod_{n=1}^{\infty} (1 - q^n)^{24}.$$

EXAMPLE 3. Here we revisit the infinite product for $E_4(z)$. Since the divisor of $E_4(z)$ is supported at a simple zero at $\tau = \omega$, Theorem 2 implies that

$$E_4(z) = 1 + 240 \sum_{n=1}^{\infty} \sigma_3(n) q^n = (1-q)^{-240}(1-q^2)^{26760} \cdots = \prod_{n=1}^{\infty} (1 - q^n)^{c(n)},$$

where

$$c(n) = 8 + \frac{1}{3n} \sum_{d \mid n} \mu(n/d) j_d(\omega).$$

By (2.4), it turns out that

$$\sum_{n=0}^{\infty} j_n(\omega) q^n = \frac{E_6(z)}{E_4(z)} = 1 - 744q + 159768q^2 - 36866976q^3 + \cdots.$$

Using these terms, Theorem 2 gives

$$\begin{aligned} c(1) &= 8 + j_1(\omega)/3 = 8 - 744/3 = -240, \\ c(2) &= 8 + \big(j_2(\omega) - j_1(\omega)\big)/6 = 26760, \\ c(3) &= 8 + \big(j_3(\omega) - j_1(\omega)\big)/9 = -4096240. \end{aligned}$$

In §2 we sketch the proof of a fundamental formula describing the action of Ramanujan's Theta-operator (see also [**4**]), and in §3 we use this result to prove Theorems 1 and 2.

2. An explicit formula for Ramanujan' s Θ-operator

Recall that Ramanujan's Theta-operator is defined by

$$\Theta\left(\sum_{n=h}^{\infty} a(n) q^n\right) := \sum_{n=h}^{\infty} n a(n) q^n. \tag{2.1}$$

We refer to Θ as Ramanujan's operator since he observed that [**6**]

$$\Theta(E_4) = (E_4 E_2 - E_6)/3 \quad \text{and} \quad \Theta(E_6) = (E_6 E_2 - E_8)/2. \tag{2.2}$$

If $f(z) = \sum_{n=h}^{\infty} a_f(n) q^n$ is a weight k meromorphic modular form on $\mathrm{SL}_2(\mathbb{Z})$, then

$$\Theta(f) = (\tilde{f} + k f E_2)/12, \tag{2.3}$$

where $\tilde{f}$ is a meromorphic modular form of weight $k+2$ on $\mathrm{SL}_2(\mathbb{Z})$. Here we give an explicit description of the form $\tilde{f}$ appearing above in terms of the modular functions $j_m(z)$.

To state these results, for every point $\tau \in \mathcal{H}$ let $H_\tau(z) := \sum_{n=0}^{\infty} j_n(\tau) q^n$. Asai, Kaneko, and Ninomiya [**1**, Th. 3] proved that

$$H_\tau(z) = \sum_{n=0}^{\infty} j_n(\tau) q^n = \frac{E_4^2(z) E_6(z)}{\Delta(z)} \cdot \frac{1}{j(z) - j(\tau)}. \tag{2.4}$$

In particular, for every τ it turns out that $H_\tau(z)$ is a weight 2 meromorphic modular form. The utility of (2.4) was already known; for example, it can be used to prove the famous denominator formula for the monster Lie algebra.

In [**4**], the following result is proved.

THEOREM 3. *If $f = \sum_{n=h}^{\infty} a_f(n)q^n$ is a nonzero weight k meromorphic modular form on $\mathrm{SL}_2(\mathbb{Z})$ for which $a_f(h) = 1$, then*

$$\frac{\Theta(f)}{f} = \frac{kE_2}{12} - f_\Theta,$$

where f_Θ is defined by

$$f_\Theta := \sum_{\tau \in \mathfrak{F}} e_\tau \operatorname{ord}_\tau(f) H_\tau(z).$$

For convenience, we recall the following straightforward fact (see [**4**, Prop. 2.1]) which explains the connection between Theorem 3 and the combinatorial properties of the infinite product exponents of a given modular form.

PROPOSITION 1. *Let $f = \sum_{n=h}^{\infty} a_f(n)q^n$ be a meromorphic function in a neighborhood of $q = 0$, and suppose that $a_f(h) = 1$. Then there are uniquely determined complex numbers $c(n)$ such that*

$$f = q^h \prod_{n=1}^{\infty} (1 - q^n)^{c(n)},$$

where the product converges in a small neighborhood of $q = 0$. Moreover, the following identity is true

$$\frac{\Theta(f)}{f} = h - \sum_{n=1}^{\infty} \sum_{d|n} c(d)\, dq^n.$$

SKETCH OF THE PROOF OF THEOREM 3. As before, let $\mathfrak{F}$ denote the standard fundamental domain for the action of $\mathrm{SL}_2(\mathbb{Z})$ on $\mathcal{H}$. We cut off $\mathcal{F}$ by a horizontal line $\mathcal{L} := \{iC - t : -\frac{1}{2} \leq t \leq \frac{1}{2}\}$ where $C > 0$ is chosen so large that all poles and zeros of f, apart from those at the cusp at infinity, are contained in $\{z \in \mathcal{H} : \mathrm{Im}(z) < C\}$.

For simplicity, suppose that $f(z)$ has no zero or pole on the boundary $\partial\mathcal{F}$ except possibly i or ω, and let γ be the closed path with positive orientation consisting of $\mathcal{L}$ and γ_1 where γ_1 is the part of $\partial\mathfrak{F}$ below $\mathcal{L}$ modified in the usual way (i.e. as in the standard proof of the "$k/12$-valence formula").

We integrate

$$\frac{1}{2\pi i} \cdot \frac{f'(z)}{f(z)} \cdot j_n(z)$$

along γ. By the residue theorem, taking into account that $j_n(z)$ is holomorphic on $\mathcal{H}$, this integral is equal to

$$\sum_{\tau \in \mathfrak{F} - \{\omega, i\}} \operatorname{ord}_\tau(f) j_n(\tau).$$

On the other hand, the integral can be evaluated separately along the different pieces of γ. If we let r tend to zero, we then find that

$$\sum_{\tau\in\mathfrak{F}-\{\omega,i\}} \operatorname{ord}_\tau(f) j_n(\tau) = -\tfrac{1}{3}\operatorname{ord}_\omega(f) j_n(\omega) - \tfrac{1}{2}\operatorname{ord}_i(f) j_n(i) + \frac{1}{2\pi i}\int_\rho \frac{F'(q)}{F(q)} J_n(q)\,dq - \frac{k}{2\pi i}\int_\sigma \frac{j_n(z)}{z}\,dz. \tag{2.5}$$

Here $F(q) = f(z)$ as before and $J_n(q) := j_n(z)$. Furthermore, ρ is a small circle around $q = 0$ with negative orientation and not containing any pole or zero of $F(q)$ except possibly 0, and σ is the part of the unit circle in the upper half-plane that connects ω and i, with positive orientation.

By Proposition 1, we have that

$$\frac{qF'(q)}{F(q)} = \frac{\Theta(f)}{f} = h - \sum_{n=1}^{\infty}\sum_{d\mid n} c(d)\,dq^n,$$

where h is the order of F at $q = 0$. Therefore we find that

$$\frac{1}{2\pi i}\int_\rho \frac{F'(q)}{F(q)} J_n(q)\,dq = \sum_{d\mid n} c(d)\,d. \tag{2.6}$$

Since (2.5) holds for $f = \Delta$, and since Δ has no zeros on $\mathcal{H}$, we obtain from (2.5) that

$$\frac{1}{2\pi i}\int_\sigma \frac{j_n(z)}{z}\,dz = 2\sigma_1(n). \tag{2.7}$$

To prove that

$$\frac{\Theta(f)}{f} = \frac{kE_2}{12} - f_\Theta,$$

where f_Θ has the claimed form, one now simply argues coefficient by coefficient with (1.2), (2.5), (2.6) and (2.7). □

3. Proof of Theorems 1 and 2

Proof of Theorem 1. By Theorem 3, if n is a positive integer, then

$$\sum_{d\mid n} c(d)d = 2k\sigma_1(n) + \sum_{\tau\in\mathfrak{F}} e_\tau \operatorname{ord}_\tau(f) j_n(\tau). \tag{3.1}$$

Therefore, to prove Theorem 1 it suffices to obtain a closed formula for

$$b(n) = \sum_{d\mid n} c(d)d \tag{3.2}$$

in terms of the coefficients of $f(z)$. In particular, a straightforward calculation reveals that it suffices to show that if $n \geq 1$, then

$$b(n) = n \sum_{\substack{m_1,\ldots,m_n\geq 0,\\ m_1+2m_2+\cdots+nm_n=n}} (-1)^{m_1+\cdots+m_n}\frac{(m_1+\cdots+m_n-1)!}{m_1!\cdots m_n!} \times a_f(h+1)^{m_1}\cdots a_f(h+n)^{m_n}. \tag{3.3}$$

To prove (3.3), observe that

$$0 = b(n) + b(n-1)a_f(h+1) + b(n-2)a_f(h+2) + \cdots + b(1)a_f(h+n-1) + na_f(h+n),$$

and then use the fact that

$$0 = s_n - s_{n-1}\sigma_1 + s_{n-2}\sigma_2 - \cdots + (-1)^{n-1}s_1\sigma_{n-1} + (-1)^n n\sigma_n.$$

Here the σ_i are the elementary symmetric functions in $X_1, \ldots, X_n$ and the s_i are the power functions in these variables (i.e. $s_i := X_1^i + \cdots + X_n^i$). By evaluating these identities at $(X_1, \ldots, X_n) = (\lambda(1, n), \ldots, \lambda(n, n))$ where the $\lambda(j, n)$ are the roots of the polynomial

$$X^n + a_f(h+1)X^{n-1} + a_f(h+2)X^{n-2} + \cdots + a_f(h+n),$$

one obtains (3.3). This final calculation requires the fact that

$$s_i = i \sum_{\substack{m_1,\ldots,m_n \geq 0,\\ m_1+2m_2+\cdots+nm_n=i}} (-1)^{m_2+m_4+\cdots} \frac{(m_1+m_2+\cdots+m_n-1)!}{m_1!m_2!\cdots m_n!} \cdot \sigma_1^{m_1}\cdots\sigma_n^{m_n}.$$

□

Proof of Theorem 2. By (3.1), it follows that

$$F(n) := \sum_{d|n} c(d)d = 2k\sigma_1(n) + \sum_{\tau\in\mathfrak{F}} e_\tau \operatorname{ord}_\tau(f) j_n(\tau).$$

By Möbius Inversion, it follows that

$$\begin{aligned} c(n) &= \frac{1}{n}\sum_{d|n} \mu(n/d)F(d) \\ &= \frac{1}{n}\sum_{d|n} \mu(n/d)\left(2k\sigma_1(d) + \sum_{\tau\in\mathfrak{F}} e_\tau \operatorname{ord}_\tau(f) j_d(\tau)\right). \end{aligned}$$

The formula for $c(n)$ follows from the elementary identity

$$n = \sum_{d|n} \mu(n/d)\sigma_1(d).$$

□

References

1. T. Asai, M. Kaneko, and H. Ninomiya, *Zeros of certain modular functions and an application*, Comm. Math. Univ. Sancti Pauli **46** (1997), 93–101.
2. R. E. Borcherds, *Automorphic forms on* $O_{s+2,2}(\mathbb{R})$ *and infinite products*, Invent. Math. **120** (1995), 161–0213.
3. R. E. Borcherds ______, *Automorphic forms on* $O_{s+2,2}(\mathbb{R})^+$ *and generalized Kac-Moody algebras*, Proc. Internat. Congress of Mathematicians (Zürich, 1994), vol. 1, 2, Birkhäuser, Basel, 1995, pp. 744–752.
4. J. H. Bruinier, W. Kohnen, and K. Ono, *Values of modular functions and divisors of modular forms*, Compositio Math. (to appear).
5. J. H. Bruinier and K. Ono, *The arithmetic of Borcherds exponents*, Math. Ann. **327** (2003), 293–303.
6. S. Ramanujan, *On certain arithmetical functions*, Trans. Camb. Phil. Soc. **22** (1916), no.18, 159–184.
7. D. Zagier, *Traces of singular moduli*, Motives, Polylogarithms and Hodge Theory, Part I (Irvine, CA, 1998), Int. Press Lect. Ser., 3, I, Somerville, MA, 2002, pp. 211–244.

Department of Mathematics, University of Wisconsin, Madison, WI 53706, USA.
E-mail address: ono@math.wisc.edu

Centre de Recherches Mathématiques
CRM Proceedings and Lecture Notes
Volume **36**, 2004

An *ABC* Construction of Number Fields

David P. Roberts

ABSTRACT. We describe a general three step method for constructing number fields with Lie-type Galois groups and discriminants factoring into powers of specified primes. The first step involves extremal solutions of the matrix equation $ABC = I$. The second step involves extremal polynomial solutions of the equation $A(x) + B(x) + C(x) = 0$. The third step involves integer solutions of the generalized Fermat equation $ax^p + by^q + cz^r = 0$. We concentrate here on details associated to the third step and give examples where the field discriminants have the form $\pm 2^a 3^b$.

1. Introduction

Let S be a finite set of primes and let G be a finite group. Let $\mathrm{NF}(S, G)$ be the set of Galois number fields $K \subset \mathbb{C}$ with $\mathrm{Gal}(K/\mathbb{Q}) \cong G$ and discriminant divisible only by primes in S. The sets $\mathrm{NF}(S, G)$ are finite, by a classical theorem of Hermite.

We are interested in the following inverse Galois problem. *For given S and G, find defining polynomials $f(x) \in \mathbb{Q}[x]$ for as many fields in* $\mathrm{NF}(S, G)$ *as possible.* This problem is most interesting when $|\,\mathrm{NF}(S, G)|$ can be expected to be small, so that one can reasonably aim for complete lists. Thus we are interested in "collecting number fields," the most valuable specimens being those which are least ramified, for a given Galois group G.

In this paper we consider only cases with $S = \{2, 3\}$. If G has a faithful permutation representation of degree ≤ 7, we have completely identified $\mathrm{NF}(S, G)$ previously with Jones [**13, 15**]. Here we supplement these complete lists with lists of fields with larger groups G, involving the simple groups $\mathrm{PSL}_2(9)$, $\mathrm{SL}_2(8)$, $\mathrm{SL}_3(3)$, $\mathrm{SU}_3(3)$, A_8, $\mathrm{W}(\mathrm{E}_6)'$, A_9, $\mathrm{W}(\mathrm{E}_7)'$, and A_{32}.

We construct all our fields by specializing three point covers. Our requirement that the field discriminant be of the form $\pm 2^a 3^b$ is extremely restrictive. In the main, it restricts consideration to three point covers with bad reduction at 2 and 3 only. To keep ramification within $\{2, 3\}$, the specialization point has to be chosen judiciously as well. Throughout, our computations are done in *Mathematica*, supplemented by the *nfdisc*, *factorpadic*, *polredabs*, and *nfisisom* commands of *Pari.*

2000 *Mathematics Subject Classification.* 11R21.
This is the final form of the paper.

Our title refers to the three step procedure we go through in order to produce our final polynomials $f(x)$. Each step centers on an ABC-equation, these equations being (1.1), (1.4), and (1.5).

First, the general ABC construction makes use of Katz's theory of rigid local systems [**16**]. This theory associates a rigid local system to a rigid solution of the matrix equation

$$ABC = 1. \tag{1.1}$$

There are motivic aspects to Katz's theory which arise when one works with characteristic zero coefficients, as we will briefly discuss at the end of this introduction. But for the ABC construction itself, we take A, B, C in some $\mathrm{GL}_n(\overline{\mathbb{F}}_\ell)$. Rigid means that the group $\widetilde{M}$ generated by A, B and C acts irreducibly on $\overline{\mathbb{F}}_\ell^n$ and that the centralizer dimensions of A, B, and C sum to the largest possible number compatible with irreducibility, namely n^2+2. Katz's theory says that if the primes dividing the orders of A, B, and C are all in S, and also the coefficient characteristic ℓ is in S, then the associated local system is guaranteed to have its bad reduction entirely in S too. So the role of this first step is to point us to three point covers guaranteed to be ramified within S. The hypergeometric family of solutions to (1.1) given in (1.6)–(1.8) suffices to cover all our examples in this paper, and so we will not be making much explicit reference to this first step.

Second, the three point covers we present in Sections 5–12 are mostly of the form $F\colon \mathbb{P}^1_x \to \mathbb{P}^1_t$. Here the subscripts distinguish different copies of the projective line by the coordinate we are using. So $\mathbb{P}^1_x$ has function field $\mathbb{Q}(x)$ while $\mathbb{P}^1_t$ has function field $\mathbb{Q}(t)$. We give F by an equation $t = -A(x)/C(x)$ with $A(x) \in \mathbb{Z}[x]$ of degree N and $C(x) \in \mathbb{Z}[x]$ of degree $\leq N$. If we write

$$f(t,x) = A(x) + tC(x) \tag{1.2}$$

then the discriminant with respect to x has the form

$$D(t) = \pm 2^A 3^B t^{e_0}(t-1)^{e_1}. \tag{1.3}$$

The simple way that t appears here reflects the fact that F ramifies only over 0, 1, and ∞, i.e. F is a three point cover. The form of the numerical coefficient reflects the fact that the bad reduction set of the cover is exactly $\{2,3\}$. These covers are calculated from extremal relatively prime solutions to the polynomial equation

$$A(x) + B(x) + C(x) = 0. \tag{1.4}$$

Here extremal means that the total number of roots of $A(x)$, $B(x)$, and $C(x)$ in $\mathbb{P}^1_x$, not counting multiplicities, is as small as possible, namely $N+2$. In contrast, the total number of roots, counting multiplicities, is $3N$. The nature of the multiplicities is determined in a group-theoretic way from the given solution to (1.1). If the group generated by A, B, C is $\widetilde{M}$ with center Z, the cover we seek has monodromy group $M = \widetilde{M}/Z$. Here we are modding out by centers because the essence of the situation is maintained in a computationally simpler context: once one has a cover $X = \mathbb{P}^1_x$ corresponding to M, it is typically easy to take the appropriate abelian cover $\widetilde{X}$ of X corresponding to $\widetilde{M}$. For roughly half of our examples, the instance of (1.4) we need to solve has already been solved in the literature. For the other half, we find the solution by standard techniques. So we will not say too much more about this second step either.

Third, we specialize the three point covers. We view (1.2) as a family of separable polynomials $f(\tau, x) \in \mathbb{Q}[x]$ indexed by $\tau \in \mathbb{Q} - \{0, 1\}$. Let K_τ be the splitting field of $f(\tau, x)$ in $\mathbb{C}$. To keep the discriminant of K_τ divisible by 2 and 3 only, we consider the generalized Fermat equation

$$ax^p + by^q + cz^r = 0. \tag{1.5}$$

Here p, q, and r are given positive integers, giving the order of local monodromy about the cusps 0, 1, and ∞ respectively in (1.2). We look for integer solutions of (1.5) with relatively prime terms and with a, b, and c divisible by 2 and 3 only. From each solution, we take $\tau = -ax^p/cz^r$ as our specialization point. It is this third step, and various issues associated with it, that we will concentrate on in this paper. It is the most purely number-theoretical of the three steps.

Of course, there are very substantial differences between the three steps, despite the formal similarity between the central ABC equations. Of the three equations, the matrix equation (1.1) has the most powerful theory associated to it. There is a hypergeometric family of solutions, treated in depth in [**3**]. There are also infinitely many other non-classical but similar families, introduced in [**16**] and discussed further in [**25**]. The hypergeometric family of solutions goes as follows. Let $u(x)$ and $w(x)$ be relatively prime degree n polynomials in $\overline{\mathbb{F}}_\ell[x]$ with non-zero constant terms. Put

$$A = m_u, \tag{1.6}$$

$$B = m_u^{-1} m_w, \tag{1.7}$$

$$C = m_w^{-1}, \tag{1.8}$$

where m_p indicates the companion matrix of p, as illustrated by (8.1). Then (A, B, C) is a rigid solution of (1.1), as one has irreducibility and A, B, and C have centralizer dimension dimension n, $n^2 - 2n + 2$, and n respectively. More precisely with respect to B, one has

$$\det(B) = \det(A)^{-1} \det(C)^{-1} = u(0)^{-1} w(0).$$

If $\det(B) \neq 1$, then B is conjugate to a diagonal matrix with diagonal entries 1, ..., 1, $\det(B)$. If $\det(B) = 1$, then B is conjugate to a matrix with diagonal entries 1, the 1–2 entry also 1, and all other entries zero.

From a practical point of view, the polynomial equation (1.4) is the most problematic. We have formulated things so far in terms of $F\colon \mathbb{P}^1_x \to \mathbb{P}^1_t$, but the general situation is $F\colon X \to \mathbb{P}^1_t$ with X a curve of arbitrary genus. In fact, consider a rigid solution (A, B, C) to (1.1) generating $\widetilde{M} \subset \mathrm{GL}_n(\overline{\mathbb{F}}_\ell)$ with center Z. Choose a faithful transitive permutation representation of $M = \widetilde{M}/Z$ into some symmetric group S_N. Then A, B, and C each determine a partition of N. Write the total number of parts as $N + 2 - 2g$. Then g is a nonnegative integer. The general polynomial ABC equation then takes place in the function field $\mathbb{Q}(X)$ of an unspecified curve of genus g, as one looks for three divisors of degree N, supported at altogether $N + 2 - 2g$ points, and a function $F \in \mathbb{Q}(X)$, with the three divisors being the zero locus of F, the zero locus of $F - 1$, and the polar locus of F. This positive genus case, in other words the great mass of the cases, presents computational problems which at present are usually insurmountable.

While we do not apply any theory to the integer equation (1.5), from our point of view it presents no problem at all. We simply carry out a computer search to find some solutions of the sort we need. Finding all the solutions, and proving that there are no more, are entirely other issues, both wrapped up with the ABC conjecture. Section 2 provides an overview of the fields constructed in this paper. Our main theorem is stated there; it says how many fields in $\mathrm{NF}(\{2,3\},G)$ we construct in this paper for the various G we consider. Sections 3 and 4 discuss three point covers and especially specialization. Sections 5–12 discuss the individual constructions, sorted approximately by $|G|$. In each case, we describe in some detail what happens in the specialization step, i.e. Step 3 of the ABC construction. Each of these sections also discusses a more theoretical issue, as indicated by the section titles.

As stated above, we will be saying very little about Steps 1 and 2 of the ABC construction. As to Step 1, our covers 6, $9a$, $10a$, $13a$, $13b$, $27a$, $27b$, $27c$, N_m, and $28b$ all come directly from a hypergeometric solution to (1.1). We will just write down a $\big(u(x),w(x)\big)$ giving this connection via (1.6)–1.8; there are sometimes quite different pairs $\big(u(x),w(x)\big)$ which would also work. Our covers $10b$, $9b$, $26c$, $28a$, and $27d$ come indirectly from a hypergeometric solution to (1.1), via a base change. We display equations indicating how these base changes go, and describe in §8 the case of $28a$ in more detail as an example, the other cases being similar. Some of our covers also come from non-hypergeometric solutions to (1.1), but we will not enter into these connections at all, save for some comments on $27d$. The coefficient characteristic is $\ell=3$ in Sections 5, 7, 8 and we use $\mathbb{F}_9=\mathbb{F}_3[i]$ with $i^2=-1$, taking $\rho=1+i$ as our standard generator of $\mathbb{F}_9^\times$. The coefficient characteristic is $\ell=2$ in Section 6 and we use models of $\mathbb{F}_8$ and $\mathbb{F}_{64}$ introduced there. In Sections 9–12 the coefficient characteristic can be generally taken to be either $\ell=2$ or $\ell=3$, as A, B, and C defined by (1.6)–1.8 generate a finite group even when one takes $\mathbb{Q}$ as the coefficient field. This finiteness even in characteristic zero is a very unusual situation, and in this sense Sections 5–8 represent the general case better than Sections 9–12 do. As to Step 2, we will limit ourselves to a brief general discussion in §5, and a few comments about the new covers.

Throughout this paper, we work as much as possible in the setting of algebraic number theory. However the reader should realize that a systematic study of the sets $\mathrm{NF}(S,G)$ would fundamentally involve both the theory of motives and the theory of automorphic representations. Our ABC construction is on the motivic side, as Katz's theory from which we begin is motivic in nature, as mentioned before. In fact, via rigid characteristic zero solutions to (1.1), all our $f(t,x)$ can be thought of as generalizations of classical division polynomials $f_m(t,x)$ associated to the m-torsion points of the elliptic curve

$$y^2=(t-1)x^3-3tx+2t$$

with J-invariant $1728t$. A great advance would be to bypass the computational techniques from general three point covers and somehow use the motivic theory to directly calculate $f(t,x)$, just as the classical division polynomials $f_m(t,x)$ can be calculated for all m by explicit recursion relations [**22**].

On the automorphic side, when G is solvable, the sets $\mathrm{NF}(S,G)$ are in principle accessible by abelian class field theory. For groups embeddable into some $\mathrm{GL}_2(\overline{\mathbb{F}}_\ell)$, Serre's conjecture applies [**27**]. Extending Serre's conjecture, Ash and Gross both have conjectural Langlands-type non-abelian class field theory for quite general groups [**1, 11**]. Gross's theory predicts in particular that for a given prime p, $\mathrm{NF}(\{p\},G)$ is non-empty for suitable non-solvable G. For example, it is predicted that $\mathrm{NF}(\{5\}, \mathrm{G}_2(5))$ is non-empty. The fields we produce in this paper seem good candidates for comparison with automorphic forms.

2. Summary of fields constructed

To keep our investigation of manageable size, we focus attention on groups G of the form $H.A$, where H is non-abelian simple and the natural map $A \to \mathrm{Out}(H)$ is injective. In passing, however, we construct interesting fields for G not in this class. For example, in Sections 5–9, some interesting solvable fields appear. Also, in Section 6 some G involving two copies of $\mathrm{SL}_2(8)$ appear, and many other such $H^n.A$ could be constructed just by specializing at non-rational points τ. Also, related to the distinction between $\widetilde{M}$ and M, many of our covers have natural lifts; for example, Cover $27d$ in Section 9 with Galois group $\mathrm{W}(\mathrm{E}_6)$ has a natural lift to a cover with Galois group $2.\mathrm{W}(\mathrm{E}_6)$.

Table 2.1 lists some groups G of the form $H.A$. Groups G are placed in the same block iff they have the same H. We give a characteristic 0 description, a characteristic 2 description, and/or a characteristic 3 description. The Atlas [**7**] sometimes gives even more descriptions, e.g. $\mathrm{SL}_3(2).2 \cong \mathrm{PGL}_2(7)$. The column N gives the degree of a minimal faithful permutation representation.

An entry under $\#_G$ of the form $\bullet x$ summarizes results from [**13**] and [**15**]. In this case, $|\,\mathrm{NF}(\{2,3\}, G)| = x$. The complete search for $\mathrm{GL}_3(2)/\mathrm{A}_7/\mathrm{S}_7$ fields took thirteen hours, but we estimate an analogous complete search for $\mathrm{GL}_3(2).2/\mathrm{A}_8/\mathrm{S}_8$ fields would take somewhere around ten thousand years [**14**]. This is why we are shifting attention to non-exhaustive but still systematic ways of constructing fields in a given $\mathrm{NF}(S, G)$.

The main focus of the present paper is the following theorem.

THEOREM 2.1. *For* $G = \mathrm{M}_{10}$, $\mathrm{PGL}_2(9)$, $\mathrm{P\Gamma L}_2(9)$, $\mathrm{SL}_2(8)$, $\Sigma\mathrm{L}_2(8)$, $\mathrm{SL}_3(3)$, $\mathrm{SL}_3(3).2$, $\mathrm{G}_2(2)'$, $\mathrm{G}_2(2)$, S_8, $\mathrm{PSp}_4(3)$, $\mathrm{SO}_5(3)$, A_9, S_9, $\mathrm{Sp}_6(2)$, *and* S_{32}, *the polynomials presented in Section* 5–12 *give* $\#_G$ *elements of* $\mathrm{NF}(\{2,3\}, G)$, *with* $\#_G$ *as on Table* 2.1.

In general, suppose given separable polynomials $f_i \in \mathbb{Q}[x]$, which one expects have distinct splitting fields K_i in some $\mathrm{NF}(S, G)$. After a variable change, one can assume all the f_i are monic polynomials in $\mathbb{Z}[x]$. To prove that the K_i are in $\mathrm{NF}(S, G)$ and distinct one has to do three things:

1. *Verify that the ramification set* S_i *of* K_i *is really in* S. Let $L_i = \mathbb{Q}[x]/f_i(x)$. Let D_i be the polynomial discriminant f_i and let d_i be the field discriminant of L_i. One has to show that the $p \notin S$ dividing D_i do not divide d_i. This may require substantial computation in general, but in the setting of three point covers we do it uniformly without computation by (3.4).

2. *Verify that the Galois group* $G_i := \mathrm{Gal}(K_i/\mathbb{Q})$ *is indeed isomorphic to* G. One can very easily compute lower bounds by means of Frobenius elements; we briefly indicate how this goes in Section 5, and then give no more details. On a heuristic level, one can expect that the lower bound obtained after moderate computation is always exact. On a rigorous level, to get upper bounds may require very substantial computation, like those we carry out for related purposes in Sections 9 and 11. In the setting of three point covers, we get the needed upper bounds uniformly without computation from (3.2).

TABLE 2.1. Lower bounds for $|\,\mathrm{NF}(\{2,3\},G)|$.

$\lvert G \rvert$	0	2	3	N	$\#_G$	§
$60 = 2^2 3\ 5$	A_5	$SL_2(4)$		5	•0	
$120 = 2^3 3\ 5$	S_5	$\Sigma L_2(4)$		5	•5	
$168 = 2^3 3\ 7$		$SL_3(2)$		7	•0	
$336 = 2^4 3\ 7$		$SL_3(2).2$		8		
$360 = 2^3 3^2 5$	A_6	$Sp_4(2)'$	$PSL_2(9)$	6	•4	
$720 = 2^4 3^2 5$	S_6	$Sp_4(2)$	$PGO_4^-(3)$	6	•27	
$720 = 2^4 3^2 5$			$PGL_2(9)$	10	4	5
$720 = 2^4 3^2 5$	M_{10}			10	15	5
$1,440 = 2^5 3^2 5$			$P\Gamma L_2(9)$	10	79	5
$504 = 2^6 3^2 7$		$SL_2(8)$		9	3	6
$1,512 = 2^6 3^3 7$		$\Sigma L_2(8)$		9	64	6
$2,520 = 2^3 3^2 5\ 7$	A_7			7	•0	
$5,040 = 2^4 3^2 5\ 7$	S_7			7	•10	
$5,616 = 2^4 3^3 13$			$SL_3(3)$	13	6	7
$11,232 = 2^5 3^3 13$			$SL_3(3).2$	26	85	7
$6,048 = 2^5 3^3 7$		$G_2(2)'$	$SU_3(3)$	28	1	8
$12,096 = 2^6 3^3 7$		$G_2(2)$	$SU_3(3).2$	28	41	8
$20,160 = 2^6 3^2 5\ 7$	A_8	$GL_4(2)$		8		
$40,320 = 2^7 3^2 5\ 7$	S_8	$SO_6^+(2)$		8	2	11, 12
$25,920 = 2^6 3^4 5$	$W(E_6)'$	$SU_4(2)$	$PSp_4(3)$	27	21	9
$51,840 = 2^7 3^4 5$	$W(E_6)$	$SO_6^-(2)$	$SO_5(3)$	27	124	9
$181,440 = 2^6 3^4 5\ 7$	A_9			9	10	10
$362,880 = 2^7 3^4 5\ 7$	S_9			9	26	10, 12
$1,451,520 = 2^9 3^4 5\ 7$	$W(E_7)'$	$Sp_6(2)$		28	34	11
$2.63 \times 10^{35} \approx 32!$	S_{32}			32	1	12

3. *Verify that the K_i are distinct.* Often the group G has up to isomorphism exactly one faithful permutation representation of a given degree N, and all given defining polynomials f_i have degree N. In this case, for each pair $i \neq i'$ one needs to find a prime p, not dividing $D_i D_{i'}$, for which the factorization partitions of f_i and $f_{i'}$ over $\mathbb{F}_p$ are distinct. In general the situation may be very slightly more complicated, for example by sextic twinning in Section 5 and projective twinning in Section 7. But even here, one needs to just find p such that the factorization partitions come from elements of different orders in the symmetric group S_N. We have done this, but do not present any details here.

In the rest of the paper we complete the statement of Theorem 2.1 by presenting the defining equations. The proof of Theorem 2.1 is completed simultaneously, since, as we have just explained, the theorem is essentially self-proving from its full statement. Of course, it would be pointless to restrict ourselves only to establishing those facts literally contributing to Theorem 2.1. Rather, we present a fuller picture of both our particular examples and the general technique of constructing fields in a given $\mathrm{NF}(S,G)$ by means of specializing three point covers.

Jones has run modest computer searches which have found an A_8 field and some more S_8 fields ramified at 2 and 3 only. From looking at mod ℓ Galois representations of various curves, we know also that there are more fields for $\mathrm{PGL}_2(9)$, $\mathrm{SL}_2(8)$, $\mathrm{SU}_3(3).2$ and $\mathrm{SO}_5(3)$ than listed here.

In [**13**], we analyzed the local behavior at 2 and 3 of low degree fields in complete detail. In particular, all $5+4+27$ fields on the top lines of Table 2.1 are wildly ramified at both 2 and 3, mostly very wildly ramified, as explained in Section 3.3 there. Similarly, all 10 S_7 fields are wildly ramified at both 2 and 3, as explained in [**15**]. In contrast, four of the fields here are wildly ramified at only one of these two primes; see (5.1), (6.1), (6.2) and (9.5). One can't have tame ramification at both primes, as a Galois field K with discriminant $\pm 2^a 3^b$ and tame ramification only has root discriminant < 6. Odlyzko's bounds [**23**], even the unconditional ones, then force $K = \mathbb{Q}$ or $K = \mathbb{Q}(\sqrt{-3})$.

3. Three point covers and specialization

Let F be a field and work with smooth projective curves over the fixed projective line $\mathbb{P}^1_F = \mathrm{Spec}(F[t]) \cup \{\infty\}$. A three point cover over F is a finite separable cover $X \to \mathbb{P}^1_F$, ramified only above the three points 0, 1, and ∞. A defining polynomial for X is a polynomial $f(t,x)$ with $F(X) = F(t)[x]/f(t,x)$. An important issue extensively addressed in the literature is how one explicitly constructs defining polynomials; throughout this paper, we mainly take the defining polynomials simply as given, and the bulk of this section is devoted to setting up notation. Standard references for three point covers include the books [**20, 21, 26, 28, 30**].

We need first some notation with respect to the universal base curve. Let $T = \mathrm{Spec}\big(\mathbb{Z}[t, 1/t, 1/(t-1)]\big)$ be the thrice-punctured projective line, so that for any field F, $T(F) = F - \{0,1\}$. The topological space $T(\mathbb{C})$ plays a central role. We work with the fundamental group

$$\pi_1 := \pi_1\big(T(\mathbb{C}), 1/2\big) = \langle \gamma_0, \gamma_1, \gamma_\infty \mid \gamma_0\gamma_1\gamma_\infty = 1\rangle.$$

Here γ_0 and γ_1 come from the counterclockwise circle with radius $\frac{1}{2}$ centered at 0 and 1 respectively. The group π_1 is free on the two generators γ_0 and γ_1. The third element $\gamma_\infty := (\gamma_0\gamma_1)^{-1}$ is introduced so as to treat the three cusps in the same way.

To keep control of Galois groups even when specializing, it is convenient to "remove from $T(\mathbb{C})$ the segments $(-\infty, 0) - i\varepsilon$ and $(1,\infty) - i\varepsilon$ with ε a positive infinitesimal." What this really means is that we define $T(\mathbb{C})^{\mathrm{cut}}$ to be the point set $T(\mathbb{C}) = \mathbb{C} - \{0,1\}$ with a stronger topology. The open sets are generated by finite intersections and arbitrary unions from the standard ones together with the new open set consisting of $\{t \in \mathbb{C} : \mathrm{Im}(t) \geq 0\} - [0,1]$. So $T(\mathbb{C})^{\mathrm{cut}}$, unlike $T(\mathbb{C})$, is simply connected; one can think of it as a spread out version of the base point $\frac{1}{2}$.

A simple partition of $T(\mathbb{Q}_p)$ into subregions plays a fundamental role in the analysis of ramification. For $\tau \in T(\mathbb{Q}_p)$, write $\tau = -A/C$ with $A, C \in \mathbb{Z}_p$ not both divisible by p. Define $B \in \mathbb{Z}_p$ by $A+B+C=0$. Define

$$\mathrm{ord}_0(\tau) := \mathrm{ord}_p(A), \quad \mathrm{ord}_1(\tau) := \mathrm{ord}_p(B), \quad \mathrm{ord}_\infty(\tau) := \mathrm{ord}_p(C).$$

The partition is

$$T(\mathbb{Q}_p) = T(\mathbb{Q}_p)^{\text{gen}} \coprod \left(\coprod_{i=1}^{\infty} T(\mathbb{Q}_p)^{0,i}\right) \coprod \left(\coprod_{i=1}^{\infty} T(\mathbb{Q}_p)^{1,i}\right) \coprod \left(\coprod_{i=1}^{\infty} T(\mathbb{Q}_p)^{\infty,i}\right).$$

Here ord_c takes the value i exactly on $T(\mathbb{Q}_p)^{c,i}$; so for c fixed and i increasing, the $T(\mathbb{Q}_p)^{c,i}$ are smaller and smaller annuli, all centered at the cusp c. The generic piece, empty for $p = 2$, is where all three ord_c take the value 0.

For the rest of this section, a degree N three point cover $X \to \mathbb{P}^1_F$ is fixed. A particular defining polynomial $f(t,x)$ is fixed too. Objects Δ, a, b, $c(t)$, and Σ below depend on f. Otherwise the objects M, G, ... depend only on X. We will use this notation systematically in the sequel, with indices when discussing particular examples. For example, once we have denoted a particular cover X_{27d}, automatically M_{27d} denotes its monodromy group.

The polynomial discriminant of $f(t,x)$, with respect to the variable x, factors uniquely as

$$D(t) = \Delta t^a (t-1)^b c(t)^2, \tag{3.1}$$

with $c(t) \in F[t]$ monic and prime to $t(t-1)$. Here $\Delta \in F^\times$ and a and b are integers. Let Σ be the subvariety of $\mathbb{P}^1_F$ corresponding to the roots of $c(t)$. Switching to a different defining polynomial $f^*(t,x)$ typically makes the corresponding $c^*(t)$ relatively prime to $c(t)$. So the factor $c(t)^2$ plays essentially no role in our situation; actually $c(t)$ is identically 1 in most of our examples.

For τ in $F - \{0,1\}$, let $X_\tau = \text{Spec}(L_\tau)$ be the corresponding fiber. So $L_\tau = F(X_\tau)$ is a finite separable algebra over F. For $\tau \notin \Sigma(F)$, $L_\tau = F[x]/f(\tau,x)$; so in this case, L_τ is a field iff $f(\tau,x)$ is irreducible.

Suppose henceforth that $F \subseteq \mathbb{C}$. Let $X(\mathbb{C})^{\text{cut}}$ be the inverse image of $T(\mathbb{C})^{\text{cut}}$ in $X(\mathbb{C})$. Let C be set of components of $X(\mathbb{C})^{\text{cut}}$, i.e. $C = \pi_0\big(X(\mathbb{C})^{\text{cut}}\big)$. So C has N components and the group π_1 acts naturally on C. The image of π_1 in the symmetric group $\text{Sym}(C)$ is called the monodromy group M. For $c \in \{0,1,\infty\}$, the image of γ_c in M is denoted m_c; its order is denoted e_c.

Let $\overline{\mathbb{Q}}$ be the algebraic closure of $\mathbb{Q}$ in $\mathbb{C}$. Suppose henceforth that $F \subseteq \overline{\mathbb{Q}}$; this is the essential case anyway. For $\tau \in F - \{0,1\}$, the branch cuts let us canonically identify the fiber $X_\tau(\mathbb{C})$ with the component set C; we use this identification without comment in the sequel. But also $X_\tau(\mathbb{C})$ is identified with the set of homomorphisms $\text{Hom}(L_\tau, \overline{\mathbb{Q}})$; for $\tau \notin \Sigma(F)$, $X_\tau(\mathbb{C})$ is identified with the set of roots $X_\tau(\mathbb{C}) \subset \overline{\mathbb{Q}}$ of $f(\tau,x)$. Either way, one sees a natural action of $\text{Gal}(\overline{\mathbb{Q}}/F)$ on $X_\tau(\mathbb{C}) = X_\tau(\overline{\mathbb{Q}})$. The image of $\text{Gal}(\overline{\mathbb{Q}}/F)$ in the symmetric group $\text{Sym}(C)$ is the Galois group G_τ.

Let M^* be the normalizer of M in the symmetric group $\text{Sym}(C)$. A basic fact is that

$$\text{each } G_\tau \text{ is contained in } M^*. \tag{3.2}$$

Another basic fact is that all the biggest G_τ coincide, and we call this common group the Galois group G.

Suppose now that $F \subset \overline{\mathbb{Q}}$ is a number field of degree d. Then one has d conjugate covers $X^\delta \to F^\delta$. Restricting scalars, one gets a single cover of degree dN over the field $\mathbb{Q}$. On the level of defining polynomials, this operation of restricting scalars is the passage from $f(t,x) \in F[t,x]$ to $\prod f^\delta(t,x) \in \mathbb{Q}[t,x]$. Since our object is to construct Galois extensions of $\mathbb{Q}$, this is a natural thing for us to do; in the rest of this section, we take $F = \mathbb{Q}$.

Let K_τ be the splitting field in $\mathbb{C}$ of L_τ. So if $\tau \notin \Sigma(\mathbb{Q})$, K_τ is the field generated by the roots of $f(\tau, x)$ in $\mathbb{C}$. The fields K_τ are the fields which we emphasized in Sections 1 and 2; we have set things up so that $G_\tau = \mathrm{Gal}(K_\tau/\mathbb{Q})$. On the other hand, for most of this paper we focus on the degree N algebras L_τ which arise naturally and are computationally more accessible. Thus, for example, we sometimes give the algebra discriminant of L_τ, as computed by *nfdisc*. Determining the discriminant of K_τ would be much harder.

Let S be the set of primes at which the cover $X \to \mathbb{P}^1_{\mathbb{Q}}$ has bad reduction. Always S contains all primes dividing $e_0 e_1 e_\infty$. Always S is contained in the primes dividing numerator or denominator of Δ. By adjusting the defining equation, one can always make S exactly the primes dividing numerator or denominator of Δ.

A basic fact is that all the fields K_τ are tamely ramified outside S. To be more precise, recall that for a Galois number field $K \subset \mathbb{C}$, tamely ramified at p, one has a conjugacy class r_p in $\mathrm{Gal}(K/\mathbb{Q})$, measuring the ramification at p. In our case, for $p \notin S$, one has an explicit formula for the ramification class $r_{\tau,p}$ as a conjugacy class in $G \supseteq G_\tau$:

$$r_{\tau,p} = \begin{cases} [1] & \text{if } \tau \in T(\mathbb{Q}_p)^{\mathrm{gen}}, \\ [m_c^i] & \text{if } \tau \in T(\mathbb{Q}_p)^{c,i}. \end{cases} \tag{3.3}$$

Thus L_τ is unramified at p iff $\tau \in T(\mathbb{Q}_p)^{\mathrm{gen}}$ or $\tau \in T(\mathbb{Q}_p)^{c,i}$ with $m_c^i = 1$. Also

$$\text{if } \tau \in T(\mathbb{Q}_p)^{c,i} \text{ then } \mathrm{ord}_p\big(\mathrm{disc}(L_\tau)\big) = N - |C/m_c^i|, \tag{3.4}$$

the last term being the number of orbits of m_c^i on the component set C.

4. Specialization lists for $S = \{2, 3\}$

Let p, q, and r be positive integers or ∞. Let S be a finite set of primes. Define $T_{p,q,r}(\mathbb{Z}^S)$ to be the set of those rational numbers $\tau \in \mathbb{Q} - \{0, 1\}$ such that

$$\tau \in T(\mathbb{Q}_p)^{\mathrm{gen}} \coprod \left(\coprod_{i=1}^{\infty} T(\mathbb{Q}_p)^{0,pi}\right) \coprod \left(\coprod_{i=1}^{\infty} T(\mathbb{Q}_p)^{1,qi}\right) \coprod \left(\coprod_{i=1}^{\infty} T(\mathbb{Q}_p)^{\infty,ri}\right)$$

for all p not in S. Here if an index is ∞, then the corresponding cuspidal summand is by definition empty.

Explicitly, and without reference to p-adic numbers, a rational number τ is in $T_{p,q,r}(\mathbb{Z}^S)$ iff there exist integers a, b, c, x, y, z as follows: a, b, and c are divisible only by primes in S;

$$\tau = -\frac{ax^p}{cz^r}; \quad ax^p + by^q + cz^r = 0.$$

The six-tuple (a, b, c, x, y, z) is then uniquely determined under the following auxiliary normalization conditions: x, y, and z are divisible only by primes not in S; $A = ax^p$, $B = by^q$, and $C = cz^r$ are relatively prime; x, y, and z are positive; two of A, B, and C are positive. In short, identifying $T_{p,q,r}(\mathbb{Z}^S)$ requires finding the solutions of the generalized Fermat equation.

The discussion at the end of Section 3 makes clear why we are interested in the sets $T_{p,q,r}(\mathbb{Z}^S)$. Namely, let $X \to \mathbb{P}^1_{\mathbb{Q}}$ be a three point cover with bad reduction set within S and local monodromy orders $p = e_0$, $q = e_1$, $r = e_\infty$. Then the specialized

TABLE 4.2. The 56 S_3-orbits of $T_h(\mathbb{Z}^{\{2,3\}})$ needed for $T_{2,3,\infty}(\mathbb{Z}^{\{2,3\}})$.

$a\cdot x^p$	$b\cdot y^q$	$c\cdot 1$
1	1	$-2\cdot 1$
$2\cdot 1$	1	$-3\cdot 1$
$3\cdot 1$	1	$-2^2\cdot 1$
$2^3\cdot 1$	1	$-3^2\cdot 1$
$3\cdot 1$	5^3	$-2^7\cdot 1$
-5^2	$2^4\cdot 1$	$3^2\cdot 1$
-5^2	$2^3 3\cdot 1$	1
5^2	$2\cdot 1$	$-3^3\cdot 1$
-7^2	$2^4 3\cdot 1$	1
7^2	$2^5\cdot 1$	$-3^4\cdot 1$
$2\cdot 11^2$	1	$-3^5\cdot 1$
-17^2	$2^5 3^2\cdot 1$	1
-131^2	5^6	$2^9 3\cdot 1$
$2\cdot 5^4$	-11^3	$3^4\cdot 1$
$2\cdot 7^2$	-5^3	$3^3\cdot 1$
11^2	-5^3	$2^2\cdot 1$
13^2	$-2\cdot 5^3$	$3^4\cdot 1$
$2^2\cdot 5^2$	-7^3	$3^5\cdot 1$
17^2	-7^3	$2\cdot 3^3\cdot 1$
-19^2	7^3	$2\cdot 3^2\cdot 1$
19^2	5^3	$-2\cdot 3^5\cdot 1$
13^2	7^3	$-2^9\cdot 1$
$3^3\cdot 7^2$	-11^3	$2^3\cdot 1$
37^2	$-2^2\cdot 7^3$	$3\cdot 1$
$-3\cdot 23^2$	11^3	$2^8\cdot 1$
35^2	-13^3	$2^2 3^5\cdot 1$
$2^2\cdot 23^2$	-13^3	$3^4\cdot 1$
-47^2	13^3	$2^2 3\cdot 1$

$a\cdot x^2$	$b\cdot y^3$	$c\cdot 1$
$-2^3\cdot 17^2$	5^3	$3^7\cdot 1$
59^2	$-3\cdot 11^3$	$2^9\cdot 1$
61^2	$3\cdot 5^3$	$-2^{12}\cdot 1$
-71^2	17^3	$2^7\cdot 1$
$-2^3 3^2\cdot 13^2$	23^3	1
11^2	23^3	$-2^{12} 3\cdot 1$
73^2	23^3	$-2^3 3^7\cdot 1$
143^2	$-3\cdot 19^3$	$2^7\cdot 1$
$2^3\cdot 73^2$	-35^3	$3^5\cdot 1$
107^2	$-3^2\cdot 17^3$	$2^{15}\cdot 1$
-215^2	19^3	$2\cdot 3^9\cdot 1$
-253^2	$2^9\cdot 5^3$	$3^2\cdot 1$
-359^2	$3\cdot 35^3$	$2^8\cdot 1$
545^2	$-2\cdot 53^3$	$3^6\cdot 1$
595^2	-73^3	$2^4 3^7\cdot 1$
$-3\cdot 389^2$	$2\cdot 61^3$	1
-827^2	73^3	$2^{15} 3^2\cdot 1$
955^2	-97^3	$2^3 3^4\cdot 1$
1871^2	$-3^2\cdot 73^3$	$2^9\cdot 1$
2359^2	47^3	$-2^5 3^{11}\cdot 1$
2681^2	-193^3	$2^4 3^4\cdot 1$
$-2\cdot 2761^2$	239^3	$3^{13}\cdot 1$
8549^2	$-3^5\cdot 67^3$	$2^3\cdot 1$
-23053^2	505^3	$2^{27} 3\cdot 1$
$2\cdot 21395^2$	-971^3	$3^8\cdot 1$
39151^2	-1153^3	$2^5 3^5\cdot 1$
$-3\cdot 48383^2$	1915^3	$2^{13}\cdot 1$
$-2\cdot 184211^2$	4079^3	$3\cdot 1$

fields K_τ have bad reduction within S exactly if $\tau \in T_{p,q,r}(\mathbb{Z}^S)$. This important statement is called the Chevalley–Weil theorem for M-curves in [**8**].

If $X \to \mathbb{P}^1_{\mathbb{Q}}$ yields $\{p,q,r\}$ with $1/p+1/q+1/r \geq 1$, then the monodromy group M is solvable, except perhaps for some composition factors isomorphic to A_5. Thus we are interested here exclusively in the hyperbolic case

$$\frac{1}{p} + \frac{1}{q} + \frac{1}{r} < 1. \tag{4.1}$$

Here the sets $T_{p,q,r}(\mathbb{Z}^S)$ are known to be finite [**9**], the proof appealing to Faltings' general finiteness theorem for curves.

TABLE 4.3. 45 more S_3-orbits in $T_h(\mathbb{Z}^{\{2,3\}})$.

$a{\cdot}x^p$	$b{\cdot}y^q$	$c{\cdot}z^r$
$2^5{\cdot}1$	$-3{\cdot}5^3$	7^3
23^2	-5^4	$2^5 3{\cdot}1$
7^2	-5^4	$2^6 3^2{\cdot}1$
-29^2	5^4	$2^3 3^3{\cdot}1$
47^2	-7^4	$2^6 3{\cdot}1$
$2^5 3{\cdot}5^2$	-7^4	1
113^2	7^4	$2^7 3^4{\cdot}1$
$-2^2{\cdot}61^2$	11^4	$3^5{\cdot}1$
239^2	$-2{\cdot}13^4$	1
287^2	-17^4	$2^7 3^2{\cdot}1$
$2{\cdot}5861^2$	$-3^2{\cdot}59^4$	7^9
$3^2{\cdot}5^3$	$2^2{\cdot}19^3$	-13^4
-37^3	$2{\cdot}29^3$	$3{\cdot}5^4$
$-2^8{\cdot}7^3$	$3{\cdot}29^3$	11^4
-71^3	23^3	$2^4 3^2{\cdot}7^4$
$-2{\cdot}203^3$	$3^3{\cdot}79^3$	43^4
2293^2	$2{\cdot}67^3$	$-3{\cdot}5^9$
1079^2	$-2^{10}{\cdot}19^3$	$3{\cdot}5^9$
$-3{\cdot}36553^2$	203^3	$2^{11}{\cdot}5^9$
-138743^2	$3{\cdot}1027^3$	$2^{13}{\cdot}5^9$
$2^3 3^2{\cdot}12515^2$	-2797^3	13^9
107567^2	$-3{\cdot}3155^3$	$2^{11}{\cdot}7^9$

$a{\cdot}x^2$	$b{\cdot}y^q$	$c{\cdot}z^8$
$-2{\cdot}3{\cdot}263^2$	29^3	5^8
$-2^2{\cdot}353^2$	$3^4{\cdot}11^3$	5^8
5239^2	$-3^2{\cdot}163^3$	$2{\cdot}7^8$
$-2^5{\cdot}4015^2$	799^3	7^8
26311^2	$2^4 3^2{\cdot}95^3$	-13^8
-32039^2	241^3	$2^5 3^4{\cdot}5^8$
-39313^2	$2{\cdot}767^3$	$3{\cdot}11^8$
31987^2	-1489^3	$2^3 3^6{\cdot}5^8$
36631^2	1679^3	$-2^6 3^5{\cdot}5^8$
-99431^2	$2^3{\cdot}1073^3$	$3^2{\cdot}5^8$
-160975^2	$2^{10} 3{\cdot}203^3$	11^8
-185039^2	1633^3	$2^6 3^4{\cdot}7^8$
$2^2{\cdot}774517^2$	-15613^3	$3^8{\cdot}11^8$
-6827035^2	$2{\cdot}28559^3$	$3^3{\cdot}13^8$
9101359^2	-43873^3	$2^7 3^7{\cdot}7^8$
-26615519^2	78913^3	$2^7 3^5{\cdot}17^8$
-30042907^2	$2^3 3^3{\cdot}16037^3$	43^8
$2{\cdot}45707519^2$	1171537^3	$-3^5{\cdot}95^8$
$2{\cdot}13^2$	$-3{\cdot}19^4$	5^8
437147^2	-21769^3	$2^9 3^4{\cdot}5^{12}$
1169^2	$2^3 3^4{\cdot}5^4$	-11^6
2591^2	$-3{\cdot}43^4$	$2{\cdot}11^6$
14089^2	-131^4	$2^{11} 3{\cdot}5^6$

In the application to specializing three point covers, only cases satisfying the condition

(4.2) all primes dividing pqr are in S

arise. Accordingly, put

$$T_h(\mathbb{Z}^S) = \bigcup_{(p,q,r)} T_{p,q,r}(\mathbb{Z}^S), \tag{4.3}$$

the union being over (p,q,r) satisfying (4.1) and (4.2). Write $(p,q,r) \mid (p',q',r')$ iff $p \mid p'$, $q \mid q'$, and $r \mid r'$, with the convention that (any positive integer) $\mid \infty$. Then

$$T_{p,q,r}(\mathbb{Z}^S) \supseteq T_{p',q',r'}(\mathbb{Z}^S)$$

if $(p,q,r) \mid (p',q',r')$. Thus (4.3) can be written as a finite union over (p,q,r) minimal with respect to divisibility, and so $T_h(\mathbb{Z}^S)$ is also finite. To avoid confusion, we should mention that it is natural in other contexts to consider a larger set $T_H(\mathbb{Z}^S)$

by letting (p,q,r) on the right side of (4.3) run over triples satisfying (4.1) but not necessarily (4.2). The ABC conjecture would say that each $T_H(\mathbb{Z}^S)$ is finite, but this finiteness is not known. The group $S_3 = \mathrm{Sym}(\{0,1,\infty\})$ acts naturally on the sets $T_h(\mathbb{Z}^S)$ and $T_H(\mathbb{Z}^S)$. If $2 \in S$, one has the three-element orbit $\{-1, 1/2, 2\}$; otherwise all orbits have six elements. A case which has received a lot of attention recently is the case $S = \emptyset$. The corresponding set $T_H(\mathbb{Z})$ is known to contain ten S_3-orbits [**2**], and conjectured not to contain any more [**8, 9**].

The set $T_{2,3,\infty}(\mathbb{Z}^{\{2,3\}})$ has been completely identified [**6**]; it has 81 elements giving rise to 56 S_3-orbits. We have carried out a several-day computer search for more elements of $T_h(\mathbb{Z}^{\{2,3\}})$, with the expectation that each τ found will be involved in the construction of infinitely many essentially distinct number fields K_τ, via our ABC construction. We have found 45 more orbits. Representatives of these 101 orbits are given in Table 4.2 and Table 4.3. The representatives τ on these tables are grouped according to their corresponding $\{p,q,r\}$. Each S_3-orbit is given in terms of a solution to a generalized Fermat equation. For example, the fifth line in the left column of Table 4.2 corresponds to $3 + 5^3 - 2^7 = 0$. The members of the corresponding S_3-orbit are all possible negative quotients of the three terms, i.e. $3/2^7$, $2^7/3$, $5^3/2^7$, $2^7/5^3$, $-3/5^3$, $-5^3/3$.

Our computer search shows that Table 4.3 is complete with respect to solutions with $|ax^p|$, $|bx^q|$, and $|cz^r|$ all $\leq 10^9$. To get larger solutions, we permuted cusps and applied the following base change maps iteratively.

$$\begin{aligned} f_2 : T_{m,m,n}(\mathbb{Z}^S) &\to T_{m,2,2n}(\mathbb{Z}^{S\cup\{2\}}), \\ \tau &\mapsto 4\tau(1-\tau), \\ (A,B,C) &\mapsto \big(4AB, (2A+C)^2, -C^2\big), \end{aligned}$$

$$\begin{aligned} f_3 : T_{2n,2,n}(\mathbb{Z}^S) &\to T_{3,2,2n}(\mathbb{Z}^{S\cup\{3\}}), \\ \tau &\mapsto \frac{(4\tau-1)^3}{27\tau}, \\ (A,B,C) &\mapsto \big((4A+C)^3, (8A-C)^2 B, -27AC^2\big), \end{aligned}$$

$$\begin{aligned} f_4 : T_{3n,3,n}(\mathbb{Z}^S) &\to T_{3,2,3n}(\mathbb{Z}^{S\cup\{2\}}), \\ \tau &\mapsto \frac{(9\tau-1)^3(1-\tau)}{64\tau}, \\ (A,B,C) &\mapsto \big(B(9A+C)^3, (27A^2+18AC-C^2)^2, 64AC^3\big). \end{aligned}$$

These maps themselves are three point covers. In fact, f_3 and f_4 describe how the modular curve $X_0(N)$ covers the j-line $X_0(1)$ for $N = 2, 3$.

The base change operations are quite efficient. For example, not using [**6**], but rather starting from just $\tau = -2$ corresponding to $2 - 3 + 1 = 0$, one gets 73 of the 101 orbits. Two more examples of low height solutions being transformed to larger height solutions are

$$f_3(3^5, -2^2 61^2, 11^4) = (-15613^3, 2^2 774517^2, 3^8 11^8), \tag{4.4}$$

$$f_2(3^3 79^3, -2^1 203^3, 43^4) = (-2^3 3^3 16037^3, 30043907^2, -43^8). \tag{4.5}$$

TABLE 4.4. Order of the specialization sets $T^*_{p,q,r}$.

		Section:		5	6	7	8	9	10	11
p	q	r	$\lvert T^*_{p,q,r}\rvert$	A_6	$SL_2(8)$	$SL_3(3)$	$SU_3(3)$	$W(E_6)'$	A_9	$W(E_7)'$
2	3	8	99	f_{10a}						
2	3	9	87		f_{9a}					
2	3	12	82							
2	3	≥ 16	81		f_{18}					
2	4	6	48							
2	4	8	45			f_{26c}				
2	4	9	45					f_{27d}		
2	4	≥ 12	44				f_{28a}			
2	6	≥ 8	36							
2	≥ 8	≥ 8	35	f_{10b}				f_{27abc}	$f_{9,1}$	f_{28b}
3	3	4	39	f_6						
3	3	≥ 6	27			f_{13a}				
3	4	≥ 6	24			f_{13b}				
3	≥ 6	≥ 6	23							
≥ 4	≥ 4	≥ 4	21							

Note that the orbits appearing in (4.4) and the right side of (4.5) are three of the 10 known orbits of $T_H(\mathbb{Z})$. Our list contains four more of these orbits; the remaining three are excluded from our consideration, as they involve the exponent 7 on primes other than 2 and 3.

In the rest of this paper, we abbreviate the part of $T_{p,q,r}(\mathbb{Z}^{\{2,3\}})$ appearing on Tables 4.2 and 4.3 by $T^*_{p,q,r}$. Table 4.4 gives the cardinality of $T^*_{p,q,r}$ in all cases, under the normalization hypothesis $p \leq q \leq r$. The block on the right illustrates that we are using the bulk of Tables 4.2 and 4.3 in Sections 5-11.

5. A_6, S_6, $PGL_2(9)$, M_{10} and $P\Gamma L_2(9)$; basics of computations

In this section, we work with the following three covers, with $\rho = 1 + i \in \mathbb{F}_9$, as explained towards the end of §1.

$$\begin{aligned}
u_6 &= (x-1)^2,\\
w_6 &= (x-\rho)(x-\rho^7) = x^2 + ix + 1,\\
\Lambda_6 &= (3A, 3B, 4A) \to (33, 3111, 42),\\
f_6(t,x) &= (x^2-2)^3 + t(3x-4)^2,\\
D_6(t) &= 2^{13}3^6t^4(t-1)^2,
\end{aligned}$$

$$\begin{aligned}
u_{10a} &= (x-1)^3,\\
w_{10a} &= (x+1)(x-\rho)(x-\rho^7) = x^3 + \rho x^2 + \rho x + 1,
\end{aligned}$$

$$\begin{aligned}
\Lambda_{10a} &= (3AB, 2D, 8A) \to (3331, 22222, 811),\\
f_{10a}(t,x) &= (x^3+12x^2+60x+96)^3x + 1728t(3x^2+28x+108),\\
D_{10a}(t) &= -2^{99}3^{42}t^6(t-1)^5,
\end{aligned}$$

$$\begin{aligned}
u^s_{10b} &= (x-1)(x-\rho),\\
w^s_{10b} &= (x-\rho^3)(x-\rho^6),\\
\Lambda^s_{10b} &= (8A, 4A, 8B) \to (811, 4411, 811) \quad \text{(genus 1)},\\
t &= -(s-1)^2/4s,\\
\Lambda_{10b} &= (8CD, 2BC, 8AB) \to (82, 22211, 811),\\
f_{10b}(t,x) &= x^8(x-3)^2 - 27t(3x^2-2x+3),\\
D_{10b}(t,x) &= 2^{29}3^{42}(t-1)^3t^8.
\end{aligned}$$

The covers in the later sections will be presented in the same format. First, if the cover comes directly from a hypergeometric solution (1.6)–(1.8) to (1.1), we simply give such a solution by giving $u_N(x)$ and $w_N(x)$, as we have done for $N=6$ and $N=10a$. There are more lines if the cover comes only indirectly from a hypergeometric solution to (1.1). In the case of $10b$, the hypergeometric solution gives a cover $X^s_{10b} \to \mathbb{P}^1_s$ where X^s_{10b} has genus one. However there is an involution on X^s_{10b} over the involution $s \leftrightarrow 1/s$ on $\mathbb{P}^1_s$. We mod out by these involutions to obtain our cover $X_{10b} = \mathbb{P}^1_x \to \mathbb{P}^1_t$. We consistently reserve the variable t for the base variable in the final three point cover; s and/or u enter as base variables in intermediate three point covers. The intermediate covers, such as $X^s_{10b} \to \mathbb{P}^1_s$ here, drop from consideration as the final covers, such as $X_{10b} \to \mathbb{P}^1_t$, specialize to give more fields. The bulk of a given section discusses specialization, and the connection with the matrix ABC equation (1.1) no longer plays a role; in these discussions, only the last three lines corresponding to each cover enter.

The line beginning Λ_N gives first the conjugacy class $[m_c]$ of m_c in M, in Atlas notation, for $c = 0, 1, \infty$. When M has non-trivial outer automorphisms, like in Cases 6 and $10a$, there may be some ambiguity in how Atlas notation is used; in Case $10a$, for example, we could just as well write $8B$ rather than $8A$. The Λ_N line gives next the orbit-partition λ_c of m_c acting on the component set C. All the covers in this paper are rigid, meaning that they are completely determined by the group-theoretical data $\big([m_0]; [m_1], [m_\infty]; M \subseteq \mathrm{Sym}(C)\big)$. The computation of the defining equation starts from this group-theoretical information.

The component set $C_6 = \pi_0\big(X_6(\mathbb{C})^{\mathrm{cut}}\big)$ of the cover $X_6 \to \mathbb{P}^1_{\mathbb{Q}}$ has six elements. The monodromy transformations $m_{6,0}$, $m_{6,1}$, and $m_{6,\infty}$ generate the alternating group $M_6 = \mathrm{Alt}(C_6)$. The Galois group G_6 is all of $\mathrm{Sym}(C_6)$. The constant field extension with Galois group G_6/M_6 is $\mathbb{Q}(\sqrt{2})$, as can be seen from considering $D_6(t)$ modulo squares. Similarly, C_{10a} can be identified with a projective line over $\mathbb{F}_9$ so that the monodromy group M_{10a} is $\mathrm{PGL}_2(9)$ and the Galois group G_{10a} is $\mathrm{P\Gamma L}_2(9)$; the constant field extension is again $\mathbb{Q}(\sqrt{2})$, although this can no longer be seen from $D_{10a}(t)$. In contrast, Cover $10b$ has $M_{10b} = G_{10b} = \mathrm{P\Gamma L}_2(9)$.

Except for f_{9a} in Section 6 and the $f^*_{N,m}$ in Section 10, the equations for all our covers have the simple form $f(t,x) = A(x)+tC(x)$. In these cases, x is a coordinate on the covering curve X, identifying it with $\mathbb{P}^1_x$. The cover $X \to \mathbb{P}^1_t$ is given by the rational function $x \mapsto -A(x)/C(x)$. Define $B(x)$ by requiring that the polynomial ABC equation (1.4) hold, i.e. $B(x) = -A(x) - C(x)$. Then the polynomials $A(x)$, $B(x)$, and $C(x)$ factor over $\mathbb{Q}$ according to λ_0, λ_1, and λ_∞. Thus, for example, $f_{10a}(1,x)$ factors as a quintic squared, according to $\lambda_{10a,1} = 22222$. In brief, one determines the coefficients of A and C by imposing some normalization conditions and demanding that B factor in the proper way. For very simple computations of this form see [**4**]; to make the computations feasible in our range of degrees, one has to use the "differentiation trick" described there systematically.

We specialize the cover X_6 at the 39-element set $T^*_{3,3,4}$. The 39 algebras $L_{6,\tau}$ are all of degree six, thus the factor fields all appear in [**13**]. The algebra $L_{6,\tau}$ is the twin [**24**] of the algebra $L_{6,1-\tau}$, so that their splitting fields in $\mathbb{C}$ are identical: $K_{6,\tau} = K_{6,1-\tau}$. Since $L_{6,1/2}$ is self-twin, it is forced to be a non-field; in fact, $L_{6,1/2}$ factors as a quintic field L_5 times $\mathbb{Q}$, with L_5 the unique quintic $\{2,3\}$-field with Galois group the Frobenius group F_5.

The 39 specialization points yield 37 isomorphism classes of algebras, the repetitions being $L_{6,2^2 19^3/13^4} \cong L_{6,2^2}$ and $L_{6,3^2 5^3/13^4} \cong L_{6,-3}$. Excluding $L_5 \times \mathbb{Q}$ and counting the repeated fields once, one has 10 S_6 twin pairs and 2 $\mathrm{C}_3^2.\mathrm{D}_4$ twin pairs. The remaining 6 twin pairs consist of a field and a non-field. Two of these have Galois group S_5, three have Galois group $\mathrm{S}_4 \times \mathrm{S}_2$, and one has Galois group $\mathrm{S}_3 \times \mathrm{S}_3$.

In [**13**], we found a rather mysterious lack of balance in the 27-element set $\mathrm{NF}(\{2,3\},\mathrm{S}_6)$. Namely, if we group these fields, and also those in $\mathrm{NF}(\{2,3\},\mathrm{S}_5)$, by their unique quadratic subfields $\mathbb{Q}(\sqrt{d})$, the decomposition is as follows:

$d:$	-6	-3	-2	-1	2	3	6
$\mathrm{NF}(\{2,3\},\mathrm{S}_5)$	1	0	0	0	2	1	1
$\mathrm{NF}(\{2,3\},\mathrm{S}_6)$	1	1	0	2	13	4	6

In each case, just under half of the fields have discriminant class $d = 2$. So one can view the existence of the cover f_6 as explaining this imbalance; it accounts for both S_5's and 10 of the 13 S_6's.

Next, we specialize the cover X_{10a} at the 99-element set $T^*_{3,2,8}$. There is a tight relation between the covers X_6 and X_{10a}. Namely $f_6(s,x)$, $f_6(1-s,x)$ and $f_{10a}(1-(2s-1)^2,x)$ have the same splitting field over $\mathbb{Q}(s)$. This accounts for the behavior of the nineteen specialization points with $1-\tau \in \mathbb{Q}^{\times 2}$.

Twelve of the elements $\tau \in T^*_{3,2,8}$ satisfy $2(1-\tau) \in \mathbb{Q}^{\times 2}$. By the form of $D_{10a}(t)$, the corresponding G_τ are in the index two subgroup M_{10} of $\mathrm{P\Gamma L}_2(9)$. The polynomials $f_{10a}(-23^3,x)$ and $f_{10a}(-799^3/7^8,x)$ each factor as $f_9 f_1$. The field $K_{10a,-23^3}$ is one member of the two-element set $\mathrm{NF}(\{2,3\},\mathrm{C}_3^2.\mathrm{C}_4)$. The field $K_{10a,-731^3/7^8}$ has Galois group $\mathrm{C}_3^2.\mathrm{Q}$, with Q being the quaternion group. One knows that $\mathrm{NF}(\{2,3\},\mathrm{Q})$ also has two elements, with defining equations

$$g_\pm(x) = x^8 \pm 12x^6 + 36x^4 \pm 36x^2 + 9.$$

The field K_- is totally real while K_+ is totally imaginary; the quaternionic subfield of $K_{10a,-731^3/7^8}$ is K_+. The remaining ten fields are all distinct, with Galois group M_{10}, and discriminants $2^a 3^b$, $24 \le a \le 34$, $10 \le b \le 18$.

For eight of the remaining sixty-eight τ, the polynomial $f_{10a}(\tau,x)$ factors as $9+1$, namely $\tau = -2^1 61^3$, $-2^1 28559^3/3^3 13^8$, $-239^3/3^{13}$, $-212567^3/3^1 11^8$, $-29^3/5^8$, $2/3^3$, $1171537^3/3^5 95^8$, and $-4079^3/3$. The splitting fields $K_{10a,\tau}$ all have Galois group of the form $\mathrm{C}_3^2.\widetilde{\mathrm{D}}_4$, with $\widetilde{\mathrm{D}}_4$ the Sylow 2 subgroup of $\mathrm{GL}_2(3)$; these eight fields are all distinct. The remaining sixty specialization points give 59 fields in $\mathrm{NF}(\{2,3\},\mathrm{P\Gamma L}_2(9))$, the unique duplication being $K_{10a,156133^3/3^8 11^8} = K_{10a,5^3/2^2}$, verified by *polredabs*. Note that none of the 59 algebras $L_{10a,\tau}$ are tame at 2, because all of them have $\mathbb{Q}(\sqrt{2})$ in their splitting field. However exactly one is tame at 3, namely

$$L_{10a,11^3/2^3} : x^{10} - 4x^9 + 6x^8 + 24x^2 + 32x + 16, \tag{5.1}$$

with discriminant $-2^{34}3^9$. Note that $f(11^3/2^3, x)$ has relatively large coefficients. The displayed polynomial is obtained from $f(11^3/2^3, x)$ by *polredabs*. We have used *polredabs* analogously below, so as to always display only polynomials with relatively small coefficients.

Next, we specialize X_{10b} at the 35-element set $T^*_{8,2,8} = T^*_{\infty,2,\infty}$. The polynomial $f_{10b}(3^5, x)$ factors into an irreducible nonic and a linear, the nonic defining the same nonic field as the nonic factor of $f_{10a}(-23^3, x)$. Otherwise the Galois fields produced from X_{10b} are distinct from those produced from X_{10a}. The polynomial $f_{10b}(2/3^3, x)$ also factors in the form $9+1$; it has splitting field of degree 72 and is tame at 3. The remaining 33 elements of $T^*_{8,2,8}$ give irreducible degree 10 polynomials. The polynomial $f_{10b}(4, x)$ has splitting field the unique field in $\mathrm{NF}(\{2,3\}, \mathrm{S}_5)$ containing $\mathbb{Q}(\sqrt{6})$. The remaining thirty-two specialization points give thirty-two distinct fields containing A_6 in their Galois group, one with Galois group A_6, four with Galois group $\mathrm{PGL}_2(9)$, five with Galois group M_{10}, two with Galois group S_6, and twenty with Galois group $\mathrm{P\Gamma L}_2(9)$. The four with Galois group $\mathrm{PGL}_2(9)$ come from $\tau = -8, -288, -1$, and 2. Their corresponding field discriminants are $-2^{33}3^{10}$, $-2^{33}3^{16}$, $-2^{26}3^{12}$, and $2^{33}3^{12}$.

As we discussed in §2, we carried out many computations with Frobenius elements to distinguish fields and get lower bounds on Galois groups. Here are some group-theoretical details for the situation of this section Let $G^\natural$ be the set of conjugacy classes in $G = \mathrm{P\Gamma L}_2(9)$. For every $N \in \{10a, 10b\}$, $p \geq 5$ and $\tau \in \mathbb{F}_p - \{0,1\}$, one has a Frobenius element $\mathrm{Fr}_{N,\tau,p}$ in $G^\natural$. As an element of the quotient group $\mathrm{C}_2 \times \mathrm{C}_2$, it is given by a pair of Jacobi symbols:

$$[\mathrm{Fr}_{10a,\tau,p}] = \left(\left(\frac{2}{p}\right), \left(\frac{-2(\tau-1)}{p}\right)\right), \tag{5.2}$$

$$[\mathrm{Fr}_{10b,\tau,p}] = \left(\left(\frac{2\tau(\tau-1)}{p}\right), \left(\frac{2(\tau-1)}{p}\right)\right). \tag{5.3}$$

The Frobenius class $\mathrm{Fr}_{N,\tau,p}$ itself is then completely determined by the factorization pattern $\lambda_{N,\tau,p}$ of $f_N(\tau, x)$ in $\mathbb{Q}_p$. The possibilities are

	$\mathrm{A}_6 = \mathrm{PSL}_2(9) = \mathrm{M}'_{10}$					$\mathrm{S}_6 - \mathrm{A}_6$			$\mathrm{PGL}_2(9) - \mathrm{PSL}_2(9)$			$\mathrm{M}_{10} - \mathrm{M}'_{10}$	
$[\mathrm{Fr}_{N,\tau,p}]$	$(1,1)$					$(-1,-1)$			$(1,-1)$			$(-1,1)$	
Atlas	$1A$	$2A$	$3AB$	$4A$	$5AB$	$2BC$	$4B$	$6AB$	$2D$	$8AB$	$10AB$	$4C$	$8CD$
#	1	45	80	90	144	30	90	240	36	180	144	180	180
$\lambda_{N,\tau,p}$	1^{10}	2^41^2	3^31	4^21^2	5^2	2^31^4	4^22	631	2^5	81^2	10	4^21^2	82

The line # gives the number of elements of $\mathrm{P\Gamma L}_2(9)$ belonging to each conjugacy class. We do not use this information to distinguish fields and get lower bounds on Galois groups. However this information is very orienting as one carries out Frobenius computations, because a field with Galois group H has its Frobenius elements distributed in $H^{\natural}$ in proportion to the analogous numbers, here tabulated for $H = \mathrm{P\Gamma L}_2(9)$.

6. $\mathbf{SL_2(8)}$ and $\mathbf{\Sigma L_2(8)}$; Shimura curves

Here we specialize two covers. For the first, we work with $\mathbb{F}_8 = \mathbb{F}_2[a]/(a^3+a+1)$. For the second, we work with $\mathbb{F}_{64} = \mathbb{F}_2[b]/(b^6+b^3+1)$. The field element a has multiplicative order 7, while b has multiplicative order 9. We regard $\mathbb{F}_{64}$ as containing $\mathbb{F}_8$ via $a = b + b^8$.

The first cover is

$$\begin{aligned}
u_{9a} &= x^2 + x + 1,\\
w_{9a} &= x^2 + ax + 1,\\
\Lambda_{9a} &= (333, 22221, 9),\\
f_{9a}(t,x) &= (x^3 - 9x^2 - 69x - 123)^3 - 2^{14}t(9x^4 - 42x^3 - 675x^2 - 1485x - 441) - 2^{28}t^2,\\
D_{9a}(t) &= 2^{140}3^{18}t^6(t-1)^4\\
&\quad \cdot (43980465111104t^3 - 5421322469376t^2 - 7496810496t + 513922401)^2.
\end{aligned}$$

The defining polynomial f_{9a} was sent to us by Elkies in 1995. This cover has genus one. This is the reason behind the extraneous cubic-squared factor in $D_{9a}(t)$, an example of a non-trivial $c(t)^2$ in (3.1). All the other covers in this paper, besides a few that we see briefly while carrying out a construction involving base change, have genus zero.

The construction of the second cover involves two base changes, the first a degree three base change $\mathbb{P}^1_s \to \mathbb{P}^1_u$, and the second a degree two base change $\mathbb{P}^1_u \to \mathbb{P}^1_t$.

$$\begin{aligned}
u^s_{9b} &= (x-b)(x-b^6),\\
w^s_{9b} &= (x-b^4)(x-b^6),\\
\Lambda^s_{9b} &= (9A, 9B, 9C) \to (9,9,9) \quad \text{(genus 4)},\\
u &= (s-2)(s+1)(2s-1)/3(s-1)s,\\
\Lambda^u_{9b} &= (3B, 3B, 9ABC) \to (33111, 33111, 9) \text{ at } u = \sqrt{-3}, -\sqrt{-3}, \infty,\\
f_{9b}(u,x) &= A(x) + u2^{13}3^4,\\
A(x) &= x^9 + 108x^7 + 216x^6 + 4374x^5 + 13608x^4 + 99468x^3 + 215784x^2\\
&\quad + 998001x + 810648,\\
D_{9b}(u) &= 2^{104}3^{50}(u^2+3)^4,\\
t &= -u^2/3,\\
\Lambda_{18} &= (2^9, 3^4 1^6, 18),\\
f_{18}(t,x) &= f_{9b}(u,x) f_{9b}(-u,x)\\
&= A(x)^2 + 2^{26}3^9 t,\\
D_{18}(t) &= -2^{460}3^{189}t^9(t-1)^8.
\end{aligned}$$

The formula $f_{9b}(u,x)$ for Cover $9b$ is from [**21**, page 193]. The monodromy and Galois group of this cover are both $\Sigma L_2(8) = SL_2(8).3$. Matzat's three point cover does not fit into our set-up because the ramification locus is $u = \sqrt{-3}$, $-\sqrt{-3}$, ∞.

Our second base change is to place the ramification points at our standard locations, $t = 0, 1, \infty$. To do this without introducing irrationalities, we double the degree of the cover. The Galois group becomes $G_{18} = \Sigma\mathrm{L}_2(8)^2.2$. The monodromy group is the unique index three normal subgroup. The constant field extension, with Galois group G_{18}/M_{18}, is $\mathbb{Q}(\cos(2\pi/9))$, with defining polynomial $x^3 - 3x + 1$; this is the unique field in $\mathrm{NF}(\{2,3\}, \mathrm{A}_3)$.

We specialize f_{9a} at the 87-element set $T^*_{3,2,9}$. The polynomial $f_{9a}(-17^3/2^7, x)$ factors as $f_8 f_1$, the degree eight factor having Galois group $\mathrm{A}_4 \times \mathrm{C}_2$. The polynomials $f_{9a}(-2^5 3^2, x)$ and $f_{9a}(3^2 17^3/2^{15}, x)$ define the same field, with defining equation $x^9 + 6x^3 - 2$ and Galois group the 54-element affine group $(\mathbb{Z}/9).(\mathbb{Z}/9)^\times$. The remaining 84 specialization points give 55 fields, all with Galois group $\Sigma\mathrm{L}_2(8)$. All these fields are wildly ramified at three. Exactly two of them are only tamely ramified at 2:

$$L_{9a,\mathcal{T}} : x^9 - 36x^6 - 162x^4 - 54x^3 - 972x^2 + 486x - 594, \tag{6.1}$$

$$L_{9a,4/3} : x^9 - 18x^3 + 27x - 6. \tag{6.2}$$

The first field arises from five specialization points, namely $\tau = 2^5/3^4$, $-505^3/2^{27}3$, $2^{10}19^3/5^9 3$, $-2^9 5^3/3^2$, and $-73^3/2^{15}3^2$. The field discriminants are $2^8 3^{26}$ and $2^6 3^{26}$, respectively.

We specialize f_{18} at the 81-element set $T^*_{2,3,18} = T^*_{2,3,\infty}$. When -3τ is not a square in $\mathbb{Q}$, we get fields which generically have Galois group containing two copies of $\mathrm{SL}_2(8)$, thus interesting, but out of our self-imposed context. The six elements τ with -3τ a square are $\tau = -u^2/3$ with $u \in \{1, 10/9, 35/18, 3, 595/108, 37\}$. The twelve algebras $L_{9b,\pm u}$ are all fields, and so is $L_{9b,0}$; they are all pairwise non-isomorphic. Exactly one of these fields appeared already as a specialization of f_{9a}, namely $L_{9b,-5\cdot 7/2\cdot 3^2}$ which coincides with the tame-at-two field $L_{9a,4/3}$. Of the twelve remaining fields, nine have Galois group $\Sigma L_2(8)$, and three have Galois group $\mathrm{SL}_2(8)$, these being

$$\begin{aligned} L_{9b,-3} &: x^9 - 12x^6 - 18x^5 + 36x^2 - 27x - 128, \\ L_{9b,37} &: x^9 - 36x^6 - 54x^5 - 324x^4 - 216x^3 - 972x^2 - 243x - 2124, \\ L_{9b,1} &: x^9 - 36x^6 - 54x^5 + 432x^3 + 324x^2 - 243x - 1152. \end{aligned}$$

These fields have discriminants $2^{14}3^{22}$, $2^{14}3^{26}$, and $2^{14}3^{26}$, respectively.

In comparison with the other examples in this paper, the most remarkable phenomenon here is that often two or more $L_{9a,\tau}$ are isomorphic. This phenomenon can be partially explained as follows. The cuspidal data identifies the base $\mathbb{P}^1$ with a minimal-area Shimura curve $X_0(1)$ associated to the cubic field $\mathbb{Q}(\cos(2\pi/9))$ and no ramification at finite places. The cover $X_{9a} \to \mathbb{P}^1$ is identified with $X_0(2) \to X_0(1)$, the ideal (2) having residual cardinality 8. But now, one has also a degree four cover $\pi : X_0(P) \to X_0(1)$, where P is the unique prime above 3. The curve $X_0(P)$ has an Atkin–Lehner-type involution W_P, and hence a second natural map $\pi \circ W_P$ to the base curve $X_0(1)$. As in the classical case, one can think of $X_0(P) \subset X_0(1) \times X_0(1)$ as being a correspondence T_P from $X_0(1)$ to $X_0(1)$ of bidegree $(4,4)$. In terms of our fixed coordinate t on $X_0(1)$, and the same coordinate s on the second copy of $X_0(1)$, the defining equation for $X_0(P)$ is unique up to scalars:

$$
\begin{aligned}
h_P(s,t) = 2^{12}st\big(&-3^{10}17^3 66383 + 2^{13}3^8 1054805(s+t) + 2^{27}3^5 211(s^2+t^2) \\
&+ 2^{13}3^6 2486119st - 2^{21}3^3 22267(s^2t+st^2) \\
&- 2^{24}55s^2t^2 + 2^{30}(s^3t^2+s^2t^3)\big) \\
&- 3^6(-3^2 17^3 + 2^{15}s)^3 - 3^6(-3^2 17^3 + 2^{15}t)^3 - 3^{12}17^9.
\end{aligned}
$$

The points giving isomorphic algebras are exactly as follows. Here $\sigma \approx \tau$, rather than just $\sigma \sim \tau$, indicates that $h_P(\sigma,\tau) = 0$; so these isomorphisms $L_{9a,\sigma} \cong L_{9a,\tau}$ are explained by the Hecke correspondence T_P.

$$
\begin{array}{rcl}
-1 & \sim & 2797^3/13^9 \\
5^3 7^3/3^5 & \sim & -4079^3/3 \\
3^2 & \sim & 5^3/3^3 \\
3^5 & \sim & -23^3 \\
-2^5 3^2 & \sim & 3^2 17^3/2^{15} \\
-1/2^3 & \sim & 3^5 67^3/2^3
\end{array}
\qquad
\begin{array}{rcccl}
-2^4/3^2 & \sim & 2^2 7^3/3 & \approx & 23^3/2^{12}3 \\
2^3/3^2 & \sim & -13^3/2^2 3 & \approx & -5^6/2^9 3 \\
-2^3 3 & \sim & 1/2^2 & \approx & 3^1 11^3/2^9 \\
-3 & \sim & 5^3/2^7 & \approx & 3^1 19^3/2^7
\end{array}
$$

$$
\begin{array}{rcccccccl}
2^5/3^4 & \sim & -505^3/2^{27}3 & \approx & 2^{10}19^3/5^9 3 & \sim & -2^9 5^3/3^2 & \approx & -73^3/2^{15}3^2 \\
-2^3 & \sim & -3^2/2^4 & \sim & 5^3/2^2 & \approx & 7^3/2^9 & \approx & 3^2 73^3/2^9 \\
3/2^2 & \sim & -11^3/2^8 & \approx & -35^3 3/2^8 & \sim & -203^3/2^{11}5^9 & \approx & 3155^3 3/2^{11}7^9 \\
-2^4 3 & \sim & 2^2 & \approx & 3^1 5^3/2^{12} & \sim & -1915^3/2^{13} & \approx & -3^1 1027^3/2^{13}5^9
\end{array}
$$

The Hecke correspondence T_P also explains two of the three cases mentioned above of unexpectedly small Galois group. First, the degree seven polynomial $h_P(t,t)$ factors as $f_3^2 f_1$, the root of the linear factor being $\tau = -17^3/2^7$. So this τ is a CM point, which forces $L_{9a,\tau}$ to be non-generic. Second, the degree three polynomial $h_P(0,t)$ factors as f_1^3, the root of the linear factor being $\tau = 3^2 17^3/2^{15}$. So this τ, like 0, is also a CM point, again forcing non-genericity in $L_{9a,\tau}$.

The cover X_{10a} from the previous section is also a minimal area Shimura curve, coming from the field $\mathbb{Q}(\sqrt{2})$. For more on Shimura curves as covers of the projective line, see [**10, 29**].

7. $\mathbf{SL_3(3)}$ and $\mathbf{SL_3(3).2}$; projective twinning

Here we use two new covers $13a$, $13b$, and a cover $26c$, which is a doubled version of one of the covers in [**18**]. The new covers are

$$
\begin{aligned}
u_{13a} &= (x-1)^3, \\
w_{13a} &= (x-\rho)(x-\rho^3)(x+1) = x^3 + 2x^2 + 2, \\
\Lambda_{13a} &= (3B, 3A, 8A) \to (33331, 3331111, 841), \\
f_{13a}(t,x) &= (x^3 - 6x + 6\sqrt{-2}x - 4 - 8\sqrt{-2})^3(x - 2 - \sqrt{-2})^3(x - 2 + 2\sqrt{-2}) \\
&\qquad - t2^2 3^2(3x - 4 + \sqrt{-2})^4(3x - 8 + \sqrt{-2}), \\
D_{13a}(t) &= 2^{112}3^{54}(1-\sqrt{-2})^{72}t^8(t-1)^6, \\
u_{13b} &= (x-1)(x^2+1), \\
w_{13b} &= w_{13a}, \\
\Lambda_{13b} &= (4A, 3A, 8A) \to (44221, 3331111, 841), \\
f_{13b}(t,x) &= (x^2 - 3\sqrt{-2}x - 3\sqrt{-2} - 3)^4(x^2 + 6\sqrt{-2} - 3)^2(x + 3 - 3\sqrt{-2}) \\
&\qquad + t2^2 3^3(1+\sqrt{-2})^8(x + 1 - \sqrt{-2})^4(3x + 5 - \sqrt{-2}),
\end{aligned}
$$

$$D_{13b}(t) = 2^{92}3^{72}(1+\sqrt{-2})^{96}t^8(t-1)^6.$$

These covers were first computed over $\mathbb{F}_{11}$, where each cover appeared with its projective twin, as described below; at this level, there is no evident relation between the equation for a cover ($\sqrt{-2} \mapsto 3 \in \mathbb{F}_{11}$) and the equation for its twin ($\sqrt{-2} \mapsto 8 \in \mathbb{F}_{11}$). We worked with the auxiliary prime 11 because it is the smallest prime besides 3 which is split in the field $\mathbb{Q}(\sqrt{-2})$. We then lifted via the 11-adics to solutions in $\mathbb{Q}(\sqrt{-2})$ as explained in [**19**]. At this level, passing from a cover to its twin is induced by complex conjugation in the ground field $\mathbb{Q}(\sqrt{-2})$.

The third cover is

$$\begin{aligned}
u^u_{13c} &= (x-\rho^5)(x-\rho^7)(x-1) = x^3+x^2+1,\\
w^u_{13c} &= w_{13a},\\
\Lambda^u_{13c} &= (8A, 8B, 2A) \to (841, 841, 222211111) \text{ at } u = -\sqrt{-8}, \sqrt{-8}, \infty,\\
f_{13c}(u,x) &= A(x) - uB(x),\\
A(x) &= (x^3-12x^2-6x-64)(x^4+16x^3-36x^2+128x-28)\\
&\qquad \cdot (x^6+12x^5+54x^4+176x^3+444x^2+624x+552),\\
B(x) &= (x^4+16x^3+72x^2+128x+188), (3x^4-4x^3+12x^2-24x-68)^2,\\
D_{13c}(u) &= 2^{160}3^{114}(u^2+8)^{10},\\
t &= -u^2/8,\\
\Lambda_{26c} &= (2B, 8AB, 4B) \to (2^{13}, 8^2 4^2 1^2, 4^4\, 2^5),\\
f_{26c}(t,x) &= f_{13a}(u,x) f_{13a}(-u,x),\\
&= A(x)^2 + 8tB(x)^2,\\
D_{26c}(t) &= -2^{777}3^{452}t^{13}(t-1)^{20}.
\end{aligned}$$

One has $M_{13c} = G_{13c} = \mathrm{SL}_3(3)$. Just as we doubled the Matzat cover in the previous section, we double the Malle cover to place the critical values at our standard positions 0, 1, and ∞ and remove irrationalities. In distinction to what happened when we doubled the Matzat cover, here the groups do not become much bigger: $M_{26c} = G_{26c} = \mathrm{SL}_3(3).2$.

To facilitate comparison with f_{26c}, we consider the degree 26 polynomials $f_{26a} = f_{13a}\bar{f}_{13a}$ and $f_{26b} = f_{13b}\bar{f}_{13b}$ in $\mathbb{Q}[t,x]$. We specialize f_{26a} at the 27-element set $T^*_{3,3,8}$ and f_{26b} at the 24-element set $T^*_{4,3,8}$. The polynomials $f_{26a}(-8,x)$, $f_{26b}(4,x)$, $f_{26b}(3/4,x)$ each factor, and the corresponding splitting fields have Galois group $3^{1+2}_+.D_4$, $3^{1+2}_+.V$, and $3^{1+2}_+.C_2$ respectively. A Frobenius computation shows that the remaining 48 polynomials have Galois group all of $\mathrm{SL}_3(3).2$. The Frobenius computation shows that these fields are all non-isomorphic except for perhaps $L_{26b,1/4}$ and $L_{26b,-8}$. We have verified that this last pair of fields is indeed isomorphic. We used the method described in Section 9, working only with the roots of $f_{13b,1/4}$ and $f_{13b,-8}$.

Now consider Cover 26c. In general, if -2τ is not a square in $\mathbb{Q}$, then $L_{26c,\tau}$ contains a subfield isomorphic to $\mathbb{Q}(\sqrt{-2\tau})$. If -2τ is a square, then one has the factorization $f_{26c}(\tau,x) = f_{13c}(u,x)f_{13c}(-u,x)$, with $u^2 = -8\tau$. Thirty-eight of the elements in the 45-element set $T^*_{2,8,4}$ are such that -2τ is a non-square. A Frobenius computation shows that all of these have Galois group all of $\mathrm{SL}_3(3).2$

and are pairwise non-isomorphic. Moreover the fields $L_{26c,\tau}$ are not isomorphic with any of the 47 fields coming from $26a$ and $26b$.

The seven elements τ with -2τ a square are $\tau = -u^2/8$ with $u = 1, 2, 7/2, 4, 8, 44$ and 10. A Frobenius computation shows generic behavior in the first six cases: the splitting field of $f_{13c}(u,x)$ has Galois group all of $\mathrm{SL}_3(3)$, and these six splitting fields are non-isomorphic. Note that the degree thirteen fields $\mathbb{Q}[x]/f_{13c}(u,x)$ and $\mathbb{Q}[x]/f_{13c}(-u,x)$ are non-isomorphic, corresponding to different permutation representations of their common Galois group. This is seen clearly in the degenerate case $\tau = -5^2/2$. In this case, $f_{13c}(-10,x)$ factors as $9+4$ and $f_{13c}(10,x)$ factors as $12+1$; the common Galois group is $3_+^{1+2}.\widetilde{\mathrm{S}_4}$.

The twinning phenomenon here is quite general, deriving from the fact that the permutation representations of a group PGL_n on the projective space $\mathbb{P}^{n-1}$ and its dual $\check{\mathbb{P}}^{n-1}$ are not isomorphic, for $n \geq 3$. A prime ℓ and a degree $n \geq 3$ being fixed, say that a number field of degree $\ell^{n-1} + \ell^{n-2} + \cdots + \ell + 1$ is projective iff its Galois group can be identified with $\mathrm{PGL}_n(\ell)$, as a permutation group. Then a projective field L has a non-isomorphic twin L^t, there being a canonical isomorphism $L = L^{tt}$.

This situation of twin projective fields contrasts in one important way with the situation of twin A_6 or S_6 sextic fields, which played a role in Section 5. Namely, the two different permutation representations induce the same linear representation only in the projective case. So twin fields have the same Dedekind zeta function only in the projective case. One can summarize by saying that A_6 and S_6 sextic fields come in fraternal twin pairs while projective fields come in identical twin pairs. In particular, twin projective fields have the same discriminant. In the cases here, one has discriminants $2^{28}3^{24}$, $2^{36}3^{26}$, $2^{28}3^{22}$, $2^{38}3^{26}$, $2^{32}3^{24}$, $2^{38}3^{20}$ for $|s| = 1, 2, 7/2, 4, 8, 44$.

8. $\mathrm{SU}_3(3)$ and $\mathrm{SU}_3(3).2$; connections with a base change

Here we work with a cover $X_{28a} \to \mathbb{P}^1_t$, which comes from a solution to the matrix *ABC* equation (1.1) as follows.

$$\begin{aligned}
u^s_{28a}(x) &= (x-1)^2(x-i) = x^3 + (1+2i)x^2 + (1+2i)x + 2i,\\
w^s_{28a}(x) &= (x+1)^2(x+i) = x^3 + (2+i)x + (1+2i)x + i,\\
\Lambda^s_{28a} &= (12A, 2A, 12B) \to (12^2 3\ 1, 2^{12}1^4, 12^2 3\ 1) \quad \text{(genus 3)},\\
t &= -(s-1)^2/4s,\\
\Lambda_{28a} &= (4D, 2B, 12AB) \to (4^6 1^4, 2^{12}1^4, 12^2 3\ 1),\\
f_{28a}(t,x) &= (x^6 - 6x^5 - 435x^4 - 308x^3 + 15x^2 + 66x + 19)^4\\
&\quad \cdot (x^4 + 20x^3 + 114x^2 + 68x + 13) - 2^2 3^9 t(x^2+4x+1)^{12}(2x+1),\\
D_{28a}(t) &= 2^{576}3^{630}t^{18}(t-1)^{12}.
\end{aligned}$$

The monodromy group of the genus three cover $X^s_{28a} \to \mathbb{P}^1_s$ is the simple group $\mathrm{SU}_3(3)$. The Galois group is $\mathrm{SU}_3(3).2$, the corresponding constant field extension being $\mathbb{Q}(i)$. The equation $f_{28a}(t,x)$ for Cover $28a$ appears in [**20**, page 412]. Its monodromy and Galois groups are both $M_{28a} = G_{28a} = \mathrm{SU}_3(3).2$.

Explicitly, the rigid solution to (1.1) we are using is

$$\begin{pmatrix} 0 & 0 & i \\ 1 & 0 & 2+i \\ 0 & 1 & 2+i \end{pmatrix} \begin{pmatrix} 1 & 0 & 1+2i \\ 0 & 1 & 0 \\ 0 & 0 & 2 \end{pmatrix} \begin{pmatrix} 1+i & 1 & 0 \\ 2+2i & 0 & 1 \\ i & 0 & 0 \end{pmatrix} = I. \tag{8.1}$$

The three displayed matrices generate $M = \mathrm{SU}_3(3)$ inside $\mathrm{GL}_3(9)$. In constrast, the faithful linear representations of the larger group $\mathrm{SU}_3(3).2$ in characteristic two and three all have dimension ≥ 6. None of these representations gives rise to a rigid solution of the matrix equation (1.1). This is why our passage from (1.1) to Cover $28a$ involves an intermediate cover.

The intermediate cover $X^s_{28a} \to \mathbb{P}^1_s$ is a pullback via the base change $\mathbb{P}^1_s \to \mathbb{P}^1_t$, $s \mapsto -(s-1)^2/4s$ of the cover $X_{28a} = \mathbb{P}^1_x \to \mathbb{P}^1_t$. Some particular points that play a special role in our discussion behave as follows.

$$1 \mapsto 0, \tag{8.2}$$

$$-1 \to 1, \tag{8.3}$$

$$0, \infty \mapsto \infty. \tag{8.4}$$

The points $s = -1, 1$ in $\mathbb{P}^1_s$ are the unique ramification points of this quadratic base change map. From (8.2), the ramification class $4D$ associated to $t = 0$ pulls back to its square $2A$, which is now associated to $s = 1$. From (8.3), the class $2B$ associated with $t = 1$ pulls back to its square, the identity class $1A$; so there is no ramification associated to $s = -1$ in $\mathbb{P}^1_s$. Finally, from (8.4), the ramification class $12AB$ associated to $t = \infty$ splits into two ramification classes, $12A$ at $s = 0$ and $12B$ at the $s = \infty$. This entire discussion explains on a conceptual level why Cover $28a$ is ramified only at 2 and 3; were it not for the connection via the base change to rigidity, one would have expected 7 to very likely ramify too.

Two of the specialization points τ in the 44-element set $T^*_{4,2,12}$ yield a factorizable polynomial. Namely $f_{28a}(2/3^3, x)$ factors as $27 + 1$ and $f_{28a}(-5^4/2^3 3^3, x)$ factors as $24 + 4$. The splitting fields of these polynomials have Galois groups $3^{1+2}_+.8.2$ and $4.\mathrm{S}_4.2$ respectively, these being maximal subgroups of $\mathrm{U}_3(3).2$. The specialization point $\tau = \frac{1}{2}$ yields a polynomial with Galois group $\mathrm{SU}_3(3)$; the remaining polynomals yield 41 distinct fields, all having Galois group all of $\mathrm{SU}_3(3).2$.

9. $\mathbf{W(E_6)'}$ and $\mathbf{W(E_6)}$; exhibiting exceptional isomorphisms

Here we work with four degree twenty-seven covers, denoted $27a$, $27b$, $27c$, $27d$. Cover $27a$ is due to Häfner [**12**], while $27b$ and $27c$ are new here. These three covers are related to algebraic hypergeometric functions with finite monodromy, corresponding to the entries 47, 45, and 48 on the Beukers–Heckman list [**3**]; they all have $M = G = \mathrm{W(E_6)}$. Cover $27d$ is also new. It is directly related to hypergeometric-like functions with finite monodromy, namely functions corresponding to the families A and D of [**25**], rather than to the family H corresponding to hypergeometric functions. Here, however we present $27d$ as related via a base change to hypergeometric functions with infinite monodromy. For this cover, $M = \mathrm{W(E_6)}'$ and $G = \mathrm{W(E_6)}$, the corresponding constant field extension being $\mathbb{Q}(\sqrt{-3})$.

$$u_{27a} = x^6 + x^3 + 1,$$

$$w_{27a} = (x^2 + x + 1)(x^2 + 1)(x + 1)(x - 1),$$

$$\begin{aligned}
\Lambda_{27a} &= (9AB, 2C, 12C) \to (9^3, 2^6\,1^{15}, 12\,6\,4^2\,1),\\
f_{27a}(t,x) &= (x^3+6x^2-8)^9 - t2^4 3^{12} x^6 (x^2+5x+4)^4 (x-2),\\
D_{27a}(t) &= 2^{414}3^{450}t^{24}(t-1)^6,
\end{aligned}$$

$$\begin{aligned}
u_{27b} &= (x^4-x^2+1)(x^2+x+1),\\
w_{27b} &= (x^4+1)(x+1)(x-1),\\
\Lambda_{27b} &= (12AB, 2C, 8A) \to (12^2\,3, 2^6\,1^{15}, 8^3\,2\,1),\\
f_{27b}(t,x) &= 2^4 x^3 (x^2-3)^{12} - t3^9(x-1)^8(x-2)(x^2-2x-1)^8,\\
D_{27b}(t) &= 2^{542}3^{270}t^{24}(t-1)^6,
\end{aligned}$$

$$\begin{aligned}
u_{27c} &= u_{27a},\\
w_{27c} &= w_{27b},\\
\Lambda_{27c} &= (9AB, 2C, 8A) \to (9^3, 2^6\,1^{15}, 8^3\,2\,1),\\
f_{27c}(t,x) &= 2^{18}(x^3+9x^2+6x+1)^9 - 3^{15}\,t\,x(2x+1)^8(x^2-2x-1)^8,\\
D_{27c}(t) &= 2^{522}3^{450}t^{24}(t-1)^6,
\end{aligned}$$

$$\begin{aligned}
u^s_{27d} &= (x-1)^5 \in \mathbb{F}_3[x],\\
w^s_{27d} &= (x+1)^5 \in \mathbb{F}_3[x],\\
\Lambda^s_{27d} &= (9A, 2A, 9A) \to (9^3, 2^{12}1^3, 9^3) \quad \text{(genus 4)},\\
t &= -4s/(s-1)^2,\\
\Lambda_{27d} &= (9A, 2B, 4A) \to (9^3, 2^{10}\,1^7, 4^6\,1^3),\\
f_{27d}(t,x) &= 4(3x^3-12x-8)^9 - t(12x^3+54x^2+63x+22)\\
&\quad \cdot (9x^6+81x^5+135x^4+276x^3+432x^2+288x+64)^4,\\
D_{27d}(t) &= 2^{518}3^{450}t^{24}(t-1)^{10}.
\end{aligned}$$

The three new covers were all found first modulo 5, and then lifted via the 5-adics to rational solutions as explained in [**19**]. The cover X_{28b} of Section 11 was also computed in this way. Of these four new covers, $27d$ was the hardest to obtain, since the $N+2$ parts of $3N$ are more evenly distributed among the cusps.

In Cases $27a$, $27b$, and $27c$, we specialize to τ in the 35-element set $T^*_{9,2,12} = T^*_{12,2,8} = T^*_{9,2,8} = T^*_{\infty,2,\infty}$. In Case $27d$, we specialize to τ in the 45-element set $T^*_{9,2,4}$. All 150 defining polynomials $f_{27*}(\tau,x)$ are irreducible, so all the algebras $L_{27*,\tau}$ are fields. A Frobenius computation proves that at least 146 of these fields have Galois group all of G or G'. Using the monodromy technique described in Section 11, we have verified that each apparent group-drop indeed occurs, and computed degree nine resolvents as follows.

(9.1) $$h_{27a,-1}(x) = x^9-3x^8+6x^7-3x^5+15x^4+6x^3+6x^2+6x+2,$$

(9.2) $$h_{27b,-1/48}(x) = x^9-12x^6-9x^5+54x^3+36x^2+18x-56,$$

(9.3) $$g_{27d,1/2}(x) = x^9-6x^7+27x^5-12x^4-174x^3-108x^2+183x+124,$$

$$(9.4) \qquad g_{27d,1/2\cdot 13^4}(x) = x^9 + 3x^8 - 16x^6 - 36x^5 + 36x^4 + 96x^3 - 108x - 36.$$

Here $h_{27a,-1}$ and $f_{27a,-1}$ have the same splitting field, and so too do $h_{27b,-1/48}$ and $f_{27b,-1/48}$; the Galois groups here have the form $3^3.(\mathrm{S}_4 \times \mathrm{C}_2)$ and $3^3.\mathrm{S}_4$ respectively. On the other hand, the Galois groups $G_{27d,1/2}$ and $G_{27d,1/2\cdot 13^4}$ each have the form $3^{1+2}.\widetilde{\mathrm{S}}_4$; the corresponding degree nine resolvents have Galois group of the form $3^2.\widetilde{\mathrm{S}}_4$. The field discriminants of the four displayed nonic polynomials are $2^{13}3^{15}$, $2^{12}3^{18}$, $-2^{22}3^{13}$, and $-2^{22}3^{15}$ respectively.

All the 150 specializations are wild at both 2 and 3 except

$$(9.5) \qquad L_{27d,-48} = \mathbb{Q}[x]/f_{27d}(-48, x),$$

which is tame at 2, with discriminant $2^{20}3^{84}$. The tameness at 2 can almost be seen from the Newton polygon at 2, which has slopes 1/5, 1, 8/5, and 5 with multiplicities 10, 1, 15, and 1 respectively. It can be seen from the full 2-adic factorization of $f_{27d}(-48, x)$ which has three factors of degree five and discriminant 2^4, one factor of degree ten and discriminant 2^8, and two linear factors.

The Frobenius computation proves that except for the possible isomorphism

$$L_{27b,-48} \cong L_{27d,32/81},$$

the 146 generic specializations are all distinct. To prove that $L_{27b,-48}$ and $L_{27d,32/81}$ are isomorphic one needs to exhibit the unique isomorphism; this is our topic for the rest of the section.

In numerical terms, one needs to find the unique bijection σ from the complex roots $A = \{\alpha_i\}$ of $f_{27b,-48}$ to the complex roots $B = \{\beta_j\}$ of $f_{27d,32/81}$ such that the the polynomial

$$(9.6) \qquad \mathrm{test}_{\sigma,r}(x) = \prod_i (x - \alpha_i - r\beta_{\sigma(i)})$$

has rational coefficients for all $r \in \mathbb{Q}$; here we introduce r just to avoid possible inseparability problems; in practice one can simply fix r at 1.

The bijection σ must intertwine complex conjugation: $c_B \circ \sigma = \sigma \circ c_A$. Since the polynomials in question have exactly three real roots, there are $3!12!2^{12} \approx 1.177 \times 10^{13}$ such bijections. The sheer size of this number is why *Pari*'s *nfisisom*, which is designed to find isomorphisms between fields, does not work here. Indeed, if we were simply given the bare polynomials $f_{27b,-48}$ and $f_{27d,32/81}$, we would not know how to find the desired isomorphism.

But we are not given the polynomials in isolation. Rather we can figure out how the monodromy operators m_c act on the roots. In the case of $f_{27b,-48}$ this action is as follows.

$$\begin{aligned} m_{27b,0} &= (1a, 15a, 7a, 4a, 3a, 2a, 2b, 3b, 4b, 7b, 15b, 1b)(5a, 6, 5b) \\ &\qquad \cdot (11, 12b, 13b, 14b, 8b, 9b, 10, 9a, 8a, 14a, 13a, 12a), \\ m_{27b,\infty} &= (1a, 1b)(2a, 3a, 4a, 5a, 5b, 4b, 3b, 2b)(6, 7a, 8a, 9a, 10, 9b, 8b, 7b) \\ &\qquad \cdot (11)(12a, 13a, 14a, 15a, 15b, 14b, 13b, 12b). \end{aligned}$$

Here we have ordered the $\mathrm{Gal}(\mathbb{C}/\mathbb{R})$-orbits from left to right, 1 through 15. If a root is in the lower half plane, we append "a"; if it is the upper half plane we append "b"; if it is on the real line we index the root simply by a number; thus $\alpha_{1a} = -(12.976\ldots) - (243.333\ldots)i$ and $\alpha_6 = .518\ldots$. The action of $m_{27b,0}$ is obtained by considering the inverse image of $[-48, 0]$ in the complex plane with coordinate x. It is the union of three "wheels," two with 12 spokes and one with 3. The action is obtained by "rotating the wheels one spoke counter-clockwise." All this can be done numerically, working visually with say the roots of $f_{27b}(-48u^{12}, x)$, with $u = 0, 1/100, 2/100, \ldots, 1$; restricting to $u \geq 20/100$ lets one work entirely in standard precision and still do the computation. The other monodromy action $m_{27b,\infty}$ is obtained similarly, i.e. again in the the spirit of Grothendieck's *dessins d'enfants* [**26**]. Repeating the procedure for the roots of $f_{27d,32/81}$ gives

$$\begin{aligned} m_{27d,0} &= (1a, 2a, 10a, 6a, 3, 6b, 10b, 2b, 1b)(9b, 8b, 7b, 5b, 4, 5a, 7a, 8a, 9a) \\ &\qquad \cdot (11a, 13a, 14a, 12a, 15, 12b, 14b, 13b, 11b), \\ m_{27d,1} &= (5a)(5b)(13a)(13b)(14a)(14b)(15)(1a, 1b)(2a, 12a)(2b, 12b) \\ &\qquad \cdot (3, 4)(6a, 9a)(6b, 9b)(7a, 8a)(7b, 8b)(10a, 11a)(10b, 11b). \end{aligned}$$

Let C be one of the two root sets A or B. We use the monodromy action to view the 27-element set C as a structured set. Namely consider the set $\mathrm{Sub}_2(C)$ of two-element subsets. Under the action of the monodromy group one has a decomposition into two orbits

$$\mathrm{Sub}_2(C) = \mathrm{Sub}_2(C)' \coprod \mathrm{Sub}_2(C)''.$$

Here the first orbit has 135 elements and the second has 216. In other words, we can view C as the vertices of a graph Γ_C, with edge-set $\mathrm{Sub}_2(C)'$. The action of complex conjugation on C extends to Γ_C. The graph-with-involution (Γ_C, i_C) has only $2^7 3^2 = 1152$ automorphisms.

The bijection $\sigma\colon A \to B$ we seek respects not only complex conjugation but also the graph structure. To find it, we work purely group-theoretically to first find some isomorphism $s\colon (A, i_A) \to (B, i_B)$ respecting the graph structure; this is easy. Then we compose it with the 1152 elements of $\mathrm{Aut}(\Gamma_B, i_B)$ to get 1152 candidates for σ. For only one of them does (9.6) appear to have rational coefficients, and we have thereby numerically determined σ. It is as follows:

roots of $f_{27b,-48}$:	$1a$	$2a$	$3a$	$4a$	$5a$	6	$7a$	$8a$	$9a$	10	11	$12a$	$13a$	$14a$	$15a$
roots of $f_{27d,32/81}$:	$7b$	$1b$	$9a$	$5b$	$8a$	4	$6a$	$2b$	$11a$	15	3	$14b$	$13a$	$12b$	$10a$

Here, since $\sigma(\alpha_{1a}) = \beta_{7b}$, we must have $\sigma(\alpha_{1b}) = \beta_{7a}$ as well, and so too for the other complex conjugate pairs.

To check rigorously that indeed σ induces an isomorphism we proceed as follows. First, we numerically solve the 27-by-27 system $Tz = B$ with $T_{ik} = \alpha_i^k$ and $B_i = \beta_{\sigma(i)}$ for the vector z, rationalizing at the end. Then

$$y = \sum_{k=0}^{26} z_k x^k, \tag{9.7}$$

as an element of the ring $\mathbb{Q}[x]/f_{27b}(-48, x)$, should satisfy $f_{27d}(32/81, y) = 0$. In our case, (9.7) takes the explicit form $126414618624y =$

$$\begin{array}{rrr}
-778448003x^{26} & -981509289x^{25} & -45939794218464x^{24} \\
1137207587245554x^{23} & -12724591174373616x^{22} & 84720990963862440x^{21} \\
-370643301489686778x^{20} & 1104517184207752350x^{19} & -2221386267735948267x^{18} \\
2770729912599438087x^{17} & -1364804638977353670x^{16} & -1659195683617968252x^{15} \\
3294135546577106040x^{14} & -1424334213106643304x^{13} & -1494872066602993428x^{12} \\
1744593287940021708x^{11} & 585147838091136391x^{10} & -818860129717879323x^{9} \\
171625804007741892x^{8} & 237402868177405098x^{7} & -68548289134461768x^{6} \\
-5005683284871398 4x^{5} & 11561622988644846x^{4} & 7711324091701110x^{3} \\
-484480924209657x^{2} & -608102864271747x & -73397655884286.
\end{array}$$

This indeed satisfies $f_{27d}(32/81, y) = 0$. The case $L_{13c,1/4} \cong L_{13c,-8}$ from §8 is substantially easier, as the root sets to be identified have only 13 elements each.

10. A_9 and S_9; cuspidal specialization

Let (A, B, C) be a rigid solution to the matrix ABC equation (1.1) generating a subgroup $\widetilde{M}$ of $\mathrm{GL}_n(\overline{\mathbb{F}}_\ell)$ with center Z. Almost always, $\widetilde{M}$ is closely related to a finite classical group, i.e. an orthogonal group, a symplectic group, a unitary group, or a general linear group. Our ABC construction requires us to choose a faithful permutation representation of $M = \widetilde{M}/Z$, i.e. an overgroup $S_N \supseteq M$, and only then do we have a three point cover for which we seek an equation. The equations we have been finding are rather complicated, because they reflect not only the fundamental datum (A, B, C), but also the choice of S_N. Accordingly, even though solutions to (1.1) come in very regularly behaving families, we have had to work one cover at a time.

In this section, we are in the very exceptional situation where the monodromy group is closely related to a symmetric group. We work with two positive integer parameters $N > m$, and the construction is uniform in these parameters. Throughout, we take $r = N - m$.

$$\begin{aligned}
u_{N,m}(x) &= (x^N - 1)/(x - 1), \\
w_{N,m}(x) &= (x^m - 1)(x^r - 1)/(x - 1), \\
\Lambda_{N,m} &= (N, 2\,1^{N-2}, mr), \\
f_{N,m}(t, x) &= m^m x^N - t(Nx - r)^m, \\
D_{N,m}(t) &= (-1)^{(N+2m)(N+1)/2} N^N (mr)^{(N-1)m} t^{N-1}(t - 1).
\end{aligned}$$

The cover $X_{N,m}$ is isomorphic to the cover $X_{N,r}$, an isomorphism being $x \mapsto mx/(Nx - r)$. So, without loss of generality, we restrict to the case $m \leq N/2$. We assume further that N and m are relatively prime, so as to make $u_{N,m}(x)$ and $w_{N,m}(x)$ relatively prime and hence (1.6)–(1.8) irreducible. In this case, the monodromy group $M_{N,m}$ and the Galois group $G_{N,m}$ are both the full symmetric group S_N.

For $m = 1$, the polynomial $f_{N,m}$ is a trinomial and these covers have been discussed in e.g. [**21**, Section II.3]. As we explain next, even the other covers $X_{N,m}$ can be given by trinomial equations, and in this guise they have appeared in many places. Define integers v and w by

$$v \in \{0, 1, \ldots, r-1\}, \quad Nv \equiv 1 \mod (r), \quad w = (Nv-1)/r.$$

Define

$$y = \left(\frac{m}{Nx - r}\right)^{w-v} x^w \in \mathbb{Q}(x).$$

One has

$$y^m = t^{w-v} x \in \mathbb{Q}(x)$$

and y is a root of

$$f^*_{N,m}(t, y) = my^N - Nt^v y^m + rt^w.$$

In summary, we have two defining polynomials for the same cover, the canonical one $f_{N,m}$ and a trinomial version $f^*_{N,m}$. We use $f^*_{N,m}$ exclusively in the sequel. We mention $f_{N,m}$ because this is the cover given by the standard three point cover algorithm sketched in Section 5: x is a coordinate on $X_{N,m}$ while y is a rational function in x, typically of degree > 1. The general trinomial $ax^N + bx^m + c$ fits in this situation via $t = (m/a)^m (b/N)^N (r/c)^r$. Thus we call the $X_{N,m}$ trinomial covers.

The bad reduction set $\mathrm{S}_{N,m}$ is the set of primes dividing Nmr; in other words, $X_{N,m}$ has bad reduction within S iff $N/m \in T_{\infty,\infty,\infty}(\mathbb{Z}^S)$. In particular, $\mathrm{S}_{N,m} \subseteq \{2, 3\}$ exactly for the four pairs $(N, m) = (2, 1)$, $(3, 1)$, $(4, 1)$, and $(9, 1)$. For comparison, we note that specializing $f_{3,1}$ just at $T^*_{\infty,2,\infty}$ already yields all 9 $\mathrm{A}_3/\mathrm{S}_3$ cubics. Similarly, specializing $f_{4,1}$ just at $T^*_{\infty,2,\infty}$ already yields all 23 $\mathrm{A}_4/\mathrm{S}_4$ quartics, except the unique totally real one; in fact, no specialization of $f_{4,1}$ is totally real.

We are interested here in the case $(N, m) = (9, 1)$, coming from the Catalan equation $2^3 + 1 = 3^2$. Here a Frobenius computation shows that nothing unexpected happens when one specializes t to τ in the 35-element set $T^*_{8,2,9} = T^*_{\infty,2,\infty}$. One gets 10 fields with Galois group A_9 and 25 with Galois group S_9. The absolute field discriminants $2^a 3^b$ are mostly near the maximum possible $2^{31}3^{26}$, with $12 \leq a \leq 31$ and $12 \leq b \leq 26$ being the exact ranges. Ref. [**17**] lets one understand a and b in terms of the 2-adic and 3-adic placement of τ, respectively.

In general, suppose $X \to \mathbb{P}^1_{\mathbb{Q}}$ is a three point cover with defining equation $f(t, x)$. It makes sense to specialize t to one of the cuspidal points $c = 0$, 1, and ∞ as well, to obtain a Galois field $K_c \subset \mathbb{C}$. Let Norm_c be the normalizer of the monodromy transformation m_c in G. Then $\mathrm{Gal}(K_c/\mathbb{Q})$ is contained in $\mathrm{Norm}_c/\langle m_c \rangle$.

For the covers $X_{N,m}$ of this section, the most interesting cusp to specialize at is $c = 1$. The Galois group $G_{N,m,1}$ is contained in a symmetric group S_{N-2}. These fields $K_{N,m,1} \subset \mathbb{C}$ have been studied in [**5**]. In our example, one has

$$f_{9,1}(1, x) = (x-1)^2 (x^7 + 2x^6 + 3x^5 + 4x^4 + 5x^3 + 6x^2 + 7x + 8).$$

The septic field defined by the degree seven factor has Galois group S_7 and discriminant $-2^{12}3^{10}$. In general, cuspidal specializations seem worthy of special attention. First, they depend only on the group-theoretic data defining the cover from which they come. Second, they serve as an aid in understanding non-cuspidal specializations which are p-adically near the cusp.

11. $\mathbf{W(E_7)'}$; computing lower degree resolvents

Here we use the following new cover with $M = G = \mathrm{W}(\mathrm{E}_7)' = \mathrm{Sp}_6(2)$:

$$\begin{aligned} u_{28b} &= (x^6 - x^3 + 1)(x + 1), \\ w_{28b} &= (x^4 - x^2 + 1)(x^2 + x + 1)(x - 1), \\ \Lambda_{28b} &= (12C, 2A, 9A) \to (12^2\, 3\, 1, 2^6\, 1^{16}, 9^3\, 1), \\ f_{28b}(t, x) &= 3^6(x^2 + 6x + 6)^{12}x^3(3x + 4) - t\, 2^{18}(x^3 + 3x^2 - 3)^9, \\ D_{28b}(t) &= 2^{540}3^{450}t^{24}(t - 1)^6. \end{aligned}$$

Here x, as a multivalued function of t, can be expressed in terms of hypergeometric functions with finite monodromy, as this cover corresponds to Entry 58 of the Beukers–Heckman list [**3**, page 353].

We specialize to τ in the 35-element set $T^*_{12,2,9} = T^*_{\infty,2,\infty}$. A Frobenius computation proves that for $\tau \neq -1$ the algebras $L_{28b,\tau}$ all have Galois group all of $\mathrm{Sp}_6(2)$. At $\tau = -1$, there is an apparent group-drop to $\mathrm{S}_8 \cong \mathrm{SO}_6^+(2) \subset \mathrm{Sp}_6(2)$. The rest of this section explains how we produce a degree eight polynomial with the same splitting field as $f_{28b}(-1, x)$, thereby proving that indeed $G_{28b,-1} \cong \mathrm{S}_8$.

In general, suppose given an irreducible degree 28 polynomial f over $\mathbb{Q}$ with Galois group S_8 or $\mathrm{W}(\mathrm{E}_7)'$. For each bijection

$$\begin{aligned} L = (\text{Two element subsets of } \{0,\ldots,7\}) &\to (\text{Complex roots of } f), \\ \{i, j\} &\to \alpha_{L(i,j)}, \end{aligned}$$

one has an octic polynomial

$$g_L(x) = \prod_{i=1}^{8}\left(x - \sum_{j \neq i} \alpha_{L(i,j)}\right).$$

Assuming f is sufficiently generic, one gets $28!/8! \approx 7.5 \times 10^{24}$ distinct octic polynomials. If the Galois group is S_8, then exactly one of these has coefficients in $\mathbb{Q}$, it being the desired polynomial.

Suppose now that f has exactly four real roots so that the desired octic has two real roots. Then it suffices to consider bijections L where the involution -1 acting on $\mathbb{Z}/8$ goes over to complex conjugation. Then one gets $(4!24!!)/(2!3!2^3) \approx 4.9 \times 10^{11}$ octic polynomials. So still this method for finding a resolvent octic is impractical.

In our case, however, we have an action of $\mathrm{W}(\mathrm{E}_7)' \subset \mathrm{S}_{28}$ on the roots of $f_{28b}(-1, x)$ and the desired g_L is among the thirty-six g_L coming from the thirty-six subgroups of $\mathrm{W}(\mathrm{E}_7)'$ isomorphic to S_8. Using monodromy techniques as described in Section 9, we can identify these 36 copies of S_8. Looking among only these, we find a labeling L giving a desired g with quite large coefficients. Applying *polredabs*, gives

$$g_8(x) = x^8 + 4x^7 + 8x^6 + 16x^5 + 22x^4 + 20x^3 + 10x^2 - 8x - 10, \tag{11.1}$$

with field discriminant $-2^{17}3^9$.

To prove rigorously that $f_{28b}(-1, x)$ and $g_8(x)$ have the same splitting field, we proceed again as in Section 9. Write

$$g_8(x) = \prod_{\alpha \in A'}(x - \alpha), \quad g_{28}(x) = \prod_{\{\alpha_1, \alpha_2\} \subset A'}(x - \alpha_1 - \alpha_2), \quad f_{28b}(-1, x) = 3^7 \prod_{\beta \in B}(x - \beta).$$

We need to numerically find the right bijection σ from $A = \mathrm{Sub}_2(A')$ to B, and then algebraically confirm that one indeed gets an isomorphism from $\mathbb{Q}[x]/g_{28}(x)$ to $\mathbb{Q}[y]/f_{28b}(-1, y)$. This unique correct bijection is indicated on the following chart, with the root-labeling convention of Section 9:

	1	$2a$	$2b$	$3a$	$3b$	$4a$	$4b$	5
1		$11b$	$11a$	$13a$	$13b$	$10a$	$10b$	2
$2a$			15	$5b$	$14b$	$12b$	$7b$	$16a$
$2b$				$14a$	$5a$	$7a$	$12a$	$16b$
$3a$					9	$4a$	$8a$	$3b$
$3b$						$8b$	$1b$	$3a$
$4a$							1	$6a$
$4b$								$6b$
5								

Thus, for example, $\{\alpha_1, \alpha_2\} = \{1, 2a\} \in A$ matches $11b \in B$. Our computation and confirmation of the resolvents (9.1), (9.2), (9.3), (9.4) were each of similar complexity.

12. S_{32}; prime-dropping specialization

Let $f(t, x) \in \mathbb{Q}[t, x]$ define a degree N three point cover, with bad reduction set S'. For $\tau \in \mathbb{Q} - \{0, 1\}$ it may happen that the specialized algebra L_τ is ramified strictly within S'. Then we call τ a prime-dropping specialization point for f. Our experience suggests that such prime-dropping specialization points are quite rare. Here we report on three related instances of this phenomenon.

Consider again the trinomial covers $X_{N,m}$ of Section 10, with $r := N - m$. Recall that the specialized algebra is given by a trinomial equation: $L_\tau = \mathbb{Q}[x]/f^*_{N,m}(\tau, x)$. Recall also that the cover has bad reduction at all primes dividing Nmr. Nonetheless, one has the following fact, i being any positive integer:

$$\text{(12.1)} \qquad \text{If } p^e || \left\{ \begin{array}{l} N \\ m \\ r \end{array} \right. \text{ and } \tau \in \left\{ \begin{array}{l} T(\mathbb{Q}_p)^{0,Nei} \\ T(\mathbb{Q}_p)^{\infty,mei} \\ T(\mathbb{Q}_p)^{\infty,rei} \end{array} \right. \text{ then } K_{N,m,\tau} \text{ is unramified at } p$$

This fact can be proved directly: for suitable rational numbers a and b, one has $af^*_{N,m}(\tau, bx) \in \mathbb{Z}[x]$, with polynomial discriminant prime to p, as in (12.2), (12.3), and (12.4) below. Alternatively, this fact is a special case of the main theorem of [**17**].

We found three new fields this way with $S = \{2, 3\}$. All three fields come from the specialization point $\tau = 2 \cdot 5^5/3^2$, associated to the *ABC* triple

$$-2 \cdot 5^5 + 79^2 + 3^2 = 0.$$

The fields come from the covers

$$f^*_{8,3}(t, x) = 3x^8 - 8t^2x^3 + 5t^3,$$
$$f^*_{9,4}(t, x) = 4x^9 - 9t^4x^4 + 5t^7,$$
$$f^*_{32,5}(t, x) = 5x^{32} - 32t^{11}x^5 + 27t^{13}.$$

Nice defining polynomials for the number fields are as follows.

$$\text{(12.2)} \qquad 3^7 5^{-16} f^*_{8,3}(2 \cdot 5^5/3^2, -5^2x/3) = x^8 + 2^5x^3 + 2^3 3,$$

$$(12.3) \qquad 2^{-2}3^{18}5^{-36}f^*_{9,4}(2\cdot 5^5/3^2, 5^4x/3^2) = x^9 - 2^2 3^4 x^4 + 2^5 3^4,$$

$$(12.4) \qquad 3^{32}5^{65}f^*_{32,5}(2\cdot 5^5/3^2, -5^2x/3) = x^{32} + 2^{16}3^5x^5 + 2^{13}3^9.$$

The polynomial discriminants are $-2^{45}3^5 79^2$, $-2^{40}3^{48}79^2$, and $-2^{563}3^{277}79^2$, respectively. The field discriminants are $-2^{31}3^5$, $-2^{14}3^{24}$, and $-2^{191}3^{111}$, respectively. Here, in the degree 32 case, we used [**17**] to compute the exponents 191 and 111. Note that the maximal absolute discriminant allowed by local considerations in the last case is $2^{191}3^{112}$.

In all three cases, the fact that 79 does not divide the field discriminant d can be seen directly from the polynomial, rather than by our usual appeal to (3.4). Namely, each polynomial factors in the form $f_{N-2}f_1^2$ over $\mathbb{F}_{79}$, with f_{N-2} separable. This implies that $\mathrm{ord}_{79}(d) \in \{0, 1\}$. However from the polynomial discriminant we know that $\mathrm{ord}_{79}(d) \in \{0, 2\}$.

References

1. A. Ash, *Galois representations and cohomology of* $\mathrm{GL}(n, \mathbb{Z})$, Séminaire de Théorie des Nombres (Paris, 1989-90), Prog. Math., vol. 102, Birkhäuser Boston, Boston, MA, 1992 pp. 9–22.
2. F. Beukers, *The diophantine equation* $Ax^p + By^q = Cz^r$, Duke Math J. **91** (1998), 61–88.
3. F. Beukers and G. Heckman, *Monodromy for the hypergeometric function* ${}_nF_{n-1}$, Invent. Math. **95** (1989), 325–354.
4. B. Birch, *Noncongruence subgroups, covers and drawings*, The Grothendieck Theory of Dessins d'Enfants (Luminy, 1993), London Math. Soc. Lecture Note Ser., vol. 200, Cambridge Univ. Press, Cambridge, 1994, pp. 25–46.
5. A. Borisov, *On some polynomials allegedly related to the abc conjecture*, Acta Arith. **84** (1998), 109–128.
6. F. B. Coghlan, *Elliptic curves with conductor* $N = 2^a3^b$, Ph.D. Thesis, Manchester University, 1967; Tables published in: *Modular functions of one variable.* IV, Proc. Internat. Summer School (Antwerp, 1972), Lecture Notes in Math. vol. 476, Springer, Berlin, 1975, pp. 123–134.
7. J. H. Conway, R. T. Curtis, S. P. Norton, R. A. Parker, and R. A. Wilson, *An atlas of finite groups*, Oxford University Press, 1985.
8. H. Darmon, *Faltings plus ε, Wiles plus ε, and the generalized Fermat equation*, C. R. Math. Rep. Acad. Sci. Canada **19** (1997), 3–14.
9. H. Darmon and A. Granville, *On the equations* $z^m = F(x, y)$ *and* $Ax^p + By^q = Cz^r$, Bull. London Math Soc. **27** (1995), 513–543.
10. N. D. Elkies, *Shimura curve computations*, Algorithmic Number Theory (Portland, 1998), Lecture Notes in Comput. Sci., vol. 1423, Springer, Berlin, 1998. pp. 1–47.
11. B. H. Gross, *Modular forms* (mod p) *and Galois representations*, Internat. Math. Res. Notices 1998, no. 16, 865–875.
12. F. Häfner, *Einige orthogonale und symplektische Gruppen als Galoisgruppen über* $\mathbb{Q}$, Math. Ann. **292** (1992), no. 4, 587–618.
13. J. W. Jones and D. P. Roberts, *Sextic number fields with discriminant* $-^j2^a3^b$, Number Theory (Ottawa, 1996), CRM Proc. Lecture Notes, vol. 19, Amer. Math. Soc., Providence, RI, 1999, pp. 141–172.
14. ———, *Timing analysis of targeted Hunter searches*, Algorithmic Number Theory (Portland, 1998), Lecture Notes in Comput. Sci., 1423, Springer, Berlin, 1998.
15. ———, *Septic number fields with discriminant* $\pm 2^a3^b$, Math. Comp. **72** (2003), no. 244, 1975–1985.
16. N. M. Katz, *Rigid local systems*, Ann. of Math. Stud., vol. 139, Princeton University Press, Princeton, NJ, 1996.
17. P. Llorente, E. Nart, and N. Vila, *Discriminants of number fields defined by trinomials*, Acta Arith. **43** (1984), 367–373.
18. G. Malle, *Polynomials for primitive nonsolvable permutation groups of degree* $d \leq 15$, J. Symbolic Comput. **4** (1987), 83–92.

19. ———, *Polynomials with Galois groups* Aut(M_{22}), M_{22}, *and* $PSL_3(\mathbb{F}_4).2_2$ *over* $\mathbb{Q}$, Math. Comp. **51** (1988), no. 184, 761–768.
20. G. Malle and B. H. Matzat, *Inverse Galois theory*, Springer Monogr. Math., Springer-Verlag, Berlin, 1999.
21. B. H. Matzat, *Konstructive Galoistheorie*, Springer Lect. Notes in Math. **1284** (1987).
22. J. McKee, *Computing division polynomials*, Math. Comp. **63** (1994), no. 208, 767–771.
23. A. M. Odlyzko, *Bounds for discriminants and related estimates for class numbers, regulators, and zeros of zeta functions: a survey of recent results*, Sém. Théor. Nombres Bordeaux **2** (1990), no. 1, 119–141.
24. D. P. Roberts, *Twin sextic algebras*, Rocky Mountain J. Math **28** (1998), no. 1, 341–368.
25. ———, *Rigid Jordan tuples*, (submitted).
26. L. Schneps (ed.), *The Grothendieck theory of dessins d'enfants*, Conference on Dessins d'Enfant (Luminy, 1993), London Math. Soc. Lecture Notes Ser. **200**, Cambridge University Press, Cambridge, 1994.
27. J.-P. Serre, *Sur les représentations modulaires de degré 2 de* Gal($\overline{\mathbb{Q}}/\mathbb{Q}$), Duke Math. J. **54** (1987), 179–230.
28. J.-P. Serre and H. Darmon, *Topics in Galois theory*, Jones and Bartlett Publishers, 1992.
29. K. Takeuchi, *Commensurability classes of arithmetic triangle groups*, J. Fac. Sci. Univ. Tokyo Sect. 1A Math. **24** (1977), no. 1, 201–212.
30. H. Völklein, *Groups as Galois groups: An introduction*, Cambridge Stud. Adv. Math. **53**, Cambridge University Press, 1996.

Division of Science and Mathematics, University of Minnesota-Morris, Morris, Minnesota, 56267, USA.

E-mail address: `roberts@mrs.umn.edu`

Centre de Recherches Mathématiques
CRM Proceedings and Lecture Notes
Volume **36**, 2004

Diophantine Approximation in Small Degree

Damien Roy

1. Introduction

This paper, partly a survey, deals with the problem of finding optimal exponents in Diophantine estimates involving one real number ξ. The prototype of such an estimate is the fact, known at least since Euler, that, for any given irrational real number ξ, there exist infinitely many rational numbers p/q with

$$\left|\xi - \frac{p}{q}\right| \le q^{-2}. \tag{1.1}$$

Here, the exponent of q in the upper bound is optimal because, when ξ has bounded partial quotients, there is also a constant $c > 0$ such that $|\xi - p/q| \ge cq^{-2}$ for all rational numbers p/q (see [**14**, Chapter I]).

Define the *height* $H(P)$ of a polynomial $P \in \mathbb{R}[T]$ as the largest absolute value of its coefficients, and the *height* $H(\alpha)$ of an algebraic number α as the height of its irreducible polynomial over $\mathbb{Z}$. Then the above estimate may be generalized in the following two ways related respectively with Mahler's and Koksma's classifications of numbers.

Consider a real number ξ which, for a fixed integer $n \ge 1$, is not algebraic over $\mathbb{Q}$ of degree $\le n$. On one hand, an application of Dirichlet's box principle shows that there exist infinitely many non-zero polynomials $P \in \mathbb{Z}[T]$ such that

$$|P(\xi)| \ll H(P)^{-n} \tag{1.2}$$

where, as in the sequel, the implied constant depends only on n and ξ. On the other hand, Wirsing showed in [**17**] that there exist infinitely many algebraic numbers α of degree at most n with

$$|\xi - \alpha| \ll H(\alpha)^{-(n+3)/2}. \tag{1.3}$$

For $n = 1$, both estimates are equivalent to (1.1), up to the values of the implied constants. For general n, Spindžuk proved in [**15**] that the exponent of $H(P)$ in the upper bound (1.2) is optimal by showing that, for each ξ outside a set of Lebesgue measure zero and for each $\varepsilon > 0$, there are only finitely many non-zero integer polynomials of degree at most n with $|P(\xi)| \le H(P)^{-n-\varepsilon}$. However, in the second

---

2000 *Mathematics Subject Classification.* Primary: 11J13; Secondary: 11J04, 11J82.
Work partially supported by NSERC and CICMA..
This is the final form of the paper.

estimate (1.3), it was conjectured by Wirsing [**17**, p. 69] and Schmidt [**14**, p. 258] that the optimal exponent for $H(\alpha)^{-1}$ is $n+1$ instead of $(n+3)/2$. Aside from the case $n=1$ where these numbers agree, this is known to be true only for $n=2$ thanks to work of Davenport and Schmidt [**5**]. Despite of several refinements by Bernik, Tishchenko and Wirsing, the optimal exponent remains unknown for any $n \geq 3$ (see [**3**, Chapter 3] for more details and references).

In 1969, Davenport and Schmidt [**6**] devised a new method based on geometry of numbers to study the second type of estimate. Assuming, for a fixed $n \geq 2$, that ξ is not algebraic over $\mathbb{Q}$ of degree $\leq n-1$, they proved that there exist infinitely many algebraic integers α of degree at most n with

$$|\xi - \alpha| \ll H(\alpha)^{-[(n+1)/2]} \tag{1.4}$$

(where, for a real number x, the notation $[x]$ stands for the integral part of x). They showed that, for $n=2$, the optimal exponent of approximation is 2 in agreement with the natural conjecture that it should, in general, be n. They also provided sharper estimates for $n=3,4$.

Their approach which we will describe in the next section leads them to establish first another Diophantine approximation result. In the case $n=3$, it is the following statement [**6**, Theorem 1a]) where

$$\gamma = \frac{1+\sqrt{5}}{2}$$

denotes the golden ratio.

THEOREM 1.1 (Davenport-Schmidt). *Suppose that $\xi \in \mathbb{R}$ is neither rational nor quadratic over $\mathbb{Q}$. Then there are arbitrarily large values of X such that the inequalities*

$$|x_0| \leq X, \quad |x_0\xi - x_1| \leq cX^{-1/\gamma}, \quad |x_0\xi^2 - x_2| \leq cX^{-1/\gamma}, \tag{1.5}$$

where c is a suitable positive number depending on ξ, have no solution in integers x_0, x_1, x_2, not all 0.

Note that, an application of Dirichlet's box principle shows that, for any $X \geq 1$, there exists a non-zero point (x_0, x_1, x_2) in $\mathbb{Z}^3$ with $|x_0| \leq X$ and $|x_0\xi^j - x_j| \leq [\sqrt{X}]^{-1}$ for $j=1,2$ (see [**14**, Theorem 1A in Chapter II]). Since $1/\gamma \simeq 0.618 > 1/2$, the condition (1.5) is a far stronger requirement. Nevertheless, it is shown in [**10,11**] that the exponent $1/\gamma$ in Theorem 1.1 is best possible. More precisely, there are countably many real numbers ξ which are neither rational nor quadratic over $\mathbb{Q}$ such that, for some other choice of $c > 0$ (also depending on ξ), the inequalities (1.5) admit a non-zero integer solution for every $X \geq 1$ [**11**, Theorem 1.1]. Because of this property, we shall call these numbers ξ *extremal.*

In Section 3 below, we will sketch a proof of Theorem 1.1 and of a criterion for a real number to be extremal. This criterion attaches to an extremal real number a sequence of approximation triples which we will show in Section 4 to be essentially unique and to satisfy a certain recurrence relation. This will allow us in Section 5 to derive a construction of extremal real numbers which generalizes that of Section 6 of [**11**].

Theorem 1.1 concerns simultaneous approximations of a real number and its square by rational numbers with the same denominator. Davenport and Schmidt looked more generally at simultaneous approximations of the first $n-1$ powers

$\xi, \dots, \xi^{n-1}$ of a real number ξ by rational numbers with the same denominator [**6**, Theorem 2a], and their result was recently improved by M. Laurent [**7**]. However, the optimal exponent for this problem is unknown for $n \geq 4$.

By an argument of duality that we will describe in the next section, Davenport and Schmidt deduce the following consequence of Theorem 1.1.

COROLLARY 1.2 (Davenport-Schmidt). *Let ξ be as in Theorem 1.1. There exist infinitely many algebraic integers α of degree at most 3 with*

$$|\xi - \alpha| \ll H(\alpha)^{-\gamma-1}.$$

Although the case of general degree is unclear, it is shown in [**12**] that here also the exponent $\gamma + 1 \simeq 2.618$ is best possible, against the natural conjecture mentioned earlier that the optimal exponent would be 3. More precisely, there are real numbers ξ which are neither rational nor quadratic over $\mathbb{Q}$ and satisfy $|\xi - \alpha| \gg H(\alpha)^{-\gamma-1}$ for all algebraic integers α of degree at most 3 [**12**, Theorem 1.1]. Such numbers have to be extremal and it would be interesting to know if this property extends to all extremal real numbers.

Considering the irreducible polynomials in $\mathbb{Z}[T]$ of the approximations α provided by the above corollary, we readily deduce:

COROLLARY 1.3. *Let ξ be as above. There exist infinitely many monic polynomials $P \in \mathbb{Z}[T]$ of degree at most 3 with*

$$|P(\xi)| \ll H(P)^{-\gamma}.$$

We will prove in Section 6 that the exponent γ in this statement is also best possible. Note that an argument of Bugeaud and Teulié [**4, 16**] shows more precisely that the inequality (1.4) has infinitely many solutions in algebraic integers α of degree exactly n under the same assumption that ξ is not algebraic of degree $\leq n-1$. In the situation of Corollary 1.2, their argument shows that we may require the algebraic integers α to have degree exactly 3. Therefore, one may as well require that the polynomials P of Corollary 1.3 have degree 3.

Denote by $\mathbb{Z}[T]_{\leq n}$ the group of integer polynomials of degree at most n. In another direction, we have the following Gel'fond type criterion in degree two [**1**] which is, in a sense, dual to Theorem 1.1.

THEOREM 1.4 (Arbour-Roy). *Let $\xi \in \mathbb{C}$. Assume that for any sufficiently large positive real number X there exists a non-zero polynomial $P \in \mathbb{Z}[T]_{\leq 2}$ of height at most X such that*

$$|P(\xi)| \leq \tfrac{1}{4} X^{-\gamma-1}. \tag{1.6}$$

Then ξ is algebraic over $\mathbb{Q}$ of degree at most 2.

Reference [**11**, Theorem 1.2] shows that the exponent of X in (1.6) is best possible. Analogous statements involving polynomials of degree at most n for a fixed integer $n \geq 3$ are well-known but the corresponding optimal exponent is not known (see [**2**, Theorem 1] for a general setting, and see [**6**, Theorem 2b] for a sharper estimate in the present context).

Finally, we will show in Section 7 that the extremal real numbers (associated with Theorem 1.1) are also characterized as those real numbers ξ for which Theorem 1.4 is optimal up to the value of the multiplicative constant.

THEOREM 1.5. *Let ξ be a real number which is not rational nor quadratic over $\mathbb{Q}$. The following conditions are equivalent:*

(a) *there exists a constant $c > 0$ such that, for any real number $X \geq 1$, there is a non-zero point $\mathbf{x} = (x_0, x_1, x_2) \in \mathbb{Z}^3$ satisfying the condition (1.5);*

(b) *there exists a constant $c > 0$ such that, for any real number $X \geq 1$, there is a non-zero polynomial $P \in \mathbb{Z}[T]_{\leq 2}$ of height at most X satisfying $|P(\xi)| \leq cX^{-\gamma-1}$.*

It would be interesting to know if a similar property holds in higher degree. Note that any real number satisfying one of the above conditions (a) or (b) is transcendental over $\mathbb{Q}$ by virtue of Schmidt's subspace theorem.

The present work originates from a colloquium talk given at the University of Ottawa in November 2002. The author thanks the editors for suggesting to include it in the proceedings of CNTA-7. He also thanks Yann Bugeaud for pointing out a number of stimulating questions in relation to the present topic, in particular concerning the status of the exponent in Corollary 1.3.

2. The method of Davenport and Schmidt

Let n be a fixed positive integer and let ξ be a fixed real number which is not algebraic over $\mathbb{Q}$ of degree at most n. The natural approach to construct algebraic approximations of ξ of degree at most n is to produce non-zero polynomials of $\mathbb{Z}[T]_{\leq n}$ with "small" value and "large" derivative at ξ. One then concludes using the fact that any non-zero polynomial $P \in \mathbb{R}[T]$ of degree at most n with $P'(\xi) \neq 0$ has at least one root α with

$$|\xi - \alpha| \leq n \frac{|P(\xi)|}{|P'(\xi)|}. \tag{2.1}$$

Define a *convex body* of $\mathbb{R}^n$ to be a compact, convex neighborhood $\mathcal{C}$ of 0 which is symmetric with respect to 0 (i.e. $\mathcal{C} = -\mathcal{C}$). According to a well-known result of Minkowski, if such a convex body $\mathcal{C}$ has volume at least 2^n, then it contains a non-zero point of $\mathbb{Z}^n$. Applying this result to the convex body of $\mathbb{R}^{n+1}$ defined, for a real number $X \geq 1$, by

$$\begin{cases} |x_0 + x_1\xi + \cdots + x_n\xi^n| \leq X^{-n} \\ |x_1| \leq X \\ \cdots \\ |x_n| \leq X \end{cases}$$

and noting that its volume is 2^{n+1}, we deduce that there exists a non-zero polynomial $P \in \mathbb{Z}[T]$ with

$$\deg(P) \leq n, \quad H(P) \leq (1 + |\xi| + \cdots + |\xi|^n)X \quad \text{and} \quad |P(\xi)| \leq X^{-n}.$$

The difficulty is to control the derivative of P at ξ. The best one can hope is, for arbitrarily large values of X, to have $|P'(\xi)| \gg X$. Then, using (2.1), one finds that there is at least one root α of P with

$$\deg(\alpha) \leq n, \quad H(\alpha) \ll X \quad \text{and} \quad |\xi - \alpha| \ll X^{-n-1} \ll H(\alpha)^{-n-1}.$$

This motivates the conjecture mentioned in the introduction. In general, one has recourse to resultants to establish lower bounds on $|P'(\xi)|$.

The approach of Davenport and Schmidt in [6] is different as they require a set $\{P_1, \dots, P_n\}$ of n linearly independent polynomials of $\mathbb{Z}[T]_{\le n-1}$, all having small absolute value at ξ. Then, taking any monic polynomial $Q \in \mathbb{R}[T]$ of degree n with $Q(\xi) = 0$ and writing it as a linear combination

$$Q = T^n + \theta_1 P_1 + \cdots + \theta_n P_n$$

with $\theta_1, \dots, \theta_n \in \mathbb{R}$, one sees that the polynomial $P \in \mathbb{Z}[T]$ given by

$$P = T^n + [\theta_1]P_1 + \cdots + [\theta_n]P_n$$

is monic of degree n, has height $H(P) \le H(Q) + \sum_{i=1}^n H(P_i)$, and satisfies

$$|P(\xi)| \le \sum_{i=1}^n |P_i(\xi)| \quad \text{as well as} \quad |P'(\xi)| \ge |Q'(\xi)| - \sum_{i=1}^n |P_i'(\xi)|.$$

So, if X denotes an upper bound for $\sum_{i=1}^n H(P_i)$, an appropriate choice of Q (with $H(Q) \ll X$ and $|Q'(\xi)| \gg X$) produces a polynomial P with

$$H(P) \ll X, \quad |P(\xi)| \le \sum_{i=1}^n |P_i(\xi)| \quad \text{and} \quad |P'(\xi)| \ge X.$$

The roots of such a polynomial are algebraic integers of degree $\le n$ and height $\ll X$ and, by (2.1), at least one of them, say α, satisfies

$$|\xi - \alpha| \le \frac{n}{X} \sum_{i=1}^n |P_i(\xi)|.$$

To construct appropriate sets of polynomials $\{P_1, \dots, P_n\}$, Davenport and Schmidt apply a result of duality of Mahler [8]. To state this result or rather a consequence of it, let $\mathcal{C}$ be a convex body of $\mathbb{R}^n$ and let $\mathcal{C}^*$ denote the set of points $(y_1, \dots, y_n) \in \mathbb{R}^n$ satisfying

$$|x_1 y_1 + \cdots + x_n y_n| \le 1$$

for all $(x_1, \dots, x_n) \in \mathcal{C}$. Then, $\mathcal{C}^*$ is again a convex body of $\mathbb{R}^n$, called the *dual* (or *polar*) convex body to $\mathcal{C}$ (the bi-dual $\mathcal{C}^{**}$ being $\mathcal{C}$ itself) and, if $\mathcal{C}$ contains no non-zero integral point, then $n!\mathcal{C}^*$ contains n linearly independent points of $\mathbb{Z}^n$.

For example, given real numbers $X, Y \ge 1$, the convex bodies of $\mathbb{R}^n$ defined by

$$\mathcal{C} : \begin{cases} |x_0| \le X \\ |x_0\xi - x_1| \le Y^{-1} \\ \cdots \\ |x_0\xi^{n-1} - x_{n-1}| \le Y^{-1} \end{cases} \quad \text{and} \quad \mathcal{K} : \begin{cases} |y_0 + y_1\xi + \cdots + y_{n-1}\xi^{n-1}| \le X^{-1} \\ |y_1| \le Y \\ \cdots \\ |y_{n-1}| \le Y \end{cases}$$

are essentially dual to each other in the sense that

$$n^{-1}\mathcal{K} \subseteq \mathcal{C}^* \subseteq \mathcal{K}.$$

So, if $\mathcal{C}$ contains no non-zero integral point, then there are n linearly independent polynomials of $\mathbb{Z}[T]_{\le n-1}$ of height $\ll Y$ whose absolute values at ξ are $\ll X^{-1}$ and therefore there exists an algebraic integer of degree $\le n$ and height $\ll Y$ whose distance to ξ is $\ll (XY)^{-1}$.

In the case $n = 3$, this explains why Theorem 1.1 implies the existence of infinitely many algebraic integers α of degree ≤ 3 with $|\xi - \alpha| \ll H(\alpha)^{-\gamma-1}$ as stated in Corollary 1.2. In general, the fact that (1.4) has infinitely many solutions in algebraic integers of degree at most $\le n$ follows from a similar statement

[**6**, Theorem 2a] showing that for $X = cY^\nu$ with $\nu = [(n-1)/2]$ and an appropriate constant $c > 0$, there are arbitrarily large values of Y for which the convex body $\mathcal{C}$ contains no non-zero integral point.

A modification of the method produces approximation by algebraic numbers or algebraic integers of degree n or even by algebraic units of degree n (see [**4, 16**]). A more general choice of convex bodies involving several derivatives still produces simultaneous approximations of a real number by conjugate algebraic integers [**13**].

3. Extremal real numbers

In this section, we present a sketch of proof of Theorem 1.1 and establish some properties of the corresponding "extremal" real numbers.

Let ξ be a fixed real number. For each point $\mathbf{x} = (x_0, x_1, x_2) \in \mathbb{Z}^3$, we define

$$\|\mathbf{x}\| = \max\{|x_0|, |x_1|, |x_2|\} \quad \text{and} \quad L(\mathbf{x}) = L_\xi(\mathbf{x}) = \max\{|x_1 - \xi x_0|, |x_2 - \xi^2 x_0|\}.$$

Identifying any such point with the corresponding symmetric matrix

$$\mathbf{x} = \begin{pmatrix} x_0 & x_1 \\ x_1 & x_2 \end{pmatrix}, \tag{3.1}$$

we define

$$\det(\mathbf{x}) = \begin{vmatrix} x_0 & x_1 \\ x_1 & x_2 \end{vmatrix} = x_0 x_2 - x_1^2.$$

Then, using the multilinearity of the determinant, one finds

$$|\det(\mathbf{x})| = \left\| \begin{matrix} x_0 & x_1 - \xi x_0 \\ x_1 & x_2 - \xi x_1 \end{matrix} \right\| \ll \|\mathbf{x}\| L(\mathbf{x}). \tag{3.2}$$

Similarly, if $\det(\mathbf{x}, \mathbf{y}, \mathbf{z})$ denotes the determinant of the 3×3 matrix whose rows are points $\mathbf{x}, \mathbf{y}, \mathbf{z} \in \mathbb{Z}^3$, one finds,

$$\begin{aligned} |\det(\mathbf{x}, \mathbf{y}, \mathbf{z})| &= \left\| \begin{matrix} x_0 & x_1 - \xi x_0 & x_2 - \xi^2 x_0 \\ y_0 & y_1 - \xi y_0 & y_2 - \xi^2 y_0 \\ z_0 & z_1 - \xi z_0 & z_2 - \xi^2 z_0 \end{matrix} \right\| \\ &\ll \|\mathbf{x}\| L(\mathbf{y}) L(\mathbf{z}) + \|\mathbf{y}\| L(\mathbf{x}) L(\mathbf{z}) + \|\mathbf{z}\| L(\mathbf{x}) L(\mathbf{y}). \end{aligned} \tag{3.3}$$

We are now ready to present a sketch of proof of Theorem 1.1. To this end, assume that ξ is neither rational nor quadratic over $\mathbb{Q}$ and that there exists a positive real number c such that the inequalities (1.5) have a non-zero solution $\mathbf{x} = (x_0, x_1, x_2) \in \mathbb{Z}^3$ for any sufficiently large real number X. We need to show that c is bounded below by some positive constant depending only on ξ.

First note that there is a sequence of points $(\mathbf{x}_i)_{i \geq 1}$ in $\mathbb{Z}^3$ with the following three properties:

- $1 \leq \|\mathbf{x}_1\| < \|\mathbf{x}_2\| < \|\mathbf{x}_3\| < \dots,$
- $L(\mathbf{x}_1) > L(\mathbf{x}_2) > L(\mathbf{x}_3) > \dots,$
- if $\mathbf{y} \in \mathbb{Z}^3$ has $1 \leq \|\mathbf{y}\| < \|\mathbf{x}_{i+1}\|$, then $L(\mathbf{y}) \geq L(\mathbf{x}_i)$.

Although this differs slightly from the construction in §3 of [**6**], we say that $(\mathbf{x}_i)_{i \geq 1}$ is a sequence of *minimal points* for ξ. In such a sequence, any point $\mathbf{x}_i$ with $i \geq 2$ is *primitive*, i.e. has relatively prime coordinates. Moreover any two consecutive points $\mathbf{x}_i$ and $\mathbf{x}_{i+1}$ with $i \geq 2$ are linearly independent. Define

$$X_i = \|\mathbf{x}_i\| \quad \text{and} \quad L_i = L(\mathbf{x}_i)$$

for each $i \geq 1$. Then, for any sufficiently large i, the hypotheses imply the existence of a non-zero point $\mathbf{x} \in \mathbb{Z}^3$ with $\|\mathbf{x}\| < X_{i+1}$ and $L(\mathbf{x}) \ll cX_{i+1}^{-1/\gamma}$. This gives

$$L_i \ll cX_{i+1}^{-1/\gamma}. \tag{3.4}$$

Davenport and Schmidt show that, for any sufficiently large i, we have $\det(\mathbf{x}_i) \neq 0$ [**6**, Lemma 2]). Since $\det(\mathbf{x}_i)$ is an integer, its absolute value is then bounded below by 1 and, using the estimate (3.2) combined with (3.4), we find

$$1 \leq |\det(\mathbf{x}_i)| \ll X_i L_i \ll cX_i X_{i+1}^{-1/\gamma}. \tag{3.5}$$

They also show that, for infinitely many i, the points $\mathbf{x}_{i-1}$, $\mathbf{x}_i$ and $\mathbf{x}_{i+1}$ are linearly independent [**6**, Lemma 5]. For these i, the same argument based on (3.3) and (3.4) then gives

$$1 \leq |\det(\mathbf{x}_{i-1}, \mathbf{x}_i, \mathbf{x}_{i+1})| \ll X_{i+1} L_i L_{i-1} \ll c^2 X_{i+1}^{1-1/\gamma} X_i^{-1/\gamma}. \tag{3.6}$$

The required lower bound on c then follows by comparing (3.5) and (3.6), upon noting that $1 - 1/\gamma = 1/\gamma^2$.

The above considerations apply in particular to any extremal real number ξ. Combining (3.5) and (3.6) then shows that, for all indices i such that $\mathbf{x}_{i-1}$, $\mathbf{x}_i$ and $\mathbf{x}_{i+1}$ are linearly independent, we have

$$\|\mathbf{x}_{i+1}\| \sim \|\mathbf{x}_i\|^\gamma, \quad L_\xi(\mathbf{x}_i) \sim \|\mathbf{x}_i\|^{-1}, \quad |\det(\mathbf{x}_i)| \sim 1, \quad |\det(\mathbf{x}_{i-1}, \mathbf{x}_i, \mathbf{x}_{i+1})| \sim 1$$

writing $a \sim b$ to mean $a \ll b$ and $a \gg b$. A further analysis shows that, by going to a subsequence $(\mathbf{y}_i)_{i\geq 1}$ of $(\mathbf{x}_i)_{i\geq 1}$, one may assume these estimates to hold for all $i \geq 1$. More precisely, we have the following equivalence [**11**, Theorem 5.1]:

THEOREM 3.1. *A real number ξ is extremal if and only if there exists a constant $c \geq 1$ and an unbounded sequence of non-zero primitive points $(\mathbf{y}_i)_{i\geq 1}$ of $\mathbb{Z}^3$ satisfying, for all $i \geq 1$,*

$$\begin{aligned}
&c^{-1}\|\mathbf{y}_i\|^\gamma \leq \|\mathbf{y}_{i+1}\| \leq c\|\mathbf{y}_i\|^\gamma,\\
&c^{-1}\|\mathbf{y}_i\|^{-1} \leq L_\xi(\mathbf{y}_i) \leq c\|\mathbf{y}_i\|^{-1},\\
&1 \leq |\det(\mathbf{y}_i)| \leq c,\\
&1 \leq |\det(\mathbf{y}_i, \mathbf{y}_{i+1}, \mathbf{y}_{i+2})| \leq c.
\end{aligned}$$

In the next section, we show that the sequence $(\mathbf{y}_i)_{i\geq 1}$ is essentially uniquely determined by ξ.

4. The sequence of approximation triples

In this section, we fix an extremal real number $\xi \in \mathbb{R}$ and a sequence of approximation triples $(\mathbf{y}_i)_{i\geq 1}$ as in Theorem 3.1. We first prove:

PROPOSITION 4.1. *There exists a constant $c_3 > 0$ such that any non-zero primitive point $\mathbf{y} \in \mathbb{Z}^3$ with*

$$L_\xi(\mathbf{y}) \leq c_3\|\mathbf{y}\|^{-1/\gamma} \tag{4.1}$$

is of the form $\mathbf{y} = \pm\mathbf{y}_i$ for some index $i \geq 1$.

PROOF. Fix a constant c_3 with $0 < c_3 \leq \|\mathbf{y}_1\|^{-1}$. Then, for any non-zero point $\mathbf{y} \in \mathbb{Z}^3$, there exists an index $i \geq 1$ such that

$$c_3\|\mathbf{y}_i\| \leq \|\mathbf{y}\| < c_3\|\mathbf{y}_{i+1}\|.$$

If (4.1) holds, we then find

$$|\det(\mathbf{y},\mathbf{y}_i,\mathbf{y}_{i+1})| \ll \|\mathbf{y}\|\|\mathbf{y}_i\|^{-1}\|\mathbf{y}_{i+1}\|^{-1} + \|\mathbf{y}_{i+1}\|\|\mathbf{y}_i\|^{-1}L_\xi(\mathbf{y}) \ll c_3 + c_3^{1/\gamma^2},$$

$$|\det(\mathbf{y},\mathbf{y}_i,\mathbf{y}_{i-1})| \ll \|\mathbf{y}\|\|\mathbf{y}_i\|^{-1}\|\mathbf{y}_{i-1}\|^{-1} + \|\mathbf{y}_{i}\|\|\mathbf{y}_{i-1}\|^{-1}L_\xi(\mathbf{y}) \ll c_3 + c_3^{1/\gamma^2}.$$

So, provided that c_3 is sufficiently small, these determinants vanish and, since $\mathbf{y}_{i-1}$, $\mathbf{y}_i$ and $\mathbf{y}_{i+1}$ are linearly independent with $\mathbf{y}_i$ primitive, we conclude that $\mathbf{y} = \pm\mathbf{y}_i$. □

Since, for all sufficiently large values of i, the point $\mathbf{y} = \mathbf{y}_i$ satisfies the condition (4.1), we deduce from this proposition that the sequence $(\mathbf{y}_i)_{i\geq 1}$ is uniquely determined by ξ up to its first terms and up to multiplication of its terms by ± 1.

In proving the inequalities (3.2) and (3.3), we used the multi-linearity of the determinant. Equivalently, we could have looked at the Taylor series expansion of $\det(\mathbf{x})$ and $\det(\mathbf{x},\mathbf{y},\mathbf{z})$ at the points $\mathbf{x} = (x_0, x_0\xi, x_0\xi^2), \dots, \mathbf{z} = (z_0, z_0\xi, z_0\xi^2)$. Proposition A.1 of the appendix generalizes this idea and, through a computer search, provided the following relations.

PROPOSITION 4.2. *For any sufficiently large index i, we have*

$$\det(\mathbf{y}_i,\mathbf{y}_{i+1},[\mathbf{y}_{i+3},\mathbf{y}_{i+3},\mathbf{y}_{i+4}]) = 0, \quad \det(\mathbf{y}_{i+1},\mathbf{y}_{i+2},[\mathbf{y}_{i+3},\mathbf{y}_{i+3},\mathbf{y}_{i+4}]) = 0, \tag{4.2}$$

where, upon identifying points $\mathbf{x},\mathbf{z} \in \mathbb{Z}^3$ with the corresponding symmetric matrices as in (3.1) *and upon denoting by* $\mathrm{Adj}(\mathbf{z})$ *the adjoint matrix of* $\mathbf{z}$, *we define*

$$[\mathbf{x},\mathbf{x},\mathbf{z}] = \mathbf{x}\mathrm{Adj}(\mathbf{z})\mathbf{x}. \tag{4.3}$$

A direct proof of these relations can be found in [**11**], as part of the proof of [**11**, Corollary 5.2]. It uses the estimates of [**11**, Lemma 3.1] to show that the above determinants (4.2) have absolute values tending to zero as i tends to infinity. As in [**12**, Proposition 2.3], we deduce:

COROLLARY 4.3. *There exists a 2×2 matrix M with relatively prime integer coefficients and an index i_0 such that the symmetric matrix corresponding to $\mathbf{y}_{i+2}$ is a rational multiple of $\mathbf{y}_{i+1}M\mathbf{y}_i$ when $i \geq i_0$ is odd, and a rational multiple of ${\mathbf{y}_{i+1}}^tM\mathbf{y}_i$ when $i \geq i_0$ is even. Such a matrix M is non-singular, non-symmetric and non-skew-symmetric.*

PROOF. Choose $i_0 \geq 2$ so that the relations (4.2) hold for $i \geq i_0 - 1$. Since $\mathbf{y}_i$, $\mathbf{y}_{i+1}$ and $\mathbf{y}_{i+2}$ are linearly independent, these relations imply that $[\mathbf{y}_{i+3},\mathbf{y}_{i+3},\mathbf{y}_{i+4}]$ is a rational multiple of $\mathbf{y}_{i+1}$ and thus, by definition of the latter symbol, since all these matrices are invertible, that $\mathbf{y}_{i+4}$ is a rational multiple of $\mathbf{y}_{i+3}\mathbf{y}_{i+1}^{-1}\mathbf{y}_{i+3}$ for $i \geq i_0 - 1$. Therefore, assuming that $\mathbf{y}_{i+3}$ is a rational multiple of $\mathbf{y}_{i+2}S\mathbf{y}_{i+1}$ for some integer matrix S and some index $i \geq i_0 - 1$, we find that $\mathbf{y}_{i+4}$ is a rational multiple of $\mathbf{y}_{i+2}S\mathbf{y}_{i+3}$, and so, by taking transpose, that $\mathbf{y}_{i+4}$ is a rational multiple of ${\mathbf{y}_{i+3}}^tS\mathbf{y}_{i+2}$. The first assertion of the corollary then follows by induction on i, upon choosing M so that it holds for $i = i_0$. The matrix M is clearly non-singular. It is not symmetric since a simple computation based for example on the formulas (2.1) and (2.2) of [**11**] gives

$$\det(\mathbf{y}_i,\mathbf{y}_{i+1},\mathbf{y}_{i+1}M\mathbf{y}_i) = \det(\mathbf{y}_i)\det(\mathbf{y}_{i+1})\,\mathrm{trace}(MJ) \quad \text{where} \quad J = \begin{pmatrix} 0 & 1 \\ -1 & 0 \end{pmatrix}, \tag{4.4}$$

while, for odd $i \geq i_0$, the above determinant is non-zero. Finally, it is not skew-symmetric, otherwise in the above notation we would have $M = \pm J$ which, for $i \geq i_0$, would imply proportionality relations

$$\mathbf{y}_{i+3} \propto \mathbf{y}_{i+2} J \mathbf{y}_{i+1} \propto \mathbf{y}_i J \mathbf{y}_{i+1} J \mathbf{y}_{i+1} \propto \mathbf{y}_i$$

and thus $\mathbf{y}_{i+3} = \pm \mathbf{y}_i$, against the fact that the norms of the $\mathbf{y}_i$'s are unbounded. □

As the sequence $(\mathbf{y}_i)_{i\geq 1}$ is uniquely determined by ξ up to its first terms and up to multiplication of its terms by ± 1, we deduce that the matrix M of the corollary is uniquely determined by ξ up to multiplication by ± 1 and up to transposition. We say that M is the matrix *associated* with ξ and, for a given M, we denote by $\mathcal{E}(M)$ the set of extremal real numbers with associated matrix M. In the next section, we present a criterion for showing that $\mathcal{E}(M)$ is not empty.

Remark that, in the notation of the corollary, the sequence of matrices $(M_i)_{i\geq 1}$ given by $M_i = \mathbf{y}_i M$ for i even and by $M_i = \mathbf{y}_i {}^t M$ for i odd satisfies

$$M_{i+2} \propto M_{i+1} M_i$$

for all $i \geq i_0$. It can therefore be viewed as a Fibonacci sequence of matrices in $\mathrm{PGL}_2(\mathbb{Q})$.

5. Construction of extremal real numbers

PROPOSITION 5.1. *Let M be a non-singular, non-symmetric 2×2 matrix with relatively prime integer coefficients. Assume that there exist non-singular 2×2 symmetric matrices $\mathbf{y}_1$, $\mathbf{y}_2$, $\mathbf{y}_3$ with relatively prime integer coefficients such that $\mathbf{y}_3$ is a rational multiple of $\mathbf{y}_2 M \mathbf{y}_1$. Extend the definition of $\mathbf{y}_i$ coherently for $i \geq 4$ by asking that $\mathbf{y}_i$ have relatively prime integer coefficients and that it be a rational multiple of $\mathbf{y}_{i-1} M \mathbf{y}_{i-2}$ for odd $i \geq 3$ and a rational multiple of $\mathbf{y}_{i-1} {}^t M \mathbf{y}_{i-2}$ for even $i \geq 4$. Assume further that the $\mathbf{y}_i$'s are unbounded and that there exist positive constants c_4 and c_5 such that*

$$|\det(\mathbf{y}_i)| \leq c_4 \quad \text{and} \quad \|\mathbf{y}_{i+2}\| \geq c_5 \|\mathbf{y}_{i+1}\| \|\mathbf{y}_i\|$$

for all $i \geq 1$. Then $(\mathbf{y}_i)_{i\geq 1}$ is a sequence of approximation triples associated with an extremal real number $\xi \in \mathcal{E}(M)$.

PROOF. For each $i \geq 1$, we have

$$\mathbf{y}_{i+2} = \rho_i \mathbf{y}_{i+1} S \mathbf{y}_i \tag{5.1}$$

for some appropriate choice of $S = M$ or ${}^t M$ and some non-zero rational number ρ_i with $|\rho_i| \leq 1$. Assuming that $\mathbf{y}_i$ and $\mathbf{y}_{i+1}$ are non-singular, this shows that $\mathbf{y}_{i+2}$ is also non-singular. Thus, by induction, all $\mathbf{y}_i$'s are non-singular, and so

$$1 \leq |\det(\mathbf{y}_i)| \leq c_4$$

for $i \geq 1$. We also deduce that

$$\mathbf{y}_{i+3} = \rho_{i+1} \mathbf{y}_{i+2} {}^t S \mathbf{y}_{i+1} = \rho_i \rho_{i+1} \mathbf{y}_{i+1} S \mathbf{y}_i {}^t S \mathbf{y}_{i+1}$$

which, starting from the fact that $\mathbf{y}_1$, $\mathbf{y}_2$ and $\mathbf{y}_3$ are symmetric, implies by induction that all $\mathbf{y}_i$'s are symmetric. The formula (4.4) with M replaced by S also gives

$$\det(\mathbf{y}_i, \mathbf{y}_{i+1}, \mathbf{y}_{i+2}) = \rho_i \det(\mathbf{y}_i) \det(\mathbf{y}_{i+1}) \operatorname{trace}(SJ).$$

Since M is non-symmetric, we have $\operatorname{trace}(SJ) = \pm\operatorname{trace}(MJ) \neq 0$ and therefore, the above determinant being an integer, it satisfies

$$1 \le |\det(\mathbf{y}_i, \mathbf{y}_{i+1}, \mathbf{y}_{i+2})| \le c_4^2 |\operatorname{trace}(MJ)|.$$

The relation (5.1) also implies that

$$\|\mathbf{y}_{i+2}\| \le c_6 \|\mathbf{y}_{i+1}\| \|\mathbf{y}_i\|,$$

where c_6 denotes the sum of the absolute values of the coefficients of M. Defining $q_i = \|\mathbf{y}_{i+1}\| \|\mathbf{y}_i\|^{-\gamma}$ and using the similar lower bound for $\|\mathbf{y}_{i+2}\|$ that comes from the hypotheses of the theorem, we deduce that

$$c_5 q_i^{-1/\gamma} \le q_{i+1} \le c_6 q_i^{-1/\gamma},$$

which, by induction, implies $c_7^{-1} \le q_i \le c_7$ with $c_7 = \max\{q_1, q_1^{-1}, c_5^{-\gamma^2}, c_6^{\gamma^2}\}$ and so

$$c_7^{-1} \|\mathbf{y}_i\|^\gamma \le \|\mathbf{y}_{i+1}\| \le c_7 \|\mathbf{y}_i\|^\gamma,$$

for all $i \ge 1$.

Denote by $[\mathbf{x}]$ the image of a non-zero point $\mathbf{x}$ of $\mathbb{R}^3$ in the projective space $\mathbb{P}^2(\mathbb{R})$, and, for any non-zero point $\mathbf{y} \in \mathbb{R}^3$, define

$$d([\mathbf{x}], [\mathbf{y}]) = d(\mathbf{x}, \mathbf{y}) = \frac{\|\mathbf{x} \wedge \mathbf{y}\|}{\|\mathbf{x}\| \|\mathbf{y}\|}$$

where $\mathbf{x} \wedge \mathbf{y}$ denotes the vector product of $\mathbf{x}$ and $\mathbf{y}$. This distance function defines the usual topology on $\mathbb{P}^2(\mathbb{R})$ (see for example [**9**, Lemma 1.16]) and it is easily proved to satisfy

$$d(\mathbf{x}, \mathbf{z}) \le d(\mathbf{x}, \mathbf{y}) + 2d(\mathbf{y}, \mathbf{z}) \tag{5.2}$$

for any three non-zero points $\mathbf{x}$, $\mathbf{y}$, $\mathbf{z}$ of $\mathbb{R}^3$. Since $\mathbb{P}^2(\mathbb{R})$ is compact, the sequence $([\mathbf{y}_i])_{i\ge1}$ has an accumulation point $[\mathbf{y}]$ for some non-zero $\mathbf{y} \in \mathbb{R}^3$. Since the points $\mathbf{y}_i$ have bounded determinant and norm tending to infinity with i, we deduce that, by continuity,

$$\det(\mathbf{y}) = 0. \tag{5.3}$$

In order to estimate the distance between two consecutive points of this sequence, we note that

$$\mathbf{y}_{i+2} J \mathbf{y}_{i+1} = \rho_i \mathbf{y}_i{}^t S \mathbf{y}_{i+1} J \mathbf{y}_{i+1} = \rho_i \det(\mathbf{y}_{i+1}) \mathbf{y}_i{}^t SJ,$$

thus

$$\max_{k,\ell=0,1} |y_{i+2,k} y_{i+1,\ell+1} - y_{i+2,k+1} y_{i+1,\ell}| \le c_4 c_6 \|\mathbf{y}_i\|,$$

and so

$$d(\mathbf{y}_{i+1}, \mathbf{y}_{i+2}) \le 2c_4 c_6 \frac{\|\mathbf{y}_i\|}{\|\mathbf{y}_{i+1}\| \|\mathbf{y}_{i+2}\|} \le c_8 \|\mathbf{y}_{i+1}\|^{-2} \tag{5.4}$$

with $c_8 = 2c_4c_6/c_5$. Using (5.2), (5.4) and the fact that the norms of the $\mathbf{y}_i$'s grow faster than any geometric series, we deduce that, for $k > i \ge 2$, we have

$$d(\mathbf{y}_i, \mathbf{y}_k) \le \sum_{j=i}^{k-1} 2^{j-i} d(\mathbf{y}_j, \mathbf{y}_{j+1}) \le c_8 \sum_{j=i}^{k-1} 2^{j-i} \|\mathbf{y}_j\|^{-2} \le c_9 \|\mathbf{y}_i\|^{-2}$$

for some constant $c_9 > 0$. As $d(\mathbf{y}_k, \mathbf{y})$ can be made arbitrarily small for a suitable choice of $k > i$, this implies

$$d(\mathbf{y}_i, \mathbf{y}) \le c_9 \|\mathbf{y}_i\|^{-2},$$

showing in particular that the sequence $([\mathbf{y}_i])_{i\ge 1}$ converges to $[\mathbf{y}]$.

We claim that the point $\mathbf{y} = (y_0, y_1, y_2)$ has $y_0 \neq 0$. Otherwise, upon denoting by k a fixed index for which $y_k \neq 0$, we would have, for $i \ge 2$,

$$|y_k y_{i,0}| = |y_k y_{i,0} - y_0 y_{i,k}| \le \|\mathbf{y}\| \|\mathbf{y}_i\| d(\mathbf{y}_i, \mathbf{y}) \le c_9 \|\mathbf{y}\| \|\mathbf{y}_i\|^{-1}.$$

As this upper bound tends to zero for $i \to \infty$, this would force the integer $y_{i,0}$ to be zero for all sufficiently large values of i, against the fact that the determinant of any three consecutive $\mathbf{y}_i$'s is non-zero.

Since $y_0 \neq 0$, we may assume without loss of generality that $y_0 = 1$. Writing $\xi = y_1$, we then deduce, by virtue of (5.3), that

$$\mathbf{y} = (1, \xi, \xi^2)$$

and so

$$L_\xi(\mathbf{y}_i) \le \|\mathbf{y} \wedge \mathbf{y}_i\| \le \|\mathbf{y}\| \|\mathbf{y}_i\| d(\mathbf{y}_i, \mathbf{y}) \le c_9 \max\{1, \xi^2\} \|\mathbf{y}_i\|^{-1}$$

for any $i \ge 2$. We also get a lower bound of the same type for $L_\xi(\mathbf{y}_i)$ by combining the estimate (3.2) with the lower bound $|\det(\mathbf{y}_i)| \ge 1$. Therefore, by Theorem 3.1, the number ξ is extremal, and $(\mathbf{y}_i)_{i\ge 1}$ is an associated sequence of approximation triples. □

To apply the above proposition for a given skew-symmetric matrix M, one has to choose $\mathbf{y}_1$ and $\mathbf{y}_2$ so that

- the product $\mathbf{y}_2 M \mathbf{y}_1$ is symmetric,
- the $\mathbf{y}_i$'s have bounded non-zero determinants,
- the $\mathbf{y}_i$'s are unbounded and the ratios $\|\mathbf{y}_{i+2}\|/(\|\mathbf{y}_{i+1}\| \|\mathbf{y}_i\|)$ are bounded below by some positive constant.

The second condition is automatically fulfilled if M, $\mathbf{y}_1$ and $\mathbf{y}_2$ have determinant ± 1, because then all $\mathbf{y}_i$'s have determinant ± 1. The third condition is also fulfilled if, for example, the coefficients of M are positive while those of $\mathbf{y}_1$ and $\mathbf{y}_2$ are non-negative.

EXAMPLE 1. If we define

$$A = \begin{pmatrix} a & 1 \\ 1 & 0 \end{pmatrix}, \quad B = \begin{pmatrix} b & 1 \\ 1 & 0 \end{pmatrix} \quad \text{and} \quad M = AB = \begin{pmatrix} ab+1 & a \\ b & 1 \end{pmatrix}$$

for a choice of distinct positive integers a and b, then

$$\mathbf{y}_1 = A, \quad \mathbf{y}_2 = ABA \quad \text{and} \quad \mathbf{y}_3 = \mathbf{y}_2 M \mathbf{y}_1 = ABAABA$$

are symmetric matrices of determinant ± 1 with non-negative entries while M has positive entries. The resulting sequence $(\mathbf{y}_i)_{i\ge 1}$ thus fulfills all requirements of Proposition 5.1. It can be shown that the corresponding extremal real number has continued fraction expansion

$$\xi = [0, a, b, a, a, b, a, \dots]$$

given by the Fibonacci word on $\{a, b\}$ (see [**10**, Theorem 2.2] or [**11**, Corollary 6.3]).

EXAMPLE 2. Take

$$\mathbf{y}_1=\begin{pmatrix}1&1\\1&0\end{pmatrix},\quad \mathbf{y}_2=\begin{pmatrix}a^3+2a & a^3-a^2+2a-1\\ a^3-a^2+2a-1 & a^3-2a^2+3a-2\end{pmatrix},\quad M=\begin{pmatrix}a&1\\-1&0\end{pmatrix}$$

for a fixed positive integer a. One readily checks that $\mathbf{y}_2M\mathbf{y}_1$ is symmetric, that $\det(\mathbf{y}_1)=\det(\mathbf{y}_2)=-1$ and that $\det(M)=1$. Thus, to ensure that the corresponding sequence $(\mathbf{y}_i)_{i\geq 1}$ defines an extremal real number, it remains only to verify the growth condition on the norms of these points. To this end, we note that if $\mathbf{x},\mathbf{y}\in\mathbb{Z}^3$ have coordinates satisfying $x_0\geq x_1\geq x_2\geq 0$ and $y_0\geq y_1\geq y_2\geq 0$ and if the product $\mathbf{z}=\mathbf{y}M\mathbf{x}$ is symmetric, then we also have $z_0\geq z_1\geq z_2\geq 0$ and moreover $z_0\geq (a-1)y_0x_0$. By recurrence on i, using $\mathbf{y}_{i+2}=\mathbf{y}_{i+1}M\mathbf{y}_i$ for odd i and $\mathbf{y}_{i+2}=\mathbf{y}_iM\mathbf{y}_{i+1}$ for even i, we deduce that $\|\mathbf{y}_i\|=y_{i,0}$ and that $\|\mathbf{y}_{i+2}\|\geq (a-1)\|\mathbf{y}_{i+1}\|\|\mathbf{y}_i\|$ for all $i\geq 1$. So, for $a\geq 2$, the required growth condition is satisfied (we have $\lim_{i\to\infty}\|\mathbf{y}_i\|=\infty$ since $\|\mathbf{y}_2\|>\|\mathbf{y}_1\|=1$). In the case where $a=1$, one finds that the points $\mathbf{x}=\mathbf{y}_3$ and $\mathbf{y}=\mathbf{y}_2$ satisfy the stronger conditions $x_0\geq 2x_1\geq 4x_2\geq 0$ and $y_0\geq 2y_1\geq 4y_2\geq 0$ and that, for such points $\mathbf{x},\mathbf{y}\in\mathbb{Z}^3$, when the product $\mathbf{z}=\mathbf{y}M\mathbf{x}$ is symmetric, we also have $z_0\geq 2z_1\geq 4z_2\geq 0$ and $z_0\geq y_0x_0/2$. By recurrence on i, this gives $\|\mathbf{y}_i\|=y_{i,0}$ and $\|\mathbf{y}_{i+2}\|\geq \frac{1}{2}\|\mathbf{y}_{i+1}\|\|\mathbf{y}_i\|$ for all $i\geq 2$. In particular, since $\|\mathbf{y}_2\|=3$ and $\|\mathbf{y}_3\|=5$, we deduce that $\|\mathbf{y}_{i+1}\|>\|\mathbf{y}_i\|>2$ for $i\geq 1$ and so $\lim_{i\to\infty}\|\mathbf{y}_i\|=\infty$. Thus, in all cases, the sequence $(\mathbf{y}_i)_{i\geq 1}$ defines an extremal real number in $\mathcal{E}(M)$. This proves the remark at the end of [**12**, §3].

6. Approximation by cubic algebraic integers

In order to show that the exponent in Corollary 1.3 is best possible, we apply the following criterion where, for a real number x, the symbol $\{x\}$ denotes the distance from x to a closest integer (compare with [**11**, Proposition 9.1]).

LEMMA 6.1. *Let ξ be an extremal real number and let $(\mathbf{y}_i)_{i\geq 1}$ be a corresponding sequence of approximation triples. Assume that there exists a constant $c_1>0$ such that*

$$\{y_{i,0}\xi^3\}\geq c_1 \tag{6.1}$$

for any sufficiently large index i. Then there exists a constant $c_2>0$ such that, for any monic polynomial $P\in\mathbb{Z}[T]_{\leq 3}$, we have

$$|P(\xi)|\geq c_2H(P)^{-\gamma}.$$

PROOF. Choose an index $i_0\geq 1$ such that (6.1) holds for each $i\geq i_0$. Multiplying P by a suitable power of T if necessary, we may assume without loss of generality that P has degree three. Writing $P(T)=T^3+pT^2+qT+r$, we find, for any $i\geq i_0$,

$$\begin{aligned}\{y_{i,0}\xi^3\}&\leq |y_{i,0}P(\xi)|+|p|\{y_{i,0}\xi^2\}+|q|\{y_{i,0}\xi\}\\ &\leq \|\mathbf{y}_i\||P(\xi)|+2H(P)L_\xi(\mathbf{y}_i)\\ &\leq \|\mathbf{y}_i\||P(\xi)|+c_3H(P)\|\mathbf{y}_i\|^{-1},\end{aligned}$$

with a constant $c_3>0$ depending only on ξ. Choosing i to be the smallest integer $i\geq i_0$ for which

$$H(P)\leq \frac{c_1}{2c_3}\|\mathbf{y}_i\|,$$

and using (6.1) this implies

$$|P(\xi)| \geq \frac{c_1}{2}\|\mathbf{y}_i\|^{-1}.$$

The conclusion follows as the above choice of i implies $\|\mathbf{y}_i\| \ll H(P)^{\gamma}$. □

Let $a \geq 1$ be an integer and, for shortness, define $\mathcal{E}_a = \mathcal{E}\left(\begin{smallmatrix} a & 1 \\ -1 & 0 \end{smallmatrix}\right)$. Example 2 of the preceding section shows that this set is not empty. Moreover, it is proved in [**12**, §4] that any extremal real number ξ in $\mathcal{E}_a$ satisfies the hypotheses of the preceding lemma (more precisely it is shown that $\lim_{j\to\infty}\{y_{i+3j,0}\xi^3\}$ exists and is a positive real number for $i = 0, 1, 2$). So, we deduce:

THEOREM 6.2. *Let $a \geq 1$ be an integer and let $\xi \in \mathcal{E}_a$. There exists a constant $c > 0$ such that, for any monic polynomial $P \in \mathbb{Z}[T]_{\leq 3}$, we have*

$$|P(\xi)| \geq cH(P)^{-\gamma}.$$

COROLLARY 6.3. *With ξ as above, there exists a constant $c' > 0$ such that, for any algebraic integer α of degree at most 3 over $\mathbb{Q}$, we have*

$$|\xi - \alpha| \geq c'H(\alpha)^{-\gamma-1}.$$

7. Proof of Theorem 1.5

Reference [**11**, Theorem 8.1] provides a sequence of polynomials in $\mathbb{Z}[T]_{\leq 2}$ with small absolute value at a given extremal real number ξ. These polynomials are constructed by taking exterior products of consecutive approximation triples of ξ. As a corollary, this result implies [**11**, Theorem 1.2] which in turn tells us that (a) implies (b) in Theorem 1.5. Our proof that (b) implies (a) will be similar. We will first prove that, for any given real number ξ which satisfies this condition (b), there is a sequence of polynomials in $\mathbb{Z}[T]_{\leq 2}$ with properties parallel to those stated in Theorem 3.1. Then, by taking exterior products of consecutive polynomials in this sequence, we will get points $\mathbf{x}$ in $\mathbb{Z}^3$ for which $L_\xi(\mathbf{x})$ is small and deduce from this that ξ is extremal. The latter step will require the following lemma.

LEMMA 7.1. *Let ξ be a real number and let $P = p_0 + p_1T + p_2T^2$ and $Q = q_0 + q_1T + q_2T^2$ be polynomials of $\mathbb{Z}[T]_{\leq 2}$ with*

$$2H(P)|Q(\xi)| \leq H(Q)|P(\xi)|.$$

Then the point

$$\mathbf{x} = P \wedge Q = (p_2q_1 - p_1q_2, p_0q_2 - p_2q_0, p_1q_0 - p_0q_1) \in \mathbb{Z}^3$$

satisfies

$$(2\max\{1, |\xi| + |\xi^2|\})^{-1}H(Q)|P(\xi)| \leq L_\xi(\mathbf{x}) \leq \tfrac{3}{2}H(Q)|P(\xi)|.$$

PROOF. The upper bound follows immediately from the relations

$$x_0\xi - x_1 = p_2Q(\xi) - q_2P(\xi) \quad \text{and} \quad x_0\xi^2 - x_2 = q_1P(\xi) - p_1Q(\xi).$$

For the lower bound, we simply note that

$$\begin{aligned} q_0P(\xi) - p_0Q(\xi) &= -\xi\big(q_1P(\xi) - p_1Q(\xi)\big) - \xi^2\big(q_2P(\xi) - p_2Q(\xi)\big) \\ &= -\xi(x_0\xi^2 - x_2) + \xi^2(x_0\xi - x_1) \end{aligned}$$

implies

$$\max\{1, |\xi| + |\xi^2|\}L_\xi(\mathbf{x}) \geq \max_{0\leq i\leq 2}|q_iP(\xi) - p_iQ(\xi)| \geq H(Q)|P(\xi)| - H(P)|Q(\xi)|. \ \square$$

We now prove Theorem 1.5 by adding one more equivalent condition (compare with [**11**, Theorem 5.1]):

THEOREM 7.2. *Let ξ be a real number. The following conditions are equivalent:*

(a) *the number ξ is extremal;*

(b) *the number ξ is neither rational nor quadratic over $\mathbb{Q}$ and, for any real number $X \geq 1$, there is a non-zero polynomial $P \in \mathbb{Z}[T]_{\leq 2}$ of height at most X satisfying $|P(\xi)| \leq c_1 X^{-\gamma-1}$ with a constant $c_1 = c_1(\xi)$;*

(c) *There exists a constant $c_2 \geq 1$ and an unbounded sequence of non-zero polynomials $(Q_k)_{k\geq 1}$ of $\mathbb{Z}[T]_{\leq 2}$ with relatively prime coefficients satisfying, for all $k \geq 1$,*

$$c_2^{-1} H(Q_k)^{\gamma} \leq H(Q_{k+1}) \leq c_2 H(Q_k)^{\gamma},$$
$$c_2^{-1} H(Q_k)^{-\gamma^3} \leq |Q_k(\xi)| \leq c_2 H(Q_k)^{-\gamma^3},$$
$$1 \leq |\operatorname{Res}(Q_k, Q_{k+1})| \leq c_2,$$
$$1 \leq |\det(Q_k, Q_{k+1}, Q_{k+2})| \leq c_2.$$

PROOF. As mentioned earlier, Reference [**11**, Theorem 1.2] shows that (a) implies (b).

Assume now that (b) is satisfied. We prove (c) by going back to the arguments of [**1**]. First of all, we recall that there is a sequence of "minimal polynomials" $(P_i)_{i\geq 1}$ in $\mathbb{Z}[T]_{\leq 2}$ with the following three properties:

- $1 \leq H(P_1) < H(P_2) < H(P_3) < \cdots$
- $|P_1(\xi)| > |P_2(\xi)| > |P_3(\xi)| > \cdots$
- if $P \in \mathbb{Z}[T]_{\leq 2}$ has $1 \leq H(P) < H(P_{i+1})$, then $|P(\xi)| \geq |P_i(\xi)|$

(see [**5**, §3] or [**1**, Lemma 5]). For each $i \geq 1$, let V_i denote the sub-$\mathbb{Q}$-vector space of $\mathbb{Q}[T]$ generated by P_i and P_{i+1}. Then, for $i \geq 2$, the polynomials P_i and P_{i+1} form a basis of the group $V_i \cap \mathbb{Z}[T]$ of integral polynomials in V_i (see the proof of Lemma 2 of [**5**]). In particular, P_i has relatively prime coefficients for $i \geq 2$. Moreover the condition (b) implies

$$|P_i(\xi)| \leq c_1 H(P_{i+1})^{-\gamma-1} \tag{7.1}$$

for any $i \geq 1$.

Let I denote the set of indices $i \geq 2$ for which P_{i-1}, P_i and P_{i+1} are linearly independent. Reference [**1**, Lemma 6] shows that I is an infinite set and the arguments in [**1**, §3] show that there exists an index $i_1 \in I$ such that $\operatorname{Res}(P_i, P_{i+1}) \neq 0$ for all $i \in I$ with $i \geq i_1$. We define a sub-sequence $(Q_k)_{k\geq 1}$ of $(P_i)_{i\geq 1}$ by putting $Q_k = P_{i_k}$ where i_k denotes the k-th element i of I with $i \geq i_1$. We claim that this sequence enjoys all properties stated in (c).

Take $i = i_k$ for some $k \geq 1$. Using [**1**, Lemmas 2 and 4] together with (7.1), we find

$$1 \leq |\det(P_{i-1}, P_i, P_{i+1})| \leq 6H(P_i)H(P_{i+1})|P_{i-1}(\xi)| \leq 6c_1 H(P_i)^{-\gamma} H(P_{i+1})$$

and

$$1 \leq |\operatorname{Res}(P_i, P_{i+1})| \leq 12H(P_i)H(P_{i+1})^2 |P_i(\xi)| \leq 12c_1 H(P_i) H(P_{i+1})^{-1/\gamma}.$$

Comparing these two sets of inequalities, we deduce

$$H(P_{i+1}) \sim H(P_i)^{\gamma}, \quad |P_{i-1}(\xi)| \sim H(P_i)^{-\gamma-1}, \quad |P_i(\xi)| \sim H(P_{i+1})^{-\gamma-1}. \tag{7.2}$$

In particular, the polynomial $Q_k = P_i$ satisfies

$$|Q_k(\xi)| \sim H(Q_k)^{-\gamma^3}. \tag{7.3}$$

Moreover, if i is large enough, the pairs $(P, Q) = (P_{i-1}, P_i)$ and $(P, Q) = (P_i, P_{i+1})$ both satisfy the hypotheses of Lemma 7.1 and so we get

$$\begin{aligned} L_\xi(P_{i-1} \wedge P_i) &\sim H(P_i)|P_{i-1}(\xi)| \sim H(P_i)^{-\gamma}, \\ L_\xi(P_i \wedge P_{i+1}) &\sim H(P_{i+1})|P_i(\xi)| \sim H(P_{i+1})^{-\gamma}. \end{aligned} \tag{7.4}$$

Adjusting the implied constants if necessary, we may assume that these estimates hold for all $i \in I$.

Consider now the next element $j = i_{k+1}$ of I. By construction, we have $V_i = V_{j-1}$ and so $P_i \wedge P_{i+1} = \pm P_{j-1} \wedge P_j$. Using (7.4), we find

$$H(P_j)^{-\gamma} \sim L_\xi(P_{j-1} \wedge P_j) = L_\xi(P_i \wedge P_{i+1}) \sim H(P_{i+1})^{-\gamma}$$

and so, by (7.2),

$$H(Q_{k+1}) = H(P_j) \sim H(P_{i+1}) \sim H(Q_k)^\gamma. \tag{7.5}$$

Moreover, since P_i and P_j are linearly independent (being primitive of distinct height), they form a basis of V_i. Since V_i contains P_i and P_{i+1} whose resultant is non-zero, we deduce that P_i and P_j also have a non-zero resultant and using [**1**, Lemma 2] together with (7.3) and (7.5), we get

$$1 \le |\operatorname{Res}(Q_k, Q_{k+1})| \ll H(Q_k)H(Q_{k+1})^2|Q_k(\xi)| \ll 1.$$

Finally, putting $\ell = i_{k+2}$, we observe that the polynomials P_i, P_j and P_ℓ are linearly independent as they span the same vector space over $\mathbb{Q}$ as P_{j-1}, P_j and P_{j+1}. Using [**1**, Lemma 4] together with (7.3) and (7.5), this gives

$$1 \le |\det(Q_k, Q_{k+1}, Q_{k+2})| \ll H(Q_{k+1})H(Q_{k+2})|Q_k(\xi)| \ll 1.$$

The last three estimates together with (7.3) show that the sequence $(Q_k)_{k\ge1}$ satisfies the condition (c).

To prove the last implication that (c) implies (b), assume the existence of a sequence of polynomials $(Q_k)_{k\ge1}$ as in condition (c). We first claim that ξ is neither rational nor quadratic over $\mathbb{Q}$. To prove this, assume on the contrary that there exists a non-zero polynomial $Q \in \mathbb{Z}[T]_{\le2}$ which vanishes at ξ. Using [**1**, Lemma 4], we then find

$$|\det(Q, Q_k, Q_{k+1})| \ll H(Q)H(Q_{k+1})|Q_k(\xi)| \ll H(Q)H(Q_k)^{-\gamma^2},$$

which implies that the integer $\det(Q, Q_k, Q_{k+1})$ is zero for all sufficiently large values of k, against the hypothesis that any three consecutive Q_k's are linearly independent. Define

$$\mathbf{y}_k = Q_k \wedge Q_{k+1}$$

for each index $k \ge 1$. We have

$$\|\mathbf{y}_k\| \ll H(Q_k)H(Q_{k+1}) \ll H(Q_k)^{\gamma^2}$$

for all $k \ge 1$ and, applying Lemma 7.1, we deduce

$$L_\xi(\mathbf{y}_k) \sim H(Q_{k+1})|Q_k(\xi)| \sim H(Q_{k+1})^{-\gamma} \ll \|\mathbf{y}_{k+1}\|^{-1/\gamma}.$$

This shows in particular that $L_\xi(\mathbf{y}_k)$ tends to zero as k tends to infinity and thus that the sequence $(\mathbf{y}_k)_{k\ge1}$ is unbounded. So, for any sufficiently large real number X, there exists an index k such that $\|\mathbf{y}_k\| \le X \le \|\mathbf{y}_{k+1}\|$ and, for such a choice of k, the previous estimate gives $L_\xi(\mathbf{y}_k) \ll X^{-1/\gamma}$. This shows that ξ is extremal. □

Appendix A. Finding new relations

Let the notation be as in Section 4. Fix an integer $k \geq 1$, a $(k+1)$-tuple of non-negative integers $\mathbf{d} = (d_0, \dots, d_k)$, an integer $p \geq 0$ and triples of indeterminates $\mathbf{u}_j = (u_{j,0}, u_{j,1}, u_{j,2})$ for $j = 0, 1, \dots, k$. To each monomial

$$\mathbf{u}_0^{\mathbf{e}_0} \cdots \mathbf{u}_k^{\mathbf{e}_k} = \prod_{j=0}^{k} \prod_{\ell=0}^{2} u_{j,\ell}^{e_{j,\ell}} \in \mathbb{Q}[\mathbf{u}_0, \dots, \mathbf{u}_k]$$

we associate a *weight* given by

$$\sum_{j=0}^{k} \sum_{\ell=0}^{2} \ell e_{j,\ell}$$

and a *multi-degree* given by

$$(|\mathbf{e}_0|, \dots, |\mathbf{e}_k|) = (e_{0,0} + e_{0,1} + e_{0,2}, \dots, e_{k,0} + e_{k,1} + e_{k,2}).$$

We denote by $E(\mathbf{d}, p)$ the sub-$\mathbb{Q}$-vector space of $\mathbb{Q}[\mathbf{u}_0, \dots, \mathbf{u}_k]$ generated by all monomials of multi-degree $\mathbf{d}$ and weight p. We also denote by $\mathbf{1}$ the element of $\mathbb{Q}^{3k+3}$ all of whose coordinates are equal to 1.

PROPOSITION A.1. *Suppose that a polynomial $P \in E(\mathbf{d}, p)$ satisfies*

$$\left(\prod_{j=0}^{k} \prod_{\ell=1}^{2} \left(\frac{\partial}{\partial u_{j,\ell}} \right)^{f_{j,\ell}} P \right) (\mathbf{1}) = 0$$

for every choice of non-negative integers $f_{0,1}, f_{0,2}, \dots, f_{k,1}, f_{k,2}$ with

$$2 \sum_{j=0}^{k} (f_{j,1} + f_{j,2}) \gamma^j \leq \sum_{j=0}^{k} d_j \gamma^j \quad \text{and} \quad f_{j,1} + f_{j,2} \leq d_j, \qquad (0 \leq j \leq k).$$

Then, we have

$$P(\mathbf{y}_i, \mathbf{y}_{i+1}, \dots, \mathbf{y}_{i+k}) = 0$$

for all sufficiently large values of i.

PROOF. We find

$$\begin{aligned} P(\mathbf{y}_i, \dots, \mathbf{y}_{i+k}) &= \xi^p y_{i,0}^{d_0} \cdots y_{i+k,0}^{d_k} P\left(1, \frac{y_{i,1}}{y_{i,0}\xi}, \frac{y_{i,2}}{y_{i,0}\xi^2}, \dots, 1, \frac{y_{i+k,1}}{y_{i+k,0}\xi}, \frac{y_{i+k,2}}{y_{i+k,0}\xi^2}\right) \\ &= \xi^p y_{i,0}^{d_0} \cdots y_{i+k,0}^{d_k} \sum \prod_{j=0}^{k} \prod_{\ell=1}^{2} \frac{1}{f_{j,\ell}!} \left(\left(\frac{y_{i+j,\ell}}{y_{i+j,0}\xi^\ell} - 1 \right) \frac{\partial}{\partial u_{j,\ell}} \right)^{f_{j,\ell}} P(\mathbf{1}), \end{aligned}$$

where the sum in the second expression ranges over all choices of non-negative integers $f_{0,1}, f_{0,2}, \dots, f_{k,1}, f_{k,2}$ with $f_{j,1} + f_{j,2} \leq d_j$ for $j = 0, \dots, k$. Since, for any such choice of integers, we have

$$\begin{aligned} \left| y_{i,0}^{d_0} \cdots y_{i+k,0}^{d_k} \prod_{j=0}^{k} \prod_{\ell=1}^{2} \left(\frac{y_{i+j,\ell}}{y_{i+j,0}\xi^\ell} - 1 \right)^{f_{j,\ell}} \right| &\ll \prod_{j=0}^{k} \|\mathbf{y}_{i+j}\|^{d_j - 2f_{j,1} - 2f_{j,2}} \\ &\ll \|\mathbf{y}_i\|^{\sum_{j=0}^{k} (d_j - 2f_{j,1} - 2f_{j,2})\gamma^j}, \end{aligned}$$

the hypothesis implies

$$|P(\mathbf{y}_i, \dots, \mathbf{y}_{i+k})| \ll \|\mathbf{y}_i\|^{-\varepsilon}$$

for some positive real number ε. Since $P(\mathbf{y}_i, \dots, \mathbf{y}_{i+k})$ is a rational number with bounded denominator, it must therefore vanish for all sufficiently large values of i. □

Empirically it seems that, for a given multi-degree $\mathbf{d}$, the dimension of $E(\mathbf{d}, p)$ is maximal with $p = |\mathbf{d}| = d_0 + \dots + d_k$. For values of p at equal distance from $|\mathbf{d}|$, that is for integers $p, q \geq 0$ with $p + q = 2|\mathbf{d}|$, the dimensions of the corresponding vector spaces $E(\mathbf{d}, p)$ and $E(\mathbf{d}, q)$ are the same. So, it is natural to look first for polynomials in $E(\mathbf{d}, |\mathbf{d}|)$.

A computer search based on the above proposition found two non-zero polynomials with the appropriate vanishing, namely

$$\det(\mathbf{u}_0, \mathbf{u}_1, [\mathbf{u}_3, \mathbf{u}_3, \mathbf{u}_4]) \quad \text{and} \quad \det(\mathbf{u}_1, \mathbf{u}_2, [\mathbf{u}_3, \mathbf{u}_3, \mathbf{u}_4])$$

where the symbol $[\mathbf{u}_3, \mathbf{u}_3, \mathbf{u}_4]$ is defined by (4.3). The first polynomial has multi-degree $(1, 1, 0, 2, 1)$ and weight 5, while the other has multi-degree $(0, 1, 1, 2, 1)$ and weight 5. They provide the relations (4.2) of Proposition 4.2.

References

1. B. Arbour and D. Roy, *A Gel'fond type criterion in degree two*, Acta Arith. **11** (2004), 97–103.
2. W. D. Brownawell, *Sequences of diophantine approximations*, J. Number Theory **6** (1974), 11–21.
3. Y. Bugeaud, *Approximation by algebraic numbers*, Cambridge Tracts in Math., Cambridge Univ. Press, 2004 (to appear).
4. Y. Bugeaud and O. Teulié, *Approximation d'un nombre réel par des nombres algébriques de degré donné*, Acta Arith. **93** (2000), 77–86.
5. H. Davenport and W. M. Schmidt, *Approximation to real numbers by quadratic irrationals*, Acta Arith. **13** (1967), 169–176.
6. ———, *Approximation to real numbers by algebraic integers*, Acta Arith. **15** (1969), 393–416.
7. M. Laurent, *Simultaneous rational approximation to the successive powers of a real number*, Indag. Math. (N.S.) **11** (2003), 45–53.
8. K. Mahler, *Ein Übertragungsprinzip für konvexe Körper*, Časopis Pěst. Mat. Fiz. **68** (1939), 93–102.
9. P. Philippon, *Critères pour l'indépendance algébrique*, Pub. Math. IHES **64** (1986), 5–52.
10. D. Roy, *Approximation simultanée d'un nombre et de son carré*, C. R. Acad. Sci., Paris, Sér. I Math. **336** (2003), 1–6.
11. ———, *Approximation to real numbers by cubic algebraic integers*. I, Proc. London Math. Soc. **88** (2004), 42–62.
12. ———, *Approximation to real numbers by cubic algebraic integers*. II, Ann. of Math. **158** (2003), 1081–1087.
13. D. Roy and M. Waldschmidt, *Diophantine approximation by conjugate algebraic integers*, Compositio Math. (to appear).
14. W. M. Schmidt, *Diophantine approximation*, Lecture Note in Math., vol. 785, Springer-Verlag, 1980.
15. V. G. Spindžuk, *A proof of Mahler's conjecture on the measure of the set of S-numbers*, Izv. Akad. Nauk SSSR Ser. Mat. **29** (1965), 379–436; English transl., Amer. Math. Soc. Transl. **51** (1966), 215–272.
16. O. Teulié, *Approximation d'un nombre réel par des unités algébriques*, Monatsh. Math. **132** (2001), 169–176.
17. E. Wirsing, *Approximation mit algebraischen Zahlen beschränkten Grades*, J. Reine Angew. Math. **206** (1961), 67–77.

Department of Mathematics and Statistics, University of Ottawa, 585 King Edward Avenue, Ottawa, ON K1N 6N5, Canada.

E-mail address: droy@uottawa.ca

Centre de Recherches Mathématiques
CRM Proceedings and Lecture Notes
Volume **36**, 2004

Compact Operators with Rational Generation

Lawren Smithline

Abstract. Let $A_{i,j}$, for i, j nonnegative, be an array of elements of an integral domain R with linear recurrence relation. That is, there is a finite set S of pairs of positive numbers (k, l) and constants $c_{k,l}$ and a finite set T of pairs of nonnegative numbers (i, j) such that for (i, j) not in T,

$$A_{i,j} = \sum_S c_{k,l} A_{i-k,j-l},$$

where A with a negative index takes the value zero.

When R has a nontrivial ultrametric norm, there are easy convergence conditions which imply the array $A_{i,j}$ is the matrix for a compact linear operator on an infinite rank module over R and is also the series of coefficients of the Taylor expansion of a rational function of two variables. We show that the product of two such operators is another of the same form.

The Atkin U operator which acts on overconvergent p-adic modular forms is the motivating example for this study. In cases where U is a compact operator with rational generation, its special combinatorial properties improve our understanding of its spectral theory and our ability to do numerical experiments.

1. Introduction and definitions

We describe a method for two dimensional generating functions. Afterwards, we apply this method to a space of overconvergent p-adic modular forms. These overconvergent p-adic modular forms generalize classical modular forms and arise from modular curves considered as p-adic rigid analytic varities. We show an example for prime $p = 3$ and tame level $N = 1$. The "tame" refers to the situation $(p, N) = 1$. The technicalities of overconvergent p-adic modular forms are important for constructing the combinatorial examples. Given the data for building the matrices, we can obtain results with little further reference to their origin. We refer the reader to [**1, 2, 4–6**] for details about overconvergent p-adic modular forms.

One of the specific open numerical problems in the area of overconvergent p-adic modular forms is the following conjecture of Gouvea and Mazur. Let p be a prime, N be an integer prime to p, k and k' be even integers. Let a be a positive integer such that $(p-1)p^n$ divides $k - k'$. Let U_k be the Atkin U operator acting

---

2000 *Mathematics Subject Classification.* Primary: 11F11, 11B37; Secondary: 11F25, 11F85.
The author thanks the referee for suggestions which improved the clarity of this paper.
This is the final form of the paper.

on overconvergent p-adic modular forms of tame level N and weight k. Let $U_{k'}$ be the analog for weight k'.

CONJECTURE 1.1 (Gouvêa–Mazur). *The valuations of the eigenvalues of U_k and $U_{k'}$ which are less than n are the same, and occur with the same multiplicity.*

Using an explicit matrix expression of U_k, we computed the valuations of approximately one hundred terms of its characteristic series. This computation, done on a desktop computer running the PARI algebra system, was more extensive than would be possible using, for instance, Koike's formula. From these valuations, we determined the slopes of the Newton polygon of the characteristic series, which are the valuations of the eigenvalues of U_k.

The data affirm a lower bound of the Newton polygon of the characteristic series $\det(1-tU_0)$ and for $p=3$, $N=1$, and suggest the theorem that the Newton polygon coincides with the lower bound infinitely often. We sketch these results in Section 4.

The data also motivate the conjecture[1] that the slopes are exactly the 3-adic valuation of

$$\frac{(2n!)(6n!)}{n!n!}$$

for $n \geq 0$.

Let R be an integral domain.

DEFINITION 1.2. An *array with recursive generation* A is a doubly indexed tableau $A_{i,j}$ determined by the data

S^A, a finite set of ordered pairs of positive integers,
a finite set of ordered pairs of nonnegative integers,
$c^A_{k,l} \in R$ for each $(k,l) \in S^A$,
$d^A_{i,j} \in R$ for each $(i,j) \in T^A$;

$$A_{i,j} = \sum_{(k,l)\in S^A} c^A_{k,l} A_{i-k,j-l} + d^A_{i,j},$$

where we evaluate A with either index negative as zero, and also $d^A_{i,j} = 0$ for $(i,j) \notin T$.

We denote the ancillary objects related to array with recursive generation A by placing A as a superscript.

The array with recursive generation generalizes the *linear recursive sequence.* When every element of S^A and T^A is a pair of the form (i,i), the sequence $a_i = A_{i,i}$ is a linear recursive sequence.

DEFINITION 1.3. The *generating function* of an array with recursive generation A is the formal power series

$$f^A(x,y) = \sum_{i=0}^{\infty}\sum_{j=0}^{\infty} A_{i,j} x^i y^j,$$

expressed as a formal rational function

$$\frac{\sum_{(i,j)\in T^A} d^A_{i,j} x^i y^j}{1 - \sum_{(k,l)\in S} c^A_{k,l} x^k y^l}.$$

---

[1]Frank Caligari produced this concise description of the slopes.

DEFINITION 1.4. The sum of two arrays with recursive generation, $A+B$ is the tableau C such that $C_{i,j} = A_{i,j} + B_{i,j}$.

DEFINITION 1.5. The matrix product of two arrays with recursive generation, AB, is the tableau C such that $C_{i,j} = \sum_{k=0}^{\infty} A_{i,k}B_{k,j}$.

The tableau C exists, because every row and column of each of A and B has a finite number of nonzero entries. Therefore, the hypothetical infinite sum $C_{i,j} = \sum_{k=0}^{\infty} A_{i,k}B_{k,j}$ has only finitely many nonzero terms.

Let R be a Banach algebra with nontrivial ultrametric norm $|\cdot|$ and M be the Banach module $R\langle t\rangle$, that is, the subring of $R[\![t]\!]$ consisting of power series with coefficients tending to norm zero.

DEFINITION 1.6. An operator with rational generation L with respect to the basis t^i is a linear operator in $\mathrm{End}(M)$ whose matrix A with respect to the basis t^i is an array with recursive generation.

A compact operator with rational generation is an operator with rational generation which is also a compact operator. When A is the matrix of an operator with rational generation, we also refer to the operator with rational generation A.

PROPOSITION 1.7. *Let L be an operator with rational generation and A be its associated array with recursive generation. If*

$$\max_{k,l} |c^A_{k,l}| < 1,$$

then the formal power series for A converges to the formal rational function for A and L is a compact operator with rational generation.

PROOF. Extend the norm $|\cdot|$ to power series by

$$\left|\sum_{k,l} c^A_{k,l} x^k y^l\right| = \max_{k,l} |c^A_{k,l}|.$$

If

$$\left|\sum_{k,l} c^A_{k,l} x^k y^l\right| < 1,$$

then the power series $f^A(x,y)$ is a geometric series with common ratio of norm less than 1. □

2. The algebra of linear recursive sequences

In Section 3, we show that for arrays with recursive generation A, B, the sum $A+B$ and product AB are also arrays with recursive generation. The simpler case of linear recursive sequences illuminates this result. As with arrays with recursive generation, we use superscripts a, b, c to denote the objects related to linear recursive sequences a, b, c.

LEMMA 2.1. *Let a and b be linear recursive sequences. The sequence $c_i = a_i + b_i$ is a linear recursive sequence.*

PROOF. Let $f^a(x) = \sum_{i=0}^{\infty} a_i x^i$. It is a standard fact about generating functions that

$$f^a(x) = D^a(x) \sum_{i=0}^{\infty} C^a(x)^i$$

for polynomials $C^a(x)$, $D^a(x)$, where $C^a(x)$ has zero constant term. Formally, $f^a(x)$ is the Taylor series at $x = 0$ for the rational function

$$\frac{D^a(x)}{1 - C^a(x)}.$$

The same recitation applies for b: there are $f^b(x)$, $D^b(x)$, $C^b(x)$ in the same relation.

Let $f^c(x) = f^a(x) + f^b(x)$. It is the generating function for c and the Taylor series for

$$\frac{D^a(x)(1 - C^b(x)) + D^b(x)(1 - C^a(x))}{1 - C^a(x) - C^b(x) + C^a(x)C^b(x)}.$$

Therefore, the sequence c is a linear recursive sequence. □

LEMMA 2.2. *Let a and b be linear recursive sequences. The sequence $c_i = a_i b_i$ is a linear recursive sequence.*

PROOF. Let $f^a(x)$ and $f^b(x)$ be the generating functions for a and b as in the previous lemma. Let K be the field of fractions of R. Let L be the splitting field of $1 - C^a(x)$ and $1 - C^b(x)$. The formal rational expressions for f^a and f^b both have partial fraction decompositions over L. That is,

$$f^a(x) = \sum_{j=1}^{m^a} \frac{\delta_j^a(x)}{(1 - \gamma_j^a x)^{e_j^a}}$$

where $\gamma_j^a \in L$ and $\delta_j^a(x)$ is a polynomial of degree less than e_j^a.

Converting the formal rational function to a geometric series shows there are polynomials $p_j^a(n)$ over L each of degree less than e_j^a such that

$$a_i = \sum_{j=1}^{m^a} p_j^a(i)(\gamma_j^a)^i.$$

The same procedure holds for f^b, i.e. there are γ_k^b, $\delta_k^b(x)$, $p_k^b(i)$ and

$$b_i = \sum_{k=1}^{m^b} p_k^b(i)(\gamma_k^b)^i.$$

Thus,

$$c_i = \sum_{j=1}^{m^a} \sum_{k=1}^{m^b} p_j^a(i) p_k^b(i)(\gamma_j^a \gamma_k^b)^i.$$

Reversing the partial fraction decomposition, we construct the formal rational function for $f^c(x)$ as the sum of fractions of the form

$$\frac{\delta_{j,k}^c(x)}{(1 - \gamma_{j,k}^c x)^{e_{j,k}^c}},$$

where

$$\gamma_{j,k}^c = \gamma_j^a \gamma_k^b, \quad e_{j,k}^c = (e_j^a - 1)(e_k^b - 1) + 1,$$

and the degree of $\delta_{j,k}^c$ is less than $e_{j,k}^c$. □

For example, the Fibonacci sequence $F_0 = 0$, $F_1 = 1$, $F_i = F_{i-1} + F_{i-2}$ has the generating function

$$f^F(x) = x/(1 - x - x^2).$$

We construct the rational function for the sequence $G_i = F_i^2$.

Let $\phi = 1 + \sqrt{5}/2$ and $\overline{\phi}$ be its conjugate. From the classical result $F_i = \sqrt{5}(\phi^i - \overline{\phi}^i)/5$, it follows that

$$G_i = \left(\phi^{2i} + 2(-1)^i + \overline{\phi}^{2i}\right)/5.$$

There is a rational function whose denominator is a cubic with roots ϕ^2, -1, $\overline{\phi}^2$, whose numerator has degree at most two and whose Taylor series expansion is $f^G(x)$. The three values $G_0 = 0$, $G_1 = 1$, $G_2 = 1$ give a solution for the numerator of the rational function. In total,

$$f^G(x) = \frac{x - x^2}{1 - 2x - 2x^2 + x^3}.$$

3. The algebra of arrays with recursive generation

PROPOSITION 3.1. *Let A and B be arrays with recursive generation. The sum $C = A + B$ is an array with recursive generation.*

PROOF. Let $f^A(x, y)$ and $f^B(x, y)$ be the generating functions for A and B. The formal rational function $f^A(x, y)$ is

$$\frac{\sum_{(i,j)\in T^A} d^A_{i,j} x^i y^j}{1 - \sum_{(k,l)\in S} c^A_{k,l} x^k y^l}.$$

The function $f_B(x, y)$ is analogous for the tableau B.

The function $f^C(x, y)$ is the sum $f^A(x, y) + f^B(x, y)$, and the parameters $c^C_{k,l}$ and $d^C_{i,j}$ appear in the expression of $f^C(x, y)$ as a rational function. □

THEOREM 3.2. *Let A and B be arrays with recursive generation. The product $C = AB$ is an array with recursive generation.*

PROOF. Let $f^A(x, y)$, $f^B(x, y)$, $f^C(x, y)$ be the generating functions of A, B, C. A priori, C has only a formal power series generating function. The functions $f^A(x, y)$ and $f^B(x, y)$ have formal expressions as rational functions, because A and B are arrays with recursive generation.

Let $g(x, y, z) = f^A(x, z) f^B(z^{-1}, y)$ be a power series in $R[\![x, y, z, z^{-1}]\!]$. The generating function $f^C(x, y)$ is the z^0 term of $g(x, y, z)$.

The function $f^A(x, z)$ is the generating function of a linear recursive sequence over $R[x]$. By Lemma 2.2, there is a partial fraction decomposition for the rational function $f^A(x, z)$ expands as follows. There is a natural number m^A and for each integer i, $1 \le i \le m^A$, there is γ^A_i algebraic over the field of fractions of $R[x]$, positive integer e^A_i, polynomial $p^A_i(n)$ of degree less than e^A_i, such that

$$f^A(x, z) = \sum_{i=1}^{m^A} \sum_{k=0}^{\infty} p^A_i(k)(\gamma^A_i)^k z^k.$$

The same procedure applied to $f^B(z^{-1}, y)$ yields m^B and for $1 \le j \le m^B$, δ^B_j algebraic over the field of fractions of $R[y]$, positive integer e^B_j, polynomial $p^B_j(n)$

such that

$$f^B(z^{-1}, y) = \sum_{j=1}^{m^B} \sum_{k=0}^{\infty} p_j^B(k)(\delta_j^B)^k z^{-k}.$$

Thus,

$$f^C(x, y) = \sum_{i=1}^{m^A} \sum_{j=1}^{m^B} \sum_{k=0}^{\infty} p_i^A(k) p_j^B(k) (\gamma_i^A \delta_j^B)^k.$$

This is the Taylor series of a rational function of all the γ_i^A and δ_j^B. The field extension L of the fraction field of $R[x, y]$ generated by the γ_i^A and δ_k^B is normal and $f^C(x, y)$ is stabilized by the automorphisms of L fixing $R[x, y]$. Therefore, $f^C(x, y) \in R[\![x, y]\!]$ and it is the Taylor series of a rational function of x, y. $\square$

4. Example application to 3-adic modular forms

Let $\mathbf{C}_3$ be the complete algebraically closed ring containing $\mathbf{Q}_3$, the 3-adic rational numbers. Let $\mathcal{O}_3$ be the ring of integers of $\mathbf{C}_3$.

We refer back to the references for overconvergent p-adic modular forms, and also [**3, 9**] for details on the following facts.

FACT 4.1. For every even integer k, there is a module $\mathcal{M}_k$ of 3-adic overconvergent modular forms of weight k and tame level 1.

FACT 4.2. The module $\mathcal{M}_0$ is a ring with the structure of $\mathcal{O}_3\langle d \rangle$ for a certain weight zero modular form

$$d = \sqrt{\Delta(q^3)/\Delta(q)},$$

where $\Delta(q)$ is the classical weight 12 level 1 cusp form.

FACT 4.3. There is a compact operator U which is a compact endomorphism of $\mathcal{M}_k$.

FACT 4.4. The evaluations of U at $1, d, d^2$ are $U(1) = 1$,

$$U(d) = 10 \cdot 3^2 d + 4 \cdot 3^7 d^2 + 3^{11} \cdot d^3,$$

and

$$U(d^2) = 3U(d)^2 + 8 \cdot 3d + 2 \cdot 3^5 d^2.$$

FACT 4.5. There is an algebraic relation between the powers d^i and $U(d^i)$:

$$U(d^{3+i}) = (3^{12}d^3 + 4 \cdot 3^8 d^2 + 10 \cdot 3^3 d) U(d^{2+i}) + (3^6 d^2 + 4 \cdot 3^2 d) U(d^{1+i}) + dU(d^i).$$

LEMMA 4.1. *Let A be the matrix realization of U with respect to the basis d^i. A is a compact operator with rational generation with*

$$c_{k,l}^A = \begin{array}{ccc|} & & 3^{12} \\ & 3^6 & 4 \cdot 3^8 \\ 1 & 4 \cdot 3^2 & 10 \cdot 3^3 \\ \hline \end{array},$$

where the bottom left corner is $c^A_{1,1}$, *and*

$$d^A_{i,j} = \begin{array}{|lll} \hline 10\cdot 3^2 & 8\cdot 3 & 1 \\ 4\cdot 3^7 & 2\cdot 3^5 & \\ 3^{11} & & \end{array}\ .$$

where the top right corner is $d^A_{1,1}$.

PROOF. This is a consequence of Facts 4.4 and 4.5. ☐

Let D be the diagonal array with recursive generation with $c^D_{1,1} = 3^{9/2}$ and $d^D_{1,1} = 1$. Let E be the diagonal array with recursive generation with $c^E_{1,1} = 3^{-3/2}$ and $d^D_{1,1} = 9$.

Let B be the array with recursive generation with data

$$c^B = \begin{array}{lll|} & & 1 \\ & 1 & 4\sqrt{3} \\ 1 & 4\sqrt{3} & 10 \\ \hline \end{array}\ , \quad \text{and} \quad d^B = \begin{array}{|lll} \hline 10 & 8\sqrt{3} & 3 \\ 4\sqrt{3} & 2 & \\ 1 & & \end{array}\ .$$

The product DBE equals A. The matrix EDB is similar to A.

A basic result of [**8**] gives a lower bound for the Newton polygon of the determinant $\det(1 - tA)$, by reading off the valuations of the diagonal ED.

PROPOSITION 4.2. *The Newton polygon of A lies on or above the Newton polygon of ED, which is the piecewise linear graph through* $\big(n, 3n(n-1)/2 + 2n\big)$.

PROOF. The matrix ED is diagonal and the matrix B has entries in $\mathcal{O}_3$. So the minimum valuation in row i of EDB is the valuation of the nonzero element in row i of ED. Serre's observation is that the coefficient of t^i in $\det\big(1 - t(EDB)\big)$ has valuation at least the sum of the lowest i valuations in rows of EDB. But ED is just the diagonal array with recursive generation with $c^{ED}_{1,1} = 3^3$, and $d^{ED}_{1,1} = 3^2$. ☐

THEOREM 4.3. *The Newton polygon of A touches the Newton polygon of ED exactly at the points* $\big(n, 3n(n-1)/2 + 2n\big)$ *for which* $n = (3^j - 1)/2$ *for some* j.

PROOF. Denote by a_n the coefficient of $3^{3n(n-1)/2+2n}t^n$ in $\det(1 - tA)$.

We show for which square finite truncations B_n of B at n rows and columns the matrix B_n has 3-adic unit determinant. In these cases, there will be a unique summand, $\det(B_n)$ of valuation zero in the sum for a_n. Otherwise, there will be no terms of valuation zero.

We consider the array with recursive generation $\overline{B}$ which is the $\mathbf{F}_3$ reduction of B. Let $c(x, y) = xy(1 + x^2 + xy + y^2)$. The generating function

$$f_{\overline{B}}(x, y) = \frac{xy - x^2y^2 + xy^3}{1 - c(x, y)}$$

may be rewritten as an infinite product

$$f_{\overline{B}}(x, y) = (xy)(1 - xy + xy^2)\prod_{j=0}^{\infty}\big(1 + c(x, y) + c(x, y)^2\big)^{3^j}.$$

In the ring $\mathbf{F}_3[\![x, y]\!]$, the exponent of 3^j distributes.

The details of the calculation, shown in [**9**] come down to: for $n = 0$, the value of $\det(B_n)$ is 1, which has valuation zero. For $n > 0$, $\det(B_n)$ has valuation zero if and only if 3 divides $n - 1$ and $\det(B_{n-1/3})$ has valuation zero. ☐

5. Other results

Given a compact operator with rational generation A representing an Atkin U operator and $n > 0$, we can compute a N such that for truncations of a A to a A_N of size at least N by N, the Newton polygons of A and A_N agree at least up to the position above ordinate n.

Therefore, we use the compact operator with rational generation for A to do numerical experiments to investigate the slopes of the U operator. The conjecture on slopes of 3-adic modular forms was mentioned earlier. Similarly, for $p = 5$, we have data for the slopes in weight zero which agrees with the 5-adic valuation of

$$3n!3n!5n!/n!^3.$$

Kilford [**7**] uses a basis for 2-adic modular forms to compute all the slopes for 2-adic modular forms. His recursion relaxes the condition on S^A and permits ordered pairs of nonnegative integers except for $(0, 0)$. This variation complicates the general multiplication property of arrays with recursive generation; however, it still holds for compact operators with rational generation. Multiplications by diagonal arrays with recursive generation and arrays with recursive generation with only finitely many nonzero entries are all that Kilford requires, and these are simpler than the general case. After deriving a compact operator with rational generation for U_2, the combinatorial analysis is independent of its origin.

The results of compact operators with rational generation so far apply to genus zero modular curves. An explicit description of U for 11-adic modular forms is evident in analogy with the others, and is more challenging because $X_0(11)$ has genus 1. We have yet to organize it for extensive computation.

References

1. R. Coleman, *p-adic Banach spaces and families of modular forms*, Invent. Math. **127** (1997), 417–479.
2. R. Coleman and B. Mazur, *The eigencurve*, London Math. Soc. Lecture Note Ser. **254** (1996), 1–113.
3. R. Coleman, G. Stevens, and J. Teitelbaum, *Numerical experiments on families of p-adic modular forms*, Stud. Adv. Math. **7** (1998), 143–158.
4. F. Gouvêa and B. Mazur, *Families of modular eigenforms*, Math. Comput. **58** (1992), 793–805.
5. ———, *On the characteristic series of the U operator*, Ann. Inst. Fourier **43** (1993), 301–312.
6. N. Katz, *p-adic properties of modular schemes and modular forms*, Modular Functions of One Variable, III, Proc. Internat. Summer School (Antwerp, 1972), Lecture Notes in Math., vol. 350, Springer, Berlin, 1973, pp. 69–190.
7. L. J. P. Kilford, *Slopes of overconvergent modular forms*, Ph.D. Thesis, Imperial College of Science, Technology and Medicine, London, 2002.
8. J. P. Serre, *Endomorphismes complètements continues des espaces de Banach p-adique*, Publ. Math. IHES **12** (1962), 69–85.
9. L. M. Smithline, *Exploring slopes of p-adic modular forms*, Ph.D. Thesis, University of California, Berkeley , 2000.

Department of Mathematics, Cornell University, Malott Hall, Ithaca, NY 14853, USA.

E-mail address: `lawren@math.cornell.edu`

Centre de Recherches Mathématiques
CRM Proceedings and Lecture Notes
Volume **36**, 2004

The Canonical Height of a Finite Étale K-Algebra

Valerio Talamanca

Abstract. In this paper we study height functions attached to adelic norms on finite dimensional vector spaces over number fields. We show that for a finite étale algebra over a number field there exists an intrinsically defined adelic norm whose associated height function can be regarded as the canonical height of A.

1. Introduction

In this paper we continue the study of heights on linear spaces that we started in [**12, 13**], with particular emphasis towards finite étale K-algebras, K being a number field. Let $\mathbf{X}$ be a vector space over number field K. We study height functions which are defined by adelic norms on $\mathbf{X}$, where an adelic norm is a family of v-adic norms on $\mathbf{X}$ satisfying certain compatibility conditions, see Section 1 for the precise definition. Adelic norms on line bundles were used by Peyre [**9**] and Zhang [**17**], among others, to define height functions on projective varieties. It is worth noting that the height functions commonly used in the literature, i.e., naive or Northcott–Weil heights, Arakelov heights, and twisted heights, can all be recovered as heights defined by suitable adelic norms on some vector space; see the examples in Section 1 for details.

Let A be a (commutative) finite étale K-algebra. In [**12**] we introduced a height function H_A on A which is invariant under K-algebra isomorphisms. The main results of this paper, Theorem 3.4 and 3.5, show that H_A can be regarded as the canonical height of A. Indeed, Theorem 3.4 exhibits a construction of H_A by an averaging process starting with any adelic norm which is compatible with the algebra structure. It must be noted that this averaging process is the exact analogue of that employed by Tate in constructing the canonical height associated to a symmetric ample line bundle on an abelian variety, see [**7**]. Theorem 3.5 shows that the only K-linear surjective maps between two finite étale K-algebras which leave the height invariant (i.e., such that $H_A = H_B \circ T$) are, up to multiplication by an invertible element of height 1, the K-algebra isomorphisms of A into B. The

---

2000 *Mathematics Subject Classification.* Primary: 11G50; Secondary: 12A80.

Research partially supported by G.N.S.G.A of Istituto Nazionale di Alta Matematica, GVA project of M.I.U.R and GTEM european training and reasearch network.

I would like to thank the referee for many corrections and suggestions.

This is the final form of the paper.

paper is organized as follows: Section 2 contains the bulk of definitions as well as some general results on heights associated to adelic norms on vector spaces. The results about finite étale K-algebras are proven in Section 3.

2. Heights and adelic norms

Let K be a number field of degree d over $\mathbb{Q}$. We denote by $\mathcal{M}_K$ the set of equivalence classes of absolute values of K; by $\mathcal{M}_K^0$ (respectively $\mathcal{M}_K^\infty$) the subset of $\mathcal{M}_K$ consisting of the equivalence classes of non-archimedean (respectively archimedean) absolute values. If $v \in \mathcal{M}_K^0$, $v \mid p$, we normalize $|\cdot|_v$ by requiring that $|p|_v = p^{-1}$; while if $v \in \mathcal{M}_K^\infty$ we normalize $|\cdot|_v$ by requiring that restricted to $\mathbb{Q}$ it coincides with the standard archimedean absolute value. Let K_v be the completion of K with respect to $|\cdot|_v$. We denote by n_v the local degree, and set $d_v = n_v/d$.

Before introducing the height functions that we will use, we need to recall a few facts about lattices and norms for vector spaces over number fields and their completions. See [**15**, §3 and §5] for more details. Given $v \in \mathcal{M}_K^0$ we denote by $\mathcal{O}_v$ the closure of $\mathcal{O}_K$ (the ring of integers of K) in K_v. Let $\mathbf{X}$ be a finite dimensional K_v-vector space. An $\mathcal{O}_v$-module $\Lambda \subset \mathbf{X}$ is called a $\mathcal{O}_v$-lattice if it is compact and open. The norm $N_\Lambda\colon \mathbf{X} \to \mathbb{R}$ associated to Λ is defined by

$$N_\Lambda(\mathbf{x}) = \inf_{\substack{\gamma \in K_v^\times \\ \gamma\mathbf{x} \in \Lambda}} |\gamma|_v^{-1}.$$

On the other hand to any $|\cdot|_v$-norm on $\mathbf{X}$ one can associate the lattice of elements of norm at most 1. Amongst all the norms associated to the same lattice, N_Λ can be characterized as the only one such that the value group of N coincides with the value group of $|\cdot|_v$.

Let $\mathbf{X}$ be a K-vector space. As customary, in order to be able to define height functions on $\mathbf{X}$ we need to provide $\mathbf{X}$ with some extra structure. In our case this extra structure is that of an adelic norm on $\mathbf{X}$. An $\mathcal{O}_K$-module $\Lambda \subset \mathbf{X}$ is called an $\mathcal{O}_K$-lattice if it is finitely generated and contains a basis of $\mathbf{X}$ over K.

An *adelic norm* on $\mathbf{X}$ is a collection $\mathcal{F} = \{N_v\colon \mathbf{X}_v = \mathbf{X} \otimes_K K_v \to \mathbb{R}, v \in \mathcal{M}_K\}$ of $|\cdot|_v$-norms, having the following properties:

(a) If $v \in \mathcal{M}_K^0$, then N_v is ultrametric, i.e., $N_v(\mathbf{x}+\mathbf{y}) \le \max\{N_v(\mathbf{x}), N_v(\mathbf{y})\}$.
(b) There exists an $\mathcal{O}_K$-lattice $\Lambda \subset \mathbf{X}$, such that N_v is the norm associated to the $\mathcal{O}_v$-lattice $\Lambda_v = \Lambda \otimes_{\mathcal{O}_K} \mathcal{O}_v$ for all but finitely many $v \in \mathcal{M}_K^0$.

The $\mathcal{O}_K$-lattice $\Lambda_\mathcal{F}$ defined by requiring that $(\Lambda_\mathcal{F})_v = \{\mathbf{x} \in \mathbf{X}_v \mid N_v(\mathbf{x}) \le 1\}$ for all $v \in \mathcal{M}_K^0$ is uniquely determined by $\mathcal{F}$, for $\mathcal{O}_K$-lattices are uniquely determined by their local completions as proven in [**15**, Theorem V.2.2].

By definition $N_v = N_{(\Lambda_\mathcal{F})_v}$ for almost all $v \in \mathcal{M}_K^0$. Thus, for a given $\mathbf{x} \in \mathbf{X}$, the set $\{v \in \mathcal{M}_K^0 \mid N_v(\mathbf{x}) \ne 1\}$ is finite. Therefore we can define $H_\mathcal{F}$, the *height* associated to $\mathcal{F}$, by setting

$$H_\mathcal{F}(\mathbf{x}) = \prod_{v \in \mathcal{M}_K^0} N_v(\mathbf{x})^{d_v},$$

for all non zero $\mathbf{x} \in \mathbf{X}$. We set by definition $H_\mathcal{F}(\mathbf{0}) = 1$.

EXAMPLES. (a) Let $\mathbf{X}$ be a K-vector space, $\Lambda \subset \mathbf{X}$ an $\mathcal{O}_K$-lattice, $\underline{b}$ a basis of $\mathbf{X}$ over K and $1 \le q \le \infty$. For each $v \in \mathcal{M}_K^\infty$, let $N_{\underline{b},v}^q$

be the ℓ^q-norm on $\mathbf{X}_v$ associated to $\underline{b}$ regarded as a K_v-basis. Then $\mathcal{E}^q_{\Lambda,\underline{b}} = \{N_{\Lambda_v}, v \in \mathcal{M}_K^0\} \cup \{N^q_{\underline{b},v}, v \in \mathcal{M}_K^\infty\}$ is an adelic norm. In particular for $\mathbf{X} = K^n$, $\Lambda = \mathcal{O}_K^n$, $\underline{b} = \underline{e}$ the canonical basis of K^n, and $q = 1, 2$ or ∞ we recover the classical naive heights as defined by Northcott ($q = 1$) [**8**], Weil ($q = \infty$) [**14**] and Schmidt ($q = 2$) [**11**].

(b) Let $\mathcal{T} = (T_v)$ be an element of $\mathrm{GL}_n(\mathbb{A}_K)$, the adele group of $\mathrm{GL}_n(K)$. Let $\Lambda_{\mathcal{T}}$ be the $\mathcal{O}_K$-lattice defined by requiring that $(\Lambda_{\mathcal{T}})_v = T_v(\mathcal{O}_v^n)$. Set $N_v^{\mathcal{T}} = N_{(\Lambda_{\mathcal{T}})_v}$ for all $v \in \mathcal{M}_K^0$, and consider the adelic norm $\mathcal{F}_{\mathcal{T}} = \{N_v^{\mathcal{T}}, v \in \mathcal{M}_K^0\} \cup \{N^{\mathcal{T}}_{\underline{e},v}, v \in \mathcal{M}_K^\infty\}$, where $N_v^{\mathcal{T}}(\mathbf{x}) = N^2_{\underline{e},v}\big(T_v(\mathbf{x})\big)$. The height $H_{\mathcal{T}} = H_{\mathcal{F}_{\mathcal{T}}}$ associated to this adelic norm was used by D. Roy and J. Thunder in [**10**] and referred to as the twisted height associated to $\mathcal{T}$.

(c) Let $\overline{E} = \big(E, \{\langle\ ;\ \rangle_v\}_{v \in \mathcal{M}_K^\infty}\big)$ be an Hermitian vector bundle over Spec $(\mathcal{O}_K)$. This means that E is a locally free projective $\mathcal{O}_K$-module of finite rank, and $\langle\ ;\ \rangle_v$ is a metric on $E_v = E \otimes_{\mathcal{O}_K} K_v$ which is a scalar product if v is real and Hermitian if v is complex. As a by-product of Arakelov theory[1] one gets a (logarithmic) height theory for the K-vector space $E_K = E \otimes_{\mathcal{O}_K} K$ which can be computed as follows (see [**4**]): given $s \in E_K = E \otimes_{\mathcal{O}_K} K$, let L_s be the Hermitian line bundle $(Ks) \cap E$ (we are regarding E as contained in E_K, via $\mathbf{x} \mapsto \mathbf{x} \otimes 1$) with the induced metric, then $h_{ar}(s) = -\widehat{\deg}(L_s)$. On the other hand to $\overline{E}$ we can associate an adelic norm on E_K, namely $\mathcal{F}_{\overline{E}} = \{N_{E_v}, v \in \mathcal{M}_K^0\} \cup \{\sqrt{\langle\ ;\ \rangle_v}, v \in \mathcal{M}_K^\infty\}$. Then, [**16**, Lemma 1] yields: $\widehat{\deg}(L_s) = [K : \mathbb{Q}] \cdot \log H_{\mathcal{F}_{\overline{E}}}(s)$.

The height function $H_{\mathcal{F}}$ enjoys the following properties:

(H1) $H_{\mathcal{F}}(\lambda \mathbf{x}) = H_{\mathcal{F}}(\mathbf{x})$ for all $\lambda \in K^\times$ and all $\mathbf{x} \in \mathbf{X}$.

(H2) Let $\mathbb{P}(\mathbf{X})$ be the set of 1-dimensional subspaces of $\mathbf{X}$. Then for all $C > 0$ the set $\{l \in \mathbb{P}(\mathbf{X}) \mid H_{\mathcal{F}}(l) \le C\}$ is finite.

The first property follows from the product formula. It implies that $H_{\mathcal{F}}(l)$ is well defined for $l \in \mathbb{P}(\mathbf{X})$. To prove (H2), recall that given an adelic norm $\mathcal{F}$ on $\mathbf{X}$, there exists an isomorphism $\iota \colon \mathbf{X} \to K^n$, $n = \dim_K \mathbf{X}$, such that

$$C_1 H_{\mathcal{E}}\big(\iota(\mathbf{x})\big) \le H_{\mathcal{F}}(\mathbf{x}) \le C_2 H_{\mathcal{E}}\big(\iota(\mathbf{x})\big)$$

for some positive constants C_1, C_2. Here $\mathcal{E} = \mathcal{E}^\infty_{\Lambda,\underline{e}}$, as defined in Example (a) above, with $\underline{e}$ the canonical basis of K^n and $\Lambda = \mathcal{O}_K^n$ (see [**13**, Lemma 2.1]). Therefore the classical Northcott's theorem [**8**] yields the conclusion.

If $\mathcal{S} \subset \mathcal{M}_K$ is a finite set we set $\mathbf{X}_{\mathcal{S}} = \prod_{v \in \mathcal{S}} \mathbf{X}_v$ and we consider $\mathbf{X}$ as embedded diagonally in $\mathbf{X}_{\mathcal{S}}$. Moreover we denote by $d_{\mathcal{F}}$ the metric on $\mathbf{X}_{\mathcal{S}}$ defined by: $d_{\mathcal{F}}(\{\alpha_v\}_{v \in \mathcal{S}}, \{\beta_v\}_{v \in \mathcal{S}}) = \sup_{v \in \mathcal{S}} N_v(\alpha_v - \beta_v)$.

LEMMA 2.1. *Let $\mathcal{F}$ be an adelic norm on* $\mathbf{X}$. *If $\mathcal{S} \subset \mathcal{M}_K$ is a finite set, then* $\mathbf{X}$ *is dense in* $\mathbf{X}_{\mathcal{S}}$ *with respect to the metric* $d_{\mathcal{F}}$.

PROOF. This follows directly from the weak approximation theorem for number fields and the fact that all the norms on a finite dimensional vector space over a complete field are equivalent. □

For the rest of this section we let $\mathcal{F} = \{N_v, v \in \mathcal{M}_K\}$ and $\mathcal{G} = \{P_v, v \in \mathcal{M}_K\}$ be adelic norms on $\mathbf{X}$ and $\mathbf{Y}$ respectively.

---

[1] For a comprehensive treatment of Arakelov theory see [**5**]; for one of the relations between Arakelov theory and heights (of cycles) on projective varieties see [**3**].

LEMMA 2.2. *Let $\mathbf{X}, \mathbf{Y}, \mathcal{F}$ and $\mathcal{G}$ be as above. Assume that $T \in \operatorname{Hom}_K(\mathbf{X}, \mathbf{Y})$ is injective. Then there exists a finite set $\mathcal{S}_T \subset \mathcal{M}_K$ such that: $T\colon (\mathbf{X}_v, N_v) \to (\mathbf{Y}_v, P_v)$ is norm preserving if and only if $v \notin \mathcal{S}_T$.*

PROOF. First note that by an abuse of notation we denote by the same letter the map that a linear transformation induces on the various completions of a vector space. Next, by the definition of an adelic norm, there exists a K-lattice $\Lambda \subset \mathbf{X}$ (respectively $\Omega \subset \mathbf{Y}$) such that N_v (respectively M_v) coincides with the norm associated to Λ_v (respectively Ω_v) for all $v \notin \mathcal{P}$, where $\mathcal{P}$ is a finite subset of $\mathcal{M}_K$ containing $\mathcal{M}_K^\infty$. Thus, for all $v \notin \mathcal{P}$, T is norm preserving if and only if $T^{-1}(\Omega_v) = \Lambda_v$, and this happens for all but finitely many $v \in \mathcal{M}_K \setminus \mathcal{P}$. □

Given $T \in \operatorname{Hom}_K(\mathbf{X}, \mathbf{Y})$ we set:

$$H^{\mathrm{op}}_{\mathcal{F},\mathcal{G}}(T) = \sup_{\mathbf{x}\in\mathbf{X}\setminus\{\mathbf{0}\}} \frac{H_{\mathcal{G}}(T(\mathbf{x}))}{H_{\mathcal{F}}(\mathbf{x})}.$$

The function $H^{\mathrm{op}}_{\mathcal{F},\mathcal{G}}\colon \operatorname{Hom}_K(\mathbf{X}, \mathbf{Y}) \to \mathbb{R}$, $T \mapsto H^{\mathrm{op}}_{\mathcal{F},\mathcal{G}}(T)$, is called the *operator height* on $\operatorname{Hom}_K(\mathbf{X}, \mathbf{Y})$ associated to $\mathcal{F}$ and $\mathcal{G}$. If $\mathbf{X} = \mathbf{Y}$ and $\mathcal{F} = \mathcal{G}$ we will use $H^{\mathrm{op}}_{\mathcal{F}}$ instead of $H^{\mathrm{op}}_{\mathcal{F},\mathcal{F}}$. On $\operatorname{Hom}_K(\mathbf{X}, \mathbf{Y})$ it is also possible to define an adelic norm associated to $\mathcal{F}$ and $\mathcal{G}$, precisely: $\mathcal{L}^{\mathcal{G}}_{\mathcal{F}} = \{M_{N_v,P_v}, v \in \mathcal{M}_K\}$, where:

$$M_{N_v,P_v}(T) = \sup_{\mathbf{x}\in\mathbf{X}_v\setminus\{\mathbf{0}\}} \frac{P_v(T(\mathbf{x}))}{N_v(\mathbf{x})}.$$

It is clear that $H_{\mathcal{L}^{\mathcal{G}}_{\mathcal{F}}}(T) \geq H^{\mathrm{op}}_{\mathcal{F},\mathcal{G}}(T)$, but in many cases the equality actually holds:

PROPOSITION 2.3. *Let $\mathbf{X}$, $\mathbf{Y}$, $\mathcal{F}$ and $\mathcal{G}$ be as above. Let $T \in \operatorname{Hom}_K(\mathbf{X}, \mathbf{Y})$ be injective. Then:*

$$H^{\mathrm{op}}_{\mathcal{F},\mathcal{G}}(T) = \prod_{v\in\mathcal{M}_K} M_{N_v,P_v}(T)^{d_v}.$$

PROOF. The argument is identical to the one used in [**13**, Proposition 3.1], where the same result is proved for the special case $\mathbf{X} = \mathbf{Y} = K^n$ and $\mathcal{F} = \mathcal{E}^2_{\mathcal{O}^n_K,\underline{e}}$, $\underline{e}$ being the canonical basis of K^n. One only needs to use Lemma 2.1 and Lemma 2.2 instead of [**13**, Lemma 3.2]. □

PROPOSITION 2.4. *Let $\mathbf{X}$, $\mathbf{Y}$, $\mathcal{F}$ and $\mathcal{G}$ be as above. Suppose T is such that there exists $\gamma \in \mathbb{R}$ such that $\gamma H_{\mathcal{F}}(\mathbf{x}) = H_{\mathcal{G}}\big(T(\mathbf{x})\big)$ for all $\mathbf{x} \in \mathbf{X} \setminus \{\mathbf{0}\}$. Then:*

(a) $\prod_{v\in\mathcal{M}_K} M_{N_v,P_v}(T)^{d_v} = \gamma$.
(b) $P_v\big(T(\mathbf{x})\big) = M_{N_v,P_v}(T) N_v(\mathbf{x})$ *for all* $\mathbf{x} \in \mathbf{X} \setminus \{\mathbf{0}\}$ *and all* $v \in \mathcal{M}_K$.

PROOF. Since $H_{\mathcal{F}}$ is constant on the subspaces generated by $\mathbf{y}$ and $\ker T$ for each $\mathbf{y} \notin \ker T$, property (H2) implies that T is injective. For ease of notation we set $M_v = M_{N_v,P_v}$. Then by Proposition 2.3 we have

$$\prod_{v\in\mathcal{M}_K} M_{N_v,P_v}(T)^{d_v} = H^{\mathrm{op}}_{\mathcal{F},\mathcal{G}}(T) = \gamma$$

proving (a). Next Lemma 2.2 implies that for all $\mathbf{x} \in \mathbf{X} \setminus \{\mathbf{0}\}$:

$$(\bullet) \qquad \gamma = \prod_{v\in\mathcal{S}_T} \left(\frac{P_v(T(\mathbf{x}))}{N_v(\mathbf{x})}\right)^{d_v}.$$

Suppose that there exists $\mathbf{x}_0 \in \mathbf{X} \setminus \{\mathbf{0}\}$ which contradicts (b), i.e., one can find $w \in S_T$ for which $P_w(T(\mathbf{x}_0)) < M_w(T)N_w(\mathbf{x}_0)$. Then ($\bullet$) yields:

$$\gamma = \prod_{v \in \mathcal{S}_T} \left(\frac{P_v(T(\mathbf{x}))}{N_v(\mathbf{x})}\right)^{d_v} \leq \left(\prod_{\substack{v \in \mathcal{S} \\ v \neq w}} M_v(T)^{d_v}\right) \left(\frac{P_w(T(\mathbf{x}))}{N_w(\mathbf{x})}\right)^{d_w} < \prod_{v \in \mathcal{S}_T} M_v(T)^{d_v} = \gamma,$$

which is a contradiction. □

COROLLARY 2.5. *Let $\mathcal{F} = \{N_v, v \in \mathcal{M}_K\}$ and $\mathcal{G} = \{P_v, v \in \mathcal{M}_K\}$ be two adelic norms on $\mathbf{X}$. Then:*

(a) *The ratio $H_{\mathcal{F}}/H_{\mathcal{G}}$ is constant on $\mathbf{X} \setminus \{\mathbf{0}\}$ if and only if the ratio N_v/P_v is constant on $\mathbf{X}_v \setminus \{\mathbf{0}\}$ for all v's.*

(b) *$H_{\mathcal{F}} = H_{\mathcal{G}}$ if and only if $N_v = \gamma_v P_v$, and $\prod_{v \in \mathcal{M}_K} \gamma_v^{d_v} = 1$.*

PROOF. Apply the proposition to the identity map. □

3. Heights on finite étale K-algebras

Let A be a K-algebra which is always assumed to be finite dimensional, commutative and with unit usually denoted by 1_A. An adelic norm $\mathcal{F} = \{N_v, v \in \mathcal{M}_K\}$ on A is said to be *compatible* (with the algebra structure) if

(C1) $N_v(ab) \leq N_v(a)N_v(b)$ for all $a, b \in A_v$ and for all $v \in \mathcal{M}_K$.

Therefore, given a compatible adelic norm $\mathcal{F}$ on A, the inequality

$$H_{\mathcal{F}}(ab) \leq H_{\mathcal{F}}(a)H_{\mathcal{F}}(b) \qquad \text{(HC1)}$$

holds for all $a, b \in A$ such that $ab \neq 0$. We will denote by $\mathcal{C}(A)$ the set of compatible adelic norms on A.

Suppose now that A is a finite étale K-algebra, i.e., that A is a product of finite extensions of K. In [**12**] we introduced the following family of v-adic norms: let $(\mathcal{X}_v, \mathcal{O}_{\mathcal{X}_v})$ be the affine K_v-scheme associated to the K_v-algebra $A_v = A \otimes_K K_v$. Let $a \mapsto \widehat{a}$ denote the canonical isomorphism $A_v \simeq \Gamma(\mathcal{X}_v, \mathcal{O}_{\mathcal{X}_v})$. We then set

$$N_v^A(a) = \sup_{x \in \mathcal{X}_v} |\widehat{a}(x)|_x$$

where $|\cdot|_x$ is the unique extension of $|\cdot|_v$ to $K_v(x)$.

LEMMA 3.1. *Let A be a finite étale K-algebra. The family $\mathcal{F}_A = \{N_v^A, v \in \mathcal{M}_K\}$ is a compatible adelic norm on A.*

PROOF. In [**12**, Proposition 2.2(b)] it was shown that property (C1) holds for $\mathcal{F}_A$. Thus it only remains to check that there exists an $\mathcal{O}_K$-lattice $\Lambda \subset A$ such that $N_v^A = N_{\Lambda_v}$ for almost all $v \in \mathcal{M}_K$. Assume first that A is simple so $A = F$ is a finite extension of K. Our choice of $\mathcal{O}_K$-lattice is $\Lambda = \mathcal{O}_F$. Let us denote by $S_0 \subset \mathcal{M}_K^0$ the subset formed by the absolute values whose associated prime ideal ramifies in the extension $F \supset K$. We want to show that for all $v \in \mathcal{M}_K^0 \backslash S$ we have $N_v^F = N_{\Lambda_v}$. A moment of reflection shows that

$$N_v^F(a) = \sup_{w \in \mathcal{M}_F^v} |a|_w,$$

where $\mathcal{M}_F^v = \{w \in \mathcal{M}_F \mid |\cdot|_w|_K = |\cdot|_v\}$. Given $a \in F$ and $\lambda \in K_v^\times$ note that γa belongs to Λ_v if and only if $|\gamma a|_w \leq 1$ for all $w \in \mathcal{M}_F^v$. Thus $\gamma a \in \Lambda_v$ implies $|a|_w \leq |\gamma^{-1}|_v$ which yields

$$|a|_w \leq \inf_{\substack{\gamma \in K_v^\times \\ \gamma a \in \Lambda_v}} |\gamma|_v^{-1} = N_{\Lambda_v}(a).$$

The above inequality yields $N_v^F(a) \leq N_{\Lambda_v}(a)$. Note that this holds also for the ramified primes. It is in the proof of the reverse inequality that we use the assumption on v. So let us assume that v is unramified in the extension $K \subset F$. Then the value groups of $|\cdot|_v$ and $|\cdot|_w$ are the same for all $w \in \mathcal{M}_F^v$. Therefore, there exists $\gamma_0 \in K_v^\times$ such that $|\gamma_0|_v = (\sup_{w \in \mathcal{M}_F^v} |a|_w)^{-1}$. It follows that $|\gamma_0 a|_w \leq 1$ for all $w \in \mathcal{M}_F^v$. Thus $\gamma_0 a \in \Lambda_v$, yielding $N_{\Lambda_v}(a) \leq N_v^F(a)$. For the general case it suffices to recall that N_v^A behaves well under direct products, i.e., if $A \simeq \prod_{i=1}^f A_i$, then $N_v^A(a) = \sup_{1 \leq i \leq f} N_v^{A_i}(\pi_i(a))$, where π_i is the canonical projection (see [**12**, Proposition 1.1(c)] for a proof). Thus, if we decompose A as the product of simple K-algebras as above, then the sought $\mathcal{O}_K$-lattice is the product of the $\mathcal{O}_{A_i}$. □

For ease of notation we set $H_A = H_{\mathcal{F}_A}$, $H_A^{\mathrm{op}} = H_{\mathcal{F}_A}^{\mathrm{op}}$. The function H_A enjoys the following properties (see [**12**, Proposition 2.2 and Corollary 2.4] for a proof):

(A1) $H_A(a) \geq 1$;
(A2) $H_A(a^k) = (H_A(a))^k$;
(A3) If $\varphi\colon A \to B$ is K-algebra isomorphism, then $H_A = H_B \circ \varphi$.

Remark. The adelic norm $\mathcal{F}_A$ has also the property that $N_v(1_A) = 1$ for all $v \in \mathcal{M}_K$, which means that (A_v, N_v^A) is a Banach algebra (either real, complex or non-archimedean) for all $v \in \mathcal{M}_K$. A compatible adelic norm enjoying this property will be called a *Banach* adelic norm.

Example. Let $K \subset F$ be a finite extension. Let $R \subseteq \mathcal{O}_F$ be an order, i.e., R is a subring of $\mathcal{O}_F$ containing 1 which is also a $\mathcal{O}_K$-lattice. Let $R_v = R \otimes_{\mathcal{O}_K} \mathcal{O}_v$, then $\mathcal{F}_R = \{N_{R_v}; v \in \mathcal{M}_K^0\} \cup \{N_v^F; v \in \mathcal{M}_K^\infty\}$, is an adelic norm on F. Set $H_R = H_{\mathcal{F}_R}$. For each $v \in \mathcal{M}_K^0$, R_v is a subring of $F \otimes_K K_v$. Thus for every pair of elements $a, b \in F \otimes_K K_v$ we have

$$\{\lambda \in K_v^\times \mid \lambda ab \in R_v\} \supseteq \{\gamma \in K_v^\times \mid \gamma = \mu\nu \text{ and } \mu a \in R_v, \nu b \in R_v\}.$$

Therefore $N_{R_v}(ab) \leq N_{R_v}(a) N_{R_v}(b)$. Since 1_A belongs to R we have $N_v(1_A) \leq 1$ for all $v \in \mathcal{M}_K^0$. Moreover, the inclusion $R \subset \mathcal{O}_F$ yields $H_{\mathcal{O}_F}(a) \leq H_R(a)$. From the proof of Lemma 3.1 we know that $H_F(a) \leq H_{\mathcal{O}_F}(a)$, so $H_F(a) \leq H_R(a)$. In particular $H_F(1_A) \geq 1$ so we must have $N_{R_v}(1_A) = 1$ for all $v \in \mathcal{M}_K^0$, i.e., $\mathcal{F}_R$ is a Banach adelic norm. Conversely, suppose that $\mathcal{F} = \{N_v; v \in \mathcal{M}_K\}$ is a Banach adelic norm on the finite étale K-algebra A. Recall that the $\mathcal{O}_K$-lattice $\Lambda_{\mathcal{F}}$ associated to $\mathcal{F}$ can be computed as (again by [**15**, Theorem V.2.2])

$$\Lambda_{\mathcal{F}} = \bigcap_{v \in \mathcal{M}_K^0} \left(\{a \in A_v \mid N_v(a) \leq 1\} \cap A\right).$$

Now, for $v \in \mathcal{M}_K^0$, given $a, b \in \Lambda_{\mathcal{F}}$ we have $N_v(ab) \leq N_v(a) N_v(b) \leq 1$. Therefore $\Lambda_{\mathcal{F}}$ is a subring of A. Moreover, $\mathcal{F}$ is Banach adelic norm and so $N_v(1_A) = 1$ for all $v \in \mathcal{M}_K^0$, therefore $1_A \in \Lambda_{\mathcal{F}}$ and so $\Lambda_{\mathcal{F}}$ is an order in A.

To better exploit the canonical nature of H_A it is convenient to bring into the picture the spectral height. In order to do so we need some definitions: first the *spectral height* $H_s\colon \mathrm{End}_K(A) \to \mathbb{R}$ is defined as follows: $H_s(T) = 1$ if T is nilpotent, otherwise we set

$$H_s(T) = \prod_{v \in \mathcal{M}_K} \rho_v(T)^{d_v}.$$

Here $\rho_v(T)$ denotes the v-adic spectral radius[2] of T. Precisely

$$\rho_v(T) = \sup_{\lambda \in \mathrm{sp}(T)} |\lambda|_{K_v(\lambda)},$$

where $\mathrm{sp}(T)$ is the set of characteristic roots of T in an algebraic closure of K_v, and $|\cdot|_{K_v(\lambda)}$ is the unique extension of $|\cdot|_v$ to $K_v(\lambda)$. Second, to any adelic norm $\mathcal{F}$ we can also associate the height function $H_{\mathcal{F}}^{\mathrm{op}}$ on $\mathrm{End}_K(A)$, defined in Section 1. Moreover, the adelic norm $\mathcal{L}_{\mathcal{F}}^{\mathcal{F}}$, also defined in Section 1, gives rise to another height function, $H_{\mathcal{L}_{\mathcal{F}}^{\mathcal{F}}}$, on $\mathrm{End}_K(A)$. In the following, given $a \in A$ we let $m_a\colon A \to A$ denote the *multiplication-by-a* map.

LEMMA 3.2. *Let A be a finite dimensional commutative K-algebra. Suppose $\mathcal{F}$ is a compatible adelic norm on A. Then for all $a \in A$,*

$$H_{\mathcal{F}}^{\mathrm{op}}(m_a) \le H_{\mathcal{L}_{\mathcal{F}}^{\mathcal{F}}}(m_a) \le H_{\mathcal{F}}(a) \le H_{\mathcal{F}}^{\mathrm{op}}(m_a) H_{\mathcal{F}}(1_A).$$

In particular if $H_{\mathcal{F}}(1_A) = 1$ we have $H_{\mathcal{F}}^{\mathrm{op}}(m_a) = H_{\mathcal{L}_{\mathcal{F}}^{\mathcal{F}}}(m_a) = H_{\mathcal{F}}(a)$.

PROOF. By definition of $H_{\mathcal{F}}^{\mathrm{op}}$ we have:

$$H_{\mathcal{F}}(1_A) H_{\mathcal{F}}^{\mathrm{op}}(m_a) \ge H_{\mathcal{F}}\big(m_a(1_A)\big) = H_{\mathcal{F}}(a).$$

On the other hand, the inequalities $H_{\mathcal{L}_{\mathcal{F}}^{\mathcal{F}}}(m_a) \ge H_{\mathcal{F}}^{\mathrm{op}}(m_a)$ holds in an even greater generality as remarked before Proposition 2.3. Since $\mathcal{L}_{\mathcal{F}}^{\mathcal{F}} = \{M_{N_v,N_v}, v \in \mathcal{M}_K\}$ and $M_{N_v,N_v}(m_a) \le N_v(a)$, it follows that $H_{\mathcal{L}_{\mathcal{F}}^{\mathcal{F}}}(m_a) \le H_{\mathcal{F}}(a)$, completing the proof of the lemma. □

LEMMA 3.3. *Let A be an étale K-algebra. Then $H_A(a) = H_s(m_a)$ for all $a \in A$.*

PROOF. Let $\mathcal{F}$ be any adelic norm on A. The limit formula

$$\text{(GBF)} \qquad H_s(m_a) = \lim_{k \to \infty} H_{\mathcal{F}}^{\mathrm{op}}(m_a^k)^{1/k}$$

is a special[3] case of [**13**, Theorem A]. Combining (GBF) with Lemma 3.2 and (A2) we obtain:

$$H_s(m_a) = \lim_{k\to\infty} H_A^{\mathrm{op}}(m_a^k)^{1/k} = \lim_{k\to\infty} H_A(a^k)^{1/k} = H_A(a). \qquad \square$$

The next theorem shows that, when A is a finite étale K-algebra, H_A can be obtained by an averaging process starting from any compatible adelic norm on A. We also prove that H_A is minimal among the heights arising from compatible adelic norms.

---

[2]For a comprehensive treatment of spectral theories for a non-archimedean Banach algebra the reader is referred to [**1**].

[3]In [**13**] the operator height associated to an adelic norm $\mathcal{F}$ was defined including $\mathbf{0}$ in the set over which we take the superior, see [**13**, Section 3]. Now it is easy to verify that for every $T \in \mathrm{End}_K(\mathbf{X})$ that is not nilpotent we have $H_{\mathcal{F}}^{\mathrm{op}}(T^k) \ge 1$ for all $k \ge \dim_K \mathbf{X}$. Therefore the discrepancy between the two definitions does not have any bearings on the limit formula.

THEOREM 3.4. *Let A be a finite étale K-algebra. Then*
(a) $\lim_{k\to\infty} H_{\mathcal{F}}(a^k)^{1/k} = H_A(a)$, *for all* $\mathcal{F} \in \mathcal{C}(A)$ *and all* $a \in A$.
(b) $H_A(a) = \min_{\mathcal{F}} \in \mathcal{C}(A) H_{\mathcal{F}}(a)$, *for all* $a \in A$.

PROOF. (a) We have:

$$\begin{aligned}\lim_{k\to\infty} H_{\mathcal{F}}(a^k)^{1/k} &= \lim_{k\to\infty} H_{\mathcal{F}}^{\mathrm{op}}(m_a^k)^{1/k} && \text{by Lemma 3.2}\\ &= H_s(m_a) && \text{by (GBF)}\\ &= H_A(a) && \text{by Lemma 3.3}\end{aligned}$$

(b) Since $\mathcal{F}_A \in \mathcal{C}(A)$ we only have to prove that $H_A(a) \leq H_{\mathcal{F}}(a)$ for all $\mathcal{F} \in \mathcal{C}(A)$. But this follows from (a) and (HC1). □

In the following $[A]$ denotes the K-vector space underlying A and $A^\times$ denotes the set of invertible elements of A.

THEOREM 3.5. *Let A and B be two finite étale K-algebras. Suppose that $T \in \mathrm{Hom}_K([A],[B])$ is surjective and such that $H_B\big(T(a)\big) = H_A(a)$ for all $a \in A$. Then A and B are isomorphic as K-algebras. Moreover, if we let $b = T(1_A)$ then $m_{b^{-1}} \circ T$ is a K-algebra isomorphism.*

PROOF. We can apply Proposition 2.4 with $\gamma = 1$ finding that T is a bijection and that for each $v \in \mathcal{M}_K$ there exists $\gamma_v \in \mathbb{R}$, such that $\gamma_v N_v^A(a) = N_v^B\big(T(a)\big)$. In particular for $v \in \mathcal{M}_K^\infty$ we have that the K_v-linear map $\gamma_v^{-1}T$ is an isometry between (A_v, N_v^A) and (A_v, N_v^B). Now A_v (respectively B_v) is either a complex Banach algebra or a real Banach algebra or a real function algebra (see [**12**]) and they have to be of the same type simultaneously, by the existence of the linear isometry. For each of these cases there is an appropriate version of the Banach–Stone theorem[4] which we can apply, as in [**12**, Proposition 1.3 and Corollary 1.5] , to this setting. We thus find that $\gamma_v^{-1}T(1_A) = c$ is invertible and that $\gamma_v^{-1}m_{c^{-1}} \circ T$ is a K_v-algebra isomorphism of A_v into B_v. Now it is straightforward to verify that $b = T(1_A) = \gamma_v c$ is invertible and that $m_{b^{-1}} \circ T$ is an isomorphism of K-algebras. □

To conclude the paper we state, as a corollary, a special case of the above proposition which generalizes [**12**, Theorem 3.8] by removing the condition of A being isotypical (i.e., of the form B^n for some simple K-algebra B).

COROLLARY 3.6. *Let A be a finite étale K-algebras. Suppose $T \in \mathrm{End}_K([A])$ is such that $H_A\big(T(a)\big) = H_A(a)$ for all $a \in A$, then $b = T(1_A)$ is invertible and $m_{b^{-1}} \circ T$ is a K-algebra automorphism.*

References

1. V. G. Berkovich, *Spectral theory and analytic geometry over non-archimedean fields*, Math. Surveys Monogr., vol. 33, Amer. Math. Soc, Providence, R.I., 1990.
2. F. Bonsall and J. Duncan, *Complete normed algebras*, Ergebnisse der Mathematik und ihrer Grenzgebiete, Band 80, Springer-Verlag, New York, 1973.
3. J. Bost, J. Gillet, and C. Soulé, *Heights of projective varieties and positive Green forms*, J. Amer. Math. Soc. **71** (1994), 903–1027.

[4]For the precise statement and the proof see [**2**, I.5.8 and I.13.7]for the case of real or complex Banach Algebras, and [**6**, Theorem 5.1.4] for the real function algebras case.

4. C. Gasbarri, *On the number of points of bounded height on arithmetic projective space*, Manuscripta Math. **98** (1999), 453–457.
5. J. Gillet and C. Soulé, *Arithmetic interSection theory*, Inst. Hautes Études Sci. Publ. Math. **72** (1990), 93–174.
6. S. H. Kulkarni and B. V. Limaye, *Real function algebras*, Monogr. Textbooks Pure Appl. Math., vol. 186, Marcel Dekker Inc. New York, 1992.
7. S. Lang, *Les formes bilinéaires de Néron et Tate*, 1963/1964 Séminaire Bourbaki, 1963/64, Fasc. 3, Exposé 274, Secrétariat mathématique, Paris.
8. D. G. Northcott, *An inequality in the theory of arithmetic on algebraic varieties*, Proc. Cambridge Philos. Soc. **45** (1949), 502–509.
9. E. Peyre, *Hauteurs et mesures de Tamagawa sur les variétés de Fano*, Duke Math. J. **79** (1995), 101–218.
10. D. Roy and J. L. Thunder, *An absolute Siegel's lemma*, J. Reine Angew. Math **470** (1990), 1–26.
11. W. M Schmidt, *On heights of algebraic subspaces and diophantine approximations*, Ann. of Math. (2) **85** (1967), 430–472.
12. V. Talamanca, *Height preserving transformations on semisimple K-algebras*, Journées Arithmétiques (Barcelona, 1995), Collect. Math. **48** (1997), no. 1-2, 217–234.
13. ———, *A Gelfand-Beurling type formula for heights on endomorphism rings*, J. Number Theory **83** (2000), 91–105.
14. A. Weil, *Arithmetic on algebraic varieties*, Ann. of Math. (2) **53** (1951), 412–444.
15. ———, *Basic number theory*, Reprint of the 2nd ed. (1973), Classics Math., Springer Verlag, Berlin, 1995.
16. E. Viada–Aehle, *Elliptic isogenies and slopes*, Ph.D. Dissertation, E.T.H. Zurich, 2001.
17. S. Zhang, *Small points and adelic metrics*, J. Algebraic Geom. **4** (1995), 281–300

DIPARTIMENTO DI MATEMATICA, UNIVERSITÀ DI L'AQUILA, VIA VETOIO SNC, COPPITO, 67010, ITALY.

E-mail address: valerio@dm.univaq.it; valerio@matrm3.mat.uniroma3.it